Nonlinear Elliptic and Parabolic Equations of the Second Order

Mathematics and Its Applications (*Soviet Series*)

N. V. Krylov
University of Moscow, U.S.S.R.

Nonlinear Elliptic and Parabolic Equations of the Second Order

D. Reidel Publishing Company
A MEMBER OF THE KLUWER ACADEMIC PUBLISHERS GROUP
Dordrecht / Boston / Lancaster / Tokyo

Library of Congress Cataloging-in-Publication Data

CIP

L.C. Data appear on a separate card

ISBN-13: 978-1-4020-0334-9 e-ISBN-13: 978-94-010-9557-0
DOI: 10.1007/978-94-010-9557-0

Published by D. Reidel Publishing Company,
P.O. Box 17, 3300 AA Dordrecht, Holland.

Sold and distributed in the U.S.A. and Canada
by Kluwer Academic Publishers,
101 Philip Drive, Norwell, MA 02061, U.S.A.

In all other countries, sold and distributed
by Kluwer Academic Publishers Group,
P.O. Box 322, 3300 AH Dordrecht, Holland.

Original title:
НЕЛИНЕЙНЫЕ ЭЛЛИПТИЧЕСКИЕ И ПАРАБОЛИЧЕСКИЕ
УРАВНЕНИЯ ВТОРОГО ПОРЯДКА
Translated from the Russian by P. L. Buzytsky

Softcover reprint of the hardcover 1st edition 1987

TABLE OF CONTENTS

EDITOR'S PREFACE

Approach your problems from the right end and begin with the answers. Then one day, perhaps you will find the final question.

'The Hermit Clad in Crane Feathers' in R. van Gulik's The Chinese Maze Murders.

It isn't that they can't see the solution. It is that they can't see the problem.

G.K. Chesterton. The Scandal of Father Brown 'The Point of a Pin'.

Growing specialization and diversification have brought a host of monographs and textbooks on increasingly specialized topics. However, the "tree" of knowledge of mathematics and related fields does not grow only by putting forth new branches. It also happens, quite often in fact, that branches which were thought to be completely disparate are suddenly seen to be related.

Further, the kind and level of sophistication of mathematics applied in various sciences has changed drastically in recent years: measure theory is used (non-trivially) in regional and theoretical economics; algebraic geometry interacts with physics; the Minkowsky lemma, coding theory and the structure of water meet one another in packing and covering theory; quantum fields, crystal defects and mathematical programming profit from homotopy theory; Lie algebras are relevant to filtering; and prediction and electrical engineering can use Stein spaces. And in addition to this there are such new emerging subdisciplines as "experimental mathematics", "CFD", "completely integrable systems", "chaos, synergetics and large-scale order", which are almost impossible to fit into the existing classificaiton schemes. They draw upon widely different sections of mathematics. This programme, Mathematics and Its Applications, is devoted to new emerging (sub)disciplines and to such (new) interrelations as exempla gratia:

- a central concept which plays an important role in several different mathematical and/or scientific specialized areas;
- new applications of the results and ideas from one area of scientific endeavour into another;
- influences which the results, problems and concepts of one field of enquiry have and have had on the development of another.

The Mathematics and Its Applications programme tries to make available a careful selection of books which fit the philosophy outlined above. With such books, which are stimulating rather than definitive, intriguing rather than encyclopaedic, we hope to contribute something towards better communication among the practitioners in diversified fields.

Because of the wealth of scholarly research being undertaken in the Soviet Union, Eastern Europe, and Japan, it was decided to devote special attention to the work emanating from these particular regions. Thus it was decided to start three regional series under the umbrella of the main MIA programme.

One area which exemplifies beautifully some of the tendencies alluded to above is that of nonlinear elliptic and parabolic partial differential equations of the second order. That sounds like part of analysis but of course, as everyone knows, these equations turn up everywhere, for example in probability and geometry. And in the case of the breakthrough represented by the general solvability results treated in this book it is a simple and fascinating fact that these were first obtained by probabilistic and geometric means. That made these results, and more importantly the underlying general principles, largely inaccessible to analysts. As it turned out, much can also be done purely analytically and important general methods thus emerged. And that is precisely what this book is about.

The unreasonable effectiveness of mathematics in science ... - Eugene Wigner.

Well, if you know of a better 'ole, go to it - Bruce Bairnsfather.

What is now proved was once only imagined - William Blake.

As long as algebra and geometry proceeded along separate paths, their advance was slow and their applications limited.
But when these sciences joined company they drew from each other fresh vitality and thenceforward marched on at a rapid pace towards perfection - Joseph Louis Lagrange.

Bussum, March 1986

MICHIEL HAZEWINKEL

PREFACE

Nonlinear second-order elliptic and parabolic equations arise in the theory of differential equations and in numerous applications. A wide class of such equations, the class of quasilinear equations, has been studied by many mathematicians almost as thoroughly as the class of linear equations. These studies have resulted in many papers and books such as those by Ladyzhenskaya and Ural'tseva [1], Ladyzhenskaya, Solonnikov, and Ural'tseva [1], Ivanov [2] and Gilbarg and Trudinger [1]. Quasilinear equations are linear with respect to the highest derivatives. For the equations nonlinear in the highest derivatives, which are studied in this book, only fragmentary results were known for a long time and no general theory could be formulated from them. The only exception is the case of equations involving two spatial variables, where owing to well-known specific properties of these equations the solvability theory was developed in the spaces $W_p^{1,2}$ and W_p^2 for any p close to 2 (see, for example, Krylov [2]).

Although a general theory was lacking, certain classes of nonlinear equations were studied within the scope of the mathematical areas where they arose. The first nonlinear elliptic equation subject to intensive study was the Monge–Ampère equation, which was basically considered from the standpoint of the geometry of convex surfaces with extensive use of geometrical notions. Until 1971 the smoothness of the generalized solutions of this equation (which were introduced by Aleksandrov [2]), was proved only for two variables (see Bakel'man [1]). Later, Pogorelov [2,3] proved internal smoothness for an arbitrary number of variables (see also, Cheng and Yau [1]). In the case of many variables, the smoothness up to the boundary was proved after a general approach to nonlinear equations was developed (see Krylov [20,22], Ivochkina [1], Caffarelli, Nirenberg, and Spruck [1]).

Much later came the appearance of the nonlinear elliptic equations, which are conventionally called the Bellman equations. They stemmed from the theory of diffusion-type controlled random processes, which, in the case of uncontrolled diffusion coefficients, is based on the theory of quasilinear equations (see Fleming and Rishel [1] and the references therein). Using probabilistic methods, this author [3 – 6] established the solvability of the general degenerate nonlinear elliptic Bellman equations in the whole space for the class of functions with bounded derivatives. These results were of paramount importance because they not only dealt with an entire class of equations but also revealed that the Monge–Ampère equation is a particular case of the Bellman equation. This permitted development of a general theory involving the Monge–Ampère equation as a special case of the Bellman equation. However, from the standpoint of differential equations, this theory had a disadvantage: it was formulated in probabilistic terms, and no other approach could be elaborated until 1979. The approach proposed in the above-mentioned works was later developed by the author, Nisio, Pragarauskas, Safonov, Lions, and Menaldi (see the Bibliography, p. 453).

Until 1979 the Bellman equations were primarily studied by probabilistic methods. In 1979 Brezis and Evans [1] proved the existence of a solution in the class $C^{2+\alpha}$ for the case of the Bellman equation with two elliptic operators, whereas the probabilistic approach yielded only the existence of a solution with bounded generalized second derivatives. The results of Evans and Friedman [1], Evans and Lions [1], and Lions [2] also date back to about this time. They were established by methods of the theory of differential equations, but the solutions possessed bounded second derivatives and so these results are no stronger than those obtained earlier by probabilistic methods.

The situation changed dramatically in 1982 when Evans [2] and Krylov [19] independently proved the solvability in $C^{2+\alpha}$ of a broad class of Bellman equations. The proof was based solely on the theory of differential equations and the fact that the Hölder condition holds for linear equations with measurable coefficients which was proved by Krylov and Safanov [1,2]. Their results were based on estimating the maximum of a solution to the parabolic equation in terms of the L_p-norm of the right-hand side. Such estimates for elliptic equations were first obtained by Aleksandrov [3,5] using geometrical considerations. This author [8,9] extended the estimates to parabolic equations. The first of these papers was based on probabilistic methods and the other used geometrical techniques. It turned out that the methods of the first paper admit of a simple interpretation in terms of differential equations (see Chapters 2 and 3), which made it possible to apply solely the theory of differential equations using neither probabilistic nor geometrical notions and to elaborate a solvability theory covering the Bellman equation, the Monge–Ampère equation, and the quasilinear equations under natural conditions on their coefficients. The purpose of this book is to construct this theory.

It should be noted that we shall not obtain in their most general form all the results of the theory of controlled diffusion processes, the theory of Monge–Ampère equations, and the theory of quasilinear equations. First, it would be absolutely impossible within the limited compass of this book and, second, our primary goal is to develop an approach which would enable us to consider these theories from a unified standpoint and to apply general techniques. Also beyond the scope of this book there remain some important topics in the theory of nonlinear equations. For instance, in Chapters 7 and 8 we consider degenerate Bellman equations including some types of Hamilton–Jacobi equations. We do not dwell on the theory of Hamilton–Jacobi equations for which the reader is referred to Kruzhkov [1,2], Crandall and Lions [1] and the references therein. Further, in the theory of controlled diffusion processes and the theory of nonlinear equations an important role is played by the so-called free-boundary problems. The reader who wishes to become acquainted with these problems is referred to Krylov [16] and Lenhart [1]. In this book we mention such problems only in Theorem 3.4.5 and Example 8.7.15. Nor does this book cover recent significant results by Safanov [7] who proved the $C^{2+\alpha}$ solvability of equations with the structure of Bellman equations under the assumption that the 'coefficients' of these equations satisfy the Hölder condition.

An important property of most nonlinear operators considered in this book is convexity with respect to the second derivatives of the unknown function. At present it is completely unclear whether this assumption is actually necessary for constructing the solvability theory. The theory of equations with two spatial variables (see Bakel'man [1] and Krylov [2]) and some results on equations (even degenerate ones) involving many variables with a 'large' coefficient in the unknown function (see Skrypnik [1], Belopol'skaya and Daletskii [1], Evans and Lions [1]) suggest that this assumption can probably be abandoned.

When presenting the subject matter, we occasionally use some well-known facts of the theory of second-order elliptic and parabolic equations. They can all be easily found in the books by Ladyzhenskaya and Ural'tseva [1], Ladyzhenskaya, Solonnikov, and Ural'tseva [1], and Friedman [1], and a phrase such as "from the theory of linear equations it is known..." always means that one of these books is referred to. Some standard notions are used in this book without explanation, for example, a domain is an open set, $\delta^{ij} = \delta_{ij}$ is Kroneker's delta, and Δ is the Laplace operator. The notions 'positive', 'negative', 'greater than', 'less than', 'increasing', 'decreasing' are always used in their inclusive form i.e., they mean 'nonnegative', etc. The constants appearing in proofs are usually denoted by N with no sub- or superscripts. Of course, each new constant N can be quite different from foregoing ones and $N = N(\ldots)$ means that N depends solely upon what is in parentheses. Also, note that $x := y$ or $y =: x$ means that "x equals y by definition".

In conclusion I would like to thank O.A. Ladyzhenskaya and N.N. Ural'tseva, on whose invitation I visited Leningrad in the spring of 1983, when preparing this book, and gave some talks there. Their interest, attention, and support were the best evidence that the effort involved in writing this book was not futile. I am also sincerely grateful to Yu.A. Rozanov whose support was of great significance for me. Finally, I would like to thank S.E. Kuznetsov for his invaluable editorial work which helped to locate and remove numerous inaccuracies in the manuscript.

Chapter 1

AUXILIARY NOTIONS

In this chapter, we introduce the basic notation and some facts (sometimes without proof) which are needed to present the subject matter. The reader who is only interested in the $C^{2+\alpha}$ theory of nonlinear nondegenerate equations can skip Section 2 except the part concerning the Aleksandrov–Busemann–Feller theorem.

1.1. Domains and Functional Spaces of Smooth Functions

Everywhere in this book, E_d denotes a d-dimensional Euclidean space equipped with a fixed orthonormal basis. If $x \in E_d$ and $x^1, \dots, x^d$ are coordinates of the point x, we write $x = (x^1, \dots, x^d)$. The points of E_d represented in coordinate form are regarded as column vectors so that if $B = (B^{ij})$ is a $d \times d$-matrix and $x \in E_d$, then Bx is a point of E_d with coordinates $(Bx)^i = B^{ij}x^j$. Here and elsewhere repeated italic sub- or superscripts symbolize summation over all their values unless a special stipulation is made. If

$$x_1,\ x_2 \in E_d,\ x_1 = (x_1^1, \dots, x_1^d),\ x_2 = (x_2^1, \dots, x_2^d),$$

$B=(B^{ij})$ is a real matrix, and r is a positive number, we set

$$(x_1,x_2) = x_1^i x_2^i,\ |x_1| = (x_1,x_2)^{1/2},\ B^* = (B^{ji}),$$

$$||B|| = (\operatorname{tr} BB^*)^{1/2} = \Big(\sum_{i,j} (B^{ij})^2\Big)^{1/2},$$

$$S_r = \{y \in E_d : |y| < r\}.$$

Some notation, here and elsewhere, cannot be considered entirely unambiguous. Indeed, the coordinates of a vector are written as a row whereas the row is regarded as a column vector. The row (2,3) can be interpreted in different ways. It can be regarded as a vector of E_2 with coordinates 2,3 and as the number 6, i.e., the inner product of two one-dimensional vectors with coordinates 2 and 3, respectively. The author hopes that common sense and the context will help the reader to understand the notation unambiguously.

If Γ is a subset of E_d, then $C(\Gamma)$ is the set of all real-valued functions $u(x)$ defined on Γ, which are continuous on Γ (in the relative topology on Γ) and have a finite norm

$$||u||_{C(\Gamma)} = \sup_{x\in\Gamma}|u(x)|.$$

Given $\alpha\in(0,1)$, let $C^{\alpha}(\Gamma)$ be a subspace of $C(\Gamma)$ consisting of all functions u such that the norm

$$||u||_{C^{\alpha}(\Gamma)} = ||u||_{C(\Gamma)} + \sup_{x_i\in\Gamma, x_1\neq x_2} \frac{|u(x_1)-u(x_2)|}{|x_1-x_2|^{\alpha}} \tag{1}$$

is finite.

The second summand in (1) is a Hölder constant of order α for the function u with respect to the distance $|x_1-x_2|$. This summand is denoted by $|[u]|_{C^{\alpha}(\Gamma)}$. It is significant to indicate the distance relative to which the Hölder constant is computed since we shall later encounter non-Euclidean distances as well.

We denote by $u_x(x)=(u_{x^i}(x))$ and $u_{xx}(x)=(u_{x^ix^j}(x))$ the gradient and the matrix of second derivatives (of course, if they exist in some sense) of a function $u(x)$. Moreover, if ξ, $\eta\in E_d$, we denote

$$u_{(\xi)} = (u_x,\xi) = u_{x^i}\xi^i, \quad u_{(\xi)(\eta)} = (u_{xx}\xi,\eta) = u_{x^ix^j}\xi^i\eta^j.$$

The expressions $u_{(\xi)}$ and $u_{(\xi)(\xi)}$ are called the first and the second derivatives of u <u>along the direction ξ</u>.

A domain of E_d is usually denoted by D with or without indices, $\bar{D}$ is the closure of D, and $\partial D=\bar{D}\setminus D$ is the boundary of D. If D is a domain in E_d and n is an integer greater than or equal to 1, then $C^n(D)$ is the space of all n-fold continuously differentiable real-valued functions u on D with the finite norm

$$||u||_{C^n(D)} = \sum_{r=1}^{n}\sum_{i_1,\ldots,i_r=1}^{d} ||u_{x^{i_1}\cdots x^{i_r}}||_{C(D)} + ||u||_{C(D)} \tag{2}$$

For $\alpha\in(0,1)$, the space $C^{n+\alpha}(D)$ is introduced as the subset of $C^n(D)$

consisting of all functions with a finite norm $||u||_{C^{n+\alpha}(D)}$ defined as the right-hand side of (2) where $C(D)$ is replaced by $C^{\alpha}(D)$. The spaces $C^{n}(\bar{D})$ and $C^{n+\alpha}(\bar{D})$ are defined as subsets of $C^{n}(D)$ and $C^{n+\alpha}(D)$ consisting of the functions continuous on $\bar{D}$ such that each of their derivatives up to and including the nth order can be continuously extended to $\bar{D}$. The norms in these spaces are defined by the same formulas that are used for the spaces $C^{n}(D)$ and $C^{n+\alpha}(D)$ respectively. It is clear that, given a $u\in C^{n+\alpha}(D)$, the function u can be defined on ∂D so that the resulting function on $\bar{D}$ belongs to $C^{n+\alpha}(\bar{D})$. In this sense, $C^{n+\alpha}(D) = C^{n+\alpha}(\bar{D})$. We denote by $C_0^{\infty}(D)$ the set of all real-valued functions infinitely differentiable on D and equal to zero outside a compactum $\Gamma\subset D$. Furthermore, $C^{\infty}(\bar{D})$ is the set of all real-valued functions infinitely differentiable on D, continuous and bounded on $\bar{D}$, and such that each of their derivatives can be extended to $\bar{D}$ as a continuous bounded function.

The set $C^{p}_{loc}(D)$ with a domain D lying in E_d is defined as the set of all functions on D belonging to $C^{p}(D_1)$ for any bounded domain D_1 such that both D_1 and $\bar{D}_1$ lie in D. If Γ is a subset of E_d with a non-empty interior D(i.e., $\bar{\Gamma}\setminus\partial\Gamma=D\neq\phi$) and $p\geq 1$, the set $C^{p}_{loc}(\Gamma)$ is introduced as the set of all functions continuous on Γ of class $C^{p}_{loc}(D)$ such that each of their derivatives up to and including the [p]th order, regarded only on D, can be continuously extended to Γ. The sets $C_{loc}(\Gamma)$ and $C^{p}(\Gamma)$ are introduced in a similar manner.

Given a function $u(x) = (u^{i}(x),\ i\in I)$ with a finite set I, the notation $u\in C^{p}(D)$ means that $u^{i}\in C^{p}(D)$ for all $i\in I$. The norm of such a function u is assumed to be the sum of respective norms of u^{i} taken over all $i\in I$.

It is often convenient to describe the domains D of E_d by using functions $\Psi\in C_{loc}(E_d)$ as $\{x\in E_d : \Psi(x)>0\}$. Let D be a domain of E_d and $p\geq 1$. We shall write $D\in C^{p}$ if there exists a function $\Psi\in C^{p}(E_d)$ such that $D = \{\Psi>0\}$ and $|\Psi_x|>0$ on ∂D. If $D\in C^{p}$ and p is fractional, it is possible to choose the function Ψ so that on the interior of D it belongs to the class $C^{[p]+1}_{loc}(D)$ and to estimate the growth rate of its ([p] + 1)th derivatives near the boundary. To state the respective assertion, we need two more definitions:

$$d(x,\Gamma) = \text{dist}(x,\Gamma) = \inf_{y\in\Gamma}|x-y|, \quad D(\kappa) = \{x\in D : d(x,\partial D) > \kappa\}.$$

1. LEMMA. Given an integer $n\geq 2$, $\alpha\in(0,1)$, $\Psi\in C^{n+\alpha}(E_d)$, a nonempty domain $D=\{\Psi>0\}$, $|\Psi_x|\geq 1$ on ∂D, and $\partial D\neq\phi$, there exists a function $\Psi_1\in C^{n+\alpha}(E_d)$ such that $D=\{\Psi_1>0\}$, $|\Psi_{1x}|\geq 1$ on ∂D, $\Psi_1\in C^{n+1}_{loc}(D)$, the norms of Ψ_1 in $C^{n+1}(D(\kappa))$ do not exceed $N\kappa^{\alpha-1}$ for $\kappa\in(0,1)$, and the norm of Ψ_1 in $C^{n+\alpha}(E_d)$ does not exceed N, where the constant N depends only upon d, n, α, and the norm of Ψ in $C^{n+\alpha}(E_d)$.

This fact is presumably well known and we shall therefore just sketch the proof. To construct the function Ψ_1 on the interior of D, consider the equation $\Delta\Psi_1-\Psi_1=-1$ in D with the boundary-value condition $\Psi_1=0$ on ∂D. From the theory of linear equations it follows that this problem has a solution $\Psi_1\in C^{n+\alpha}(D)$, the norm of Ψ_1 in $C^{n+\alpha}(\bar D)$ being estimated as needed. Using the maximum principle, we can see that $\Psi\leq N\Psi_1$ in $\bar D$ and $1\leq N|\Psi_{1x}|$ on ∂D, where N is the norm of $\Delta\Psi-\Psi$ in $C(D)$. The function Ψ_1 can be conventionally extended from D to E_d. To estimate the $(n+1)$th derivatives of Ψ_1 it obviously suffices to denote the nth derivative of Ψ_1 by u and make use of the fact that, if $\Delta u-u=0$ in S_κ and $u\in C^\alpha(S_\kappa)$, then

$$|u_x(0)|\leq N(d,\alpha)\kappa^{\alpha-1}\|u\|_{C^\alpha(S_\kappa)} \tag{3}$$

In turn, estimate (3) can be derived from the following well known fact by use of dilatation. If $u,f\in C^\alpha(\bar S_1)$ and $\Delta u=f$ in S_1, then

$$|u_x(0)|\leq N(d,\alpha)[\|f\|_{C(S_1)}+|[u]|_{C^\alpha(S_1)}].$$

As deducing the boundary estimates for solutions of an equation, we conventionally 'straighten' the boundary. Thus, we shall need the following lemma, before which we introduce the notation

$$D_R=\{x\in E_d: |x^i|<R \text{ for } i=1,\dots,d\},$$

$$D^+_R=\{x\in E_d: 0<x^1<2R,\ |x^i|<R \text{ for } i=2,\dots,d\},$$

$$\Sigma_R=\{x\in E_d: x^1=0,\ |x^i|<R \text{ for } i=2,\dots,d\}.$$

2. LEMMA. Given an integer $n \geqq 2$, $\Psi \in C^n(E_d)$, a nonempty domain $D = \{\Psi > 0\}$, $\partial D \neq \phi$, $|\Psi_x| \geqq 1$ on ∂D, and $x_0 \in \partial D$, there exists a number $\rho_0 > 0$, which depends only on d, n, and the norm of Ψ in $C^n(E_d)$, such that, for every $\rho \in (0,\rho_0]$, there is a mapping f with the following properties:

(a) $f\colon \overline{D_3^+} \to (x_0 + S_{3\rho d}) \cap \bar{D}$, $f(D_1^+) \supset (x_0 + S_\rho) \cap D$, $f(\Sigma_3) \subset \partial D$,

(b) f is continuous on $\overline{D_3^+}$, the inverse mapping g is defined on $f(\overline{D_3^+})$, f is an n-fold continuously differentiable function on D_3^+, the function g is n-fold continuously differentiable on $f(D_3^+)$,

$$\|f^i_{x^{i_1}\dots x^{i_n}}\|_{C(D_3^+)} \leqq N\rho^n, \quad \|g^i_{x^{i_1}\dots x^{i_n}}\|_{C(f(D_3^+))} \leqq N\rho^{-1}$$

for all $i, i_1, \dots, i_n = 1, \dots, d$, where N depends only upon d, n, and the norm of Ψ in $C^n(E_d)$,

(c) $\frac{1}{2}|x_1-x_2|\rho \leqq |f(x_1)-f(x_2)| \leqq 2|x_1-x_2|\rho$ for all $x_1, x_2 \in D_3^+$,

(d) if $\alpha \in (0,1)$ and $\Psi \in C^{n+\alpha}(E_d)$, the (n+1)th derivatives of f and g in $D_3^+ \cap \{x^1 > \kappa\}$ and $f(D_3^+ \cap \{x^1 > \kappa\})$ do not exceed $N\rho^{n+1}\kappa^{\alpha-1}$ and $N\rho^{-1}\kappa^{\alpha-1}$, respectively, where $\kappa \in (0,1)$ and N depends only on d, n, α, and the norm of Ψ in $C^{n+\alpha}(E_d)$.

Both the lemma and its proof are common knowledge, so we recall only basic points of the proof. Without loss of generality one can assume that $x_0 = 0$, $\Psi_{x^1}(0) \geqq 1$, $\Psi_{x^1}(0) = 0$ for $i = 2, \dots, d$. Further, define a function $\tilde{g}(x) = (\tilde{g}^1(x), \dots, \tilde{g}^d(x))$ by the formulas $\tilde{g}^1 = |\Psi_x(0)|^{-1}\Psi(x)$, $\tilde{g}^i = x^i$ for $i \geqq 2$. Using the implicit function theorem it can be checked that, for any small ρ, the mapping $\tilde{g}$ realizes a diffeomorphism between $D_{3\rho}^+$ and the intersection of a neighbourhood of zero with D. Now g is constructed from the formula $\rho^{-1}\tilde{g}$, f is defined as the inverse of g, the differential properties of functions f and g are studied using the implicit function theorem and employing Lemma 1 to prove assertion (d) (where Ψ is replaced by Ψ_1 throughout).

The following well known fact can be proved using the Lagrange therem and straightening the boundary whenever the points x_1 and x_2 are close to each other and to ∂D.

3. LEMMA. Under the conditions of the previous lemma, for any $u \in C^1(D)$ and any $x_1, x_2 \in D$, $\alpha \in (0,1]$, the inequality

$$|u(x_1) - u(x_2)| \leq N|x_1 - x_2|^{\alpha}||u||_{C^1(D)}$$

holds, where N depends only on n, d, and the norm of Ψ in $C^n(E_d)$.

The above notation suits the theory of elliptic equations. Parabolic equations call for a different approach. To begin with, it should be said that the derivative of the desired function with respect to t, enters the equation with a different sign as compared with the conventional case. For example, we normally consider the equation $u_t + \Delta u + f = 0$. The conventional notation is basically changed because these equations look more natural from the standpoint of the theory of diffusion-type random processes. Of course, these equations can easily be converted to the conventional form by replacing t with (-t). The reader who is accustomed to the conventional form will readily obtain it by drawing a figure and directing the t-axis downward.

When considering the parabolic equations, we mainly deal with the space E_{d+1}, where the points are represented as pairs (t, x) such that $t \in E_1$, $x \in E_d$. If $z_1 = (t_1, x_1)$, $z_2 = (t_2, x_2) \in E_{d+1}$, then the number

$$\rho(z_1, z_2) := |x_1 - x_2| + |t_1 - t_2|^{1/2}$$

is said to be the <u>parabolic distance</u> between z_1 and z_2. Given $z = (t,x) \in E_{d+1}$ and a set $\Gamma \subset E_{d+1}$, the <u>parabolic distance</u> between z and Γ is defined as

$$\rho(z,\Gamma) = \inf\{\rho(z, z_1) : z_1 = (t_1, x_1) \in \Gamma,\ t_1 > t\},\ \inf \phi := \infty.$$

Given $\Gamma \subset E_{d+1}$ and $\alpha \in (0,1)$, the spaces $C(\Gamma)$, $C^{\alpha}(\Gamma)$, their norms and the values $|[u]|_{C^{\alpha}(\Gamma)}$ are defined as above except that $|x_1 - x_2|$ in (1) is replaced by $\rho(x_1, x_2)$ (note that $x_1, x_2 \in \Gamma \subset E_{d+1}$). By $u_x(t,x)$ and $u_{xx}(t,x)$ we mean the gradient and the matrix of second derivatives with respect to x. If $\xi, \eta \in E_d$, the notations $u_{(\xi)}(t,x)$ and $u_{(\xi)(\eta)}(t,x)$ are defined as above. The derivative of u with respect to t is denoted by u_t.

A domain in E_{d+1} is usually denoted by Q with or without indices, $\bar{Q}$ and ∂Q are conventional closure and boundary of Q, respectively. If D is a domain of E_d, $T > 0$, and $Q = (0,T) \times D$ is a cylinder in E_{d+1}, then we introduce the notation

$$\partial_t Q = \{(t,x): t=T,\ x \in \bar{D}\},\quad \partial_x Q = (0,T) \times \partial D,$$

$$\partial_{tx} Q = \partial_t Q \cap \overline{\partial_x Q} = \{(t,x): t = T,\ x \in \partial D\}.$$

If ∂ is regarded as a symbol corresponding to the limit of increments, the notation $\partial_t Q$, $\partial_x Q$, $\partial_{tx} Q$ can easily be interpreted. If $D \in C^2$, the parabolic boundary of Q is introduced as $\partial' Q = \partial_t Q \cup \partial_x Q$ (also see Definition 3.3.4).

If Q is a domain of E_{d+1}, then $C^2(Q)$ is the space of all real-valued functions $u(t,x)$ which are defined on Q and have continuous (in (t,x)) derivatives of the form u_t, u_x, u_{xx} such that the norm

$$\|u\|_{C^2(Q)} = \|u_t\|_{C(Q)} + \sum_{i=1}^{d} \|u_{x^i}\|_{C(Q)} + \\ + \sum_{i,j=1}^{d} \|u_{x^i x^j}\|_{C(Q)} + \|u\|_{C(Q)} \tag{4}$$

is finite.

For $\alpha \in (0,1)$, the space $C^{2+\alpha}(Q)$ is defined as the subset of $C^2(Q)$ consisting of all functions whose norm $\|u\|_{C^{2+\alpha}(Q)}$ defined on the right of (4) is finite, where the space $C(Q)$ is replaced by $C^\alpha(Q)$. The spaces $C^2(\bar{Q})$, $C^{2+\alpha}(\bar{Q})$, $C^2_{loc}(Q)$, $C^{2+\alpha}_{loc}(Q)$, $C^2(\Gamma)$, $C^{2+\alpha}_{loc}(\Gamma)$, where $\Gamma \subset E_{d+1}$ and $\bar{\Gamma} \setminus \partial\Gamma \neq \phi$, are defined in the same manner as those following formula (2). The only difference is that by the derivatives up to the second order we mean the derivatives u_t, u_{xx}, and u_x.

Given $T, R > 0$, we denote $Q_{T,R} = (0,T) \times D_R$. The following fact can be proved in a straightforward manner. It can also be derived indirectly from Theorem 5.4 in Chapter 4 of Ladyzhenskaya, Solonnikov, and Ural'tseva [1] (using their definitions of norms).

4. LEMMA. Let $\alpha \in (0,1)$, $T > 0$, $u \in C^{2+\alpha}(Q_{T,2})$, and let K be a norm of u in $C^{2+\alpha}(Q_{T,2})$. Then $|u_x(t_1,x) - u_x(t_2,x)| \leqq N(d,\alpha)K|t_1-t_2|^\beta$ for any (t_1,x), $(t_2,x) \in Q_{T,1}$, where $\beta = (1/2)(1+\alpha)$. Here both domains $Q_{T,1}$ and $Q_{T,2}$ can be replaced by $(0,T) \times D$ if $D = \{x: \Psi(x) > 0\}$, $\Psi \in C^{2+\alpha}(E_d)$, $|\Psi_x| \geqq 1$ on ∂D, and N is assumed to depend also on the norm of Ψ in $C^{2+\alpha}(E_d)$.

Also, denote

$$Q^+_{T,R} = (0,T) \times D^+_R,\qquad \Sigma_{T,R} = [0,T) \times \Sigma_R.$$

5. LEMMA. Let $T > 0$ and let u be a continuous function defined on $\Sigma_{T,2}$ with first derivatives continuous on $\Sigma_{T,2}$ with respect to $x^2, \ldots, x^d$. Moreover, suppose that, given an $\alpha \in (0,1)$ and a constant $K > 0$, the inequalities $|u(z_1)| \leq K$, $|u(t_1,x_1) - u(t_2,x_1)| \leq K|t_1-t_2|^{(1+\alpha)/2}$, and $|u_{x^i}(z_1)-u_{x^i}(z_2)| \leq K\rho^{\alpha}(z_1,z_2)$ hold for all $i=2, \ldots, d$ and $z_1 = (t_1,x_1)$, $z_2 = (t_2,x_2) \in \Sigma_{T,2}$. Then there exists a function v defined on $\bar{Q}_{T,1}$ which is continuous together with its derivative v_x, belongs to the class $C^2_{loc}(Q^+_{T,1})$, such that it coincides with u on $\Sigma_{T,1}$ and, for all $z = (t,x)$, $z_1 = (t_1,x_1) \in Q^+_{T,1}$, satisfies the inequalities

$$|v_t(z)|,\ ||v_{xx}(z)|| \leq N(x^1)^{\alpha-1},\quad |v(z)|,\ |v_x(z)| \leq N,$$

$$|v_x(z) - v_x(z_1)| \leq N\rho^{\alpha}(z,z_1),$$

where N depends only upon K, α, and d.

This lemma, like Lemma 1, is assumed to be well known. To prove the lemma, it suffices first to find a function $\tilde{u}$ which coincides with u on $\Sigma_{T,1}$, is defined on $\Sigma_{\infty,\infty}$, and on $\Sigma_{\infty,\infty}$, satisfies the conditions imposed on u. Then v can be chosen as a solution to the equation $v_t + \Delta v = 0$ with $t \geq 0$, $x^1 > 0$, and the boundary condition $v = \tilde{u}$ on $\Sigma_{\infty,\infty}$. The function v can be expressed in terms of u by use of a well known explicit formula which is studied in Section 2 of Chapter 4 in Ladyzhenskaya, Solonnikov, and Ural'tseva [1].

We make one more assertion without proof which will be frequently used below. This assertion might be proved by the use of barrier functions. For a real number a we set

$$a_+ = (1/2)(|a| + a),\quad a_- = (1/2)(|a| + a).$$

6. LEMMA. Let D be a domain of E_d such that $D \in C^2$, $T > 0$, and $Q = (0,T) \times D$, and let real-valued functions $a^{ij}(t,x)$, $b^i(t,x)$, and $c(t,x)$, $i,j=1, \ldots, d$, be defined on Q. Suppose that a^{ij} and b^i are bounded, c is bounded from above, $a^{ij} = a^{ji}$, and $a^{ij}\xi^i\xi^j \geq \varepsilon|\xi|^2$ for all $\xi \in E_d$ on Q, where the constant ε is strictly positive. Finally, let $u \in C(\bar{Q} \setminus \partial_{tx}Q) \cap C^2_{loc}(Q)$ and let

$$Lu := u_t + a^{ij}u_{x^ix^j} + b^iu_{x^i} + cu$$

be bounded from below on Q. Then, for any point $z_0 \in \partial_{tx}Q$, we obtain

$$\varlimsup_{\substack{z \to z_0 \\ z \in Q}} u(z) \leqq \varlimsup_{\substack{z \to z_0 \\ z \in \partial' Q \setminus \partial_{tx} Q}} (u(z))_+ \qquad (5)$$

In addition, if Lu and c are bounded on Q, then $(u(z))_+$ on the right of (5) can be replaced by $u(z)$.

We complete this section by introducing some more notation.

$$a_1 \wedge \dots \wedge a_n = \min(a_1, \dots, a_n)\ a_1 \vee \dots \vee a_n = \max(a_1, \dots, a_n),$$

$$C_{\tau,\rho} = \{(t,x) \in E_{d+1}: 0 < t < \tau,\ |x| < \rho\},\ C_\rho = E_1 \times S_\rho.$$

1.2. The Sobolev Spaces, Measures, and Lattices

Let D be a domain of E_d, $p \geqq 1$. We denote by $\mathcal{L}_p(D)$ the space of all real-valued functions u on D with the finite norm

$$||u||_{\mathcal{L}_\rho(D)} = \left(\int_D |u(x)|^p dx\right)^{1/p}$$

If $u \in C^\infty(\bar{D})$, we set

$$||u||_{W^2_p(D)} = \sum_{i,j=1}^{d} ||u_{x^i x^j}||_{\mathcal{L}_\rho(D)} + \sum_{i=1}^{d} ||u_{x^i}||_{\mathcal{L}_\rho(D)} + \\ + ||u||_{C(D)} \qquad (1)$$

For a bounded D this expression is always finite. We next define $W^2_\rho(D)$ as the completion of the set of all functions from $C^\infty(\bar{D})$ such that the right-hand side of (1) is finite, the set being completed in the norm defined by this right-hand side.

The resulting space differs from the ordinary Sobolev space since in (1) we use the $C(D)$-norm of u rather than the $\mathcal{L}_\rho(D)$-norm, and $C^\infty(\bar{D})$ is taken as an initial set of functions. In this book however, we will primarily use the spaces $W^2_p(D)$ only for the case of bounded domains with infinitely differentiable boundaries and $p \geqq d$ when (even for $p > (1/2)\,d$) our spaces $W^2_p(D)$ are well known to coincide with the ordinary Sobolev spaces. When considering the general type of domains, we shall need, to a greater extent, the spaces $W^2_{p,loc}(D)$ which are defined in an obvious way.

The basic facts of the Sobolev space theory used in the theory of the second-order elliptic and parabolic equations are presented in the books of Ladyzhenskaya and Ural'tseva [1], and Ladyzhenskaya, Sollonikov, and Ural'tseva [1]. In using various results of Sobolev space theory, we always keep in mind that one of these books is referred to.

The notion of the Sobolev space is tightly related to the Sobolev derivative. In the framework of the theory of generalized functions any function u, which is locally summable over D, is identified with a generalized function, and like any generalized function it has all generalized derivatives which are generalized functions as well. When a generalized derivative of u can be identified with a function locally summable over D, the latter is called the Sobolev derivative. All functions u of $W^2_p(D)$ are known to have the Sobolev derivatives u_x and u_{xx}.

Later in the book we will also need the notion of the derivative which is a measure on D. By a measure on D we mean any σ-additive real-valued function defined and finite on every Borel bounded subset of D which lies in D together with its closure. By Λ we denote Lebesgue measure on E_d (and E_{d+1}). If μ is a measure on D, then $\mu = \mu^s + \mu^a$ is its expansion into the sum of the absolutely continuous part μ^a and the singular part μ^s with respect to Λ. We set $\mu^{(o)}(x) = \mu^a(dx)/dx$. The density $\mu^{(o)}(x)$ of measure μ^a with respect to Λ is defined uniquely to within a set of Lebesgue measure zero. It is convenient to assume that, for any measure μ on D, the values of $\mu^{(o)}(x)$ at any point $x \in D$ are somehow chosen and fixed. We will consider various measures on D, hence it should be stressed that the expression "almost everywhere" (without pointing out a specific measure) will always mean "almost everywhere with respect to Lebesgue measure".

For a measure μ on D we denote by $|\mu|$ the variation of μ, $\mu_+ = (1/2)(|\mu| + \mu)$, $\mu_- = (1/2)(|\mu| - \mu)$, and by $f \circ \mu$ the indefinite integral* of a function f which is locally summable on D with respect to μ. In speaking just of a locally summable function, we always mean that it is locally summable with respect to the Lebesgue measure.

Of course, if u is a locally summable function on D, μ is a measure on D, and for any $\varphi \in C_0^\infty(D)$,

$$\int \varphi_{x^{i_1} \dots x^{i_n}} u \, dx = (-1)^n \int \varphi \mu(dx).$$

when $i_1, \dots, i_n$ are fixed, then μ is called a generalized derivative of u with respect to $x^{i_1} \dots x^{i_n}$. In such a case we write $u_{x^{i_1} \dots x^{i_n}} =$

* $f \circ \mu$ is a measure: $f \circ \mu(\Gamma) := \int_\Gamma f\mu(dx)$.

$= \mu$. It is obvious that the Sobolev derivative $u_{x^{i_1} \dots x^{i_n}}$ exists if and only if the measure $u_{x^{i_1} \dots x^{i_n}}$ exists and is absolutely continuous with respect to Λ.

The measures specified by derivatives will be basically considered in the case of convex functions. Let us discuss this notion as applied to the convex case in some more detail. If a function u is convex on a segment $[a,b]$, then considering its graph, we immediately conclude that

$(u(b) - u(x_2))(b-x_2)^{-1} \geq (u(x_2) - u(x_1))(x_2-x_1)^{-1} \geq (u(x_1)-u(a))(x_1-a)^{-1}$

for $a < x_1 < x_2 < b$. In particular the Lipschitz constant of the function u can be estimated on any segment $[a_1, b_1] \subset (a,b)$ using only $(a_1-a)^{-1} \vee (b-b_1)^{-1}$ and the oscillation of u on $[a,b]$. We now suppose that the function u is convex on $\bar{S}_R$. Then estimating the value $u(x_2)-u(x_1)$, we can consider only the case when u is defined on the straight line passing through the points x_2 and x_1. Thus, for any $R_1 < R$ the Lipschitz constant of u on S_{R_1} can be estimated in terms of $(R-R_1)^{-1}$ and the oscillation of u on $\bar{S}_R$.

From the theory of Sobolev spaces it is known that all first Sobolev derivatives of a Lipschitz continuous function do not exceed the Lipschitz constant (a.e.) Therefore, a function u convex on S_R has locally bounded generalized derivatives of the form u_x. To study the properties of second derivatives, we consider a function $\zeta \in C_o^\infty(E_d)$ such that $\zeta \geq 0$, $\zeta = \zeta(|x|)$, $\zeta = 0$ for $|x| \geq 1$, $\int \zeta dx = 1$. Let $\zeta^\tau(x) = \tau^{-d}\zeta(\tau^{-1} x)$, where $\tau > 0$, and let $f^{(\tau)} := f*\zeta^\tau$ for a function f.

It is clear that if u is convex on S_R, then $u^{(\tau)}$ is convex on $S_{R-\tau}$. Moreover, $u^{(\tau)}$ is infinitely differentiable on $S_{R-\tau}$, therefore $u^{(\tau)}_{(l)(l)} \geq 0$ on $S_{R-\tau}$ for any $l \in E_d$. Hence, for any $\tau < \tau_o < R$ and any nonnegative $\eta \in C_o^\infty(S_{R-\tau})$, which is equal to 1 on $S_{R-\tau_o}$, we obtain

$$\int \eta_{(l)(l)} u^{(\tau)} dx = \int \eta u^{(\tau)}_{(l)(l)} dx \geq \int_{S_{R-\tau_o}} u^{(\tau)}_{(l)(l)} dx \qquad (2)$$

Consequently, the positive measures $u^{(\tau)}_{(l)(l)} \circ \Lambda$ considered on $S_{R-\tau_o}$ with $\tau < \tau_o$ are uniformly bounded in variation. It is possible to select a subsequence of these measures which corresponds to $\tau_n \downarrow 0$ and weakly converges on any $S_{R-\tau_o}$. The first equality in (2) implies that the limit measure is the second generalized derivative of u along l. We denote it by $u_{(l)(l)}$. The first equality in (2) also implies that $u_{(l)(l)}$ is a a quadratic form in l which is uniquely defined by u. Again by virtue of (2), the measures $u_{x^ix^j}$, which are coefficients of l^il^j in the quadratic form, are the derivatives/measures of u. Thus, for any convex on S_R function u the measures $u_{x^ix^j}$ do exist.

In Section 2.2 we shall need the following Aleksandrov—Busemann—Feller theorem (see Aleksandrov [1] and Appendix 2).

1. THEOREM. let u be a convex function on S_R. Then, for almost every point $x_o \in S_R$ letting $x \in x_o$, we obtain

$$u(x) = u(x_0) + u_{x^i}(x_0)(x^i-x_0^i) + (1/2)u^{(o)}_{x^ix^j}(x_0)(x^i-x_0^i)(x^j-x_0^j) + + o(|x-x_0)|^2).$$

We will also use some results based on the fact that the sets of measurable functions and the sets of measures are complete lattices. The results presented below can partly be found in Dunford and Schwartz [1], partly they are obvious. In a domain D of a Euclidean space E_d, we consider the set of all functions onto $[-\infty,\infty]$ with the order relation $f_1 \leqq f_2$ if $f_1(x) \leqq f_2(x)$ almost everywhere on D. This order relation converts our set to a lattice. It turns out that any subset M of our lattice has the greatest lower bound and the least upper bound. These bounds are denoted, respectively, by

$$\bigwedge_{f\in M} f \quad \text{and} \quad \bigvee_{f\in M} f$$

We shall sketch the proof of the existence, for instance, of the least upper bound.

First, applying the transformation $f \to \tan^{-1} f$, we make all functions of M satisfy the inequality $|f| \leqq 1$. Second, by partitioning D into subsets, it is not difficult to reduce the proof to the case when D is bounded. Further, denote by λ the least upper bound of the integrals over D of all functions $f_1\chi_{\Gamma_1} + \ldots + f_n\chi_{\Gamma_n}$ *), the least upper bound being taken for all n, $f_1, \ldots, f_n \in M$, and the Borel sets $\Gamma_i \subset D$ such that $\Gamma_i \cap \Gamma_j = \emptyset$ if $i \neq j$, and $\Gamma_1 \cup \ldots \cup \Gamma_n = D$. It is not difficult to select an increasing sequence of such functions so that their integrals

*) $\chi_\Gamma(x)$ is a characteristic function of the set Γ, i.e., a function which is equal to unity on Γ and zero outside Γ.

converge to λ. The limit of this sequence is the desired least upper bound of M.

Specifically, this well known construction shows that the set M can be replaced by its suitable countable subset without changing the least upper bound. If M is a countable set, the least upper bound of M in the lattice coincides (a.e.) with $\sup\{f(x): f \in M\}$.

The set of all measures on D can also be equipped with the following obvious order relation: $\mu_1 \leq \mu_2$ if $\mu_1(\Gamma) \leq \mu_2(\Gamma)$ for all bounded Borel sets $\Gamma \subset D$ such that $\bar{\Gamma} \subset D$. Any subset M of the measure lattice bounded from above (below) turns out to have the least upper (greatest lower) bound. These bounds are denoted as

$$\bigvee_{\mu\in M} \mu \quad \text{or} \quad \bigvee_{\mu\in M} \mu(dx) \quad \text{and} \quad \bigwedge_{\mu\in M} \mu \quad \text{or} \quad \bigwedge_{\mu\in M} \mu(dx).$$

We now discuss the existence of the least upper bound of M. Let ν be a measure on D such that $\mu \leq \nu$ for all $\mu \in M$. We fix a measure $\mu_o \in M$ and denote $\tilde{\mu} = (\mu - \mu_o)_+ + \mu_o (= \mu \vee \mu_o)$. It is not difficult to prove that seeking for the least upper bound of μ over M is equivalent to seeking for the least upper bound of $\tilde{\mu}$ over $\mu \in M$. Since all $\tilde{\mu}$ have a common summand, namely μ_o, it remains to find the least upper bound only for $(\mu - \mu_o)_+$. Thus, the problem reduces to the case when $\mu \geq 0$ for all $\mu \in M$. The inequality $\mu \leq \nu$ then implies that $\mu \geq \nu$ and the non-negative functions $f_\mu := \mu(dx)/\nu(dx)$ are bounded by unity. Of course, the family of functions f_μ has a least upper bound, say f, in a lattice of functions where elements are compared to within 'almost everywhere' with respect to the measure ν. It is clear that $0 \leq f \leq 1$ (a.e. ν) on D, f is locally integrable with respect to ν, and $f \circ \nu$ is the least upper bound of the set M.

It follows from the above argument that, computing the least upper bound of the measure lattice, we can always pass to the least upper bound of the densities of these measures in an appropriate measure and, moreover, the set of measures can be replaced by its countable subset without changing the least upper bound. Note that the latter directly implies that, if $\bar{\mu} = \vee \{\mu : \mu \in M\}$, then $\bar{\mu}^a = \vee \{\mu^a: \mu \in M\}$, $\bar{\mu}^s = \vee \{\mu^s: \mu \in M\}$, $\bar{\mu}^{(0)} = \vee \{\mu^{(0)}: \mu \in M\}$.

When considering parabolic equations, we also deal with the spaces $W_p^{1,2}(Q)$ and $W_{p,loc}^{1,2}(Q)$, where the latter is a standard 'localization' of the former (cf. Section 1), while the former is defined for a domain $Q \subset E_{d+1}$ like $W_p^2(D)$ departing from the norm

$$\|u\|_{W_p^{1,2}(Q)} = \|u_t\|_{L_p(Q)} + \sum_{i,j=1}^{d} \|u_{x^ix^j}\|_{L_p(Q)} + \sum_{i=1}^{d} \|u_{x^i}\|_{L_p(Q)} + \|u\|_{C(Q)}$$

1.3. Conditional existence theorems for the solution of a nonlinear equation

In this section we prove a theorem on the existence of a solution to a nonlinear parabolic equation, given an a priori estimate of the solution in $C^{2+\alpha}$. A similar theorem is true for nonlinear elliptic equations. Thus, we shall see that the problem of proving the existence of a solution reduces to the problem of constructing a priori estimates. This reduction is well known in the theory of differential equations thanks, first of all, to the classical works of S.N. Bernstein. He was also the first to suggest the parametric extention method which allows us to establish the solvability from a priori estimates.

We now introduce some notation. Let numbers $T, K, K_1, \gamma, \varepsilon > 0$, $\alpha \in (0,1)$ and a function $\Psi \in C^{2+\alpha}(E_d)$ be fixed such that $D := \{x \in E_d: \Psi(x) > 0\}$ is a nonempty domain, $|\Psi_x| \geqq \varepsilon$ on ∂D (if $\partial D \neq \phi$). Also, let

$$Q = (0,T) \times D, \quad P(M) = P(M,Q)$$

$$= \{(u_{ij}, u_i, u, z): \sum_{i,j=1}^{d} |u_{ij}| + \sum_{i=1}^{d} |u_i| + |u| \leqq M, \ u_{ij} = u_{ji}, \ z \in Q\}.$$

Let, for $z \in Q$ and arbitrary real-valued u_{ij}, u_i $(i, j=1, \ldots, d)$, and u, a real-valued function $F(u_{ij}, u_i, u, z)$ be defined. Suppose that the function F, its first derivatives with respect to all arguments and its second derivatives with respect to (u_{ij}, u_i, u, x) are bounded on $P(K+\gamma)$ and satisfy the Hölder condition of order α relative to the distance

$$\sum_{i,j} |u^1_{ij} - u^2_{ij}| + \sum_i |u^1_i - u^2_i| + |u^1 - u^2| + |x_1 - x_2| + |t_1 - t_2|^{1/2}. \tag{1}$$

Also, suppose that, for all $\lambda \in E_d$ and $(u_{ij}, u_i, u, z) \in P(K+\gamma)$, the inequality

$$\varepsilon|\lambda|^2 \leq F_{u_{ij}} \lambda^i\lambda^j.$$

holds.

By virtue of the above conditions, the function F is uniformly continuous on $P(K+\gamma)$ and it can be continuously extended to the closure of this set. The above derivatives of F can also be extended to the closure.

1. THEOREM. Let $\varphi \in C^{2+\alpha}(\bar{Q})$, $(\varphi_{x^ix^j}(z), \varphi_{x^i}(z), \varphi(z), z) \in P(K)$ for $z \in Q$,

$$\varphi_t + \Delta\varphi = 0,\ \varphi_t + F(\varphi_{x^ix^j}, \varphi_{x^i}, \varphi, z) = 0 \text{ on } \partial_{tx}Q. \tag{2}$$

(Any two functions are supposed to coincide on the empty set.)

Let $\alpha \geqq \beta > 0$ and let, for any $\eta \in [0,1]$ and any function u such that $u \in C^{2+\beta}(\bar{Q})$, u_x, u_t, $u_{xx} \in C^2_{loc}(Q)$, and

$$u_t(z) + \eta F(u_{x^ix^j}(z), u_{x^i}(z), u(z), z) + (1-\eta)\Delta u(z) = 0 \text{ in } Q \tag{3}$$

$$u = \varphi \quad \text{on } \partial'Q, \tag{4}$$

we have

$$\sum_{i,j} |u_{x^ix^j}| + \sum_i |u_{x^i}| + |u| \leqq K \text{ in } Q, \quad ||u||_{C^{2+\beta}(Q)} \leqq K_1. \tag{5}$$

Then, for $\eta=1$, equation (3) with boundary conditions (4) has a solution $u \in C^{2+\beta}(\bar{Q})$ such that u, u_x, u_t, $u_{xx} \in C^{2+\alpha}_{loc}(Q)$.

In the second hypothesis of the theorem we assumed the availability of a priori estimates (5) if problem (3) and (4) has a sufficiently regular solution. This is a conditional assumption which means that, for any $\eta \in [0,1]$, the set of sufficiently regular solutions to problem (3)-(4) has to belong to the set of functions satisfying (5). For a given $\eta \in [0,1]$, this assumption is automatically fulfilled if the former of the sets mentioned turns out to be empty.

To prove this theorem we need two lemmas.

2. LEMMA. Let $\eta \in [0,1]$, $u \in C^{2+\beta}_{loc}(Q)$, u satisfy (3), and

$$\sum_{i,j} |u_{x^ix^j}| + \sum_i |u_{x^i}| + |u| \leqq K$$

in Q. Then u, u_x, u_t, and $u_{xx} \in C^{2+\alpha}_{loc}(Q)$.

Proof. Except for some detail, we reproduce the proof of Theorem 13 presented in Chapter 3 of Friedman [1]. Suppose that $z_0 \in Q$, $\rho \in (0,1)$, $G_i := z+C_{i^2\rho^2, i\rho}$, $i=1,2$, and $\bar{G}_2 \subset Q$. From (3) we subtract the same equation considered at the point $z+\zeta$, where $\zeta = (\tau,\xi) \in E_{d+1}$ is a chosen point. Then, for

$$u^{\zeta}(z) := \frac{1}{|\tau| + |\xi|}\,[u(z+\zeta)-u(z)]$$

and sufficiently small $|\tau| + |\xi|$ (such that $\zeta+G_2 \subset Q$), it follows from the Hadamard formula that on G_2

$$u^{\zeta}_t + (1-\eta)\Delta u^{\zeta} + \eta a^{ij}u^{\zeta}_{x^ix^j} + \eta b^i u^{\zeta}_{x^i} + \eta c u^{\zeta} + \eta f = 0,$$

where

$$a^{ij} = a^{ij}(z) = \int_0^1 F_{u_{ij}}\Big(\theta v_{x^rx^s} + (1-\theta)u_{x^rx^s},\ \theta v_{x^r} + (1-\theta)u_{x^r},$$
$$\theta v + (1-\theta)u,\ \theta(z+\zeta) + (1-\theta)z\Big)d\theta,$$

and $v=v(z)=u(z+\zeta)$. Replacing $F_{u_{ij}}$ by F_{u_i}, F_u, and $(|\tau| + |\xi|)^{-1}F_{(\zeta)}$ respectively, we obtain b^i, c, and f. Here, for the sake of brevity, obvious values of the arguments are omitted.

It follows from the conditions on F that a^{ij}, ..., f satisfy a Hölder condition with respect to $z \in G_2$ with the exponent $\delta = \beta\alpha$ and a Hölder constant independent of ζ. Therefore, it follows from the theory of linear equations that, for sufficiently small ζ, the norms of u^{ζ} in $C^{2+\delta}(G_1)$ are bounded by a constant which is independent of ζ. Choosing $\zeta \| (1,0)$, $(0,e_i)$, where e_i is a basic vector of E_d, and letting $\zeta \to 0$, we conclude that u_t, $u_x \in C^{2+\delta}(G_1)$. Since z_0 is an arbitrary point, it follows that u_t and u_x belong to $C^{2+\delta}_{loc}(Q)$. In particular, this yields an upper bound uniform in ζ for $|(u_x)_{xx}|$ and $|(u_t)_{xx}|$ on G_2 and for $|(v_x)_{xx}|$ and $|(v_t)_{xx}|$ on G_2 for sufficiently small ζ. It follows that for the functions u_{xx} and v_{xx} the constants of the first-order Hölder condition with respect to $z \in G_2$ are bounded. Now, the functions a^{ij}, ..., f involved in (6) satisfy the Hölder condition of order α (when

$\zeta = (\tau,0)$, the order exactly equals α) with a constant which is independent of ζ. Hence, $u_t, u_x \in C^{2+\alpha}_{loc}(Q)$.

We next fix the rth coordinate axis of E_d, set $w=u_{x^r}$, and differentiate (3) with respect to x^r. In view of the above discussion, we can do this. For w in Q, we obtain

$$w_t + F_{u_{ij}} w_{x^i x^j} + F_{u_i} w_{x^i} + F_u w + F_{x^r} = 0.$$

We regard this relationship as an equation in w. By virtue of the above consideration, its coefficients and constant term are differentiable with respect to x, moreover, the coefficients themselves and their derivatives with respect to x satisfy a Hölder condition with respect to $z \in G_2$ of order α. From the theory of linear equations it follows that $w_x \in C^{2+\alpha}(G_1)$, $w_x \in C^{2+\alpha}_{loc}(Q)$ and recalling the definition of w, we conclude that $u_{xx} \in C^{2+\alpha}_{loc}(Q)$ as well. This proves the lemma.

3. LEMMA. Let $N \geqq 0$, let a function $\phi(\xi,z)$ be defined for $\xi \in \{\sigma \in E_{d_1} : |\sigma| \leqq N\}$ and $z \in \bar{Q}$, and let this function be bounded together with its first derivatives with respect to ξ. Let, for any ξ, the functions ϕ and ϕ_ξ, as considered on $\bar{Q}$, satisfy a Hölder condition of order β with a constant which is independent of ξ. Also suppose that ϕ_ξ is uniformly continuous in (ξ,z). Then $\phi : \xi(z) \to \phi(\xi(z),z)$ is a continuous operator from $C^\beta(\bar{Q}) \cap \{\xi: ||\xi||_{C(Q)} \leqq N\}$ into $C^\beta(\bar{Q})$.

Proof. Suppose that our assertion is not true. Then there exist $\delta > 0$ and $\xi_n \in C^\beta(\bar{Q}) \cap \{\xi: ||\xi||_{C(Q)} \leqq N\}$, n=0, 1, 2, ..., such that

$$||\xi_n - \xi_o||_{C^\beta(Q)} \to 0, \quad ||\phi(\xi_n) - \phi(\xi_o)||_{C^\beta(Q)} \geqq 2\delta. \tag{7}$$

The function ϕ is uniformly continuous in ξ with respect to z, therefore, $\phi(\xi_n) \to \phi(\xi_o)$ uniformly on $\bar{Q}$. Thus, inequality (7) implies that, for some points $z_n', z_n'' \in \bar{Q}$, we obtain

$$|[\phi(\xi_n(z_n'), z_n') - \phi(\xi_o(z_n'), z_n')] - -[\phi(\xi_n(z_n''), z_n'') - \phi(\xi_o(z_n''), z_n'')]| \geqq \delta\rho_n^\beta \tag{8}$$

where $\rho_n = \rho(z_n', z_n'')$, $\rho_n \to 0$. In what follows, we shall omit the subscript n. Note that from (7) it follows that

$$\mu := [\xi(z') - \xi_0(z')] - [\xi(z'') - \xi_0(z'')] = o(\rho^\beta),$$

$$\nu := \xi_0(z') - \xi_0(z'') = O(\rho^\beta).$$

From this, by Hadamard's formula, we obtain

$$[\phi(\xi(z'), z') - \phi(\xi(z''), z')] - [\phi(\xi_0(z'), z') - \phi(\xi_0(z''), z')]$$

$$= \mu^i \int_0^1 \phi_{\xi^i}(\theta\xi(z') + (1-\theta)\xi(z''), z')\, d\theta +$$

$$+ \nu^i \int_0^1 [\phi_{\xi^i}(\theta\xi(z') + (1-\theta)\xi(z''), z') -$$

$$- \phi_{\xi^i}(\theta\xi_0(z') + (1-\theta)\xi_0(z''), z')]d\theta = o(\rho^\beta).$$

In addition,

$$[\phi(\xi(z''), z') - \phi(\xi_0(z''), z')] - [\phi(\xi(z''), z'') - \phi(\xi_o(z''), z'')]$$

$$= (\xi^i(z'') - \xi_0^i(z'')) \int_0^1 [\phi_{\xi^i}((\theta\xi + (1-\theta)\xi_0)(z''), z') -$$

$$- \phi_{\xi^i}((\theta\xi + (1-\theta)\xi_0)(z''), z'')]d\theta = o(\rho^\beta),$$

since $\xi(z'') - \xi_0(z'') \to 0$. Adding together the derived relationships, we come to a contradiction with (8). This proves the lemma.

4. Proof of Theorem 1. (see Section 2.3 in Chapter 4 of Agmon, Douglis, and Nirenberg [1]) Let I be the set of all $\eta \in [0,1]$ such that problem (3)-(4) has a solution $u = u^\eta \in C^{2+\beta}(\bar{Q})$ which possesses the property $u, u_x, u_t, u_{xx} \in C^{2+\alpha}_{loc}(Q)$. Zero is known to belong to I. We shall prove that $I = [0,1]$ (i.e., the theorem will be proved), if I is shown to be closed and open in the relative topology on $[0,1]$.

From Lemma 2 it follows that I is closed. Indeed, if $\eta_n \in I$ and $\eta_n \to \eta$, then from (5) it follows that the norms of u^{η_n} in $C^{2+\beta}(Q)$ are bounded by a constant K_1 and it is possible to select a subsequence uniformly converging on $\bar{Q}$ of the sequence of functions u^{η_n}, $u_x^{\eta_n}$, $u_t^{\eta_n}$, $u_{xx}^{\eta_n}$. This subsequence converges to u, u_x, u_t, u_{xx}, where $u \in C^{2+\beta}(\bar{Q})$ is a solution to problem (3)-(4) which satisfies inequalities (5). Moreover, from Lemma 2 it follows that $u, u_x, u_t, u_{xx} \in C^{2+\alpha}_{loc}(Q)$, i.e. $\eta \in I$.

In order to prove that I is open in [0,1], we take an $\eta' \in I$ and denote $u' = u^{\eta'}$. Assuming that w is a known function, consider the following problem with respect to the function v:

$$\begin{aligned} & v_t + (1-\eta)\Delta v + \eta(a^{ij}v_{x^ix^j} + b^i v_{x^i} + cv) + \\ & + \eta\left[F\left(w_{x^kx^l}, w_{x^k}, w, z\right) - a^{ij}w_{x^ix^j} - b^i w_{x^i} - cw\right] = 0 \text{ in } Q, \end{aligned} \tag{9}$$

$$v = \varphi \text{ on } \partial' Q, \tag{10}$$

where

$$(a^{ij}, b^i, c) = \left(F_{u_{ij}}, F_{u_i}, F_u\right)\left(u'_{x^kx^l}, u'_{x^k}, u', z\right). \tag{11}$$

Our hypothesis implies that u' satisfies inequalities (5), and since the second derivatives of F with respect to (u_{ij}, u_i, u) are bounded on P(K), while $u' \in C^{2+\beta}(Q)$, then $a^{ij}, b^i, c \in C^{\beta}(Q)$. Moreover, there exists a small $\delta_0 > 0$ such that if

$$w \in \mathbf{A}(\delta_0) := \left\{w \in C^{2+\beta}(\bar{Q}) : ||w - u'||_{C^{2+\beta}(Q)} \leqq \delta_0, \right.$$
$$\left. w = \varphi \text{ on } \partial' Q\right\},$$

then the condition $(w_{x^ix^j}, w_{x^i}, w, z) \in P(K+\gamma)$ holds for all $z \in Q$. For such functions w, the last multiplier of η in (9) belongs to $C^{\beta}(Q)$, the function φ satisfies the first-order matching condition and hence linear equation (9) with boundary condition (10) has a unique solution $v \in C^{2+\beta}(\bar{Q})$. Therefore, for every $\eta \in [0,1]$, a mapping $\Psi_\eta : w \to v$ from $\mathbf{A}(\delta_0)$ in to $C^{2+\beta}(\bar{Q})$ is defined.

We further want to show that, for a $\delta > 0$ and all η which are sufficiently close to η' (not necessarily belonging to [0,1]), Ψ_η is a contracting mapping from $\mathbf{A}(\delta)$ into $\mathbf{A}(\delta)$ in the metric of $C^{2+\beta}(Q)$.

It is not difficult to see that, by virtue of (9),

$$\begin{aligned} & (v-u')_t + (1-\eta)\Delta(v-u') + \\ & + \eta\left[a^{ij}(v-u')_{x^ix^j} + b^i(v-u')_{x^i} + c(v-u')\right] + \\ & + \eta\left[F\left(w_{x^kx^l}, w_{x^k}, w, z\right) - F\left(u'_{x^kx^l}, u'_{x^k}, u', z\right) - \right. \\ & \left. - a^{ij}(w-u')_{x^ix^j} - b^i(w-u')_{x^i} - c(w-u')\right] + \end{aligned}$$

$$+ (\eta'-\eta)\left[\Delta u' - F\left(u'_{x^kx^l}, u'_{x^k}, u', z\right)\right] = 0. \tag{12}$$

Further, using the Hadamard formula, we transform

$$F\left(w_{x^kx^l}, w_{x^k}, w, z\right) - F\left(u'_{x^kx^l}, u'_{x^k}, u', z\right),$$

recall definition (11) of the functions a^{ij}, b^i, and c and the conditions on F. From Lemma 3 it follows that if $w \in \mathbf{A}(\delta)$, then the $C^{\beta}(\bar{Q})$ - norm of the last multiplier of η in (12) is estimated in terms of $\chi(\delta)\delta$, where $\chi(\delta) \to 0$ as $\delta \downarrow 0$, and $\chi(\delta)$ is independent of w. This along with (12) imply that, by the theorem on estimates of solutions to linear equations,

$$||v - u'||_{C^{2+\beta}(\bar{Q})} \leqq N[\chi(\delta)\delta + |\eta' - \eta|],$$

where N does not depend upon w. Thus, there exists a $\delta \in (0,\delta_0]$ such that for all η, which are sufficiently close to η', Ψ_η is a mapping from $\mathbf{A}(\delta)$ into $\mathbf{A}(\delta)$.

If $v = \Psi_\eta w$ and $\tilde{v} = \Psi_\eta \tilde{w}$, while $w, \tilde{w} \in \mathbf{A}(\delta)$, it follows from (9) that

$$(v-v)_t + (1-\eta)\Delta(v-\tilde{v}) + \eta\left(a^{ij}(v-\tilde{v})_{x^ix^j} + b^i(v-\tilde{v})_{x^i} + c(v-\tilde{v})\right) +$$

$$+ \eta\left[F\left(w_{x^kx^l}, w_{x^k}, w, z\right) - F\left(\tilde{w}_{x^kx^l}, \tilde{w}_{x^k}, \tilde{w}, z\right) -\right.$$

$$\left. - a^{ij}(w-\tilde{w})_{x^ix^j} - b^i(w-\tilde{w})_{x^i} - c(w-\tilde{w})\right] = 0.$$

Hence, it can be seen that

$$||v-\tilde{v}||_{C^{2+\beta}(\bar{Q})} \leqq N\chi(\delta)||w-\tilde{w}||_{C^{2+\beta}(\bar{Q})},$$

and reducing, if necessary, δ, we can make Ψ_η a contracting mapping of the set $\mathbf{A}(\delta)$ (for example, simultaneously for all $\eta \in [0,1]$).

Thus, from the fixed point theorem for a contracting mapping it follows that, for every $\eta \in [\eta'+\delta, \eta'+\delta]$, there exists a $v \in \mathbf{A}(\delta)$ such that $v = \Psi_\eta v$. This v obviously satisfies (3) and (4). Moreover, Lemma 2 implies that $v, v_x, v_t, v_{xx} \in C^{2+\alpha}_{loc}(Q)$ and $(\eta'-\delta, \eta'+\delta) \cap [0,1] \subset I$. This proves the theorem.

5. Remark. The assumption concerning the second derivatives of F is used twice, first, to prove the continuity of the operator Ψ_η and second, to increase the inner smoothness of a solution to equation (3) by use of

Lemma 2. The latter seems to be of minor importance. Indeed, if we remove the assumption u_x, u_t, $u_{xx} \in C^2_{loc}(Q)$, from the conditions preceding (3) to (5), the hypothesis would be stronger, while the proof simpler. In particular, Lemma 2 would actually be excessive. However, in this book, we can check this strong assumption only if the condition u_x, u_t, $u_{xx} \in C^2_{loc}(Q)$ holds.

1.4. Comments

The Aleksandrov–Busemann–Feller theorem 2.1 and its extensions to the functions which depend on t are presented in Appendix 2. In addition, Aleksandrov [1] shows that u_{x^i} are differentiable almost everywhere and the partial derivatives of u_{x^i} with respect to x^j coincide with $u^{(0)}_{x^i x^j}$ almost everywhere. This fact and its extensions to arbitrary monotone mappings are also proved in the author's paper [18].

Chapter 2

NONLINEAR EQUATIONS WITH CONSTANT 'COEFFICIENTS' IN THE WHOLE SPACE

In this brief chapter the solvability of Bellman equations with constant 'coefficients' in the whole space is proved. On the one hand, we easily obtain a broad class of nonlinear equations which enable us to establish the solvability by elementary means. On the other hand, the results of this chapter are essentially used in the next chapter for the derivation of the Aleksandrov estimates.

2.1. Constructing the resolvent

Given a set Ω and an $\varepsilon > 0$, let a nonnegative symmetric $d \times d$ - matrix $a(\omega) = (a^{ij}(\omega))$, a d-dimensional vector $b(\omega) = (b^i(\omega))$, and real-valued functions $c(\omega)$ and $f(\omega,x)$ be defined for all $\omega \in \Omega$ and $x \in E_d$. We shall need three functional spaces. The space $C^2(E_d)$ was introduced in Section 1.1. The space $BC(\Omega,E_d)$ is defined as the set of all bounded functions $g(\omega,x)$ on $\Omega \times E_d$ with supremum norm which satisfy the condition

$$\lim_{y\to 0} \sup_{\omega,x} |g(\omega, x+y) - g(\omega,x)| = 0.$$

The space $C_u(E_d)$ is defined as the subset of $BC(\Omega,E_d)$ consisting of functions which are independent of ω.

In this section we always assume that $a(\omega)$, $b(\omega)$, and $c(\omega)$ are bounded on Ω, f, f_{x^i}, $f_{x^ix^i} \in BC(\Omega,E_d)$, and

$$\sup_\omega \|f(\omega)\|_{C^2(E_d)} < \infty, \quad \sup_\omega c(\omega) \leqq -\varepsilon,$$

where $f(\omega)$ is the function $f(\omega,x)$ with $\omega \in \Omega$. This abbreviation is also used for different functions defined on $\Omega \times E_d$.

We set

$$F(u_{ij}, u_i, u, x) = \sup_{\omega\in\Omega}\left[a^{ij}(\omega)u_{ij} + b^i(\omega)u_i + c(\omega)u + f(\omega,x)\right],$$

$$F[u](x) = F\left(u_{x^ix^j}(x), u_{x^i}(x), u(x), x\right)$$

and consider the following problem of finding a sufficiently smooth function $v = v(x)$ such that, in E_d,

$$F[v] = 0. \tag{1}$$

We now explain in its general form the idea underlying the solution of this problem. We denote

$$L(\omega)u(x) = a^{ij}(\omega)u_{x^ix^j}(x) + b^i(\omega)u_{x^i}(x) + c(\omega)u(x),$$

and let $\mathfrak{A}$ be the set of all functions on E_d taking their values in Ω. Suppose (which is absolutely unrealistic) that, for any $\lambda \geqq 0$, $\alpha \in \mathfrak{A}$, and bounded g, we can solve the equation

$$L(\alpha(x))u(x) - \lambda u(x) + g(x) = 0$$

in the class of sufficiently smooth functions. Denote the solution by $\tilde{R}^{\alpha}_{\lambda}g$. From (1) it follows that

$$L(\alpha(x))v(x) - \lambda v(x) + f(\alpha(x),x) + \lambda v(x) \leqq 0 \tag{2}$$

and the maximum principle implies that $v \geqq \tilde{R}^{\alpha}_{\lambda}(\tilde{f}^{\alpha} + \lambda v)$ for any $\alpha \in \mathfrak{A}$ where $\tilde{f}^{\alpha}(x) := f(\alpha(x),x)$ if $\alpha \in \mathfrak{A}$.

On the other hand, from the definition of F it follows that one can choose $\alpha \in \mathfrak{A}$ such that the left-hand side of (2) is arbitrarily close to zero. Hence it looks plausible that

$$v = \sup_{\alpha\in\mathfrak{A}} \tilde{R}^{\alpha}_{0}\tilde{f}^{\alpha}, \tag{3}$$

and in the general case for any $\lambda \geqq 0$,

$$v = \sup_{\alpha\in\mathfrak{A}} \tilde{R}^{\alpha}_{\lambda}(\tilde{f}^{\alpha} + \lambda v). \tag{4}$$

To fulfill our plan of studying (1), we have basically to determine v from formula (3) and then to prove that v satisfies (1). We have already mentioned that it seems impossible to construct $\tilde{R}^{\alpha}_{0}$ for arbitrary $\alpha \in \mathfrak{A}$ with no assumptions concerning the nondegeneracy of a. Therefore, we use the fact that the equations

$$\tilde{R}^{\alpha}_{0} = \tilde{R}^{\alpha}_{\lambda}(1 + \lambda\tilde{R}^{\alpha}_{0}), \quad \tilde{R}^{\alpha}_{0} = \sum_{n=1}^{\infty} (\tilde{R}^{\alpha}_{\lambda})^n \lambda^{n-1}$$

hold (cf. (4)) if $\tilde{R}^{\alpha}_{0}$ exists.

It is not difficult to see that in 'favourable' circumstances for large λ we obtain a 'good' approximation to $\tilde{R}^{\alpha}_{0}$ if, in the latter formula,

$\tilde{R}^{\alpha}_{\lambda}$ is replaced by an operator setting a correspondence between the function $g(x)$ and a function $R^{\omega}_{\lambda}g(x)|_{\omega=\alpha(x)}$, where R^{ω}_{λ} is defined as $\tilde{R}^{\alpha}_{\lambda}$, with α being independent of x and equal to ω. The operator R^{ω}_{λ} can be specified by an explicit formula (see (5)) and it can be used instead of $\tilde{R}^{\alpha}_{\lambda}$ in formula (4) which now becomes an equation. Since R^{ω}_{λ} approximates $\tilde{R}^{\alpha}_{\lambda}$ 'well' for $\omega = \alpha(x)$, the solution to this equation should give the desired function v as $\lambda \to \infty$. To prove that v is a solution to equation (1), it is again convenient to use (4) (but only as far as the operators $\tilde{R}^{\alpha}_{\lambda}$ are concerned) by carrying v to the right, multiplying by λ, letting $\lambda \to \infty$, and using the fact that, in a sense, $\lambda\tilde{R}^{\alpha}_{\lambda}\tilde{f}^{\alpha} \to f$ and $\lambda(\lambda\tilde{R}^{\alpha}_{\lambda}v - v) \to L(\alpha)v$.

To implement this plan, we need the following well known properties of the resolvent of linear equations.

1. LEMMA. If $\omega \in \Omega$, $\lambda \geq 0$, then there exist linear continuous operators $R^{\omega}_{\lambda} : C(E_d) \to C(E_d)$ such that

(a) $\lambda R^{\omega}_{\lambda}u = R^{\omega}_{\lambda}L(\omega)u + u$ for $u \in C^2(E_d)$;

(b) for $x \in E_d$, $g \in C(E_d)$, the following inequality holds:
$$|R^{\omega}_{\lambda}g(x)| \leq R^{\omega}_{\lambda}|g|(x) \leq \frac{1}{\varepsilon+\lambda}||g||_{C(E_d)};$$

(c) $R^{\omega}_{\lambda} = R^{\omega}_{\mu}(1+R^{\omega}_{\lambda}(\mu-\lambda))$ if $\lambda,\mu \geq 0$;

(d) $R^{\omega}_{\lambda}U^y = U^y R^{\omega}_{\lambda}$ for any $y \in E_d$, where the operator U^y is defined as $U^y g(x) = g(x+y)$;

(e) if $\lambda \geq 0$, $\mu \geq \nu \geq 0$, g, $u \in C(E_d)$, and $u \geq R^{\omega}_{\lambda+\mu}(g+\mu u)$ on E_d, then $u \geq R^{\omega}_{\lambda+\nu}(g+\nu u)$ on E_d.

Proof. Denote by $\sigma(\omega)$ the square root of the matrix $2a(\omega)$ and set

$$R^{\omega}_{\lambda}g(x) = \int_0^{\infty} e^{(c(\omega)-\lambda)t} \frac{1}{(2\pi t)^{d/2}} \int g(x+\sigma(\omega)y + b(\omega)t)e^{-|y|^2/(2t)}\,dy\,dt. \tag{5}$$

The reader can easily check that R^{ω}_{λ} possesses the properties (a) to (d). In order to prove (e), we note that, by virtue of (b) and (c),

$$R^{\omega}_{\lambda+\nu} = \sum_{n=0}^{\infty} (\mu-\nu)^n \left(R^{\omega}_{\lambda+\mu}\right)^{n+1},$$

and the series converges in the operator norm. Applying both sides of this equality to $g+\nu\cdot u$ and noting that

$$R^{\omega}_{\lambda+\mu}(g+\nu u) = R^{\omega}_{\lambda+\mu}(g+\mu u) - (\mu-\nu)R^{\omega}_{\lambda+\mu}u \leqq u - (\mu-\nu)R^{\omega}_{\lambda+\mu}u,$$

we obtain

$$R^{\omega}_{\lambda+\nu}(g+\nu u) \leqq \sum_{n=0}^{\infty} (\mu-\nu)^n (R^{\omega}_{\lambda+\mu})^n (u - (\mu-\nu)R^{\omega}_{\lambda+\mu}u) = u.$$

This proves the lemma.

We now make the first step towards constructing a solution to (1). For $\lambda \geqq 0$ and $g \in BC(\Omega,E_d)$, we denote

$$T_\lambda g(x) = \sup_{\omega\in\Omega} R^{\omega}_{\lambda} g(\omega)(x).$$

Since the relations $y \in E_d$ and $g \in BC(\Omega,E_d)$ imply that $U^y T_\lambda g = T_\lambda U^y g$, we obtain

$$\begin{aligned} |U^y T_\lambda g - T_\lambda g| &\leqq \sup_{\omega} |R^{\omega}_{\lambda} \sup_{\omega} |U^y g(\omega) - g(\omega)| \leqq \\ &\leqq \frac{1}{\varepsilon+\lambda} \sup_{\omega,x} |g(\omega, x+y) - g(\omega,x)|. \end{aligned} \tag{6}$$

Note that, generally, if $g_1, g_2 \in BC(\Omega,E_d)$, then

$$|T_\lambda g_1 - T_\lambda g_2| \leqq T_\lambda |g_1 - g_2| \leqq \frac{1}{\varepsilon+\lambda} \sup_{\omega,x} |g_1(\omega,x) - g_2(\omega,x)|. \tag{7}$$

We further consider the equation

$$u = T_{\lambda+\mu}(g+\mu u) \tag{8}$$

with $\lambda, \mu \geqq 0$ and $g \in BC(\Omega,E_d)$, whose solution will be sought for in the Banach space $C_u(E_d)$. From (6) and (7) it follows that $T_{\lambda+\mu}(g+\mu\cdot)$ is a contracting operator from $C_u(E_d)$ into $C_u(E_d)$. Therefore, there exists a unique solution to equation (8) which can be found by iteration starting from any $u_0 \in C_u(E_d)$. Let the solution to equation (8) be denoted by $S^{\mu}_{\lambda}g$.

2. LEMMA. Let $\omega \in \Omega$, $g, g_1, g_2 \in BC(\Omega,E_d)$, and $\lambda, \nu, \mu \geqq 0$. Then

(a) $|S^{\mu}_{\lambda}g| \leqq \frac{1}{\varepsilon+\lambda} \sup_{\omega,x} |g(\omega,x)|$;

(b) $S_\lambda^{\nu+\mu} g = S_{\lambda+\mu}^{\nu}(g+\mu S_\lambda^{\nu+\mu} g)$;

(c) if $v \in C_u(E_d)$ and $v \leqq (\geqq)\ T_{\lambda+\nu}(g+\nu\cdot v)$, then $v \leqq (\geqq) S_\lambda^\nu g$;

(d) $S_\lambda^\mu g \geqq R_{\lambda+\mu}^{\omega}(g(\omega)+\mu S_\lambda^\mu g)$;

(e) if $\mu \geqq \nu$ then $S_\lambda^\mu g \geqq T_{\lambda+\nu}(g+\nu S_\lambda^\mu g)$, $S_\lambda^\mu g \geqq S_\lambda^\nu g$;

(f) if $u, u_x, u_{xx} \in C_u(E_d)$, then $S_\lambda^\mu(g+\lambda u) = u+S_\lambda^\mu(g+Lu)$, where Lu is the function $L(\omega)u(x)$ defined on $\Omega \times E_d$;

(g) if $g_1 \geqq g_2$, then $S_\lambda^\mu g_1 \geqq S_\lambda^\mu g_2$, specifically, if $g \geqq 0$, then $S_\lambda^\mu g \geqq S_\lambda^\mu 0 = 0$, and if $g \leqq 0$, then $S_\lambda^\mu g \leqq 0$;

(h) $S_\lambda^\mu g_1 - S_\lambda^\mu g_2 \leqq S_\lambda^\mu(g_1-g_2)$;

(i) $U^y S_\lambda^\mu = S_\lambda^\mu U^y$, $\nu S_\lambda^\mu g = S_\lambda^\mu \nu g$, and the family $\{S_\lambda^\mu g : \lambda, \mu \geqq 0\}$ is equicontinuous in x on E_d.

Proof. (a) We set $u = S_\lambda^\mu g$. Then from (8) we obtain

$$|u| \leqq \frac{1}{\varepsilon+\lambda+\mu}\Big(\sup_{\omega,x}|g(\omega,x)| + \mu \sup_x |u(x)|\Big),$$

hence assertion (a) is true.

In order to prove (b), it suffices to set $u_1 := S_\lambda^{\nu+\mu} g$, $u_2 := S_{\lambda+\mu}^{\nu}(g+\mu S_\lambda^{\nu+\mu} g) = S_{\lambda+\mu}^{\nu}(g+\mu u_1)$, to take note that

$$u_1 = T_{\lambda+\mu+\nu}(g+\mu u_1+\nu u_1), \quad u_2 = T_{\lambda+\mu+\nu}(g+\mu u_1+\nu u_2),$$

and to make use of the uniqueness of the solution to the equation $u = T_{\lambda+\mu+\nu}(g+\mu u_1+\nu u)$.

Assertion (c) follows from the fact that the iterative method with an initial function v yields an increasing (decreasing) functional sequence converging to $S_\lambda^\nu g$. Assertion (d) immediately follows from the definitions of T_λ and S_λ^μ. Assertion (e) follows from (d), Lemma 1 (e), and (c).

To prove assertion (f), we denote $u_1 = S_\lambda^\mu(g+\lambda u)-u$. Then from the definition of S_λ^μ and assertion (a) of Lemma 1 it follows that

$$u + u_1 = T_{\lambda+\mu}(g+\lambda u+\mu u+\mu u_1) = T_{\lambda+\mu}(g+Lu+\mu u_1) + u.$$

Hence $u_1 = S_\lambda^\mu(g+Lu)$, which proves the statement.

Assertion (g) immediately follows from (c) since $T_{\lambda+\mu}(g_1+\mu\cdot u) \geqq T_{\lambda+\mu}$

$(g_2+\mu u)$. Assertion (h) also follows from (c) since for $u_1 = S^{\mu}_{\lambda}(g_1-g_2)$ and $u_2 = S^{\mu}_{\lambda}g_2$, we have

$$u_1 + u_2 = T_{\lambda+\mu}(g_1-g_2+\mu u_1) + T_{\lambda+\mu}(g_2+\mu u_2) \geqq T_{\lambda+\mu}(g_1+\mu u_1+\mu u_2).$$

Finally, $U^y T_{\lambda+\mu} = T_{\lambda+\mu}U^y$ and $\nu T_{\lambda+\mu} = T_{\lambda+\mu}\nu$ by virtue of Lemma 1(d), hence $U^y S^{\mu}_{\lambda} = S^{\mu}_{\lambda}U^y$ and $\nu S^{\mu}_{\lambda} = S^{\mu}_{\lambda}\nu$ (because the solution to (8) is unique), and from (h) and (a) it follows that

$$U^y S^{\mu}_{\lambda}g - S^{\mu}_{\lambda}g \leqq S^{\mu}_{\lambda}(U^y g-g) \leqq \frac{1}{\varepsilon} \sup_{\omega,x}|g(\omega, x+y) - g(\omega, x)|,$$

the latter expression tending to zero as $y\to 0$. This proves assertion (i) and completes the proof of the lemma.

From assertions (a), (e), and (i) of Lemma 2 it follows that, for any $g \in BC(\Omega,E_d)$ and $\lambda \geqq 0$, the function $S^{\mu}_{\lambda}g(x)$ increases in μ, is bounded, and, as $\mu\to\infty$, uniformly converges in x, which lies in a bounded domain, to a function of the class $C_u(E_d)$. We denote

$$P_{\lambda}g(x) = \lim_{\mu\to\infty} S^{\mu}_{\lambda}g(x).$$

Note that $P_{\lambda}(f+\lambda v)$ is precisely the desired counterpart of the right-hand side of (4). In what follows, we shall see that $P_{\lambda}f$ is a solution to the equation $\lambda u - F[u] = 0$. Therefore, it is natural to call P_{λ} the resolvent of the operator F.

3. THEOREM. Let $\omega \in \Omega$, g, g_1, $g_2 \in BC(\Omega,E_d)$, and let λ, $\mu \geqq 0$. Then

(a) $|P_{\lambda}g| \leqq \frac{1}{\varepsilon+\lambda} \sup_{\omega,x}|g(\omega,x)|$;

(b) if u, u_x, $u_{xx} \in C_u(E_d)$, then $P_{\lambda}(g+\lambda u) = u+P_{\lambda}(g+Lu)$;

(c) $P_{\lambda}g_1 - P_{\lambda}g_2 \leqq P_{\lambda}(g_1-g_2)$, and if $g_1 \geqq g_2$, then $P_{\lambda}g_1 \geqq P_{\lambda}g_2$, specifically, if $g \leqq 0$, then $P_{\lambda}g \leqq P_{\lambda}0 = 0$;

(d) $U^y P_{\lambda} = P_{\lambda}U^y$, $\mu P_{\lambda}g = P_{\lambda}\mu g$;

(e) $P_{\lambda}g \geqq R^{\omega}_{\lambda+\mu}(g(\omega)+\mu P_{\lambda}g)$;

(f) $P_{\lambda}g = P_{\lambda+\mu}(g+\mu P_{\lambda}g)$;

(g) $\lambda P_{\lambda}g \to \sup_{\omega} g(\omega)$ uniformly on E_d when $\lambda\to\infty$;

(h) $P_{\lambda}(\sup_{\omega} g) \geqq P_{\lambda}g \geqq -P\,(-\sup_{\omega} g)$, specifically, if $\sup_{\omega} g \geqq 0$, then $P_\lambda g \geq -P_\lambda 0 = 0$;

(i) if $x_0 \in E_d$, $u \in C_u(E_d)$ and there exist numbers u_i, u_{ij} ($i,j = 1, \ldots, d$) such that $u(x) = u(x_0) + u_i(x-x_0)^i + \frac{1}{2}u_{ij}(x-x_0)^i (x-x_0)^j + o(|x-x_0|^2)$ as $x \to x_0$, then $\lambda(P_\lambda(f+\lambda u)-u)(x_0) \to F(u_{ij}, u_i, u(x_0), x_0)$ as $\lambda \to \infty$.

Proof. Assertions (a) to (d) immediately follow from Lemma 2 and the definition of P_λ. Assertion (e) follows from the fact that, for $\nu \geq \mu$, Lemma 2(e) implies $S_\lambda^\nu g \geq R_{\lambda+\mu}^\omega(g(\omega)+\mu S_\lambda^\nu g)$ and it is easy to pass to the limit if one uses the integral representation (5).

From (e) and Lemma 2(c) we obtain $P_\lambda g \geq T_{\lambda+\mu+\nu}(g+\mu P_\lambda g+\nu P_\lambda g)$, $P_\lambda g \geq S_{\lambda+\mu}^\nu(g+\mu P_\lambda g)$. Hence, if $\nu \to \infty$, then $P_\lambda g \geq P_{\lambda+\mu}(g+\mu P_\lambda g)$. In order to prove the opposite inequality, i.e., assertion (f), we merely note that from Lemma 2(b) it follows that

$$S_\lambda^{\nu+\mu} g \leq S_{\lambda+\mu}^\nu(g+\mu P_\lambda g) \leq P_{\lambda+\mu}(g+\mu P_\lambda g).$$

We now prove (g). Let $\zeta \in C_0^\infty(E_d)$, $\zeta \geq 0$, $\zeta(x) = 0$ for $|x| \geq 1$, and $\int\zeta(x)dx = 1$. For any $n \geq 1$, we set $g_n(\omega,x) = g(\omega,x)*n^d\zeta(nx)$ and observe that

$$|g-g_n| \leq \int_{|y|\leq 1} \left|g(\omega,x) - g(\omega,x-\tfrac{1}{n}y)\right| \zeta(y)dy \leq$$

$$\leq \sup_{|y|\leq\frac{1}{n}} \sup_{\omega,x} |g(\omega,x) - g(\omega,x+y)| =: \gamma_n,$$

and $\gamma_n \to 0$ as $n \to \infty$. This along with (c) and (a) imply

$$\left|\lambda P_\lambda g - \sup_\omega g\right| \leq \lambda|P_\lambda g - P_\lambda g_n| + \left|\lambda P_\lambda g_n - \sup_\omega g_n\right| +$$

$$+\left|\sup_\omega g - \sup_\omega g_n\right| \leq 2\gamma_n + \left|\lambda P_\lambda g_n - \sup_\omega g_n\right|.$$

Therefore, to prove (g), it suffices to show that the latter summand tends to zero uniformly on E_d for every n as $\lambda \to \infty$. Since g_n and all its derivatives belong to $BC(\Omega,E_d)$, the proof of assertion (g) is reduced to the case when g, g_x, $g_{xx} \in BC(\Omega,E_d)$.

In this case assertion (e) with $\mu = 0$ together with Lemma 1(a) imply

$$\lambda P_\lambda g - \sup_\omega g \geqq \sup_\omega \lambda R_\lambda^\omega g(\omega) - \sup_\omega g =$$

$$= \sup_\omega\left(g + R_\lambda^\omega L(\omega) g(\omega)\right) - \sup_\omega g \geqq \inf_\omega R_\lambda^\omega L(\omega) g(\omega) \geqq$$

$$\geqq -\frac{1}{\lambda} \sup_{\omega,x} |Lg| .$$

Thus, $(\lambda P_\lambda g - \sup_\omega g)_- \to 0$ uniformly on E_d. Moreover,

$$\left(\lambda P_\lambda g - \sup_\omega g\right)_+ \leqq \left(\lambda P_\lambda \sup_\omega g - \sup_\omega g\right)_+$$

by virtue of (c). Therefore, it remains to prove (g) in the case when g is independent of ω. Besides, as shown above, we can assume that g, g_x, $g_{xx} \in C_u(E_d)$. As this takes place, from (b) and (a) we conclude that $|\lambda P_\lambda g - g| = |P_\lambda Lg| \leqq \frac{1}{\lambda} \sup_{\omega,x}|Lg|$, which proves assertion (g).

The inequality $P_\lambda(\sup_\omega g) \geqq P_\lambda g$ involved in assertion (h) follows from (c). The inequality $P_\lambda g \geqq -P_\lambda(-\sup_\omega g)$ will be proved if we show that $P_\lambda g + P_\lambda(\gamma - \sup_\omega g) \geqq 0$ for any constant $\gamma > 0$. Since $P_\lambda g_1 + P_\lambda g_2 \geqq P_\lambda(g_1+g_2)$ by virtue of (c), then $P_\lambda g + P_\lambda(\gamma - \sup_\omega g) \geqq P_\lambda(\gamma+g - \sup_\omega g)$ and thus, to prove (h), it remains to show that $P_\lambda g \geqq 0$ if $\sup_\omega g \geqq \gamma$ for a constant $\gamma > 0$.

For such a g, we choose μ large enough to obtain $P_{\lambda+\mu} g \geqq 0$. This can be done by virtue of (g). We now consider the equation $u = P_{\lambda+\mu}(g+\mu u)$, which is similar to equation (8). The uniqueness of the solution can readily be established for this equation. The solution can be found by iteration starting from any initial function $u_0 \in C_u(E_d)$. If $u_0 = 0$, then from the choice of μ and assertion (c) it immediately follows that all iterates are nonnegative. Thus, the solution is nonnegative, and from (f) it now follows that the solution coincides with $P_\lambda g$. Assertion (h) is proved.

In order to prove (i), we take a function $\eta \in C_0(E_d)$, which is equal to unity in a neighbourhood of x_0, and set $v(x) = \eta(x)\,[u(x_0) + u_i(x-x_0)^i + (1/2)u_{ij}(x-x_0)^i(x-x_0)^j]$, $w = u - v$.

Note that $w = o(|x-x_0|^2)$ as $x \to x_0$. Further, from (c) and (b) it follows that

$$|\lambda(P_\lambda(f+\lambda u) - u)(x_0) - F(u_{ij}, u_i, u(x_0), x_0)|$$
$$\leq \lambda^2 P_\lambda|w|(x_0) + |\lambda(P_\lambda(f+\lambda v) - v)(x_0) - F[v](x_0)|$$
$$= \lambda^2 P_\lambda|w|(x_0) + |\lambda P_\lambda(f+Lv)(x_0) - F[v](x_0)|,$$

where the latter summand tends to zero by virtue of (g). It remains to show that $\lambda^2 P_\lambda u(x_0) \to 0$ when $\lambda \to \infty$ provided $u \in C_u(E_d)$ and $u = o(|x-x_0|^2)$ as $x \to x_0$ and $u \geq 0$.

For this function u and for every $\delta > 0$, there exists a function u^δ such that u^δ, u^δ_x, $u^\delta_{xx} \in C_u(E_d)$, $u^\delta \geq u$ on E_d, and $u^\delta(x) = \delta|x - x_0|^2$ for those x which are sufficiently close to x_0. We obtain

$$0 \leq \overline{\lim_{\lambda\to\infty}} \lambda^2 P_\lambda u(x_0) \leq \lim_{\lambda\to\infty} \lambda P_\lambda \lambda u^\delta(x_0)$$
$$= \lim_{\lambda\to\infty} \lambda P_\lambda L u^\delta(x_0) = \sup_\omega L(\omega)u^\delta(x_0) = 2\delta \sup_\omega \operatorname{tr} a(\omega).$$

As $\delta \downarrow 0$, it follows that the desired assertion concerning $\lambda^2 P u(x_0)$ is true. This proves the theorem.

4. Remark. It is not difficult to see that $S^\mu_\lambda = R^\omega_\lambda$ and $P_\lambda = R^\omega_\lambda$ when Ω consists of a single point ω. Specifically, for R^ω_λ with any ω readily available counterparts of assertions (g) and (i) of Theorem 3 hold true.

2.2. The Existence Theorem for the Solution and the Estimates of its Derivatives

Our next discussion is based on the assumptions introduced at the beginning of Section 1 before formula (1.1). Denote $v = P_0 f$.

1. LEMMA. There exists a constant N such that $|v(x) - v(y)| \leq N|x-y|$ for all $x, y \in E_d$ and the function $v(x) + N|x|^2$ is convex on E_d.

Proof. From Theorem 1.3 it follows that, for any $x, y \in E_d$,

$$|v(x) - v(x+y)| = |P_0 f(x) - P_0 U^y f(x)| \leq P_0|f - U^y f|(x_0)$$
$$\leq \frac{1}{\varepsilon} \sup_{\omega,x} |f(\omega, x) - f(\omega, x+y)| \leq N|y|,$$

since the first derivatives of f with respect to x are uniformly bounded. Further,

$$v(x+y) - 2v(x) + v(x-y) \geq P_0(U^y+U^{-y})f(x) - P_0 2f(x)$$

$$\geq -P_0(2-U^y-U^{-y})f(x). \qquad (1)$$

However, from the Newton–Leibnitz formula it follows that

$$(2 - U^y - U^{-y})f(\omega, x) = 2f(\omega, x) - f(\omega, x+y) - f(\omega, x-y)$$

$$= -\int_0^1 (f(\omega, x+ty))'_t \, dt - \int_0^1 (f(\omega, x-ty))'_t \, dt$$

$$= -\int_0^1 dt \int_{-t}^{t} f_{(y)(y)}(\omega, x+sy) \, ds \leq N|y|^2,$$

where N is independent of ω, y, and x since the second derivatives of f with respect to x are uniformly bounded. This along with (1) and Theorem 1.3 imply $v(x+y) - 2v(x) + v(x-y) \geq -\varepsilon^{-1}N|y|^2$.

It remains to note that $|x+y|^2 - 2|x|^2 + |x-y|^2 = 2|y|^2$ and, therefore, the second differences of the continuous function $v + N|x|^2$ are nonnegative if N is appropriately chosen. This proves the lemma.

The first Sobolev derivative of a function satisfying the Lipschitz condition is known to exist (see Section 1.2) and be bounded by the Lipschitz constant, while the second generalized derivatives of convex functions are measures which allow of writing the second-order Taylor formula at almost every point $x_0 \in E_d$. Since v is a difference of two convex functions ($v + N|x|^2$ and $N|x|^2$), the above comments on convex functions remain true for this function too. This proves the first three assertions of the following theorem.

2. <u>THEOREM</u>. The Sobolev derivatives v_{x^i} exist and are bounded on E_d. For i, j = 1, ..., d, there exist the measures $v_{x^ix^j}(dx)$ which are finite on every bounded Borel set. These measures are the generalized derivatives of v with respect to x^ix^j. If $x \to x_0$, then

$$v(x) = v(x_0) + v_{x^i}(x_0)(x^i - x_0^i) +$$

$$+ \frac{1}{2} v^{(0)}_{x^ix^j}(x_0)(x^i - x_0^i)(x^j - x_0^j) + o(|x - x_0|^2)$$

for almost every $x_0 \in E_d$. Almost everywhere in E_d,

$$F\left(v^{(0)}_{x^ix^j}(x),\quad v_{x^i}(x),\quad v(x),\ x\right) = 0. \tag{2}$$

The fourth assertion of this theorem immediately follows from the third one and Theorem 1.3(i), if we note that $P_\lambda(f + \lambda v) - v = 0$, by virtue of Theorem 1.3(f).

Equation (2) provides little information on the function v. In view of Cantor's staircase function it is not difficult to devise an example of the linear equation (2) whose solution set would contain a large number of bounded functions possessing the properties of v which are stated in Lemma 1 and Theorem 2. The following property of v along with those already established makes it possible to characterize the function v uniquely using properties of its derivatives. Here, however, we do not need this uniqueness property. It will be discussed much later in Section 3.7.

3. THEOREM. For any $\omega \in E_d$ we have

$$a^{ij}(\omega)v_{x^ix^j}(dx) + [b^i(\omega)v_{x^i} + c(\omega)v + f(\omega)]\ dx \leqq 0. \tag{3}$$

<u>Proof</u>. Let $\omega \in \Omega$, $v_n(x) = v(x) * n^d\zeta(nx)$, and $f_n = f * n^d\zeta(nx)$, where ζ is taken from the proof of Theorem 1.3. For any $\lambda \geqq 0$ and $n \geqq 1$ the equation $L(\omega)u - \lambda u+f_n(\omega) + \lambda v_n = 0$ has a unique solution u_n in the class of functions whose every derivative is bounded in E_d (cf. Lemma 1.1(a) and formula (1.5)). By virtue of Lemma 1.1(a), this solution is $R^{\omega}_{\lambda}(f_n(\omega) + \lambda v_n)$. On introducing the operator $L^*(\omega)$ as

$$L^*(\omega)u = a^{ij}(\omega)u_{x^ix^j} - b^i(\omega)u_{x^i} + c(\omega)u,$$

and integrating by parts for any $\eta \in C_0^\infty(E_d)$, we obtain

$$\begin{aligned} 0 &= \int \eta(L(\omega)u_n - \lambda u_n + f_n(\omega) + \lambda v_n)\ dx \\ &= \int R^{\omega}_{\lambda}(f_n(\omega) + \lambda v_n)L^*(\omega)\eta\ dx + \\ &\quad + \int \eta[f_n(\omega) + \lambda(v_n - R^{\omega}_{\lambda}(f_n(\omega) + \lambda v_n))]\ dx. \end{aligned}$$

In the latter equation it is easy to pass to the limit by sending n to infinity, since $f_n(\omega) \to f(\omega)$, $v_n \to v$, and $R^{\omega}_{\lambda}(f_n(\omega) + \lambda v_n) \to R^{\omega}_{\lambda}(f(\omega) + \lambda v)$ uniformly on E_d. We now pass to the limit and use the fact that $v \geqq R_\lambda^\omega(f(\omega) + \lambda v)$ by virtue of Theorem 1.3(e). Then for every nonnegative $\eta \in C_0^\infty(E_d)$ we obtain

$$0 \geqq \int (R_\lambda^\omega(f(\omega) + \lambda v)L^*(\omega)\eta + \eta f(\omega))\, dx.$$

Hence, as $\lambda \to \infty$, it follows from Remark 1.4 that

$$0 \geqq \int (vL^*(\omega)\eta + \eta f(\omega))\, dx$$

for any nonnegative $\eta \in C_0^\infty(E_d)$.

This is exactly equivalent to the theorem statement in view of the definition and the existence of the measures $v_{x^ix^j}$ and the Sobolev derivatives v_{x^i} and, also, because of the arbitrary way in which the function η was selected. The theorem is proved.

The assertion of Lemma 1 concerning the convexity of $v + N|x|^2$ can be interpreted as estimation from below of the pure second derivatives of v. In some cases that are significant for the discussion of Chapter 3, it is also possible to obtain an estimate from above for the pure second derivatives of v, using Theorem 3. The idea of deriving such an estimate is as follows. Suppose that $a^{11}u_{x^1x^1} + a^{22}u_{x^2x^2} \leqq 1$ and $u_{x^1x^1}$, $u_{x^2x^2} \geqq -1$. Then $a^{11}u_{x^1x^1} \leqq 1 - a^{22}u_{x^2x^2} \leqq 1 + a^{22}$, $u_{x^1x^1} \leqq (a^{11})^{-1}(1 + a^{22})$. In other words, if we want to estimate the derivative $v_{(1)(1)}$ then we should leave it on the left of (3), while all remaining derivatives are transposed to the right. It is required that the coefficient of $v_{(1)(1)}$ be nonzero for at least one ω, i.e., the derivative $u_{(1)(1)}$ should 'actually' enter the operator $L(\omega)u$ for at least one ω or it should 'actually' enter the operator $F[u]$.

Note that when Ω consists of a single point ω, then our assumption $(f, f_x, f_{xx} \in C_u(E_d))$ imply that all the ordinary second derivatives $v_{(1)(1)}$ are bounded on E_d whether or not the derivative $u_{(1)(1)}$ enters the operator $L(\omega)$. The situation radically changes if Ω contains at least two points. To illustrate this fact we consider the following example.

4. Example. Let $d = 2$, $\Omega = \{-1\}\cup\{+1\}$, $f(\omega, x) = (x^1)^2(1 + (x^1)^2)^{-1}$, and let $L(\omega)u = \omega u_{x^1} + u_{x^2x^2} - u$. In this case it can be shown (but cf. Section 3.7) that

$$v(x^1, x^2) = e^{|x^1|} \int_{|x^1|}^{\infty} \frac{t^2}{1+t^2} e^{-t}\, dt.$$

It is obvious that v has no second derivative with respect to x^1 along the axis of ordinates, while the measure $v_{x^1x^1}(dx)$ has a singular component concentrated on this axis and does not allow of estimating from above via the Lebesgue measure.

We denote

$$F_1(u_{ij}) = \sup_{\omega} a^{ij}(\omega)u_{ij} \left(= \lim_{t\to\infty} \frac{1}{t} F(tu_{ij}, u_i, u, x)\right), \tag{4}$$

$$\mu(l) = \inf_{\xi:(l,\xi)=1} F_1(\xi^i\xi^j), \quad l \in E_d, \quad l \neq 0, \tag{5}$$

$$\mu = \inf_{\xi:|\xi|=1} F_1(\xi^i\xi^j) = \inf_{|\xi|=1} \sup_{\omega\in\Omega} a^{ij}(\omega)\xi^i\xi^j \left(= \inf_{|l|=1} \mu(l)\right). \tag{6}$$

If $l \neq 0$, we assume $\mu(l)$ to be a <u>measure of the involvement</u> of $u_{(l)(l)}$ in the operator $F[u]$. From Example 4, where $\mu(l) > 0$ only if $l \parallel (0,1)$, it can be seen that this definition is quite natural.

5. LEMMA. Let $u = (u_{ij})$ be a symmetric $d \times d$ matrix and let ψ be a number. Suppose that $(u\xi, \xi) \geq \psi|\xi|^2$ for all $\xi \in E_d$. Then for all unit vectors l

$$\mu(l)(ul, l) \leq F_1(u_{ij}) + \psi_- F_1(\delta_{ij}),$$

$$\mu(ul, l) \leq F_1(u_{ij}) + \psi_- F_1(\delta_{ij}).$$

<u>Proof</u>. We first suppose that $u > 0$ and take $\eta = \sqrt{u}l$, $\zeta = |\eta|^{-2}\eta$, $\xi = \sqrt{u}\zeta$. Then $(\xi, l) = (\zeta, \eta) = 1$, $a^{ij}u_{ij} = \operatorname{tr} au = \operatorname{tr} \sqrt{u}a\sqrt{u}$ and if $\tilde{a} := \sqrt{u}a\sqrt{u}$, then obviously $\operatorname{tr} \tilde{a} \geq |\zeta|^{-2}(\tilde{a}\zeta, \zeta) = (ul, l)(a\xi, \xi)$.

Thus, $\operatorname{tr} a(\omega)u \geq (ul, l)(a(\omega)\xi, \xi)$. Passing to the least upper bounds on both sides of this inequality and recalling that $(l, \xi) = 1$, we conclude that

$$F_1(u_{ij}) \geq (ul, l)F_1(\xi^i\xi^j) \geq (ul, l)\mu(l) \geq (ul, l)\mu.$$

These inequalities prove the lemma for $\psi > 0$, and the case $\psi \geq 0$ readily follows from the continuity.

We now consider the case $\psi < 0$. From $(u_{ij} - \psi\delta_{ij}) \geq 0$ it follows that

$$F_1(u_{ij}) \geqq F_1(U_{ij} - \psi\delta_{ij}) + \inf_{\omega} \psi \operatorname{tr} a(\omega) \geqq$$

$$\geqq \mu(l)[(ul, l) - \psi] - \psi_- \sup_{\omega} \operatorname{tr} a(\omega) \geqq$$

$$\geqq \mu\ [(ul, l) - \psi] - \psi_- F_1(\delta_{ij}).$$

Hence it is clear that assertions of the lemma are also true if $\psi > 0$. The lemma is proved.

6. THEOREM. (a) Let $\mu(l) > 0$ for a unit vector l. Then there exists the Sobolev derivative $v_{(l)(l)}$ which is bounded on E_d. Moreover, given a constant $K \geqq 0$ such that $v + K|x|^2$ is convex on E_d, the inequalities

$$\mu(l)v_{(l)(l)} \leqq 2K \sup_{\omega} \operatorname{tr} a(\omega) + |v_x|\sup_{\omega} b(\omega)| + \tag{7}$$

$$+ |v|\sup_{\omega}|c(\omega)| + \sup_{\omega}|f(\omega)|, \quad v_{(l)(l)} \geqq -2K$$

hold almost everywhere.

(b) If $\mu > 0$, then there exist all the Sobolev derivatives $v_{x^ix^j}$ which are bounded on E_d.

<u>Proof</u>. Following Lemma 1, we find a constant K, such that $v + K|x|^2$ is convex on E_d, and choose a measure ν, with respect to which all the measures $v_{x^ix^j}(dx)$ and the Lebesgue measure are absolutely continuous. We define functions $v_{ij}(x)$ and $\rho(x)$ so that

$$v_{ij} = v_{x^ix^j}(dx)/\nu(dx), \quad \rho = dx/\nu(dx).$$

From Theorem 3 it follows that, for any $\omega \in \Omega$,

$$a^{ij}(\omega)v_{ij} \leqq (|v_x|\sup_{\omega}|b(\omega)| + |v|\sup_{\omega}|c(\omega)| +$$

$$+ \sup_{\omega}|f(\omega)|)\rho =: g\rho. \tag{8}$$

almost everywhere with respect to the measure ν.

Furthermore, the set of matrices $\{a(\omega) : \omega \in \Omega\}$ is separable as a subset of an appropriate d^2-dimensional Euclidean space. In this set, we choose a countable, everywhere dense subset $\{a(\omega_n)\}$. Note that

$F_1(u_{ij}) = \sup_n a^{ij}(\omega_n)u_{ij}$ for any matrix u_{ij}.

Inequalities (8), which are true on a set of full ν-measure for every $\omega = \omega_n$, are also true on a set of full ν-measure for all ω_n at once, since the countable intersection of sets of full measure is a set of full measure. Considering (8) and passing to the least upper bound on the set of full ω_n-measure, we obtain $F_1(v_{ij}) \leqq g\rho(\text{a.e.}\nu)$. On the other hand, it is clear that $v_{(l)(l)}(dx) + 2K\,dx \geqq 0$ and $v_{ij}l^il^j \geqq -2K\rho$ (a.e.ν). Thus, from Lemma 5 it follows that

$$-2K\rho \leqq v_{ij}l^il^j \leqq \mu^{-1}(l)g\rho + \mu^{-1}(l)\ 2K\rho \sup_\omega \operatorname{tr} a(\omega)\,(\text{a.e.}\nu).$$

Passing from the densities with respect to the measure ν to the initial measures, we conclude that

$$-2K\,dx \leqq v_{x^ix^j}(dx)\ l^il^j \leqq \mu^{-1}(l)\left[g + 2K \sup_\omega \operatorname{tr} a(\omega)\right] dx. \qquad (9)$$

Hence the measure $v_{x^ix^j}(dx)l^il^j$ is absolutely continuous (with respect to dx) and since the density of this measure is a Sobolev derivative of the form $v_{(l)(l)}$, the existence of the latter is proved. Since inequalities (9) are now equivalent to (7), assertion (a) is proved.

To prove (b), it suffices to note that all Sobolev derivatives of the form $v_{(l)(l)}$ for $|l| = 1$ exist and are uniformly bounded on E_d, which is true by virtue of (a) and (7). Moreover,

$$v_{x^ix^j} = (1/2)\left[v_{(l_+^{ij})(l_+^{ij})} - v_{(l_-^{ij})(l_-^{ij})}\right],$$

where

$$l_\pm^{ij} = 2^{-1/2}(e_i \pm e_j)$$

and e_i is the ith basic vector. This proves the theorem.

7. <u>Remark</u>. From the definition of μ it follows that $\mu > 0$ if

$$\sup_\omega \inf_{|\xi|=1} a^{ij}(\omega)\xi^i\xi^j > 0$$

or there exists $\omega \in \Omega$ such that

$$\inf_{|\xi|=1} a^{ij}(\omega)\xi^i\xi^j > 0. \qquad (10)$$

The latter two conditions cannot be expressed in terms of the function F generating equation (1.1). At the same time the latter condition is essentially equivalent to the unequality $\mu > 0$.

Indeed, let $\mu > 0$. Since $a(\omega)$ are bounded with respect to ω, there exists a constant K such that $|a^{ij}(\omega)\xi^i\xi^j - a^{ij}(\omega)\eta^i\eta^j| \leqq K|\xi - \eta|$ for all $\omega \in \Omega$ and $|\xi| = |\eta| = 1$. We take a finite $4^{-1}\mu K^{-1}$-net $\{\xi_1, \ldots, \xi_n\}$ on the unit sphere of E_d and, for every $k = 1, \ldots, n$, we define ω_k such that $a^{ij}(\omega_k)\xi^i_k\xi^j_k \geqq (1/2)\mu$. Then for $\bar{a} := (a(\omega_1) + \ldots + a(\omega_n))/n, \xi \in E_d$, and $|\xi| = 1$, we obtain

$$\bar{a}^{ij}\xi^i\xi^j \geqq \frac{1}{n}(a^{ij}(\omega_k)\xi^i\xi^j - a^{ij}(\omega_k)\xi^i_k\xi^j_k) + \frac{1}{n}a^{ij}(\omega_k)\xi^i_k\xi^j_k$$

$$\geqq -\frac{1}{n}\frac{1}{4}\mu + \frac{1}{n}\frac{1}{2}\mu = \frac{1}{4n}\mu, \tag{11}$$

where ξ_k is chosen in such a way that $|\xi - \xi_k| \leqq 4^{-1}\mu K^{-1}$. We now add an element $\bar{\omega} \notin \Omega$ to the set Ω and set $a^{ij}(\bar{\omega}) = \bar{a}^{ij}$, $b^i(\bar{\omega}) = n^{-1} \times (b^i(\omega_1) + \ldots + b^i(\omega_n))$, $c(\bar{\omega}) = n^{-1}(c(\omega_1) + \ldots + c(\omega_n))$, and $f(\bar{\omega}) = n^{-1}(f(\omega_1) + \ldots + f(\omega_n)) - 1$.

It is clear that, by virtue of (11), after replacing Ω by $\{\bar{\omega}\} \cup \Omega$ condition (10) is satisfied and the function F remains unchanged. Hence, if $\mu > 0$, it is possible, without changing equation (1.1), to extend the set Ω so that condition (10) holds on the extended set.

8. Remark. To compute $F(u_{ij}, u_i, u, x)$, it suffices to know $u_1, \ldots, u_d$ and $u_{ij}l^il^j$ for all $l \in \{l : \mu(l) > 0\}$. In other words, if $u^1_{ij}l^il^j = u^0_{ij}l^il^j$ for all $l \in \{l : \mu(l) > 0\}$, then $F(u^1_{ij}, u_i, u, x) = F(u^0_{ij}, u_i, u, x)$. Thus, the condition $\mu(l) > 0$ describes all the second derivatives $u_{(l)(l)}$ which 'actually' enter the operator $F[u]$.

To prove this fact, it suffices to show that if $u^1_{ij}l^il^j = u^0_{ij}l^il^j$ for $\mu(l) > 0$, then $a^{ij}(\omega)u^1_{ij} = a^{ij}(\omega)u^0_{ij}$ for all $\omega \in \Omega$. We set $u_{ij} = u^1_{ij} - u^0_{ij}$ and take an $\omega \in \Omega$. It is obvious that

$$\mu(l) \geqq \inf\{a^{ij}(\omega)\xi^i\xi^j : \xi \in E_d, (l, \xi) = 1\},$$

and it is not difficult to see that the infimum is always strictly positive whenever $l \in \sqrt{a(\omega)}E_d$, $l \neq 0$. Therefore, $u_{ij}l^il^j = 0$ on

$\sqrt{a(\omega)}E_d$, specifically, $u_{ij}l^i_k l^j_k = 0$ for all eigenvectors l_k of the matrix $a(\omega)$ corresponding to nonzero eigenvalues λ_k. Since $a(\omega)$ can be expressed in terms of l_k and λ_k using the equation $a^{ij}(\omega) = \lambda_k l^i_k l^j_k$, therefore, $a^{ij}(\omega)u_{ij} = 0$, $a^{ij}(\omega)u^1_{ij} = a^{ij}(\omega)u^0_{ij}$, which is the required result.

In conclusion, we consider some examples of equation (1.1) whose solvability is established. If Ω consists of two points, equation (1.1) is equivalent to the set of three relations

$$L(\omega_1)u + f(\omega_1) \leqq 0, \quad L(\omega_2)u + f(\omega_2) \leqq 0,$$

$$(L(\omega_1)u + f(\omega_1))(L(\omega_2)u + f(\omega_2)) = 0.$$

In a more specific case when $L(\omega_2) = -\varepsilon$, we obtain a set of relations which are typical of free-boundary problems arising in the theory of differential equations and the problems of optimal stopping of a diffusion process well known in the theory of random processes. In the one-dimensional case, it is not difficult to see that if the operator $L(\omega_1)$ is nondegenerate, and $L(\omega_2) = -\varepsilon$, then $\mu > 0$ and the second derivatives of v are generally discontinuous though bounded, as follows from Theorem 6.

The equation $|u_x| - \varepsilon u = f$ also is a particular case of equation (1.1) since $|u_x| = \sup(\xi^i u_{x^i} : |\xi| = 1)$. For this equation Lemma 1 provides only the estimate from below for the second derivatives of v. The set of three relations

$$\Delta u - \varepsilon u + f_1 \leqq 0, \quad |u_x| - \varepsilon u \leqq f_2,$$

$$(\Delta u - \varepsilon u + f_1)(|u_x| - \varepsilon u - f_2) = 0$$

can easily be reduced to a single equation (1.1). In this case $\mu > 0$ and all the second generalized derivatives of v are bounded.

All these examples are used only as an illustration. In Section 3.2, the reader will see the most important applications of the results presented in this section to the theory under consideration.

2.3. Comments

Section 1. In many cases the right-hand side of (3) can be thought of as really providing a solution to equation (1). Sometimes it is convenient to use this fact for finding upper bounds of $\tilde{R}^{\alpha}\tilde{f}^{\alpha}$ which are uniform in α. It suffices to solve equation (1) and to estimate v from above. This is the very technique which will be used in Section 3.3 for establishing the Aleksandrov estimates that play a crucial role in the theory under investigation. The construction of operator P_λ is basically taken from the theory of controlled diffusion processes (cf. Krylov [10, Chapter 3]). Assertion (f) of Theorem 3 is called the Bellman principle in terms of this theory.

Section 2. The derivation of equation (2) is the so-called traditional derivation of the Bellman equation in the theory of controlled diffusion processes (cf. Krylov [15]). The other constructions of this section are also traditional from the point of view of this theory (cf. Krylov [10, Chapter 4]). The condition $\mu > 0$ of Theorem 6 is called the weak nondegeneracy condition for the operator F. Assertion (b) of Theorem 6 can be found in Krylov [5] (cf. Remark 7) and Krylov [6].

Chapter 3

A PRIORI ESTIMATES IN L_p FOR SOLUTIONS OF NONLINEAR ELLIPTIC AND PARABOLIC EQUATIONS

This chapter is devoted to developing general problems of the theory of nonlinear equations. The attention is focused primarily on the Aleksandrov estimates, the maximum principle, uniqueness theorems, and theorems on passing to the limit under the action of a nonlinear operator. All the considerations of this chapter are based on Sections 1 and 2, where elliptic and parabolic equations with Monge—Ampère operators in the whole space are examined using the results of the previous chapter. The reader who is interested only in the $C^{2+\alpha}$-theory of nondegenerate nonlinear equations can skip Sections 4 to 7.

3.1. Some Properties of λ-concave functions

Given a number $\lambda > 0$ and a nonnegative function $u \in C^{\infty}_{loc}(E_d)$, the function u is said to be λ-concave if $(\lambda u(x)\delta_{ij} - u_{x^i x^j}(x)) \geqq 0$ for all $x \in E_d$. In this section the function u is always assumed to be λ-concave.

The notion of the λ-concave function is a generalization of the concave function. It is important for our considerations that there exist sufficiently many λ-concave bounded functions whereas the class of bounded concave functions on E_d consists only of constant functions. Geometrically, the convexity of a function can be, in a well known fashion, interpreted by considering the chords of its graph. It was found that the λ-concavity can be interpreted in a similar manner except that, instead of line segments, one should consider segments of the graph of the function $c_1 e^{t\sqrt{\lambda}} + c_2 e^{-t\sqrt{\lambda}}$ $(t \in (-\infty, \infty))$ which is a solution to the equation $u_{tt} - \lambda u = 0$.

1. LEMMA. (a) Given $x, y \in E_d$, $x \neq y$, and $r = (1/2)|x - y|$, the inequality

$$u\left(\frac{x+y}{2} + t\,\frac{y-x}{2r}\right) \geqq u(x)\,\frac{\sinh(r-t)\sqrt{\lambda}}{\sin(2r\sqrt{\lambda})} + u(y)\,\frac{\sinh(r+t)\sqrt{\lambda}}{\sinh(2r\sqrt{\lambda})}, \tag{1}$$

is satisfied for $t \in [-r, r]$. Specifically, (if $t = 0$)

$$u\left(\frac{x+y}{2}\right) \geqq \frac{u(x) + u(y)}{2 \cosh (r \sqrt{\lambda})};$$

(b) $u(y) \geqq u(x)e^{-\sqrt{\lambda}|x-y|}$ for all $x, y \in E_d$;

(c) $|\mathrm{grad}\, u(x)| \leqq \sqrt{\lambda}u(x)$ for all $x \in E_d$;

(d) if $u(y) = 0$ for some y, then $u \equiv 0$.

Proof. We denote the functions on the left and on the right of (1) by $\varphi(t)$ and $\Psi(t)$ respectively and observe that

$$\lambda\varphi - \varphi_{tt} = (\lambda u\delta_{ij} - u_{x^ix^j}) \frac{y^i - x^i}{|y - x|} \frac{y^j - x^j}{|y - x|} \geqq 0,$$

$\lambda\Psi - \Psi_{tt} = 0$, $\varphi(\pm r) = \Psi(\pm r)$. From the maximum principle it follows that $\varphi \geqq \Psi$ on $[-r, r]$, which proves (a).

Suppose that y in (1) is the point $x + s \frac{y - x}{|y - x|}$ and t is the number $|y - x| - (1/2)\, s$ with $s \geqq |y - x|$. Then we obtain

$$u(y) \geqq u(x) \frac{\sinh (s - |y - x|) \sqrt{\lambda}}{\sinh (s \sqrt{\lambda})} +$$

$$+ u(x + s \frac{y - x}{|y - x|}) \frac{\sinh |y - x| \sqrt{\lambda}}{\sinh (s \sqrt{\lambda})}.$$

Since the latter summand is nonnegative, letting $s \to \infty$, we obtain (b). From (b), for any unit vector $l \in E_d$, it follows that

$$u_{(l)}(x) = \lim_{t\downarrow 0} \frac{u(x) - u(x - lt)}{t}$$

$$\leqq u(x) \lim_{t\downarrow 0} \frac{1 - e^{-t\sqrt{\lambda}}}{t} = \sqrt{\lambda}\, u(x).$$

Hence $|\mathrm{grad}\, u(x)| \leqq \sqrt{\lambda}\, u(x)$. Finally, (d) trivially follows from (b). This proves the lemma.

2. Remark. It is not difficult to show that if (2) is satisfied for all $x, y \in E_d$ and if u is smooth, then the matrix $(\lambda u\delta_{ij} - u_{x^ix^j})$ is positive.

In what follows we shall need some inequalities involving functions of the matrix $(\lambda u\delta_{ij} - u_{x^ix^j})$. We now derive these inequalities.

For any numbers w_{ij} $(i, j = 1, \ldots, d)$, and $m = 0, \ldots, d$, we introduce the following notation:

$$D(w_{ij}) = \det(w_{ij}), \quad c^m(w_{ij}) = \frac{1}{(d-m)!}\frac{\partial^{d-m}}{\partial\mu^{d-m}} D(\mu\delta_{ij} - w_{ij})\Big|_{\mu = 0},$$

$$c^m_{ij}(w_{rs}) = \frac{\partial}{\partial w_{ij}} c^m(w_{rs}), \quad c^m[u] = c^m(u_{x^i x^j}),$$

$$c^m_{ij}[u] = c^m_{ij}(u_{x^r x^s}).$$

We now outline the simplest properties of the functions introduced.

3. LEMMA.

(a) $D(\lambda\delta_{ij} - w_{ij}) = \sum_{m=0}^{d} \lambda^{d-m} c^m(w_{ij})$.

(b) For any $\nu \in (-\infty, \infty)$,

$$c^m(w_{ij}) = \sum_{k=0}^{m} \binom{d-k}{m-k} (-1)^{m-k}\nu^{m-k} c^k(-\nu\delta_{ij} + w_{ij}).$$

(c) If the matrix (w_{ij}) is symmetric and negative, the matrices $(c^m_{ij}(w_{rs}))$ are also symmetric and negative.

(d) $D_{w_{ir}w_{kj}} = -D_{w_{ij}w_{kr}}$ for any $i, j, k, r = 1, \ldots, d$.

Proof. (a) follows from the fact that $D(\lambda\delta_{ij} - w_{ij})$ is a polynomial of degree d in λ. To prove (b), it suffices to differentiate with respect to μ the following equation which is true by virtue of (a):

$$D(\mu\delta_{ij} - w_{ij}) = D((\mu - \nu)\delta_{ij} + (\nu\delta_{ij} - w_{ij}))$$

$$= \sum_{k=0}^{d} (\mu - \nu)^{d-k} c^k(-\nu\delta_{ij} + w_{ij}).$$

We now prove (c). If $m = 0$, it is obvious that $c^m_{ij} = 0$. If $m \geq 1$, $c^m(w_{ij})$ is known to be the sum of the m-order principal minors of the matrix $(-w_{ij})$. Furthermore, recall that the matrix with entries $\frac{\partial}{\partial v_{ij}} \det(v_{rs})$ is positive if (v_{rs}) is symmetric and positive (since multiplying it by $\det(v_{rs})^{-1}$ with $\det(v_{rs}) > 0$, we obtain the inverse matrix of (v_{rs})). Hence considering each m-order principal minor of the matrix $(-w_{ij})$, which like all its principal minors is positive, we obtain

$$(c^m_{ij}(w_{rs})) = \left(\frac{\partial}{\partial w_{ij}} c^m(w_{rs})\right) = -\left(\frac{\partial}{\partial(-w_{ij})} c^m(w_{rs})\right) \leqq 0.$$

Assertion (d) easily follows from the definition of the matrix determinant. The lemma is proved.

4. LEMMA. Let the sequence $a_0, \ldots, a_d$ be such that

$$\sum_{i=0}^{n} \binom{n}{i}(-1)^i \, a_{d-n+i} \geqq 0 \tag{3}$$

for $n = 0, \ldots, d - 1$. Then, for every $x \in E_d$, the symmetric matrix with entries

$$\sum_{m=0}^{d} a_m \lambda^{d-m} u^{d-m} c^m_{ij}[u] = \sum_{m=1}^{d} a_m \lambda^{d-m} u^{d-m} c^m_{ij}[u]$$

is negative definite.

Proof. For $k = 1, \ldots, d$, we set

$$b_k = \sum_{m=k}^{d} \binom{d-k}{m-k} (-1)^{m-k} a_m.$$

By hypothesis, $b_k \geqq 0$. Moreover, from Lemma 3(b) it follows that

$$\sum_{m=0}^{d} a_m u^{d-m} \lambda^{d-m} c^m_{ij}[u]$$

$$= \sum_{m=1}^{d} \sum_{k=1}^{m} a_m u^{d-m} \lambda^{d-m} \binom{d-k}{m-k} (-1)^{m-k} (\lambda u)^{m-k} c^k_{ij}(- \lambda u \delta_{rs} + u_{x^r x^s})$$

$$= \sum_{k=1}^{d} b_k (\lambda u)^{d-k} c^k_{ij}(- \lambda u \delta_{rs} + u_{x^r x^s}).$$

It remains to apply Lemma 3(c). This proves the lemma.

5. Remark. Given an infinite sequence of numbers $a_0, a_1, \ldots,$ such that

$$\sum_{i=0}^{n} \binom{n}{i} (-1)^i a_{k+i} \geqq 0$$

for all n, k ≥ 0, this sequence is said to be completely monotone (see Feller [1], Chapter 7). By the Hausdorff theorem, for any completely monotone sequence $a_0, a_1, \ldots$, there exists a positive measure F on [0, 1] such that $a_m = \int t^m F(dt)$ for all $m \geq 0$. In this case the assertion of the lemma readily follows from the formula

$$\sum_{m=0}^{d} a_m \lambda^{d-m} u^{d-m} c_{ij}^m[u] = -\int t^d D_{w_{ij}}(\lambda t^{-1} u \delta_{rs} - u_{x^r x^s}) F(dt).$$

6. Remark. Let $a_n = a(n)$, where $n = 0, \ldots, d$ and $a \in C^{d-1}([1, d])$, and let

$$(-1)^n a^{(n)}(t) \geq 0 \tag{4}$$

for $n = 0, 1, \ldots, d-1$, $t \in [d-n, d]$. Then the sequence a_n satisfies the condition of Lemma 4.

Indeed, if $n = 0$, inequalities (3) and (4) are equivalent. If $n \geq 1$, the left-hand side of (3) multiplied by $(-1)^n$ yields the nth difference for a(t) on the segment [d − n, d]. Thus, by the well known formula

$$\sum_{i=0}^{n} \binom{n}{i} (-1)^i a_{d-n+i} = (-1)^n \int_0^1 dt_1 \ldots \int_0^1 a^{(n)} (d - n + t_1 + + \ldots + t_n)\, dt_n \geq 0,$$

we obtain inequalities (3).

The following lemma follows from Lemma 4 and Remark 6 since the functions $a(t) = (p - t - 1)/(t + 1)$ and $1/(t(t + 1))$ satisfy condition (4) if $p \geq d + 1$.

7. LEMMA. If $p \geq d + 1$, $p_1 \geq d$, and $u > 0$, the matrices with entries

$$\sum_{m=1}^{d} \frac{p-m-1}{m+1} \lambda^{d-m} u^{p-m-2} c_{ij}^m[u],$$

$$\sum_{m=1}^{d} \frac{1}{m(m+1)} \lambda^{d-m} u^{p-m-1} c_{ij}^m[u],$$

$$\sum_{m=1}^{d} \frac{p_1-m}{m} \lambda^{d-m} u^{p_1-m-1} c_{ij}^m[u].$$

are negative definite.

To prove the basic result of this section, we shall need one more assertion, which is a specific case of the lemma in Section 12, Chapter 5 of Dunford and Schwartz [1].

8. LEMMA. Let $w(x)$ be a thrice continuously differentiable function on E_d. Then for all $m = 0, \ldots, d$ and $i = 1, \ldots, d$ the condition

$$\frac{\partial}{\partial x^j} c^m_{ij}[w] = \frac{\partial}{\partial x^j} c^m_{ji}[w] = 0. \tag{5}$$

holds true on E_d.

Proof. It is clear that

$$\sum_{m=0}^{d} \mu^{d-m} \frac{\partial}{\partial x^j} c^m_{ij}\,[w] = \frac{\partial}{\partial x^j} D_{w_{ij}} (\mu\delta_{rs} - w_{x^r x^s})$$

$$= \frac{\partial}{\partial x^j} D_{w_{ij}} ((v-w)_{x^r x^s})$$

for $\mu \in (-\infty, \infty)$ and $v = (1/2)\mu|x|^2$. Thus, it suffices to prove (5) for $m = d$. By lemma 3(d),

$$\frac{\partial}{\partial x^j} D_{w_{ij}} w_{x^p x^q} = D_{w_{ij} w_{kr}} w_{x^k x^r x^j}$$

$$= -D_{w_{ir} w_{kj}} w_{x^k x^j x^r} = -\frac{\partial}{\partial x^r} D_{w_{ir}}\left(w_{x^p x^q}\right).$$

This asserts the equality between the first and the last terms in (5). The other equality can be proved in the same manner. The lemma is proved.

We now come to major results of this section.

9. THEOREM. Let T and K be positive constants and let $u(t, x)$ be a function on E_{d+1} which is infinitely differentiable with respect to (t, x), λ-concave in x (for any t), equal to zero for $t \geqq T$ and such that

$$\lambda u - u_t \geqq 0, \quad |u| + |u_t| \leqq K e^{-\frac{1}{K}|x|},$$
$$\sum_{i=1}^{d} |u_{x^i t}| + \sum_{i,j=1}^{d} \left(|u_{x^i x^j}| + |u_{x^i x^j t}|\right) \leqq K. \tag{6}$$

Let $p \geqq d + 1$. Then for any $(t, x) \in E_{d+1}$

$$p\lambda^{-d} e^{\lambda pt} \int_t^\infty ds \int_{E_d} e^{-\lambda ps} u^{p-d-1} (\lambda u - u_s) D(\lambda u \delta_{ij} - u_{x^i x^j}) dx$$

$$\geqq \| u(t, \cdot) \|^p_{L_p(E_d)} \geqq \kappa_d d! (p\sqrt{\lambda})^{-d} u^p(t,x), \tag{7}$$

where κ_d is the volume of S_1 and $u^0 := 1$ (for $p = d + 1$).

Proof. We set $v(t, x) = u(t, x) e^{-\lambda t}$. Then the left-hand side of (7) becomes $p\lambda^{-d} d^{\lambda pt} A$, where

$$A = - \int_t^\infty ds \int_{E_d} v_s v^{p-d-1} D(\lambda v \delta_{ij} - v_{x^i x^j}) dx$$

$$= - \sum_{m=0}^{d} \int_{E_d} \int_t^\infty v_s v^{p-m-1} \lambda^{d-m} c^m[v] \, ds \, dx =: \sum_{m=0}^{d} A_m. \tag{8}$$

Transforming every term of the latter sum individually, we perform integration by parts with respect to s and make use of the fact that, by virtue of (6), we obtain an absolutely convergent integral with respect to ds dx which can be written as a multiple integral in any order. Thus, we have

$$A_m = \frac{1}{p-m} \int_{E_d} \lambda^{d-m} v^{p-m} (t, x) \, c^m[v(t, x)] \, dx$$

$$+ \frac{1}{p-m} \int_t^\infty ds \int_{E_d} \lambda^{d-m} v^{p-m} v_{x^i x^j s} c^m_{ij}[v] \, dx.$$

We now twice integrate the latter integral by parts with respect to x^i and x^j, transposing the differentiation from $v_{x^i x^j s}$ to $v^{p-m} c^m_{ij}[v]$. Here we use Lemma 8 which asserts that there is no need to differentiate $c^m_{ij}[v]$ with respect to x^i and x^j, along with assumption (6) from which it follows that the terms outside the integral disappear, and the Euler theorem on homogeneous functions showing that $v_{x^i x^j} c^m_{ij}[v] = mc^m[v]$. Thus, we obtain

$$A_m = \frac{1}{p-m} \int_{E_d} \lambda^{d-m} v^{p-m}(t, x)\ c^m[v(t, x)]\ dx - mA_m$$

$$+ (p-m-1) \int_t^\infty ds \int_{E_d} \lambda^{d-m} v^{p-m-2} v_s v_{x^i} v_{x^j} c^m_{ij}[v]\ dx. \tag{9}$$

To avoid misunderstanding, we draw the reader's attention to the fact that if $p \leqq m+2$, then, by Lemma 1(c), $v^{p-m-2}|v_{x^i} v_{x^j}| \leqq \lambda v^{p-m}$ for s such that $v(s, \cdot) \not\equiv 0$, while if s is such that $v(s, \cdot) \equiv 0$, the latter integral over E_d is assumed to be zero.

Expression (9) is an equation in A_m. We find A_m from this equation and substitute the result in (8). In addition, we use Lemma 7 which along with the inequality $v_s \leqq 0$ yields

$$\sum_{m=1}^{d} \frac{p-m-1}{m+1} \lambda^{d-m} v^{p-m-2} v_s v_{x^i} v_{x^j} c^m_{ij}[v] \geqq 0.$$

Hence

$$A \geqq \sum_{m=0}^{d} \frac{1}{(p-m)(m+1)} \int_{E_d} \lambda^{d-m} v^{p-m}(t, x)\ c^m[v(t, x)]\ dx. \tag{10}$$

On replacing $c^m[v]$ by $m^{-1} c^m_{ij}[v]\ v_{x^i x^j}$ for $m \geqq 1$, integrating by parts with respect to x^i, and using the inequality

$$- \sum_{m=1}^{d} \frac{1}{m(m+1)} \lambda^{d-m} v^{p-m-1} v_{x^i} v_{x^j} c^m_{ij}[v] \geqq 0.$$

which follows from Lemma 7, we can see that A is greater than the summand corresponding to m = 0 in (10), i.e., $A \geqq p^{-1}\lambda^d ||v(t, \cdot)||^p_{L_p(E_d)}$.

We have proved the first inequality in (7). The second one follows from Lemma 1(b) since

$$\int_{E_d} u^p(t, y)\ dy \geqq u^p(t, x) \int_{E_d} e^{-p\sqrt{\lambda}|x-y|}\ dy = u^p(t, x)\kappa_d d!\,(p\sqrt{\lambda})^{-d}.$$

This proves the theorem.

10. THEOREM. Suppose that $p \geqq d$, $K > 0$, the function u is λ-concave and

$$u(x) \leqq Ke^{-\frac{1}{K}|x|}, \quad \sum_{i,j} |u_{x^i x^j}(x)| \leqq K \tag{11}$$

for all $x \in E_d$. Then for any $x \in E_d$,

$$\int_{E_d} u^{p-d} D\left(\lambda u\delta_{ij} - u_{x^i x^j}\right) dx \geqq \lambda^d \int_{E_d} u^p \, dx \geqq u^p(x)\lambda^{d/2}\kappa_d d!p^{-d}, \tag{12}$$

where (for $p = d$) $u^0 := 1$.

Proof. We first consider the case when $p > d$. This condition along with (11) provide the existence of the desired integrals and enables us, while performing integration by parts, to discard the terms outside the integral. Thus, we obtain

$$\int_{E_d} u^{p-d} D\left(\lambda u_{\delta ij} - u_{x^i x^j}\right) dx = \sum_{m=0}^{d} \int_{E_d} \lambda^{d-m} u^{p-m} c^m[u] dx =: \sum_{m=0}^{d} A_m.$$

Here, as in the above proof,

$$A_m = \frac{1}{m} \int_{E_d} \lambda^{d-m} u^{p-m} u_{x^i x^j} c^m_{ij}[u] \, dx$$

$$= -\frac{p-m}{m} \int_{E_d} \lambda^{d-m} u^{p-m-1} u_{x^i} u_{x^j} c^m_{ij}[u] \, dx$$

for $m \geqq 1$.
Hence, by Lemma 7, we obtain

$$\int_{E_d} u^{p-d} D\,(\lambda u\delta_{ij} - u_{x^i x^j})\, dx$$

$$= \lambda^d \int_{E_d} u^p \, dx - \int_{E_d} \sum_{m=1}^{d} \frac{p-m}{m} \lambda^{d-m} u^{p-m-1} u_{x^i} u_{x^j} c^m_{ij}[u] \, dx$$

$$\geqq \lambda^d \int_{E_d} u^p \, dx.$$

As in the above proof, u(x) is estimated in terms of the latter integral, therefore it remains to examine the case $p = d$. Of course, we can assume that $\int D\left(\lambda u\delta_{ij} - u_{x^ix^j}\right) dx < \infty$, $u \not\equiv 0$. If $u \not\equiv 0$, then $u(x) > 0$ for all x and, as easily follows from the Lebesgue dominated convergence theorem,

$$\int D\left(\lambda u\delta_{ij} - u_{x^ix^j}\right) dx = \lim_{p\downarrow d} \int u^{p-d} D\left(\lambda u\delta_{ij} - u_{x^ix^j}\right) dx.$$

Therefore, it is possible to pass to the limit in (12) as $p\downarrow d$, which proves the theorem.

3.2. Estimation of Solutions of Special Nonlinear Equations

Here we consider estimates of the solutions to elliptic and parabolic equations (12) and (19) which are counterparts of the Monge–Ampère equations for the λ-concave functions. The existence of solutions for these equations will be proved by use of the results established in Chapter 2. The estimates to be derived here play a crucial role in subsequent sections of this chapter.

We fix a number $\lambda > 0$ and denote by Ω the set of all nonnegative symmetric $d \times d$ matrices. Let $\Omega_0 = \{\omega \in \Omega : \operatorname{tr}\omega \leq 1\}$, $\Omega_1 = \{\omega \in \Omega : \operatorname{tr}\omega = 1\}$. If $\omega \in \Omega_0$, let $r(\omega) := 1 - \operatorname{tr}\omega$.

1. LEMMA. (a) If $\omega, \omega_1 \in \Omega$, then

$$\frac{1}{d} \operatorname{tr}\omega\omega_1 \geq \sqrt[d]{\det \omega}\, \sqrt[d]{\det \omega_1}. \tag{1}$$

(b) If $\omega_1 \in \Omega$, then

$$\sqrt[d]{\det \omega_1} = \frac{1}{d} \inf_{\Omega \cap \{\det \omega = 1\}} \operatorname{tr}\omega\omega_1, \tag{2}$$

and $\sqrt[d]{\det \omega}$ is a concave function on Ω;

(c) If $\omega_1 \in \Omega$, $i = 1, \ldots, d - 1$, and $c_i(\omega_1)$ is the determinant of the submatrix which results from crossing out the first i rows and i columns of the matrix ω_1, then

$$\det \omega_1 \leq \omega_1^{11} \ldots \omega_1^{ii} c_i(\omega_1) , \tag{3}$$

where there is no summation over i.

Proof. (a) Denote by $\sqrt{\omega}_1$ the symmetric nonnegative square root of the matrix ω_1. Note that, for every matrix $\omega \in \Omega$, we have $\operatorname{tr}\omega \geq d\sqrt[d]{\det\omega}$, by virtue of the inequality between the arithmetic and geometric means. Hence

$$\operatorname{tr}\omega\omega_1 = \operatorname{tr}\sqrt{\omega_1}\omega\sqrt{\omega_1} \geq d\sqrt[d]{\det(\sqrt{\omega_1}\omega\sqrt{\omega_1})} = d\sqrt[d]{\det\omega\det\omega_1}.$$

This proves (a). We now prove (b). The concavity of $\sqrt[d]{\det\omega}$ follows from formula (2) which represents this function as the greatest lower bound of linear functions. It immediately follows from (1) that the left-hand side of (2) is less than its right-hand side. The opposite inequality holds true since

$$\lim_{\varepsilon\downarrow 0}\operatorname{tr}\left[(\omega_1 + \varepsilon I)^{-1}\sqrt[d]{\det(\omega_1 + \varepsilon I)}\,\omega_1\right] =$$

$$= \sqrt[d]{\det\omega_1}\lim_{\varepsilon\downarrow 0}\operatorname{tr}[(\omega_1 + \varepsilon I)^{-1}\omega_1] \leq d\sqrt[d]{\det\omega_1},\quad I = (\delta^{ij}).$$

The proof of (c) can evidently use induction on i and, therefore, it suffices to examine the case $i = 1$. Let Ω' be the set of all matrices $\omega \in \Omega$ such that $\det\omega = 1$ and $\omega^{1i} = 0$ if $i \neq 1$. By virtue of (2),

$$\inf_{\Omega'}\operatorname{tr}\omega\omega_1 = \inf_{\omega^{11}>0}\ \inf_{c_1(\omega)=(\omega^{11})^{-1}}\left(\omega_1^{11}\omega^{11} + \sum_{i,j\geq 2}\omega_i^{ij}\omega^{ij}\right) =$$

$$= \inf_{\omega^{11}>0}\ \inf_{c_1(\omega)=1}\left(\omega_1^{11}\omega^{11} + (\omega^{11})^{-\frac{1}{d-1}}\sum_{i,j\geq 2}\omega_1^{ij}\omega^{ij}\right) =$$

$$= \inf_{\omega^{11}>0}\left(\omega_1^{11}\omega^{11} + (d-1)(\omega^{11})^{-\frac{1}{d-1}}\sqrt[d-1]{c_1(\omega_1)}\right) =$$

$$= d\sqrt[d]{\omega_1^{11}c_1(\omega_1)}.$$

The first of these expressions and, consequently, the last one are greater than $d\sqrt[d]{\det\omega_1}$ by virtue of (2). This proves the lemma.

The following lemma relates equations of the Monge—Ampère type to the equations from Chapter 2 and points out a direction along which we will develop the arguments underlying the solvability of these equations.

2. LEMMA. Let u be a symmetric d × d matrix and let f and u_0 be some numbers. Then

(a) If $d \geqq 2$, then

$$\max_{\omega \in \Omega_1} \left(\operatorname{tr} \omega u + \sqrt[d]{\det \omega f} \right) = 0, \tag{4}$$

if and only if $u \leqq 0$, $(f_+)^d = d^d \det(-u)$; this assertion is also true if $d = 1$ and $f \geqq 0$;

(b) The equality

$$\max_{\omega \in \Omega_0} \left(r(\omega) u_0 + \operatorname{tr} \omega u + \sqrt[d+1]{r(\omega) \det \omega f} \right) = 0, \tag{5}$$

holds if and only if $u_0 \leqq 0$, $u \leqq 0$, and

$$(f_+)^{d+1} = (d+1)^{d+1} (-u_0) \det(-u).$$

Proof. Without loss of generality we assume that u is a diagonal matrix, and first consider the case $f > 0$. In this case $\operatorname{tr} \omega u = \omega^{ii} u^{ii}$ and, on the one hand, $\det \omega \leqq \omega^{11} \ldots \omega^{dd}$, by Lemma 2(c), so that the maxima in (4) and (5) do not exceed the maxima over diagonal matrices $\omega \in \Omega_1$, $\omega \in \Omega_0$ respectively, on the other hand, the opposite inequality between these maxima clearly holds. It can be seen that, for $f > 0$, assertions (a) and (b) will be proved if we prove that, for $n \geqq 1$,

$$\max\left(\sum_{i=1}^{n} \omega_i u_i + \left(\prod_{i=1}^{n} \omega_i \right)^{\frac{1}{n}} f: \ \omega_i \geqq 0, \ \sum_{i=1}^{n} \omega_i = 1 \right) = 0 \tag{6}$$

if and only if

$$u_i \leqq 0, \quad f^n = n^n \prod_{i=1}^{n} (-u_i). \tag{7}$$

We prove that (6) is equivalent to (7) (if $f > 0$). When $n = 1$, these two conditions are obviously equivalent even if $f \geqq 0$. Therefore, we first assume that $n \geqq 2$ and deduce (7) from (6). If $\omega_1 = 1$ and $\omega_i = 0$ for $i \geqq 2$, it follows from (6) that $u_1 \leqq 0$. Similarly, $u_i \leqq 0$ for all i. Were u_1 equal to zero, the left-hand side of (6) would be strictly positive as can readily be shown by comparing the rates of convergence to zero of the two expressions

$$\sum_{i=2}^{n} tu_i, \left[(1-(n-1)t)\prod_{i=2}^{n} t\right]^{1/n} f,$$

as $t \downarrow 0$.

Thus $u_1 < 0$ and $u_i < 0$ for all i. It immediately follows that if $\bar{\omega}_i$ provides the maximum in (6), then $\bar{\omega}_i > 0$ for all i. Finally, the inequality between the arithmetic and geometric means of $\bar{\omega}_i u_i$ implies

$$0 = \sum_{i=1}^{n} \bar{\omega}_i u_i + \left(\prod_{i=1}^{n} \bar{\omega}_i\right)^{1/n} f$$

$$\leqq \left(\prod_{i=1}^{n} \bar{\omega}_i\right)^{1/n} \left(f - n\left(\prod_{i=1}^{n} (-u_i)\right)^{1/n}\right). \tag{8}$$

This provides the second expression in (7) with an inequality of a certain sense instead of the equality. The inequality of the opposite sense can be obtained from (6) if ω_i is assigned $u_i^{-1}(\Sigma u_i^{-1})^{-1}$.

We now deduce (6) from (7). From the inequality in (8) and from (7) it follows that the left-hand side of (6) is negative. On the other hand, $u_i < 0$ for all i, since $f > 0$, and, for $\bar{\omega}_i = u_i^{-1}(\Sigma u_i^{-1})^{-1}$, the inequality in (8) turns into an equality which implies that the left-hand side of (6) is positive.

We now consider the case $f \leqq 0$. Taking diagonal matrices ω with only one nonzero diagonal entry, we find from (4), as above, that $u^{ii} \leqq 0$. It is clear that $\max u^{ii} = 0$ since the assumption $u^{ii} < 0$ readily yields a contradition. Hence, $\det(-u) = 0$ and, thus, we have proved the necessity in (a). The necessity of (b) can be proved in a similar manner. The sufficiency of (a) and (b) for $f \leqq 0$ is almost obvious. The lemma is proved.

3. THEOREM. Let $f \in C^2(E_d)$, $f \geqq 0$, and f be a function with compact support. Then there exists a nonnegative function $v(x)$ defiend on E_d such that:

(a) $v \leqq Ne^{-\frac{1}{N}|x|}$ for all x and a constant N; the generalized derivatives of the form v_{x^i} and $v_{x^i x^j}$, $i, j = 1, \ldots, d$, exist and are bounded on E_d;

(b) for any $\omega \in \Omega$(a.e.)

$$-\lambda v \operatorname{tr} \omega + \omega^{ij} v_{x^i x^j} + \sqrt[d]{\det \omega}\, f \leqq 0, \quad |v_x| \leqq \sqrt{\lambda} v,$$

$$\left(\lambda v\delta_{ij} - v_{x^ix^j}\right) \geqq 0; \tag{9}$$

(c) for any $x \in E_d$ and $p \geqq d$,

$$||v||_{L_p(E_d)} \leqq \frac{1}{\lambda d}||f||_{L_p(E_d)},$$

$$v(x) \leqq \lambda^{\frac{d}{2p} - 1} d^{-1}(\kappa_d d!p^{-d})^{-\frac{1}{p}} ||f||L_p(E_d). \tag{10}$$

<u>Proof</u>. Returning to Section 2.1, we replace Ω_1 and λ by Ω and ε respectively and, for $\omega \in \Omega_1$, denote

$$a(\omega) = \omega,\ b(\omega) = 0,\ c(\omega) = -\lambda \text{ tr } \omega = -\lambda,$$

and

$$f(\omega) = f\sqrt[d]{\det \omega}.$$

We set $v = P_0 f$ and prove that v possesses properties (a) to (c). Note that $v \geqq 0$ by Theorem 2.1.3.

We now make use of the results established in Section 2.2. In our case, $F_1(u_{ij})$ involved in (2.2.4) is the greatest eigenvalue of the matrix (u_{ij}) and, therefore, (see (2.2.6))

$$\mu = \inf_{|\xi|=1} F_1(\xi^i\xi^j) = \inf_{|\xi|=1} |\xi|^2 = 1.$$

By Theorems 2.2.2 and 2.2.6, all Sobolev's derivatives v_{x^i} and $v_{x^ix^j}$ exist and are bounded on E_d. Moreover,

$$\max_{\omega\in\Omega_1}\left(\omega^{ij}v_{x^ix^j} - \lambda v + \sqrt[d]{\det \omega}\ f\right) = 0 \text{ (a.e.)}. \tag{11}$$

This along with homogeneity in ω prove the first inequality in (9), hence for $\xi \in E_d$ and $\omega^{ij} = \xi^i\xi^j$, it follows that the matrix $(\lambda v\delta_{ij} - v_{x^ix^j})$ is nonnegative a.e.

To prove the second inequality in (9), we take $\zeta \in C_0(E_d)$ such that $\zeta \geqq 0$, $\int\zeta\, dx = 1$, and we shall also use the expression $u^{(\delta)}(x) = u(x) * \delta^{-d}\zeta(\delta^{-1}x)$ for any locally summable function u and $\delta > 0$. It is clear that

$$0 \leqq (\lambda v \delta_{ij} - v_{x^i x^j})^{(\delta)} = (\lambda v^{(\delta)} \delta_{ij} - v^{(\delta)}_{x^i x^j})$$

and, therefore, $v^{(\delta)}$ is λ-concave. By Lemma 1.1 (c), we obtain

$$|v_x^{(\delta)}| \leqq \sqrt{\lambda} v^{(\delta)}.$$

Thus, as $\delta \downarrow 0$, the inequality $|v_x| \leqq \sqrt{\lambda}\, v$ holds almost everywhere, and assersion (b) is proved.

Furthermore, it is not difficult to see that there exists a constant N_0 such that

$$f(x) \leqq N_0 \lambda \cosh^{-3} (|x|\sqrt{\lambda}),$$

$$f(x) \leqq N_0 \sqrt{\lambda} \cosh^{-2}(|x|\sqrt{\lambda}) \frac{\sinh(|x|\sqrt{\lambda})}{|x|}$$

for all $x \in E_d$. For $u(x) := N_0 \cosh^{-1}(|x|\sqrt{\lambda})$, the straightforward computation with $\omega \in \Omega_1$ yields

$$L(\omega)u + f(\omega) = \frac{\lambda N_0}{\cosh^3(|x|\sqrt{\lambda})} \times$$

$$\times \left[\sinh^2(|x|\sqrt{\lambda}) \frac{\omega^{ij} x^i x^j}{|x|^2} - \cosh^2(|x|\sqrt{\lambda})\right] +$$

$$+ f \sqrt[d]{\det \omega} - \lambda N_0 \frac{1}{\cosh^3(|x|\sqrt{\lambda})} \frac{\omega^{ij} x^i x^j}{|x|^2} -$$

$$- N_0 \sqrt{\lambda} \frac{1}{\cosh^2(|x|\sqrt{\lambda})} \times$$

$$\times \frac{\sinh(|x|\sqrt{\lambda})}{|x|} \left(1 - \frac{\omega^{ij} x^i x^j}{|x|^2}\right).$$

Here the expression between braces is nonnegative since $|\sinh t| \leqq \cosh t$ and $\operatorname{tr} \omega = 1$. In view of the choice of N_0, the remaining expressions on the right do not exceed

$$f\left[\sqrt[d]{\det \omega} - \frac{\omega^{ij} x^i x^j}{|x|^2} - \left(1 - \frac{\omega^{ij} x^i x^j}{|x|^2}\right)\right] = f\left(\sqrt[d]{\det \omega} - 1\right) \leqq 0,$$

since $\sqrt[d]{\det \omega} \leq d^{-1} \operatorname{tr} \omega = d^{-1} \leq 1$. Thus, $L(\omega)u + f(\omega) \leq 0$ and from Theorem 2.1.3 (b) it follows that $v = P_0 f = P_0(f + Lu) \leq u$. We have proved assertion (a).

Since $\omega^{ij} v_{x^i x^j} - \lambda v = \omega^{ij}(v_{x^i x^j} - \lambda v \delta_{ij})$ in (11), from Lemma 2 it follows that

$$\det\left(\lambda v \delta_{ij} - v_{x^i x^j}\right) = d^{-d} f^d \text{ (a.e.).} \tag{12}$$

For $\delta > 0$ we set

$$f_\delta = d \sqrt[d]{\det\left(\lambda v^{(\delta)} \delta_{ij} - v^{(\delta)}_{x^i x^j}\right)}.$$

The consideration above shows that the functions f_δ are uniformly bounded, $f_\delta \to f$ (a.e.) as $\delta \downarrow 0$, and, for any $\delta \in (0,1]$, there exists a constant N such that for all $x \in E_d$

$$v^{(\delta)}(x) \leq N e^{-\frac{1}{N}|x|}. \tag{13}$$

By Theorem 1.10,

$$d^{-d} \int_{E_d} (v^{(\delta)})^{p-d} f_\delta^d \, dx \geq \lambda^d \int_{E_d} (v^{(\delta)})^p \, dx \geq (v^{(\delta)}(x))^p \lambda^{d/2} \kappa_d d! p^{-d}.$$

Let $\delta \downarrow 0$ and $p > d$. Then using (13) and the Lebesgue dominated convergence theorem we obtain

$$d^{-d} \int_{E_d} v^{p-d} f^d \, dx \geq \lambda^d \int_{E_d} v^p \, dx, \quad \int_{E_d} v^p \, dx \geq v^p(x) \lambda^{-d/2} \kappa_d d! p^{-d}.$$

Applying the Hölder inequality to the integral $\int v^{p-d} f^d \, dx$, we immediately obtain from the first inequality an estimate of $\int v^p \, dx$ in terms of $\int f^p \, dx$ and hence the first inequality in (10). The second relationship along with an estimate of $\int v^p \, dx$ yields the second inequality in (10). We have thus proved (c) for $p > d$. If $p = d$, it suffices to pass to the limit in (10). This proves the theorem.

4. THEOREM. Let $f(t, x) \geq 0$ be a function on E_{d+1} which is equal to zero if $t \geq T$ or $|x| > R$ for some constants T and R. Suppose that f and all its derivatives with respect to t and x up to and including the second order lie in $C_u(E_{d+1})$. Then on E_{d+1} there exists a nonnegative

function $v(t, x)$ such that:

(a) all Sobolev derivatives of the form v_t, v_{x^i}, $v_{x^i x^j}$ exist and are bounded on E_{d+1} and $v \leqq Ne^{-\frac{1}{N}|x|}$ for all t, x and a constant N;

(b) for any $\omega \in \Omega$ and $r \geqq 0$,

$$rv_t + \omega^{ij} v_{x^i x^j} - \lambda(r + \operatorname{tr}\omega)v + \sqrt[d+1]{r \det \omega}\, f \leqq 0,$$

$$v_t - \lambda v \leqq 0, \quad \left(\lambda v \delta_{ij} - v_{x^i x^j}\right) \geqq 0, \quad |v_x| \leqq \sqrt{\lambda} v \ (\text{a.e.}); \tag{14}$$

(c) for any $x \in E_d$, $t \in (-\infty, \infty)$, and $p \geqq d+1$,

$$v(t, x) \leqq \lambda^{\frac{d+2}{2p} - 1} p^{\frac{d}{p}} (\kappa_d d!)^{-\frac{1}{p}} (d + 1)^{\frac{1}{p} - 1} \times \\ \times \left(\int_0^\infty ds \int_{E_d} e^{-\lambda(d+1)s} f^p (t + s, x)\, dx\right)^{\frac{1}{p}}, \tag{15}$$

$$||v(t, \cdot)||^{d+1}_{L_p(E_d)} \leqq \lambda^{-d}(d+1)^{-d} \int_0^\infty e^{-\lambda(d+1)s} ||f(t+s, \cdot) \times \\ ||^{d+1}_{L_p(E_d)}\, ds, \tag{16}$$

$$||v||_{L_p(E_{d+1})} \leqq \frac{1}{\lambda(d+1)} ||f||_{L_p(E_{d+1})}. \tag{17}$$

Proof. The argument below is very similar to that of the previous proof. In section 2.1, let λ be ε and t be the zeroth coordinate of the point (t, x). Moreover, for $\omega \in \Omega_0$ denote $a^{ij}(\omega) = \omega^{ij}$, $i, j \geqq 1$, $a^{0i}(\omega) = a^{i0}(\omega) = 0$, $i \geqq 0$, $b^0 = r(\omega) = 1 - \operatorname{tr}\omega$, $b^i(\omega) = 0$ for $i \geqq 1$, $c(\omega) = -\lambda$, $f(\omega) = \sqrt[d+1]{r(\omega) \det \omega}\, f$. We put $v(t, x) = P_0 f(t, x)$ and prove that v possesses properties (a) to (c).

If the matrix $(u_{ij}, i, j = 0, \ldots, d)$ is symmetric, it is not difficult to see that $F_1(u_{ij})$ in (2.2.4) is the greatest eigenvalue of the matrix $(u_{ij}, i,j = 1, \ldots, d)$. It obviously follows that, if $l \in E_{d+1}$, $l^0 = 0$, and $|l| = 1$, then $\mu(l)$ in (2.2.5) is equal to 1 and from Theorems 2.2.2 and 2.2.6 it follows that all Sobolev derivatives of the form v_t, v_{x^i}, and $v_{x^i x^j}$ $(i, j \geqq 1)$ exist, are bounded on E_{d+1}, and

$$\max_{\omega\in\Omega_0}\left(r(\omega)(v_t - \lambda v) + \sum_{i,j=1}^{d} \omega^{ij}(v_{x^ix^j} - \lambda\delta_{ij}v) + \right.$$
$$\left. + \sqrt[d+1]{r(\omega)\det\omega}\, f\right) = 0 \qquad (18)$$

almost everywhere on E_{d+1}. This relationship is similar to (11) and it is possible to derive assertion (b) from it just as above. The inequality $v \leqq Ne^{-\frac{1}{N}|x|}$ can be proved exactly in the same manner as in the previous proof (employing the same auxiliary function u).

It remains to show that (c) is true. First let $p > d + 1$. Note that $v(t, x) = 0$ for $t \geqq T'$ and $x \in E_d$ for a constant T'. Indeed, we can choose $g \in C^\infty(E_1)$ such that conditions $g(t) \geqq f(t, x)$ and

$$u(t) := e^{\lambda t}\int_t^\infty e^{-\lambda s} g(s)\, ds \geqq \frac{1}{\lambda} f(t, x)$$

are satisfied for all (t, x). This choice is possible since $f = 0$ for $t \geqq T$. It is clear that, when $\omega \in \Omega$, we obtain

$$1 = r(\omega) + \operatorname{tr}\omega \geqq \sqrt[d+1]{r(\omega)\det\omega}$$

and

$$L(\omega)u + f(\omega) = -r(\omega)g - \lambda(1 - r(\omega))\, u + \sqrt[d+1]{r(\omega)\det\omega}\, f \leqq 0.$$

Hence, by Theorem 2.1.3 (b), we conclude that

$$v = P_0 f = u + P_0(f + Lu) \leqq u$$

and $v(t, x) = 0$ for $t \geqq T'$ if T' is such that $g(t) = 0$ for $t \geqq T'$.

By Lemma 2, it follows from (18) that

$$(\lambda v - v_t)\det\left(\lambda\delta_{ij}v - v_{x^ix^j}\right) = (d + 1)^{-d-1} f^{d+1} \quad \text{(a.e.)}. \qquad (19)$$

Further, we take $\zeta(t, x) \in C_0^\infty(E_{d+1})$ such that $\zeta \geqq 0$, $\int\zeta\, dt\, dx = 1$, and for $\delta > 0$ we set

$$v^{(\delta)}(t, x) = v(t, x) * \delta^{-d-1}\zeta(\delta^{-1}t, \delta^{-1}x),$$

$$f_\delta = (d+1)\left[\left(\lambda v^{(\delta)} - v_t^{(\delta)}\right)\det\left(\lambda\delta_{ij}v^{(\delta)} - v^{(\delta)}_{x^ix^j}\right)\right]^{1/(d+1)}.$$

In view of the above argument, the functions f_δ are uniformly bounded and $f_\delta \to f$ (a.e.) as $\delta \downarrow 0$. Moreover, from relationships like $v^{(\delta)}_{x^i t} = \delta^{-d-3} v * \zeta_{x^i t}(\delta^{-1}t, \delta^{-1}x)$ and the estimate $v \leq Ne^{-\frac{1}{N}|x|}$ it follows that for every $\delta > 0$ the function $v^{(\delta)}$ satisfies the conditions of Theorem 1.9. By virtue of this theorem,

$$(d + 1)^{-d-1} p\lambda^{-d} e^{\lambda pt} \int_t^\infty ds \int_{E_d} e^{-\lambda ps} (v^{(\delta)})^{p-d-1} f_\delta^{d+1}\, dx$$

$$\geq \|v^{(\delta)}(t, \cdot)\|^p_{L_p(E_d)} \geq \kappa_d d!(p\sqrt{\lambda})^{-d} (v^{(\delta)}(t, x))^p. \qquad (20)$$

Passing to the limit, as in the previous proof, we can show that these inequalities remain valid if $v^{(\delta)}$ and f_δ are replaced by v and f. We carry out this replacement and set

$$w(t) = e^{-\lambda t}\|v(t, \cdot)\|_{L_p(E_d)}, \quad h(t) = e^{-\lambda t}\|f(t, \cdot)\|_{L_p(E_d)},$$

$$K = p\lambda^{-d}(d + 1)^{-d-1}.$$

Furthermore, we apply the Hölder inequality to the first integral with respect to dx in the first term of (20):

$$\int_{E_d} v^{p-d-1} f^{d+1}\, dx \leq e^{\lambda ps} w^{p-d-1} h^{d+1}.$$

We obtain

$$w^p(t) \leq K \int_t^\infty w^{p-d-1}(s) h^{d+1}(s)\, ds.$$

We now make use of the standard technique for transforming such inequalities. To this end, we take $\delta > 0$ and $T_1 \in (-\infty, \infty)$ and introduce an operator ϕ_δ acting on nonnegative measurable functions defined on $[T_1, \infty)$ as

$$\phi_\delta \psi(t) = \left(K \int_t^\infty \psi^{p-d-1}(s) h^{d+1}(s)\, ds + \delta\right)^{1/p}.$$

It is not difficult to see that ϕ_δ is a monotone operator, i.e. if $0 \leq \Psi^1 \leq \Psi^2$ on $[T_1, \infty)$, then $0 \leq \phi_\delta \Psi^1 \leq \phi_\delta \Psi^2$ on $[T_1, \infty)$. Moreover, since $h(t)$ is a bounded function with compact support, we obtain $\lim \phi_\delta \Psi^n = \phi_\delta \lim \Psi^n$ if the nonnegative functions Ψ^n are uniformly bounded and have limits for every t. Finally, for the function $\Psi(t) \equiv N$, we obtain $\phi_\delta \Psi \leqq \Psi$ whenever N is sufficiently large.

These properties and the boundedness of w(t) on $[T_1, \infty)$ imply that, on choosing N_1 such that $w(t) \leqq N_1$ on $[T_1, \infty)$, we obtain

$$w(t) \leqq \phi_\delta w(t) \leqq \dots \phi_\delta^n w(t) \leqq N_1.$$

Thus, $\lim_n \phi_\delta^n w$ exists and, denoting it by w_δ, we obtain $w \leqq w_\delta$, $w_\delta = \phi_\delta w_\delta$. The latter relationship is an equation in w_δ, which can easily be solved using the inequality $w_\delta \geqq \delta^{1/p}$. Hence we obtain

$$(w_\delta(t))^{d+1} = \frac{K(d+1)}{p} \int_t^\infty h^{d+1}(s)\, ds + \delta^{(d+1)/p}.$$

This, along with the inequality $w \leqq w_\delta$, as $\delta \downarrow 0$, yields

$$w^{d+1} \leqq \frac{K(d+1)}{p} \int_t^\infty h^{d+1}(s)\, ds \tag{21}$$

for $t \in [T_1, \infty)$ and since T_1 is chosen arbitrarily, this is true for all $t \in (-\infty, \infty)$. Inequality (21) coincides with (16).

By virtue of the Minkowski inequality, (21) implies

$$\left(\int_{-\infty}^\infty e^{\lambda pt} w^p\, dt\right)^{1/p}$$

$$\leqq \left(\frac{K(d+1)}{p}\right)^{\frac{1}{d+1}} \left(\int_{-\infty}^\infty e^{\lambda pt}\left(\int_0^\infty h^{d+1}(t+s)\, ds\right)^{\frac{p}{d+1}} dt\right)^{\frac{1}{p}}$$

$$\leqq \left(\frac{K(d+1)}{p}\right)^{\frac{1}{d+1}} \left(\int_0^\infty ds \left(\int_{-\infty}^\infty e^{\lambda pt} h^p(t+s)\, dt\right)^{\frac{d+1}{p}}\right)^{\frac{1}{d+1}}$$

$$= \left(\frac{K(d+1)}{p}\right)^{\frac{1}{d+1}} \left(\int_{-\infty}^\infty e^{\lambda pt} h^p(t)\, dt\right)^{\frac{1}{p}} \left(\int_0^\infty e^{-\lambda(d+1)s}\, ds\right)^{\frac{1}{d+1}}.$$

The inequality between the first and the last terms of this computation is merely another form of inequality (17). Finally, from (21) by the Hölder inequality, it follows that

$$w^p(t) \leqq \left(\frac{K(d+1)}{p}\right)^{\frac{p}{d+1}} \left(\int_t^\infty e^{-\lambda(d+1)s} (e^{\lambda s} h(s))^{d+1} \, ds\right)^{\frac{p}{d+1}}$$

$$\leqq \left(\frac{K(d+1)}{p}\right)^{\frac{p}{d+1}} \left(\int_t^\infty e^{-\lambda(d+1)s} \, ds\right)^{\frac{p-d-1}{d+1}} \int_t^\infty e^{-\lambda(d+1)s} (e^{\lambda s} h(s))^p \, ds$$

Combining this inequality with the second inequality in (20), we immediately obtain (15). We have proved (c) under the condition $p > d + 1$, which allows us to ignore the case 0^0. When $p = d + 1$, we obtain inequalities (15)-(17) letting $p \downarrow p + d$. This proves the theorem.

3.3. The Aleksandrov Estimates

A.D. Aleksandrov [4, 5] established the C-norm estimates of a solution to the Dirichlet problem for the elliptic equation $Lu = f$ in terms of the L_p-norm of f. Although among the estimates established in this section, only one (see Remark 12) was derived by A.D. Aleksandrov himself, nevertheless, his noteworthy idea of using the Monge—Ampère equations plays a major role in the proof of other estimates of this section. That is why we have collected these estimates together in this section entitled "the Aleksandrov estimates". Also, note that the analytical methods which are used here are unfortunately difficult to comprehend on a first reading, unlike the clear geometric techniques used by Aleksandrov himself.

According to the agreement adopted in Section 1.1, points of E_{d+1} are denoted by the pairs (t, x) with $t \in (-\infty, \infty)$ and $x \in E_d$. Let us fix a domain $Q \subset E_{d+1}$ and a number $R > 0$ and assume that

$$Q \subset C_R := \{(t, x) : x < R\}.$$

Suppose that, for any $(t, x) \in Q$, we are given a symmetric nonnegative $d \times d$ matrix $a(t, x) = (a^{ij}(t, x))$, a d-dimensional vector $b(t, x) = (b^i(t, x))$ and numbers $r(t, x) \geqq 0$ and $c(t, x)$. Suppose that a, b, c and r are measurable and denote

$$Lu = ru_t + a^{ij}u_{x^i x^j} + b^i u_{x^i} + cu.$$

Throughout this section constants K_1, $K \geqq 0$, and $\varepsilon \in (0, R^{-1})$ are supposed to exist such that

$$c_K \leqq 0, \quad |b| \leqq \varepsilon \operatorname{tr} a + K_1\left[a^{ij}\frac{x^i x^j}{|x|^2} + |c_K|\right] \tag{1}$$

where $(t, x) \in Q$, $x \neq 0$, and $c_K := -Kr + c$.

1. LEMMA. Let $K = 0$. Then there exists a function $\Psi \in C^\infty_{loc}(E_d)$, which depends only on R, K_1, ε, and x, such that $\Psi \geqq 0$ and $L\Psi + |b| + \operatorname{tr} a \leqq 0$ for $(t, x) \in Q$ and $x \neq 0$.

Proof. Since $L(R^2 - |x|^2) + \operatorname{tr} a = -\operatorname{tr} a - 2(b, x) + c(R^2 - |x|^2) \leqq 2|b|R$, we should only construct a function $\Psi \in C^\infty_{loc}(E_d)$ such that $\Psi \geqq 0$, $L\Psi + |b| \leqq 0$ for $(t, x) \in Q$, $x \neq 0$.

If $\alpha > 0$ and $\beta > \cosh(\alpha R)$, we set $\Psi = \beta - \cosh(\alpha|x|)$, and for $x \neq 0$, denote

$$\lambda = \frac{x}{|x|}, \quad \rho = |x|, \quad f = \varepsilon \operatorname{tr} a + K_1(a^{ij}\lambda^i\lambda^j + |c|).$$

Simple computation shows that if $x \neq 0$, it follows from (1) that $|b| \leqq f$ and

$$\begin{aligned} I &:= (1 + \alpha \sinh \alpha|x|)^{-1}(-L\Psi - f) \\ &= (1 + \alpha \sinh \alpha|x|)^{-1}\{-c(\beta - \cosh \alpha\rho) + \alpha(\sinh \alpha\rho)(b,\lambda) + \\ &\quad + \alpha^2(\cosh \alpha\rho)(a\lambda, \lambda) + \frac{\alpha}{\rho}(\sinh \alpha\rho)[\operatorname{tr} a - (a\lambda, \lambda)] - f\} \\ &\geqq -c\,\frac{\beta - \cosh \alpha\rho}{1 + \alpha \sinh \alpha\rho} + (a\lambda,\lambda)\,\frac{\alpha^2 \cosh \alpha\rho}{1 + \alpha \sinh \alpha\rho} + \\ &\quad + \frac{\alpha \sinh \alpha\rho}{1 + \sinh \alpha\rho}\,\frac{1}{\rho}\,[\operatorname{tr} a - (a\lambda, \lambda)] - f. \end{aligned} \tag{2}$$

Note that $\cosh \alpha\rho \geqq 1$, $\cosh \alpha\rho \geqq \sinh \alpha\rho$, $\alpha \sinh \alpha\rho \geqq \alpha^2\rho$, and for $(t, x) \in Q$ we obtain $\rho \leqq R$. Hence,

$$\frac{\beta - \cosh \alpha\rho}{1 + \alpha \sinh \alpha\rho} \geqq \frac{\beta - \cosh \alpha R}{1 + \alpha \sinh \alpha R},$$

$$\frac{\alpha^2 \cosh \alpha\rho}{1 + \alpha \sinh \alpha\rho} \geqq \frac{\alpha^2 \cosh \alpha\rho}{\cosh \alpha\rho + \alpha \cosh \alpha\rho} = \frac{\alpha^2}{1 + \alpha},$$

$$\frac{1}{\rho}\;\frac{\alpha \sinh \alpha\rho}{1 + \alpha \sinh \alpha\rho} \geqq \frac{1}{\rho}\,\frac{\alpha^2\rho}{1 + \alpha^2\rho} \geqq \frac{\alpha^2}{1 + \alpha^2 R}.$$

Consequently, from (2) it follows that

$$I \geqq -c\,\frac{\beta - \cosh \alpha R}{1 + \alpha \sinh \alpha R} + (a\lambda, \lambda)\,\frac{\alpha^2}{1 + \alpha} +$$

$$+ \frac{\alpha^2}{1 + \alpha^2 R}\,[\operatorname{tr} a - (a\lambda, \lambda)] + f.$$

We now take a sufficiently large α so that $\frac{\alpha^2}{1 + \alpha^2 R} \geqq \varepsilon$ and $\frac{\alpha^2}{1 + \alpha} \geqq K_1 + \varepsilon$. Then we choose a sufficiently large number β such that

$$\frac{\beta - \cosh \alpha R}{1 + \alpha \sinh \alpha R} \geqq K_1.$$

Thus, it can be seen that

$$I \geqq -cK_1 + (a\lambda, \lambda)(K_1 + \varepsilon) + \varepsilon[\operatorname{tr} a - (a\lambda, \lambda)] - f = 0.$$

Hence, $L\Psi + f \leqq 0$ and $L\Psi + |b| \leqq 0$ which proves the lemma.

The following lemma plays a dominant role in this section.

2. LEMMA. Let $K = 0$ and $Q = C_R$. Let all coefficients of L be bounded in Q and for a constant $\delta > 0$ and all $\lambda \in E_d$, $(t, x) \in Q$ the inequalities $r \geqq \delta$ and $(a\lambda, \lambda) \geqq \delta|\lambda|^2$ hold. Also, let $u \in W^{1,2}_{d+1}(Q)$ and $u(t, x) \leqq 0$ for $|x| = R$. Then for $(t_0, x_0) \in Q$ we have

$$u(t_0, x_0) \leqq N\|(r \det a)^{-\frac{1}{d+1}}(Lu)_-\|_{L_{d+1}(Q \cap \{t > t_0\})}, \tag{3}$$

where N depends only on R, K_1, ε, and d.

Proof.
1. It is easy to pass to the limit in (3), therefore we can assume that $u \in C^\infty(\bar{Q})$ and $u = 0$ for sufficiently large $|t|$. Furthermore, it is conventient to assume that $b(t, x) = 0$ for sufficiently small $|x|$. This assumption can easily be abandoned by means of passing to the limit. It is this assumption that allows us to preserve inequalities (1) if a, b, c, and r are replaced by their Sobolev means and K_1 is replaced by $2K_1$. Summarizing, we conclude that it suffices to prove the lemma under the auxiliary condition that $a, b, c, r \in C^\infty(E_{d+1})$, $u \in C^\infty(\bar{Q})$, and $u = 0$ for sufficiently large $|t|$. Moreover, if we consider $r^{-1}Lu$ in place of Lu, we can see that it suffices to prove (2) under the condition $r \equiv 1$.

One more step towards further simplification of the problem in question is as follows. We take a $\zeta \in C_0^\infty(E_{d+1})$ such that $\zeta \geqq 0$, and $\int \zeta \, dt \, dx = 1$. For every locally summable on E_{d+1} function $f(t, x)$ and $\gamma > 0$ we put $f^{(\gamma)}(t, x) = f(t, x) f * \gamma^{-d-1} \zeta(\gamma^{-1}t, \gamma^{-1}x)$; when $\gamma = 0$, let $f^{(0)}(t, x) = f(t, x)$. We denote $g = (Lu)_-$ and extend g outside Q so that it becomes a positive continuous function on E_{d+1} with compact support. For $\gamma \geqq 0$, we define functions u_i^γ and u_2 as solutions to the following problems

$$Lu_1^\gamma = -g^{(\gamma)} \text{ in } Q, \quad u_1^\gamma \Big|_{|x| = R} = 0, \tag{4}$$

$$Lu_2 = (Lu)_+ \text{ in } Q, \quad (u_2 - u)\Big|_{|x| = R} = 0. \tag{5}$$

Notethat, in any case, the functions $(Lu)_\pm$ satisfy the Lipschitz condition on $\bar{Q}$ and are zero for sufficiently large t. From the theory of linear equations it is known that there exist solutions to problems (4) and (5) in the class $C^2(\bar{Q})$ which are equal to zero for sufficiently large t. Moreover, $u_1^\gamma \in C^\infty(\bar{Q})$ for $\gamma > 0$ and $u_1^\gamma \to u_1^0$, as $\gamma \downarrow 0$, at every point of Q.

From the uniqueness theorems and the maximum principle it follows that $u = u_1^0 + u_2$, $u_2 \leqq 0$, and $u \leqq u_1^0$, in Q. Hence, if for $\gamma > 0$ we prove that

$$u_1^\gamma(t_0, x_0) \leqq N \| (\det a)^{-1/(d+1)} (Lu_1^\gamma)_- \|_{L_{d+1}(Q \cap \{t > t_0\})},$$

where N depends only on R, K_1, ε, and d, then, on passing to the limit as $\gamma \downarrow 0$, we shall prove (3) for u.

2. Thus, it remains to prove the case when $r \equiv 1$, a^{ij}, b^i, $c \in C^\infty(E_{d+1})$, $u \in C^\infty(\bar{Q})$, $u = 0$ for sufficiently large t, $u = 0$ for $|x| = R$, $Lu \leqq 0$ on Q, $Lu = 0$ for large $|t|$, and Lu can be extended to a negative function of $C_0^\infty(E_{d+1})$ on E_{d+1}. We define the function f on Q as

$$Lu = -(\det a)^{1/(d+1)} f. \tag{6}$$

Let us extend f to E_{d+1} so that $f \in C_0^\infty(E_{d+1})$ and $f \geqq 0$. In this case u^γ denotes the solution from class $C^\infty(\bar{Q})$ of the problem

$$Lu^\gamma = -(\det a)^{1/(d+1)} (e^t f)^{(\gamma)} \cdot e^{-t}$$

in Q and $u^\gamma = 0$ for $|x| = R$ such that $u^\gamma = 0$ for sufficiently large t. Furthermore, let v be the function which is associated with $\lambda = 1$ and

$e^t f(t,x)$ by Theorem 2.4,

$$K(\gamma) := \sup\{v^{(\gamma)}(t, x)e^{-t}: \ t \geqq t_0, \ x \in E_d\}.$$

Note that from (2.15) we immediately obtain that $K(\gamma)$ are finite,

$$\overline{\lim_{\gamma \downarrow 0}} K(\gamma) \leqq N||f||_{\mathcal{L}_{d+1}(E_{d+1} \cap \{t > t_0\})}. \tag{7}$$

Furthermore, inequality (2.14) holding true for all constant r and ω after convolution with $\gamma^{-d-1}\zeta(\gamma^{-1} \ t, \ \gamma^{-1} \ x)$ yields a similar inequality which is also true for all constant r and ω.

The latter inequality will obviously be preserved if r and ω are variable. Hence, by virtue of Lemma 1 with $\gamma > 0$ and $t > t_0$, we obtain

$$L(e^{-t}v^{(\gamma)} - K(\gamma)\psi) + (\det a)^{1/(d+1)} e^{-t}(e^t f)^{(\gamma)}$$

$$= e^{-t}\Big[v_t^{(\gamma)} + a^{ij}v^{(\gamma)}_{x^i x^j} - (1 + \operatorname{tr} a)v^{(\gamma)} +$$

$$+ (\det a)^{1/(d+1)} (e^t f)^{(\gamma)}\Big] + e^{-t}v^{(\gamma)}\operatorname{tr} a +$$

$$+ e^{-t}b^i v^{(\gamma)}_{x^i} + ce^{-t}v^{(\gamma)} + K(\gamma) L\psi$$

$$\leqq e^{-t}v^{(\gamma)}(\operatorname{tr} a + |b|) - K(\gamma).(\operatorname{tr} a + |b|) \leqq 0,$$

if $x \neq 0$ and since the functions involved in the expressions are continuous, this is also true for $x = 0$.

Since $v^{(\gamma)} \geqq 0$ and, specifically, $v^{(\gamma)} \geqq u^\gamma$ for $|x| = R$, the maximum principle implies that

$$u^\gamma(t_0, x_0) \leqq e^{-t_0} v^{(\gamma)} (t_0, x_0) + K(\gamma)\Psi(x_0) \leqq K(\gamma)(1 + \Psi(x_0)).$$

As $\gamma \downarrow 0$, this along with (7) yield

$$u(t_0, x_0) \leqq N||f||_{\mathcal{L}_{d+1}(E_{d+1} \cap \{t > t_0\})}.$$

The latter inequality has been proved for any nonnegative $f \in C_0^\infty(E_{d+1})$ such that (6) holds on Q. It is not difficult to construct a decreasing sequence of functions f_n such that, in addition, $f_n \downarrow 0$ outside $\bar{Q}$ and then

$$u(t_0, x_0) \leqq N\left\|\lim_{n\to\infty} f_n\right\|_{L_{d+1}(E_{d+1}\cap\{t > t_0\})}$$

$$= N\|(\det a)^{-1/(d+1)}(Lu)_-\|_{L_{d+1}(Q\cap\{t > t_0\})}.$$

This proves the lemma.

We now extend the assertion of Lemma 2 to the case when d+1 in (3) is replaced by p+1 with $p \geqq d$.

3. LEMMA. Let $K = 0$, $Q = C_R$, all coefficients of L be bounded in Q, and for a constant $\delta > 0$ and all $\lambda \in E_d$, $(t, x) \in Q$ the following inequalities hold: $c \leqq -\delta$, $r \geqq \delta$, $(a\lambda, \lambda) \geqq \delta|\lambda|^2$. Let $u \in W^{1,2}_{p+1}(Q)$, $p \geqq d$, $u(t, x) \leqq 0$ for $|x| = R$. Then, for $(t_0, x_0) \in Q$, we obtain

$$u(t_0, x_0) \leqq N\left\||c|^{\frac{d-p}{p+1}}(r \det a)^{-\frac{1}{p+1}}(Lu)_-\right\|_{L_{p+1}(Q\cap\{t > t_0\})},$$

where N depends only on R, K_1, ε, d.

<u>Proof</u>. Just as above, the proof reduces to the case when a^{ij}, b^i, $c \in C^\infty(E_{d+1})$, $r \equiv 1$, $u \in C^\infty(\bar{Q})$, $u = 0$ for sufficiently large t, $u = 0$ for $|x| = R$, and $Lu \leqq 0$. In this case we denote $f = -Lu$ and define v as a function of $C^2(\bar{Q})$ which is equal to zero for sufficiently large t and such that $Lv = -|c|^{(d-p)/(d+1)} \cdot f^{(p+1)/(d+1)}$ in Q, $v = 0$ for $|x| = R$.

We now show that $u \leqq v^{(d+1)/(p+1)}$ in Q. For any constant $\gamma > 0$, using Young's inequality

$$\varphi\Psi \leqq \frac{1}{p}\varphi^p + \frac{1}{q}\Psi^q,$$

where $p^{-1} + q^{-1} = 1$, p and $q > 1$, and $\varphi, \Psi \geqq 0$, we obtain

$$L\left(\gamma\frac{d+1}{p+1} v + \frac{p-d}{p+1}\gamma^{-\frac{d+1}{p-d}}\right)$$

$$= -\gamma\frac{d+1}{p+1}|c|^{\frac{d-p}{d+1}} f^{\frac{p+1}{d+1}} - |c|\frac{p-d}{p+1}\gamma^{-\frac{d+1}{p-d}} \leqq -f = Lu.$$

Hence, by the maximum principle, $u \leqq \gamma\frac{d+1}{p+1} v + \frac{p-d}{p+1}\gamma^{-(d+1)/(p-d)}$. On computing the greatest lower bound with respect to $\gamma > 0$ and using Lemma 2, we conclude

$$u(t_0, x_0) \leq v^{\frac{d+1}{p+1}}(t_0, x_0)$$

$$\leq N^{\frac{d+1}{p+1}} \left\| |c|^{\frac{d-p}{d+1}} (\det a)^{-\frac{1}{d+1}} (Lu)_-^{\frac{p+1}{d+1}} \right\|_{L_{d+1}(Q \cap \{t > t_0\})}^{\frac{d+1}{p+1}}$$

$$= N^{\frac{d+1}{p+1}} \left\| |c|^{\frac{d-p}{p+1}} (\det a)^{-\frac{1}{p+1}} (Lu)_- \right\|_{L_{p+1}(Q \cap \{t > t_0\})},$$

where N is the constant involved in Lemma 2. Since $N^{\alpha} \leq N+1$ for $\alpha \in [0,1]$, the lemma is proved.

In order to extend the assertion of Lemma 3 to curvilinear domains $Q \subset C_R$, we need some definitions.

4. Definition. Given a domain $G \subset E_{d+1}$, the (right) parabolic boundary of G is the set of all points (t_0, x_0) belonging to the boundary of G (as a set of E_{d+1}) each of which is associated with a continuous function $x_t \in E_d$ and a number $\delta > 0$ such that $x_{t_0} = x_0$ and $(t, x_t) \in G$ for $t \in [t_0 - \delta, t_0)$. The parabolic boundary is denoted by $\partial' G$.

5. Definition. Given a domain $G \subset E_{d+1}$ and a point $(t_0, x_0) \in \bar{G}$, the set dominating (t_0, x_0) in G is the set of all points $(t_1, x_1) \in G$ such that $t_1 > t_0$ and on $[t_0, t_1]$ there exists a continuous function $x_t \in E_d$ such that $x_{t_0} = x_0$, $x_{t_1} = x_1$, $(t, x_t) \in G$ for $t \in (t_0, t_1]$. The set dominating (t_0, x_0) in the domain G is denoted by $G(t_0, x_0)$.

It is easy to see that $G(t_0, x_0)$ is an open set and $\partial' G(t_0, x_0) \subset \partial' G$.

6. LEMMA. Let $K = 0$, all coefficients of L be bounded in Q, the domain Q be bounded, and for a constant $\delta > 0$ and all $\lambda \in E_d$, $(t, x) \in Q$ the inequalities $c \leq -\delta$, $r \geq \delta$, $(a\lambda, \lambda) \geq \delta|\lambda|^2$ hold. Let $p \geq d$, $u \in W^{1,2}_{p+1}(Q)$, and $u \leq 0$ on $\partial' Q$. Then

$$u \leq N \left\| |c|^{(d-p)/(p+1)} (r \det a)^{1/(p+1)} (Lu)_- \right\|_{L_{p+1}(Q)}, \tag{8}$$

on $\bar{Q}$, where N depends only on R, K_1, ε, and d.

Proof. As in part 1 of the proof of Lemma 2, we can assume that $r \equiv 1$, a^{ij}, b^i, $c \in C^\infty(E_{d+1})$, and $u \in C^\infty(\bar{Q})$. We put $f = (Lu)_-\chi_Q$ and define the function v as a solution of the problem $Lv = -f$ in $(-\infty, T) \times S_R$ and $v = 0$ on $\partial'(-\infty, T) \times S_R$, where T is chosen in such a way that $f(t, x)$ is equal to zero when $t \geqq T-1$. From the theory of linear equations it is known that v exists and belongs to $W^{1,2}_{p+1}((t_0, T) \times S_R)$ for any t_0.

It is clear that $v(t, x) = 0$ for $t \in [T-1, T]$. We extend $v(t, x)$ so that it is zero for $t \geqq T$ and keep the same notation for the extended function. Then $v \in W^{1,2}_{p+1}(C_R \cap \{t > t_0\})$ for any t_0. Applying Lemma 3 to the product of $(-v)$ by the function of t which is equal to zero for $t \leqq t_0-1$ and 1 for $t \geqq t_0$, we find that $v(t_0, x_0) \geqq 0$ for all $(t_0, x) \in C_R$.

We denote $w = v-u$ and prove that $w \geqq 0$ on Q. Suppose that the contrary is true. Then there exists $\xi > 0$ such that $\Gamma := \{(t, x) \in \bar{Q}: w(t, x) \leqq -\xi\} \neq \phi$. The set Γ is closed and bounded on the right with respect to t. Therefore, there exists the rightmost point $(t_0, x_0) \in \Gamma$ with respect to the axis t, i.e., a point such that $w(t_0, x_0) \leqq -\xi$ and $w(t, x) > -\xi$ if $t > t_0$ and $(t, x) \in \bar{Q}$. There are three alternative locations of the point (t_0, x_0) in $\bar{Q}$:

(a) $(t_0, x_0) \in \overline{\partial' Q}$;

(b) $(t_0, x_0) \in \partial Q \setminus \overline{\partial' Q}$;

(c) $(t_0, x_0) \in Q$.

In our case situation (a) is impossible since $w \geqq v \geqq 0$ on $\partial' Q$ and $\overline{\partial' Q}$.

Let $(t, x) \in \partial Q \setminus \overline{\partial' Q}$. Then $G_\gamma := \{(t, x) : t_0 < t < t_0 + \gamma$ and $|x - x_0| < \gamma\} \subset Q$ for a sufficiently small $\gamma > 0$. Indeed, if for any γ there exists $(t_\gamma, x_\gamma) \in G_\gamma \setminus Q$, then one can define the points $(\tilde{t}_\gamma, \tilde{x}_\gamma) \in Q$ such that $\tilde{t}_\gamma < t_\gamma$ and $(\tilde{t}_\gamma, \tilde{x}_\gamma) \to (t_0, x_0)$. Then it follows that on the line segment connecting the points $(\tilde{t}_\gamma, \tilde{x}_\gamma)$ and (t_γ, x_γ) there is obviously at least one point from ∂Q which also belongs to $\partial' Q$ and tends to (t_0, x_0) as $\gamma \downarrow 0$. On the other hand, $(t_0, x_0) \notin \overline{\partial' Q}$.

We fix a γ_0 such that $G_{\gamma_0} \subset Q$. For G_{γ_0} we have

$$Lw = Lv - Lu = -(Lu)_- - Lu = -(Lu)_+.$$

The function $(Lu)_+$ satisfies the Lipschitz condition on G_{γ_0}. By the theorem of increased smoothness, $w \in C^2(\bar{G}_\gamma)$ for $\gamma < \gamma_0$. Moreover, at the point $(t_0, x_0) \in \overline{G}_\gamma$ the function w attains its minimum which is strictly negative by the definition of (t_0, x_0), and $Lw \leqq 0$. According to the strong Nirenberg maximum principle this is impossible and, therefore, the point (t_0, x_0) cannot belong to $\partial Q \setminus \partial' Q$.

The remaining alternative $(t_0, x_0) \in Q$ is also impossible since in this case the appropriate cylinder G_γ obviously exists and this, as above, contradicts the strong maximum principle.

Thus, on Q we obtain $w \geqq 0$ and, by Lemma 3,

$$u \leqq v \leqq N \left\| |c|^{(d-p)/(p+1)} (\det a)^{-1/(p+1)} (Lv)_- \right\|_{L_{p+1}(C_R)}.$$

It remains to note that $Lv = -(Lu)_- \chi_Q$. This proves the lemma.

In the following lemma we weaken the assumptions of the previous lemma concerning the coefficients of L.

7. LEMMA. Let the domain Q be bounded, $K = 0$, $r + \operatorname{tr} a + |c| > 0$ almost everywhere on Q, $p \geqq d$, $u \in W^{1,2}_{p+1}(Q)$, and $u \leqq 0$ on $\partial' Q$. Then the assertion of Lemma 6 is true with $0^0 = 1$ and $0/0 = 1$ if such indeterminate forms occur on the right of (8).

Proof. We first note that the norm in (8) does not change if all coefficients of L are multiplied by the same positive function. Therefore, without loss of generality we can assume that

$$r + \operatorname{tr} a + c = 1 \quad (\text{a.e. } Q) \tag{9}$$

by virtue of the condition $r + \operatorname{tr} a + |c| > 0$.

If $\delta > 0$, we set

$$L^\delta = L + \delta\left(\frac{\partial}{\partial t} + \Delta - 1\right).$$

Let $a(\delta)$, $b(\delta)$, $c(\delta)$, and $r(\delta)$ be coefficients of L^δ. It is clear that

$$|b(\delta)| = |b| \leqq \varepsilon \operatorname{tr} a(\delta) + K_1\left[a^{ij}(\delta)\frac{x^i x^j}{|x|^2} + |c(\delta)|\right]$$

if $x \neq 0$. It is easy to see that the other conditions of Lemma 6 are also satisfied (for a different δ). Hence, denoting

$$g(\delta) = |c - \delta|^{p-d}(r + \delta) \det(a + \delta I),$$

we obtain from Lemma 6

$$w \leq N||g^{-1/(p+1)}(\delta)[Lw + \delta(w_t + \Delta w - w)]_-||_{L_{p+1}(Q)} \tag{10}$$

in $\bar{Q}$ for any $w \in W^{1,2}_{p+1}(Q)$ such that $w \leq 0$ on $\partial' Q$. We substitute $w = u^\gamma := u + \gamma\psi_1$ into (10), where $\gamma > 0$,

$$\psi_1 = -e^{-t} - \psi,$$

and ψ is taken from Lemma 1. It is clear that $\psi_1 \leq 0$ and, by virtue of (9),

$$L\psi_1 = re^{-t} + |c|e^{-t} - L\psi \geq re^{-t} + |c|e^{-t} + \operatorname{tr} a > 0$$

on Q.

Moreover, we set

$$g = \lim_{\delta\downarrow 0} g(\gamma), \quad \Gamma = \{(t, x) \in Q: \; g(t, x) = 0\},$$

$$f^\gamma = u^\gamma_t + \Delta u^\gamma - u^\gamma.$$

Then from (10) we obtain

$$u^\gamma \leq N||g^{-1/(p+1)}(\delta)[Lu \; + \delta f^\gamma]_-||_{L_{p+1}(\Gamma)} +$$

$$+ N||g^{-1/(p+1)}(\delta)[Lu^\gamma + \delta f^\gamma]_- \; ||_{L_{p+1}(Q\setminus\Gamma)}. \tag{11}$$

If $Lu < 0$ on a subset of Γ of positive measure, then the right-hand side of (8) obviously equals infinity and the proof is trivial. Therefore, we assume that $Lu \geq 0$ (a.e.Γ). Hence, $Lu^\gamma > 0$ (a.e.Γ). Since $g(\delta) \geq \delta^{p+1}$, the first norm in (11) does not exceed

$$\left|\left|\left[\frac{1}{\delta} Lu^\gamma + f^\gamma\right]_-\right|\right|_{L_{p+1}(\Gamma)}.$$

This expression tends to zero as $\delta \downarrow 0$ by virtue of the Lebesgue dominated convergence theorem and the fact that

$$Lu^\gamma > 0 \text{ (a.e.}\Gamma), \quad [\tfrac{1}{\delta} Lu^\gamma + f^\gamma]_- \leq |f^\gamma| \in L_{p+1}(Q).$$

To estimate the second norm in (11), we note that $g > 0$ on $Q\setminus\Gamma$ and, specifically, $\det a > 0$ and

$$1 \geq \frac{\delta^{p+1}}{g(\delta)} = \frac{\delta^{p-d}}{|c-\delta|^{p-d}} \cdot \frac{\delta}{r+\delta} \cdot \frac{\delta^d}{\det\,(a+\delta)} \to 0$$

on $Q \setminus \Gamma$ as $\delta \downarrow 0$. Furthermore, we make use of the fact that $g(\delta) \geq g$ and

$$[Lu^\gamma + \delta f^\gamma]_- \leq (Lu^\gamma)_- + \delta|f^\gamma| \leq (Lu)_- + \delta|f^\gamma|.$$

Then using the monotone convergence theorem and the Lebesgue dominated convergence theorem, we can easily prove that the limit of the second norm in (11), as $\delta \downarrow 0$, does not exceed

$$\lim_{\delta\downarrow 0}\left\{\left\|g(\delta)^{-\frac{1}{p+1}}(Lu)_-\right\|_{\mathcal{L}_{p+1}(Q\setminus\Gamma)} + \left\|\delta g(\delta)^{-\frac{1}{p+1}} f^\gamma\right\|_{\mathcal{L}_{p+1}(Q\setminus\Gamma)}\right\}$$

$$= \left\|g^{-\frac{1}{p+1}}(Lu)_-\right\|_{\mathcal{L}_{p+1}(Q\setminus\Gamma)} = \left\|g^{-\frac{1}{p+1}}(Lu)_-\right\|_{\mathcal{L}_{p+1}(Q)}.$$

Attention should be paid to the fact that those are the very relationships which imply the correctness of our stipulation on the indeterminate forms in (8). Finally, from (11) we obtain

$$u^\gamma \leq N\|g^{-1/(p+1)}(Lu)_-\|_{\mathcal{L}_{p+1}(Q)}.$$

Letting $\gamma \downarrow 0$, we come to (8). This proves the lemma.

We now weaken our condition on u.

8. LEMMA. Let Q be a bounded domain, $K = 0$, $r + \operatorname{tr} a + |c| > 0$ almost everywhere on Q, $p \geq d$, $u \in W^{1,2}_{p+1,loc}(Q) \cap C(\bar{Q})$, and $u \leq 0$ on $\partial' Q$. Then the assertion of Lemma 7 is true.

Proof. First, we can assume that $u \leq -\gamma$ on $\partial' Q$ for a constant $\gamma > 0$. Indeed, for an arbitrary function u satisfying conditions of the theorem, we have $u-\gamma \leq -\gamma$ on $\partial' Q$ and if estimate (8) can be applied to $u-\gamma$, then using $L(u-\gamma) = Lu-\gamma c \geq Lu$ and $(L(u-\gamma))_- \leq (Lu)_-$ and letting $\gamma \downarrow 0$, we obtain (8) for the initial function.

Thus, let us assume that $u \leq -\gamma$ on $\partial' Q$, take a point $(t_0, x_0) \in \bar{Q}$ and prove that $u(t_0, x_0)$ does not exceed the right-hand side of (8). When doing this, we can, of course, assume that $(t_0, x_0) \in Q$ since $u \in C(\bar{Q})$ by hypothesis. We can also assume that $u(t_0, x_0) > 0$ because otherwise the proof is trivial.

For $\delta > 0$ we denote by $Q^{(\delta)}$ the set of all points of Q whose distance from ∂Q is strictly greater than δ. Let

$$Q_1^\delta = Q^{(\delta)} \cap \{(t, x): \; u(t, x) > 0\}.$$

It is obvious that $\overline{Q_1^\delta} \subset Q$ and $u \in W^{1,2}_{p+1}(Q_1^\delta)$. Furthermore, we can take the number $\delta_0 > 0$ such that $[t_0, t_0 + 3\delta_0] \times \{|x - x_0| \leqq \delta_0\} \subset Q$. Reducing δ_0 if necessary, we can ensure the inequality $u \leqq -\frac{1}{2}\gamma$ for the part of the δ_0-neighbourhood of $\partial'Q$ which lies in Q. This $\delta_0 > 0$ can be chosen since u is continuous in $\bar{Q}$ and $u \leqq -\gamma$ on $\partial'Q$.

We now show that for $\delta \in (0, \delta_0)$ the inequality $u \leqq 0$ holds on $\partial'Q_1^\delta(t_0 + 2\delta, x_0)$. Assume the contrary. Then, by Definition 4, there exists a point (t_1, x_1) lying on the boundary of $Q^{(\delta)}$ such that $u(t_1, x_1) > 0$, and there exists a continuous curve (t, x_t), $t \in [t_0 + 2\delta, t_1]$, such that $(t, x_t) \in Q_1^\delta$ for $t \in [t_0 + 2\delta, t_1)$, $x_{t_0+2\delta} = x_0$, and $x_{t_1} = x_1$. Since $(t, x_t) \in Q^{(\delta)}$ for $t < t_1$, while $(t_1, x_{t_1}) \in \partial Q^{(\delta)}$, the trajectory (t, x_t) can be shifted, as an entity, through a distance of δ so that the new trajectory lies in Q at all time instants except for that corresponding to the shifted point (t_1, x_1) and at this latter instant it lies on ∂Q. If the initial point of the shifted trajectory transfers to the point (s_0, y_0), then obviously, $s_0 \in [t_0 + \delta, t_0 + 3\delta]$ and $|x_0 - y_0| \leqq \delta$.

Let us connect the points (t_0, x_0) and (s_0, y_0) by a line segment. By virtue of the choice of δ_0, this segment lies in Q. The composed trajectory which consists of the segment mentioned and becomes the shifted trajectory (t, x_t) by the time s_0, has the form $(t, \tilde{x}_t)$, starts at the point (t_0, x_0), and first leaves Q at the point to which (t_1, x_1) is shifted. This means specifically that (t_1, x_1) is shifted to the point lying on $\partial'Q(t_0, x_0)$ and the distance from (t_1, x_1) to $\partial'Q(t_0, x_0)$ is δ_0. However, by virtue of the choice of δ_0 the inequality $u(t_1, x_1) \leqq -\frac{1}{2}\gamma$ must then hold in contradiction to the inequality $u(t_1, x_1) > 0$ assumed above.

Thus, $u \in W^{1,2}_{p+1}(Q_1^\delta) \cap W^{1,2}_{p+1}(Q_1^\delta(t_0 + 2\delta, x_0))$ and $u \leqq 0$ on $\partial'Q_1^\delta(t_0 + 2\delta, x_0)$ for $\delta \in (0, \delta_0)$. By Lemma 7, if $\delta \in (0, \delta_0)$,

$$\begin{aligned} u(t_0 + 2\delta, x_0) &\leqq N\left\| |c|^{\frac{d-p}{p+1}} (r \det a)^{-\frac{1}{p+1}} (Lu)_- \right\|_{L_{p+1}(Q_1^\delta(t_0+2\delta, x_0))} \\ &\leqq N\left\| |c|^{\frac{d-p}{p+1}} (r \det a)^{-\frac{1}{p+1}} (Lu)_- \right\|_{L_{p+1}(Q)}. \end{aligned}$$

Letting $\delta \downarrow 0$ in the inequality between the extreme terms, we obtain that $u(t_0, x_0)$ does not exceed the right-hand side of (8). This proves the lemma.

We now formulate the basic results of this section.

9. THEOREM. Under the assumptions adopted at the beginning of this section, let the domain Q be bounded with respect to t on the right and let $r + \operatorname{tr} a + |c_K| > 0$ almost everywhere on Q and $p \geqq d$. Let $(t_0, x_0) \in Q$ and $u \in W^{1,2}_{p+1,loc}(Q) \cap C(\bar{Q})$. Then

$$u(t_0, x_0) \leqq \|e^{m(t-t_0)}u_+\|_{C(\partial' Q(t_0, x_0))} + N\left\| e^{K(t-t_0)} |c_K|^{\frac{d-p}{p+1}} (r \det a)^{-\frac{1}{p+1}} (Lu)_- \right\|_{L_{p+1}(Q^+(t_0, x_0))}, \tag{12}$$

where the constant m is chosen so that $m \cdot r \geqq c$ in $Q(t_0, x_0)$

$$Q^+(t_0, x_0) = Q(t_0, x_0) \cap \left\{(s, x) : e^{ms}u(s, x) > \|e^{mt}u_+\|_{C(\partial' Q(t_0, x_0))}\right\},$$

the constant N depends only on R, K_1, ε and d, $0^0 = 1$ and $0/0 = 0$ if these indeterminate forms occur on the right of (12).

<u>Proof</u>. Denote

$$L_K v = e^{Kt}L(e^{-Kt}v),$$

$$w(s, x) = e^{Ks}u(s, x) - e^{(K-m)s}\|e^{mt}u_+\|_{C(\partial' Q(t_0, x_0))} \tag{13}$$

and suppose that the inequality

$$w(t_0, x_0) \leqq N\left\| |c_K|^{\frac{d-p}{p+1}} (r \det a)^{-\frac{1}{p+1}} (L_K w)_- \right\|_{L_{p+1}(Q^+(t_0, x_0))}. \tag{14}$$

has already been proved. We prove (12). Note that

$$L_K e^{(K-m)t} = e^{(K-m)t}(c - mr) \leqq 0$$

in $Q(t_0, x_0)$.

Hence

$$L_K w \geqq e^{Kt} Lu, \quad (L_K w)_- \leqq e^{Kt}(Lu)_-.$$

If now $(L_K w)_-$ is replaced by the obtained estimate in (14), $w(t_0, x_0)$ is substituted with its expression from (13), and the resulting inequality is multiplied by e^{-Kt_0}, then we come to (12).

Thus, it remains to prove (14). Note that the coefficient c in the operator L_K is equal to c_K. Therefore, we can and will assume that the theorem hypothesis is true for $K = 0$.

Furthermore, for the sake of brevity we denote $Q^+ = Q^+(t_0, x_0)$. It is obvious that the domain Q^+ is bounded and

$$Q^+ = Q(t_0, x_0) \cap \{(t, x) : w(t, x) > 0\}.$$

If $w(t_0, x_0) \leqq 0$, there is no need to prove inequality (14). Thus, we may assume that $w(t_0, x_0) > 0$ and $(t_0, x_0) \in \bar{Q}^+$. Moreover, $w \leqq 0$ on $\partial' Q(t_0, x_0)$. Consequently, if $(t_1, x_1) \in \partial' Q^+$ and $(t_1, x_1) \in \partial' Q(t_0, x_0)$, then $w(t_1, x_1) \leqq 0$. But if $(t_1, x_1) \in \partial' Q^+$ and $(t_1, x_1) \notin \partial' Q(t_0, x_0)$, then, of course, $(t_1, x_1) \notin \partial Q(t_0, x_0)$, however, $(t_1, x_1) \in \bar{Q}(t_0, x_0)$ and therefore $(t_1, x_1) \in Q(t_0, x_0)$, $(t_1, x_1) \in \partial\{w > 0\}$, and $w(t_1, x_1) = 0$. Thus, $w \leqq 0$ on $\partial' Q^+$. Moreover, $w \in W^{1,2}_{p+1,loc}(Q^+) \cap C(\bar{Q}^+)$. Therefore, (14) follows from Lemma 8. This proves the theorem.

We now extend Theorem 9 to the case of domains which are unbounded on the right.

10. THEOREM. Let the domain Q be unbounded on the right, let $r + \mathrm{tr}\, a + |c_K| > 0$ almost everywhere on Q, and let $p \geqq d$. Let $(t_0, x_0) \in Q$ and let

$$u \in W^{1,2}_{p+1,loc}(Q) \cap C_{loc}(\bar{Q}),$$

$$\lim_{t\to\infty} \sup\left\{u_+(s, x)e^{ms} : \quad (s, x) \in Q,\ s = t\right\} = 0, \tag{15}$$

where m is introduced as in Theorem 9. Then the assertion of Theorem 9 is true.

To prove this theorem, it suffices to apply Theorem 9, where Q is replaced by $Q \cap \{t < t_1\}$, and then to let $t_1 \to \infty$.

We now prove similar results for elliptic operators.

11. THEOREM. Let a, b, and c be independent of t, $c \leqq 0$, and

$$L_1 v = a^{ij} v_{x^i x^j} + b^i v_{x^i} + cv.$$

Let $D \subset S_R$, $\varepsilon \in (0, R^{-1})$, $K_1 \geqq 0$, and

$$|c| + \text{tr}\, a > 0, \quad |b| \leqq \varepsilon\, \text{tr}\, a + K_1\left[a^{ij}\frac{x^i x^j}{|x|^2} + |c|\right] \tag{16}$$

for $x \in D$ and $x \neq 0$. Finally, let $p \geqq d$ and $v \in W^2_{p,loc}(D) \cap C(\bar{D})$. Then

$$v \leqq ||v_+||_{C(\partial D)} + N\left|\left||c|^{(d-p)/p}(\det a)^{-1/p}(L_1 v)_-\right|\right|_{\mathcal{L}_p(D^+)}, \tag{17}$$

in $\bar{D}$, where $D^+ = \{x \in D : v(x) > ||v_+||_{C(\partial D)}\}$, N depends only on R, K_1, ε, and d, $0^0 = 1$ and $0/0 = 0$ if such indeterminate forms are encountered.

Proof. It obviously suffices to prove (17) only on D^+. Setting $w_0 = v - ||v_+||_{C(\partial D)}$ and using the inequalities $L_1 w_0 \geqq L_1 v$ and $(L_1 w_0)_- \leqq (L_1 v)_-$, which are true by virtue of $c \leqq 0$, we conclude that it suffices to prove the inequality

$$w \leqq N\left|\left||c|^{(d-p)/p}(\det a)^{-1/p}(L_1 w)_-\right|\right|_{\mathcal{L}_p(D^+)} \tag{18}$$

on D^+ for any $w \in W^2_{p,loc}(D^+) \cap C(\bar{D}^+)$ such that $w \leqq 0$ on ∂D^+.

If w is such a function and we prove that, for any $\gamma > 0$,

$$w - \gamma \leqq N\left|\left||c|^{(d-p)/p}(\det a)^{-1/p}(L_1(w-\gamma))_-\right|\right|_{\mathcal{L}_p(D^+ \cap \{w > \gamma\})},$$

on $D^+ \cap \{w > \gamma\}$, then using $(L_1(w - \gamma))_- \leqq (L_1 w)_-$ and $D^+ \cap \{w > \gamma\} \subset D^+$ and letting $\gamma \downarrow 0$, we obtain (18). Since the closure of the domain $D^+ \cap \{w > \gamma\}$ lies in D, it follows that $w - \gamma$ is a function of the class W^2_p on $D^+ \cap \{w > \gamma\}$, and it remains to prove that

$$w \leqq N\left|\left||c|^{(d-p)/p}(\det a)^{-1/p}(L_1 w)_-\right|\right|_{\mathcal{L}_p(G)}. \tag{19}$$

on G if is a a domain $G \subset D$, $w \in W^2_p(G)$, and $w \leqq 0$ on ∂G.

We first prove (19) under additional assumptions. Suppose that the coefficients of L_1 are bounded, $c \leqq -\delta$, and $(a\lambda, \lambda) \geqq \delta|\lambda|^2$ for a constant $\delta > 0$. Here using Theorem 2.3 instead of Theorem 2.4, we could prove the respective counterpart of Lemma 2 and then use arguments similar to those from the proofs of Lemmas 3 and 6. In our opinion, however, it would be more instructive to obtain (19) in the specific case under consideration from Theorem 10. It is clear that one more assumption $w \in C^{\infty}(\bar{G})$ does not cause any loss of generality since the case $w \in W^2_p(G)$ can be studied merely by passing to the limit.

Denote $u(t, x) = e^{-t}w(x)$, $r = w^{-1}(L_1w)_-$ for $w > 0$ and $r = 1$ for $w \leqq 0$, $Lu = ru_t + L_1u$, $Q = (-\infty, \infty) \times G$, and $K = 0$. It is clear that the constant m from Theorems 9 and 10 is negative for our operator L and condition (15) holds. Moreover, $Q^+(0, x_0) = (0, \infty) \times G^+$, where $G^+ = G \cap \{w > 0\}$ and

$$Lu = e^{-t}(-rw + L_1w) = e^{-t}(-(L_1w)_- + L_1w) \geqq -2e^{-t}(L_1w)_-,$$

$$|c|^{\frac{d-p}{p+1}}(r \det a)^{-\frac{1}{p+1}}(Lu)_- \leqq 2e^{-t}|c|^{\frac{d-p}{p+1}}(r \det a)^{-\frac{1}{p+1}}(L_1w)_-$$

$$= 2e^{-t}|c|^{\frac{d-p}{p+1}}(\det a)^{-\frac{1}{p+1}}(L_1w)_-^{\frac{p}{p+1}} w^{\frac{1}{p+1}}.$$

on $Q^+(0, x_0)$.

Hence, by virtue of Theorem 10,

$$\|w_+\|_{C(G)} \leqq N_1\left\||c|^{\frac{d-p}{p+1}}(r \det a)^{-\frac{1}{p+1}}(Lu)_-\right\|_{L_{p+1}((0,\infty)\times G^+)}$$

$$\leqq 2(p+1)^{-\frac{1}{p+1}}N_1\|w_+\|^{\frac{1}{p+1}}_{C(G)}\left\||c|^{\frac{d-p}{p}}(\det a)^{-\frac{1}{p}}(L_1w)_-\right\|^{\frac{p}{p+1}}_{L_p(G^+)},$$

$$\|w_+\|_{C(G)} \leqq 2^{\frac{p+1}{p}}(p+1)^{-\frac{1}{p}}N_1^{\frac{p+1}{p}}\left\||c|^{\frac{d-p}{p}}(\det a)^{-\frac{1}{p}}(L_1w)_-\right\|_{L_p(G)}.$$

It remains to note that the multiplier of the latter norm can be estimated regardless of p, which makes it independent of p.

In order to prove (19) without additional assumptions on L_1, we can assume, without loss of generality, that $\operatorname{tr} a + |c| = 1$ on G. Then we denote $L_1^{\delta} = L_1 + \delta(\Delta - 1)$ and apply the inequality obtained in the specific case to the fuction $w^{\gamma} = w - \delta(\Psi + 1)$ and operator L_1^{δ}, where γ,

$\delta > 0$, and Ψ is taken from Lemma 1. Then following almost literally the appropriate parts of the proof of Lemma 7, we obtain the desired assertion associated with (19) in its most general form. This proves the theorem.

12. Remark. A.D. Aleksandrov [4, 5] proves estimate (17) in the case $p = d$ for a broader class of functions than $W^2_{d,loc}(D) \cap C(\bar{D})$ when assumption (16) is replaced by the following one: $\det a > 0$ (a.e. D) and $(\det a)^{-1}|b|^d \in L_1(D)$. It was not assumed in Theorem 11 that $\det a > 0$ and if, for example, $(\det a)^{-1}|b|^d \leqq K_2^d$ with a constant K_2, then the second inequality in (16) is also satisfied. Indeed, if $\mu_1 \ldots \mu_d$ are eigenvalues of a, then (for every $\varepsilon > 0$)

$$(\det a)^{\frac{1}{d}} = \mu_1^{\frac{1}{d}}(\mu_2\cdot\ldots\cdot\mu_d)^{\frac{1}{d}} \leqq \varepsilon^{-d+1}\mu_1 + (\varepsilon\mu_2\cdot\ldots\cdot\varepsilon\mu_d)^{\frac{1}{d-1}}$$

$$\leqq \varepsilon^{-d+1}\mu_1 + \varepsilon(\mu_2 + \ldots + \mu_d) \leqq \varepsilon \operatorname{tr} a + \varepsilon^{-d+1}a^{ij}\frac{x^i x^j}{|x^2|},$$

$$|b| \leqq \varepsilon \operatorname{tr} a + K_2^{d-1}\varepsilon^{1-d}a^{ij}\frac{x^i x^j}{|x|^2}.$$

It can be seen that the above-mentioned result of Aleksandrov [4, 5] is partly covered by Theorem 11. Note also that since (16) involves c, this condition can be satisfied even if $(\det a)^{-1}|b|^d \notin L_1(D)$.

3.4. The Maximum Principle and the Uniqueness of a Solution in the Sobolev Classes

In its conventional form the maximum principle for an elliptic operator L means that, given that $Lu \geqq 0$ and $L1 \leqq 0$ on a domain, the function u does not exceed the maximum value of u_+ on the boundary. Since L is linear, this assertion is equivalent to the following one. If $Lu_1 \geqq Lu_2$ and $L1 \leqq 0$, the difference $u_1 - u_2$ in the domain does not exceed the maximum value of $(u_1 - u_2)_+$ on the boundary. For non-linear operators, this second version of the maximum principle is stronger and we focus our attention on it in this section.

The maximum principle for linear operators in the Sobolev classes can easily be derived from Theorems 3.9 and 3.11. In studying non-linear operators, we use the well-known technique based on the Hadamard formula

$$F(u_{ij}, u_i, u, x) - F(v_{ij}, v_i, v, x)$$

$$= a^{ij}(u_{ij} - v_{ij}) + b^i(u_i - v_i) + c(u - v), \tag{1}$$

where

$$(a^{ij}, b^i, c) = \int_0^1 \left(F_{u_{ij}}, F_{u_i}, F_u\right)(tu_{pq} + (1 - t)v_{pq}, tu_p + + (1 - t)v_p, tu + (1 - t)v, x)\, dt.$$

Note that this formula is a particular case of the Newton–Leibnitz formula applied to $F(tu_{ij} + (1-t)v_{ij}, tu_i + (1-t)v_i, tu+(1-t)v, x)$. In order that (1) be valid, it is usually required that F be continuously differentiable with respect to (u_{ij}, u_i, u). This requirement turns out to be extremely inconvenient in applications; for example, as applied to the problems of optimal control of diffusion processes. Therefore, we consider operators in the form $\inf_\omega \sup_\pi F^{\omega\pi}$, where $F^{\omega\pi}$ are smooth functions of (u_{ij}, u_i, and u). When $F^{\omega\pi}$ does not depend on ω and π, we obtain an operator as discussed above, but if $F^{\omega\pi}$ depends on ω and π and is linear in u_{ij}, u_i, and u, we obtain a typical operator of the theory of controlled diffusion processes.

Generally speaking, $\inf\limits_\omega \sup\limits_\pi F^{\omega\pi}(u_{x^ix^j}, u_{x^i}, u, x)$ may be an unmeasurable function of x. In order to avoid this troublesome situation, we consider the infimum over ω and the supremum over π in the lattice (of equivalence classes with respect to the Lebesgue measure) of measurable functions (cf. Section 1.2) and write $\bigwedge\limits_\omega \bigvee\limits_\pi$ instead of inf sup. Of course, if ω and π assume only values from a countable set, then the pointwise infimum and supremum over ω and π a.e. coincide with respective bounds in the lattice of measurable functions.

We might present different examples when such a coincidence occurs and thus justify passing to the lattice. However, we would like to note that sometimes it is the nature of the problem that dictates considering the upper and lower bounds in a lattice.

For example, consider a bounded function $\bar{\omega}(x) \geqq 0$ on [0, 1], denote by Ω the set of all measurable functions ω on [0, 1] which satisfy the inequality $0 \leqq \omega(x) \leqq \bar{\omega}(x)$ for all $x \in [0, 1]$, and for $\omega \in \Omega$ denote by $u^\omega(x)$ the value assumed by the solution of the equation $u'' + \omega u' + 1 = 0$ on (0, 1) with the boundary conditions $u(0) = u(1) = 0$ at the point x. Finally, let us put $v(x) = \sup\{u^\omega(x) : \omega \in \Omega\}$. It can be shown (cf. Krylov [1]) that v satisfies the equation

$$v'' + \bigvee_{\omega\in\Omega} (\omega v') + 1 = 0 \tag{2}$$

and for unmeasurable $\bar{\omega}$ the function v does not satisfy the equation

$$v''(x) + \sup_{\omega\in\Omega} (\omega(x)v'(x)) + 1 = 0, \tag{3}$$

which obviously coincides with the equation $v'' + \bar{\omega}(v')_+ + 1 = 0$. Therefore, if we wish to characterize v as a unique solution of an equation in the general case, we have to consider equation (2) rather than (3).

Throughout this section (unless stated otherwise) we consider constants $R > 0$, K_1, $K \geqq 0$, and $\varepsilon \in (0, R^{-1})$, a domain $D \subset E_d$ and $D \subset S_R$, a bounded domain Q of $E_{d+1} = \{(t, x) : t \in (-\infty, \infty), x \in E_d\}$ and $Q \subset C_R$.

Denote by A the set of all operators L of the form

$$L = r\frac{\partial}{\partial t} + a^{ij}\frac{\partial^2}{\partial x^i \partial x^j} + b^i\frac{\partial}{\partial x^i} + c$$

with coefficients which are measurable with respect to (t, x), defined on Q, and satisfy the conditions

$$r \geqq 0, \quad a^* = a, \quad Kr - c \geqq 0, \quad \operatorname{tr} a + Kr - c + r = 1,$$

$$|b| \leqq \varepsilon \operatorname{tr} a + K_1\left[a^{ij}\frac{x^i x^j}{|x|^2} + Kr - c\right]$$

almost everywhere in Q, where $a = (a^{ij})$ and $b = (b^i)$. For $u \in W^{1,2}_{d+1,loc}(Q)$ we put

$$Ju = \bigvee_{L\in A} Lu \tag{4}$$

where $\vee$ stands for the supremum in the lattice of measurable functions.

1. LEMMA. For any function $u \in W^{1,2}_{d+1,loc}(Q)$ there exists an operator $L = L^u \in A$ such that $Ju = Lu$ (a.e. Q).

Proof. From the definition of the supremum in the lattice of measurable functions it readily follows that, if $Q_1 \subset Q$, then almost everywhere on Q_1 the supremum over Q coincides with the supremum over Q_1. It follows that the desired operator L can be constructed on every internal subdomain of Q. Hence, it can be assumed that $u \in W^{1,2}_{d+1}(Q)$.

As we already know (see Section 1.2), for any set of measurable functions there exists a countable subset with the same supremum in the lattice of measurable functions. Moreover, the supremum of a countable set of measurable functions in the lattice coincides with the pointwise supremum. Recalling these facts, we fix a $u \in W^{1,2}_{d+1}(Q)$ and introduce a set $\{L_1, L_2, \dots\} \subset A$ such that

$$Ju(t, x) = \sup_n L_n u(t, x) \quad (\text{a.e.}Q).$$

It is clear that

$$Ju(t, x) = \lim_{n\to\infty} \max_{i\leq n} L_i u(t, x) \quad (\text{a.e.}Q)$$

and we easily find the operator $L^n \in A$ such that $L^n u = \max_{i\leq n} L_i u$. Therefore, $L^n u \to Ju$ (a.e. Q) as $n \to \infty$. Note that the coefficients of all operators $L \in A$ are uniformly bounded, e.g., for a constant N, we have

$$|Lu| \leq N\left(\sum |u_{x^i x^j}| + \sum |u_{x^i}| + |u_t| + |u|\right) \quad (\text{a.e.}Q) \tag{5}$$

for all $L \in A$. Hence it immediately follows that $\pm$ Ju does not exceed the right-hand side of (5) and $Ju \in L_{d+1}(Q)$. From the Lebesgue dominated convergence theorem we obtain $L^n u \to Ju$ in $L_{d+1}(Q)$ as $n \to \infty$.

Moreover, since the coefficients on L^n are uniformly bounded, from the sequences of these coefficients one can select subsequence weakly convergent in $L_2(Q)$. Lest one should introduce new numbering, suppose that the coefficients of the operators L^n weakly converge to those of an operator L^0. Selecting, if necessary, one more subsequence according to the Banach—Saks theorem, we make the coefficients of the operators

$$\frac{1}{n}\sum_{i=1}^{n} L^i$$

converge to coefficients of L^0 strongly in $L_2(Q)$. In particular, the coefficients converge in measure.

It is clear that $L^0 \in A$ and, since $L^n u$ converges to Ju in $L_{d+1}(Q)$, we conclude, by the Lebesgue dominated convergence theorem, that

$$||Ju - L^0 u||_{\mathcal{L}_{d+1}(Q)} = \lim_{n\to\infty} \left|\left| Ju - \frac{1}{n}\sum_{i=1}^{n} L^i u \right|\right|_{\mathcal{L}_{d+1}(Q)}$$

$$\leq \lim_{n\to\infty} \frac{1}{n}\sum_{i=1}^{n} ||Ju - L^i u||_{\mathcal{L}_{d+1}(Q)} = 0.$$

Thus, $L^0 u = Ju$ which proves the lemma.

We now formulate the maximum principle for parabolic equations. Let Ω be a set of indices and, for any $\omega \in \Omega$, let a set of indices $\Pi(\omega)$ be defined. Let for any $\omega \in \Omega$ and $\pi \in \Pi(\omega)$ functions $F^{\omega\pi}(u_0, u_{ij}, u_i, u, t, x)$ be defined, which are measurable with respect to $(t, x) \in E_{d+1}$ for any u_0, u_{ij}, u_i, u $(i, j = 1, \ldots, d)$ and continuously differentiable with respect to (u_0, u_{ij}, u_i, u) for any fixed $(t, x) \in E_{d+1}$. We denote

$$a^{\omega\pi ij} = F^{\omega\pi}_{u_{ij}}, \quad b^{\omega\pi i} = F^{\omega\pi}_{u_i}, \quad r^{\omega\pi} = F^{\omega\pi}_{u_0},$$

$$c^{\omega\pi} = F^{\omega\pi}_{u}, \quad a^{\omega\pi} = (a^{\omega\pi ij})^d_{i,\, j=1}, \quad b^{\omega\pi} = (b^{\omega\pi i})^d_{i=1}.$$

We put

$$F^{\omega\pi}[u](t, x) = F^{\omega\pi}(u_t(t, x), u_{x^i x^j}(t, x), u_{x^i}(t, x), u(t, x), t, x)$$

and introduce $a^{\omega\pi}[u], \ldots$ in a similar manner. Finally, let

$$Iu = \bigwedge_{\omega\in\Omega} \bigvee_{\pi\in\Pi(\omega)} F^{\omega\pi}[u].$$

2. THEOREM. Let $u, v \in W^{1,2}_{d+1,loc}(Q) \cap C(\bar{Q})$. We denote

$$\tilde{a}^{\omega\pi} = \int_0^1 a^{\omega\pi}[su + (1 - s)v]\, ds$$

and introduce $\tilde{b}^{\omega\pi}$, $\tilde{c}^{\omega\pi}$, and $\tilde{r}^{\omega\pi}$ in a similar manner. Suppose that for any ω and π

$$\tilde{r}^{\omega\pi} \geq 0, \quad \tilde{a}^{\omega\pi} \geq 0, \quad K\tilde{r}^{\omega\pi} - \tilde{c}^{\omega\pi} \geq 0, \tag{6}$$

$$|\tilde{b}^{\omega\pi}| \leq \varepsilon \operatorname{tr} \tilde{a}^{\omega\pi} + K_1\left[\tilde{a}^{\omega\pi ij} \frac{x^i x^j}{|x|^2} + K\tilde{r}^{\omega\pi} - \tilde{c}^{\omega\pi}\right], \tag{7}$$

$$\bigwedge_{\omega\in\Omega} \bigwedge_{\pi\in\Pi(\omega)} [\operatorname{tr} \tilde{a}^{\omega\pi} + K\tilde{r}^{\omega\pi} - \tilde{c}^{\omega\pi}] > 0. \tag{8}$$

almost everywhere on Q.

Let $Iu \geqq Iv$ (a.e.Q) and the sets $\{(t, x) : Iu = Iv = \infty\}$, $\{(t, x) : Iu = Iv = -\infty\}$ be of measure zero. Then:

(a) if $(t_0, x_0) \in Q$ and $u \leqq v$ on $\partial' Q(t_0, x_0)$, then $u(t_0, x_0) \leqq v(t_0, x_0)$;

(b) if $(t_0, x_0) \in Q$ and $\tilde{c}^{\omega\pi} \leqq 0$ (a.e.$Q(t_0, x_0)$), then

$$u(t_0, x_0) - v(t_0, x_0) \leqq \sup_{\partial' Q(t_0, x_0)} [u - v]_+ .$$

Proof. We set $w = u - v$ and take the operator J from formula (4). We first prove that $Jw \geqq 0$ (a.e.Q).

Suppose that the contrary is true. Then on a set of a positive measure the inequality $Jw \leqq -\delta$ holds for a positive number δ. Denote

$$\tilde{L}^{\omega\pi} = \tilde{r}^{\omega\pi} \frac{\partial}{\partial t} + \frac{1}{2} (\tilde{a}^{\omega\pi ij} + \tilde{a}^{\omega\pi ji}) \frac{\partial^2}{\partial x^i \partial x^j} + \tilde{b}^{\omega\pi i} \frac{\partial}{\partial x^i} + \tilde{c}^{\omega\pi}. \tag{9}$$

It is clear that

$$[\operatorname{tr} \tilde{a}^{\omega\pi} + (K + 1)\tilde{r}^{\omega\pi} - \tilde{c}^{\omega\pi}]^{-1} \tilde{L}^{\omega\pi} \in A.$$

It follows that almost everywhere on the set G

$$\tilde{L}^{\omega\pi} \omega \leqq -\delta[\operatorname{tr} \tilde{a}^{\omega\pi} + (K + 1)\tilde{r}^{\omega\pi} - \tilde{c}^{\omega\pi}]$$

$$\leqq -\delta \bigwedge_{\omega\in\Omega} \bigwedge_{\pi\in\Pi(\omega)} [\operatorname{tr} \tilde{a}^{\omega\pi} + (K + 1)\tilde{r}^{\omega\pi} - \tilde{c}^{\omega\pi}] \doteq: -\delta f,$$

where $f > 0$ (a.e.Q) by hypothesis. Finally, by the Hadamard formula,

$$F^{\omega\pi}[u] = F^{\omega\pi}[v] + \tilde{L}^{\omega\pi}\omega \leqq F^{\omega\pi}[v] - \delta f \ \text{(a.e.}G)$$

and it follows that $Iu \leqq Iv - \delta f$ (a.e.G) in contradiction to the hypothesis of the theorem.

Thus, $Jw \geqq 0$ (a.e.Q). From Lemma 1 it follows that there exists an operator $L \in A$ such that $Lw \geqq 0$ (a.e.Q). Applying Theorem 3.9 to L and w, we immediately obtain both assertions of this theorem. Theorem 2 is proved.

3. Corollary. Let the conditions of the theorem be satisfied and let $Iu = Iv$ (a.e.Q) along with $u = v$ on $\partial' Q$. Then $u = v$ in $\bar{Q}$.

To prove the corollary, it suffices to exchange u and v in assertion (a) and make use of the continuity of u and v in $\bar{Q}$.

4. Remark. In the case of a linear equation, it is easy to present an example showing that condition (8) is necessary. Let $Q = (0, T) \times S_R$, and Γ be a subset of Q of nonzero measure. We solve the equation $u_t + \Delta u - u + \chi_\Gamma = 0$ with the boundary condition $u = 0$ on $\partial' Q$. From the theory of linear equations it is known that $u \in W^{1,2}_{d+1}(Q)$, $u \geqq 0$ in Q (see, e.g., Theorem 3.9), and $u \not\equiv 0$. We take

$$L = \chi_{Q\setminus\Gamma}\left(\frac{\partial}{\partial t} + \Delta - 1\right).$$

Then it is obvious that $Lu = L0 = 0$, however, if $v = 0$, assertions (a) and (b) are not true.

The chosen form of the maximum principle in Theorem 2 enables us to establish not only the uniqueness of the solution to the equation $Iu = f$ (cf. Corollary 3), but also the uniqueness of the solution of the problem with two free boundaries.

5. THEOREM. Let $u, v \in W^{1,2}_{d+1,loc}(Q) \cap C(\bar{Q})$, $u = v$ on $\partial' Q$, and two finite-valued functions $\Psi_1 \geqq \Psi_2$ be given on Q. Let $\tilde{a}^{\omega\pi}$, ... be introduced as in Theorem 2, conditions (6)-(8) be satisfied, and Iu, $Iv < \infty$ almost everywhere in Q. Finally, suppose that $\Psi_1 \geqq u \geqq \Psi_2$, $\Psi_1 \geqq v \geqq \Psi_2$ (a.e.Q), $(-1)^i Iu \geqq 0$ (a.e.) on the set $\{(t, x) : (-1)^i (u - \Psi_i) \geqq 0\}$, $i = 1,2$, and $(-1)^i Iv \geqq 0$ (a.e.) on the set $\{(t, x) : (-1)^i (v - \Psi_i) \geqq 0\}$, $i = 1,2$. Then $u = v$ in $\bar{Q}$.

Proof. For $\gamma \geqq 0$ and $\delta \geqq 0$ we denote

$$F^{(\omega,\gamma)(\pi,\delta)}[w] = \frac{1}{\gamma + \delta + 1}\left[F^{\omega\pi}[w] - \gamma(w - \psi_1) - \delta(w - \psi_2)\right],$$

$$\tilde{I}w = \bigwedge_{\substack{\omega\in\Omega \\ \gamma\geqq 0}} \bigvee_{\substack{\pi\in\Pi(\omega) \\ \delta\geqq 0}} F^{(w,\gamma)(\pi,\delta)}[w].$$

It is not difficult to check that conditions (6) to (8) hold for u, v, and the functions $F^{(\omega,\gamma)(\pi,\delta)}$. By Corollary 3, it now suffices to show that $\tilde{I}u = \tilde{I}v = 0$ (a.e.Q). On computing the supremum consecutively with respect to π and δ, we obtain

$$\bigvee_{\substack{\pi\in\Pi(\omega)\\ \delta\geqq 0}} \frac{1}{\gamma+\delta+1}[F^{\omega\pi}[u] - \gamma(u - \psi_1) - \delta(u - \psi_2)]$$

$$= \max\left(\psi_2 - u, \frac{1}{\gamma+1}\left(\bigvee_{\pi\in\Pi(\omega)} F^{\omega\pi}[u] - \gamma(u - \psi_1)\right)\right).$$

Hence

$$\tilde{I}u = \inf_{\gamma\geqq 0} \max\left(\psi_2 - u, \frac{1}{\gamma+1}(Iu - \gamma(u - \psi_1))\right)$$

$$= \max(\psi_2 - u, \min(Iu, \psi_1 - u)),$$

and the latter equality is true in view of $Iu < \infty$. A simple analysis of this representation of $\tilde{I}u$ and the assumptions $\Psi_1 \geqq u \geqq \Psi_2$ and $(-1)^i Iu \geqq 0$ on $\{(-1)^i(u - \Psi_i) > 0\}$ $i = 1,2$ enables the reader to prove that $\tilde{I}u = 0$ (a.e.Q). Similarly, $\tilde{I}v = 0$ (a.e.Q). This proves the theorem.

We now prove the maximum principle for parabolic equations in a band.

6. THEOREM. Let $T \in (0, \infty)$, $u, v \in W^{1,2}_{d+1,loc}((0, T) \times E_d)$, and u and v be continuous on $[0, T] \times E_d$. For every $R > 0$ let there exist constants $\varepsilon \in (0, R^{-1})$, $K_1 \geqq 0$, and $K \geqq 0$ such that inequalities (6) to (8) hold almost everywhere on $C_{T,R}$. Suppose that $Iu \geqq Iv$ almost everywhere on $(0, T) \times E_d$, $u(T, x) \leqq v(T, x)$ on E_d, and the sets

$$\{(t, x) : Iu = Iv = \infty\}, \quad \{(t, x) : Iu = Iv = -\infty\}$$

are of measure zero. Finally, let there exist a function $\Psi \geqq 0$ such that $\Psi \in W^{1,2}_{d+1,loc}((0, T) \times E_d)$, Ψ is continuous on $[0, T] \times E_d$, let $\tilde{L}^{\omega\pi}_{\Psi} \leqq 0$ (see (9)) almost everywhere on $(0, T) \times E_d$ for any ω, π, and for any R_1, $\delta > 0$ let there exist $R > R_1$ such that $u \leqq v + \delta\Psi$ whenever $|x| = R$. Then $u \leqq v$ on $[0, T] \times E_d$.

<u>Proof</u>. Consider a point $(t_0, x_0) \in (0, T) \times E_d$ and a constant $\delta > 0$ and determine $R > |x_0|$ such that $u - (v + \delta\Psi) \leqq 0$ on $\partial' C_{T,R}$. Since $\tilde{L}^{\omega\pi}\Psi \leqq 0$, the inequality $J(u - (v + \delta\Psi)) \geqq 0$ (a.e. $(0,T) \times E_d$) can be proved by the same argument as in the proof of Theorem 2. By Theorem 2, $u - (v + \delta\Psi) \leqq 0$ in $C_{T,R}$, and specifically, $u(t_0, x_0) \leqq v(t_0, x_0) +$

$\delta\Psi(t_0, x_0)$. Recalling that δ and t_0, x_0 are arbitrary, the proof is completed.

The following theorem on the maximum principle for elliptic operators is stated without proof since its proof closely parallels that of Theorem 2 except that Theorem 3.11 is used instead of Theorem 3.9.

7. THEOREM. Let $F^{\omega\pi}$ be independent of t and u_0, and let $u, v \in W^2_{d,loc}(D) \cap C(\bar{D})$, $D \subset S_R$, $K_1 \geqq 0$, and $\varepsilon \in (0, R^{-1})$. Following Theorem 2, introduce the functions $\tilde{a}^{\omega\pi}$, ... Suppose that

$$\tilde{c}^{\omega\pi} \leqq 0, \quad \tilde{a}^{\omega\pi} \geqq 0, \quad \bigwedge_{\omega\in\Omega} \bigwedge_{\pi\in\Pi(\omega)} [|\tilde{c}^{\omega\pi}| + \operatorname{tr} \tilde{a}^{\omega\pi}] > 0,$$

$$|\tilde{b}^{\omega\pi}| \leqq \varepsilon \operatorname{tr} \tilde{a}^{\omega\pi} + K_1\left[\tilde{a}^{\omega\pi ij} \frac{x^i x^j}{|x|^2} + |\tilde{c}^{\omega\pi}|\right].$$

almost everywhere on D for any ω and π.

Let $Iu \geqq Iv$ (a.e.D) and let the sets

$$\{x : Iu = Iv = +\infty\}, \quad \{x : Iu = Iv = -\infty\}$$

be of measure zero. Then $u - v \leqq \max\{(u - v)_+ (x) : x \in D\}$ in D.

It is clear that theorems similar to Theorem 5 and 6 are valid for elliptic operators. These theorems are not formulated here, in order to avoid encumbering the exposition. We discuss Theorem 7 using the two following examples.

8. Example. Let $p > d$ and let the domain D be such that the Sobolev theorem on imbedding $W^2_p(D)$ into $C^1(D)$ holds on it.

We assert that, if $f > 0$ (a.e.D), then the equation

$$\det\left(u_{x^i x^j} - u_{x^i} u_{x^j} - \delta_{ij} u\right) = f \text{ (a.e.D)} \tag{10}$$

can have only one solution from $W^2_p(D)$ such that the matrix $P[u] :=$ $(u_{x^i x^j} - u_{x^i} u_{x^j} - \delta_{ij} u)$ is nonnegative almost everywhere on D and the solution is equal to the prescribed function on ∂D.

Indeed, let u and v be two such functions. Suppose that in Theorem 7 Ω and $\Pi(\omega)$ are single-point sets and $F^{\omega\pi} = F$, where

$$F(u_{ij}, u_i, u) := \det(u_{ij} - u_i u_j - \delta_{ij} u).$$

By hypothesis, $P[u] \geqq 0$, $P[v] \geqq 0$, and $\det P[u] = \det P[v] = f > 0$. It follows that $P_s := P[su + (1 - s)v] \geqq sP[u] + (1 - s)P[v]$ (a.e.D, $s \in [0, 1]$). Denote by M_s the matrix composed of cofactors of the entries of P_s ($M_s = P_s^{-1}\det P_s$). Then (omitting superfluous superscripts) we obtain

$$\tilde{a} = \int_0^1 M_s \, ds > 0, \quad \tilde{c} = -\operatorname{tr} \tilde{a}, \quad \tilde{b} = -2 \int_0^1 M_s (su_x + (1 - s) v_x) \, ds$$

$$|\tilde{b}| \leqq 2(\|u_x\|_{C(D)} + \|v_x\|_{C(D)}) \int_0^1 \|M_s\| \, ds$$

$$\leqq N \int_0^1 \operatorname{tr} M_s \, ds = N|\tilde{c}| \quad (\text{a.e.}D).$$

The conditions of Theorem 7 are satisfied and, therefore, $u - v \leqq 0$, $v - u \leqq 0$, and $u = v$ in D. In this discussion, the assumption that f is strictly positive is needed only to provide the inequality $|\tilde{c}| + \operatorname{tr} \tilde{a} > 0$ (a.e. D).

If we use a different argument, the uniqueness of the solution to (10) in the same class of functions can be proved under the condition $f \geqq 0$ (a.e.D). Here we use Lemma 2.2 which implies that equation (10) in the class under consideration is equivalent to the equation

$$\inf\left(a^{ij}u_{x^i x^j} - a^{ij}u_{x^i}u_{x^j} - u - d\sqrt[d]{f \det a}\right) = 0 \quad (\text{a.e.}D), \tag{11}$$

where the infimum is taken over the set of all symmetric nonnegative matrices a such that tr a = 1. It is clear that the infimum can also be taken over any countable everywhere dense subset of this set, and since the pointwise infimum of a countable set of measurable functions coincides with the infimum in the lattice of measurable functions, the sign inf in (11) can be replaced by $\wedge$. Now it is not difficult to use Theorem 7.

9. Example. Let p and D be the same as above and let us consider the equation

$$\det\left(u_{x^i x^j} + u_{x^i}u_{x^j} - \delta_{ij}u\right) = f \quad (\text{a.e.}D) \tag{12}$$

in the class of functions $u \in W_p^2(D)$ such that $P[u] := (u_{x^i x^j} + u_{x^i x^j} - \delta_{ij}u) \geqq 0$. Here the argument using the single-point sets Ω, $\Pi(\omega)$ and the functions

$$F^{\omega\pi} = F(u_{ij}, u_i, u, x) = \det(u_{ij} + u_i u_j - \delta_{ij}u)$$

can no longer be employed since in the general case the inequalities $P[u] \geqq 0$ and $P[v] \geqq 0$ do not imply $P[su + (1-s)v] \geqq 0$, $\tilde{a} \geqq 0$.

Thus, we represent equation (12) with $f \geqq 0$, like (11), in the form

$$\wedge\left(a^{ij}u_{x^i x^j} + a^{ij}u_{x^i}u_{x^j} - u - d\sqrt[d]{f \det a}\right) = 0. \tag{13}$$

Equation (13) can have only one solution in the class of functions $u \in W^2_p(D)$ which are equal to a prescribed function on ∂D. This can easily be proved using Theorem 7.

There is nothing surprising in the fact that Theorem 7 can be used in different ways for proving the uniqueness of (12) and its equivalent (13) (when $P[u] \geqq 0$). Indeed, given single-point sets Ω and Π, the condition $\tilde{a} \geqq 0$ is associated with the values of F at the points (u_{ij}, u_i, u, x) which do not necessarily lie on the surface $\{F(u_{ij}, u_i, u, x) = 0\}$. Therefore, there can be different F specifying the same surface and, consequently, the same differential equation so that for some F the condition $\tilde{a} \geqq 0$ is satisfied, while for others it does not hold.

Sometimes it is convenient to use the Bony maximum principle (see Bony [1]).

10. THEOREM. Let bounded, measurable functions $a^{ij}(x)$, $b^i(x)$, and $c(x)$ be defined in D for $i, j = 1, \ldots, d$ such that $c \leqq 0$, the matrix (a^{ij}) is symmetric and nonnegative. Let $u \in W^2_d(D)$ and u attain its nonnegative maximum on D at a point $x_0 \in D$. Then the essential greatest lower bound of the expression $Lu := a^{ij}u_{x^i x^j} + b^i u_{x^i} + cu$ over D is nonpositive, i.e.,

$$\operatorname{ess\,inf}(Lu, D) \leqq 0.$$

Proof. We can assume that $c \leqq -\delta$ for a constant $\delta > 0$. If this is not the case, one could consider the operator $L-\delta$ and pass to the limit letting $\delta \downarrow 0$ and recalling that the function u is bounded on every domain $D' \subset \overline{D'} \subset D$ and the inequality $\operatorname{ess\,inf}(Lu, D') \leqq 0$ is stronger than the desired one. Moreover, we can assume that $u(x_0) > 0$ replacing u by $u + \gamma$ with $\gamma > 0$ and passing to the limit as $\gamma \downarrow 0$. Finally, we can assume that the strict maximum is attained at x_0 since we can take $u(x) - \sigma|x - x|^2$ instead of u and pass to the limit. We now consider any neighbourhood of x_0 as D' such that $\bar{D}' \subset D$ and apply Theorem 3.11 with $p = d$, $v = u$, and $L_1 = L$. It can be seen that the second summand in (3.17) is strictly positive which, in view of the adopted conditions on (3.17), is possible only if $Lu < 0$ on a set of positive measure in D', and, consequently, in D. This proves the theorem.

It is obvious that the same assertion is also true for parabolic operators.

3.5. Passage to the Limit for Nonlinear Operators in the Sobolev Classes

The passage to the limit is an important technique employed in studying linear and nonlinear equations. It is frequently necessary to pass to the limit under action of an unbounded operator when the corresponding functional sequence converges to the limit in a weak sense. For linear operators this difficulty can be more or less easily avoided. As for nonlinear operators, examples can readily be demonstrated showing that the limiting process is generally impossible. In this section we shall see that nonlinear elliptic and parabolic operators are, in a sense, an exception.

The method which will be used is demonstrated as applied to nonlinear operators in a Banach space.

1. Definition. Let X be a Banach space and let $F[x]$ be an operator defined on $D_F \subset X$ with values in X. The operator F is said to be antipositive, if there exists a constant $K \geqq 0$ such that

$$\lambda||x - y|| \leqq K||\lambda(x - y) - (F[x] - F[y])||.$$

for all $\lambda \geqq K$ and $x, y \in D_F$.

2. THEOREM. Let F be an antipositive operator into X, let D_F be a linear subspace dense in X, and let $F[x + ty]$ be continuous with respect to $t \in (-\infty, \infty)$ for any $x, y \in D_F$. Let $x_n, f_n \in X$, $n = 0, 1, 2, \ldots,$ $F[x_n] = f_n$ for $n \geqq 1$, $x_n \to x_0$, $f_n \to f_0$, and $x_0 \in D_F$. Then $F[x_0] = f_0$.

Proof. From the inequality

$$\lambda||x_n - y|| \leqq K||\lambda(x_n - y) - (F[x_n] - F[y])||$$

as $n \to \infty$ we find

$$\lambda||x_0 - y|| \leqq K||\lambda(x_0 - y) - (f_0 - F[y])||$$

for any $y \in D_F$. We put $y = x_0 + tz$ and $\lambda = t^{-1}$, where $K^{-1} \geqq t > 0$ and $z \in D_F$. Then $||z|| \leqq K||z + f_0 - F[x_0 + tz]||$. Letting $t \downarrow 0$, we conclude that $||z|| \leqq K||z + f_0 - F(x_0)||$ for any $z \in D_F$, and since D_F is dense in X, this is also true for any $z \in X$. When $z = -(f_0 - F[x_0])$, it follows that $||z|| = 0$ and $F[x_0] = f_0$. This completes the proof.

To apply this method to nonlinear differential operators, we introduce some notation and definitions. Let constants $K, T, R > 0$ and a bounded domain $Q \subset E_{d+1}$ be fixed.

3. Definition. Let $\varepsilon \geqq 0$ and let an operator $F: W^{1,2}_{d+1}(Q) \to \mathcal{L}_{d+1}(Q)$ be given. The operator F is said to be ε-nondegenerate of L-type in Q if for any $u, v \in W^{1,2}_{d+1}(Q)$ there exists a linear operator

$$L = r \frac{\partial}{\partial t} + a^{ij} \frac{\partial^2}{\partial x^i \partial x^j} + b^i \frac{\partial}{\partial x^i} + c \tag{1}$$

with measurable coefficients in Q such that $F[u] - F[v] = L(u - v)$, $a = a^*$, $r + \operatorname{tr} a + |b| + |c| \leqq K$, $r \geqq \varepsilon$, and $(a\lambda, \lambda) \geqq \varepsilon|\lambda|^2$ (a.e.Q) for all $\lambda \in E_d$, where $a = (a^{ij})$, $b = (b^i)$. When $\varepsilon = 0$, the operator F is merely called an L-type operator in Q.

Examples of L-type operators can be constructed using the following lemma. Before formulating the lemma, we note that the set of all ε-nondegenerate L-type operators is obviously convex.

4. LEMMA. (a) If F is an ε-nondegenerate L-type operator in Q, then $(-F[-u])$ is also an ε-nondegenerate L-type operator in Q and there exists a constant N which depends only on K and d and is such that

$$||F[u] - F[v]||_{\mathcal{L}_{d+1}(Q)} \leqq N||u - v||_{W^{1,2}_{d+1}(Q)}.$$

(b) If L is an operator of the form (1), whose coefficients are measurable in Q, $r + \operatorname{tr} a + |b| + |c| \leqq K$, $r \geqq \varepsilon$, $(a\lambda, \lambda) \geqq \varepsilon|\lambda|^2$ (a.e.Q) for all $\lambda \in E_d$, and $f \in \mathcal{L}_{d+1}(Q)$, then $F: u \to Lu+f$ is an ε-nondegenerate L-type operator.

(c) Given a real-valued function $F(u_0, u_{ij}, u_i, u, t, x)$ which is defined for all $(t, x) \in Q$ and real-valued u_0, u_{ij}, u_i, u $(i, j = 1, \ldots, d)$, measurable with respect to (t, x), continuously differentiable with respect to (u_0, u_{ij}, u_i, u) for all $(t, x) \in Q$ such that $F(0, 0, 0, 0, t, x) \in \mathcal{L}_{d+1}(Q)$, $F_{u_0} + \delta^{ij}F_{u_{ij}} + (F_{u_i}F_{u_i})^{1/2} + |F_u| \leqq K$, $F_{u_0} \geqq \varepsilon$, and $F_{u_{ij}}\lambda^i\lambda^j \geqq \varepsilon|\lambda|^2$ for all values of arguments and $\lambda \in E_d$, then the operator $F[u] := F(u_t, u_{x^ix^j}, u_{x^i}, u, t, x)$ is an ε-nondegenerate L-type operator in Q.

(d) Given a set of indices Ω such that for every $\omega \in \Omega$ an ε-nondegenerate L-type operator F^ω is defined in the domain Q and $\bigwedge_{\omega\in\Omega} F^\omega[0] \in \mathcal{L}_{d+1}(Q)$, then $\bigwedge_{\omega\in\Omega} F^\omega[u]$ is an ε-nondegenerate L-type operator in Q. A similar assertion is true for $\bigvee_{\omega\in\Omega} F^\omega[u]$.

Proof. Assertion (a) is trivial. Unlike Definition 3, (b) does not require that a is symmetric. However, it is clear that for $u \in W^{1,2}_{d+1}(Q)$ the expression of Lu does not change (almost everywhere in Q) if a is replaced in L by $(1/2)(a + a^*)$. To prove (c), note that the transformation of $F[u] - F[0]$ by the Hadamard formula immediately shows that $F[u] \in L_{d+1}(Q)$ if $u \in W^{1,2}_{d+1}(Q)$. Applying this formula to $F[u] - F[v]$, we obtain the operator L from Definition 3.

We now come to the proof of (d). Let $B(\varepsilon)$ be the set of all operators L of the form (1) whose coefficients satisfy the conditions of Definition 3. There obviously exists a constant such that the absolute values of the coefficients of each operator $L \in B(\varepsilon)$ do not exceed this constant almost everywhere in Q. This together with a formula like $F^{\omega}[u] = F^{\omega}[0] + L^{\omega}u$ with $L^{\omega} \in B(\varepsilon)$ imply that $F[u] := \bigwedge_{\omega\in\Omega} F^{\omega}[u]$ is an operator from $W^{1,2}_{d+1}(Q)$ into $L_{d+1}(Q)$. We further take $u, v \in W^{1,2}_{d+1}(Q)$ and the operator L^{ω} associated with u, v, and F^{ω} by Definition 3. It is clear that

$$\bigwedge_{L\in B(\varepsilon)} L(u - v) \leqq \bigwedge_{\omega\in\Omega} L^{\omega}(u - v) \leqq F[u] - F[v]$$

$$\leqq \bigvee_{\omega\in\Omega} L^{\omega}(u - v) \leqq \bigvee_{L\in B(\varepsilon)} L(u - v). \tag{2}$$

Following the proof of Lemma 4.1, one can prove the existence of operators $L', L'' \in B(\varepsilon)$ such that

$$L'(u - v) = \bigwedge_{L\in B(\varepsilon)} L(u - v), \quad L''(u - v) = \bigvee_{L\in B(\varepsilon)} L(u - v) \tag{3}$$

(they depend on u and v).

Finally, consider $s(t, x) = 0$ at those points, where $L'(u - v) = L''(u - v)$, while at other points of Q

$$s = \frac{L''(u - v) - (F[u] - F[v])}{(L'' - L')(u - v)}.$$

From (2) and (3) it follows that $s \in [0, 1]$, $L := sL' + (1 - s)L'' \in B(\varepsilon)$, and $L(u - v) = F[u] - F[v]$. This proves the lemma.

The relationship between ε-nondegenerate L-type operators and anti-positive operators is established in the following lemma.

5. LEMMA. Let $\overset{0}{W}{}^{1,2}_{d+1}(C_{T,R})$ be a subset of $W^{1,2}_{d+1}(C_{T,R})$ consisting of all functions which are equal to zero on $\partial' C_{T,R}$. Let $\varepsilon > 0$ and F be an

ε-nondegenerate L-type operator in $C_{T,R}$. Then F defined on $\overset{0}{W}{}^{1,2}_{d+1}(C_{T,R})$ is an antipositive operator in $\mathcal{L}_{d+1}(C_{T,R})$. Moreover, there exists a constant N which depends only on K, d, and ε and such that

$$\lambda||(u - v)_{\pm}||_{\mathcal{L}_{d+1}(C_{T,R})}$$
$$\leqq N||(\lambda(u - v) - (F[u] - F[v]))_{\pm}||_{\mathcal{L}_{d+1}(C_{T,R})} \tag{4}$$

for all $\lambda \geqq N$, $u, v \in \overset{0}{W}{}^{1,2}_{d+1}(C_{T,R})$.

Proof. By Definition 3, it suffices to prove (4), where $F[u] - F[v]$ is replaced by $L(u - v)$. Then passing to $w = \pm(u - v)$, we conclude that, to prove the lemma, it suffices to show that there exists a constant N which depends only on K, d, and ε and such that

$$\lambda||w_+||_{\mathcal{L}_{d+1}(C_{T,R})} \leqq N||(\lambda w - Lw)_+||_{\mathcal{L}_{d+1}(C_{T,R})} \tag{5}$$

for any $\lambda \geqq N$, $w \in \overset{0}{W}{}^{1,2}_{d+1}(C_{T,R})$, and operators of the form (1) whose coefficients satisfy the conditions $a = a^*$, $r + \operatorname{tr} a + |b| + |c| \leqq K$, $r \geqq \varepsilon$, and $(a\xi, \xi) \geqq \varepsilon|\xi|^2$ for all $\xi \in E_d$ (a.e.$C_{T,R}$).

Since in (5) it is easy to pass to the limit, we assume that r, a^{ij}, b^i, $c \in C^{\infty}(E_{d+1})$. Moreover, from the theory of linear equations it is known that for any fixed $\lambda \geqq 0$ a function $w \in \overset{0}{W}{}^{1,2}_{d+1}(C_{T,R})$ can be represented as $w_{(+)} - w_{(-)}$, where $w_{(\pm)} \in \overset{0}{W}{}^{1,2}_{d+1}(C_{T,R})$ and $\lambda w_{(\pm)} - Lw_{(\pm)} = (\lambda w - Lw)_{\pm}$ (a.e.$C_{T,R}$).

From the maximum principle (Theorem 4.2 with $F^{\omega\lambda}[u] := Lu - \lambda u$) we obtain $w_{(\pm)} \geqq 0$ in $C_{T,R}$ and consequently (5) will be proved if we prove that

$$\lambda||w_+||_{\mathcal{L}_{d+1}(C_{T,R})} \leqq N||\lambda w - Lw||_{\mathcal{L}_{d+1}(C_{T,R})} \tag{6}$$

under one further assumption that $\lambda w - Lw \geqq 0$ (a.e.$C_{T,R}$). Finally, approximating $\lambda w - Lw$ and w by smooth functions f and the solutions of problems $\lambda u - Lu = f$ of the class $\overset{0}{W}{}^{1,2}_{d+1}(C_{T,R})$ respectively, we conclude that it suffices to prove (6) in the case when $\lambda w - Lw = f$ (a.e.$C_{T,R}$), where $f \geqq 0$ and $f \in C_0^{\infty}(E_{d+1})$.

In this case we use Theorem 2.4, where λ is replaced by μ which is the root of the equation $K\sqrt{\mu} = \lambda$. Let v be the function in Theorem 2.4. Then

$$Lv - \lambda v = rv_t + a^{ij}v_{x^ix^j} + b^iv_{x^i} + cv - \lambda v$$

$$\leq (\mu(r + \operatorname{tr} a) + |b|\sqrt{\mu} - \lambda)v - \sqrt[d+1]{r \det a}\; f$$

$$\leq -\delta f \quad (\text{a.e.} C_{T,R}),$$

where δ depends only on d and ε and $\delta > 0$. By the maximum principle, it follows that $v \geq \delta w$ in $C_{T,R}$ and

$$\delta||w_+||_{L_{d+1}(C_{T,R})} \leq ||v||_{L_{d+1}(C_{T,R})}$$

$$\leq \frac{1}{\mu(d+1)}||f||_{L_{d+1}(E_{d+1})}.$$

Here the inequality between the first and the last terms holds whatever nonnegative extension of f, which belongs to the class $C_0^\infty(E_{d+1})$, to E_{d+1} is considered. Therefore,

$$\delta||w_+||_{L_{d+1}(C_{T,R})} \leq \frac{1}{\mu(d+1)}||f||_{L_{d+1}(C_{T,R})}$$

$$= \frac{1}{\mu(d+1)}||\lambda w - Lw||_{L_{d+1}(C_{T,R})}.$$

We complete the proof by noting that $\mu \geq (2K)^{-1}\lambda$ if $\lambda \geq 2K$.

The following theorem serves as a landmark, and in what follows its assumptions will be made less restrictive.

6. THEOREM. Let $\varepsilon > 0$ and let F be an ε-nondegenerate L-type operator in $C_{T,R}$, $u^n \in \overset{0}{W}{}^{1,2}_{d+1}(C_{T,R})$, $n = 0, 1, 2, \ldots$, and $u^n \to u^0$ in $L_{d+1}(C_{T,R})$ as $n \to \infty$. Then

(a) if $F[u^n] \to f$ in $L_{d+1}(C_{T,R})$ as $n \to \infty$, then $F[u^0] = f$ (a.e.$C_{T,R}$);

(b) if $\sup_n F^n[u] \in L_{d+1}(C_{T,R})$, then $\overline{\lim}\, F[u^n] \geq F[u^0]$ (a.e.$C_{T,R}$);

(c) if $\inf_n F^n[u] \in L_{d+1}(C_{T,R})$, then $F[u^0] \geq \underline{\lim}_{n\to\infty} F\, u^n$ (a.e.$C_{T;R}$).

Proof. Assertion (a) follows from Lemmas 4 and 5 and Theorem 2, assertion (c) follows from (b) if $F[u]$ is replaced by $(-F[-u])$ (see Lemma 4). To prove (b), we put $f^n = F[u^n]$, $f = \overline{\lim}\, f^n$ and make use of the proof of Theorem 2. Omitting the subscripts of norms, by Lemma 5, we obtain

$$\lambda \| (u - u^n)_+ \| \le N \| (\lambda(u - u^n) - F[u] + f^n)_+ \|$$

with $\lambda \ge N$ and $u \in \overset{0}{W}{}^{1,2}_{d+1}(C_{T,R})$. We now pass to the limit and apply the Fatou lemma to the right-hand side; this is legitimate since $f^n \le \sup_n f^n \in L_{d+1}(C_{T,R})$. Then

$$\lambda \| (u - u^0)_+ \| \le N \| (\lambda(u - u^0) - F[u] + f)_+ \| .$$

For sufficiently large λ we substitute $u = u^0 + \lambda^{-1} w$ into this inequality with $w \in \overset{0}{W}{}^{1,2}_{d+1}(C_{T,R})$ and let $\lambda \to \infty$. It is then obvious that $\|w_+\| \le N \| (w - F[u^0] + f)_+ \|$ for all $w \in \overset{0}{W}{}^{1,2}_{d+1}(C_{T,R})$ and hence for all $w \in L_{d+1}(C_{T,R})$. For $w = F[u^0] - f \vee k$ this yields $w_+ = 0$, $w \le 0$, and $F[u^0] \le f \vee k$ (a.e.$C_{T,R}$) for any k. This completes the proof.

The most awkward and unnatural assumption of Theorem 6 is that u^n equals zero on $\partial' C_{T,R}$. In order to remove this assumption, we need the following lemma.

7. LEMMA. Let the operator L have the form (1) and let in $C_{T,R}$ $c < 0$, $r + \operatorname{tr} a + |b| \le K_1 |c|$, $r \ge 0$, and $a = a^* \ge 0$, where K_1 is a constant. Consider a function $u \in W^{1,2}_{d+1}(C_{T,R})$ such that $Lu \ge 0$ (a.e.$C_{T,R}$) and define a number μ as the positive root of the equation $1 = K_1(\mu + \mu^2)$. Then in $C_{T,R}$,

$$u(t, x) \le \frac{\cosh \mu|x|}{\cosh \mu R} \sup_{\partial_x C_{T,R}} u_+ + e^{K_1^{-1}(t-T)} \sup_{\partial_t C_{T,R}} u_+ . \tag{7}$$

Proof. By the maximum principle, it suffices to check that the operator L, as applied to the right-hand side of (7), yields a negative function on $C_{T,R}$. Since $\sinh \mu|x| \le \cosh \mu|x|$ and $\sinh \mu|x| \le \mu|x| \cosh \mu|x|$, we obtain

$$L \cosh \mu|x| = c \cosh \mu|x| + \mu \sinh \mu|x| b^i \frac{x^i}{|x|} +$$

$$+ \mu^2 \cosh \mu|x| \, a^{ij} \frac{x^i x^j}{|x|^2} +$$

$$+ \mu^2 \frac{1}{\mu|x|} \sinh \mu|x| [\operatorname{tr} a - a^{ij} \frac{x^i x^j}{|x|^2}]$$

$$\leqq (c + \mu|b| + \mu^2 \operatorname{tr} a) \cosh \mu|x| \leqq 0.$$

Moreover,

$$Le^{K_1^{-1}(t - T)} = e^{K_1^{-1}(t - T)} (K_1^{-1} r + c) \leqq 0.$$

This proves the lemma.

8. LEMMA. Let $\varepsilon > 0$ and let F be an ε-nondegenerate L-type operator in $C_{T,R}$, $\tau \in [0,T]$, and $\rho \in [0,R]$. Then there exists a constant N which depends only on K, d, and ε and such that, for any $\lambda \geqq N$ and $u, v \in W^{1,2}_{d+1}(C_{T,R})$,

$$\lambda||(u - v)_{\pm}||_{L_{d+1}(C_{\tau,\rho})}$$

$$\leqq N||(\lambda(u - v) - (F[u] - F[v]))_{\pm}||_{L_{d+1}(C_{T,R})} +$$

$$+ \lambda||\varphi_\lambda||_{L_{d+1}(C_{\tau,\rho})}||(u - v)_{\pm}||_{C(\partial' C_{T,R})},$$

where

$$\varphi_\lambda(t, x) = \frac{\cosh \mu|x|}{\sinh \mu R} + \exp((\frac{\lambda}{K} - 1)(t-T))$$

and μ is the positive root of the equation $\lambda - K = K(\mu + \mu^2)$.

Proof. As in the proof of Lemma 5, it suffices to consider the case when F is a linear operator L of the form (1) with smooth coefficients, and to show that

$$\lambda||w_+||_{L_{d+1}(C_{\tau,\rho})}$$

$$\leqq N||(\lambda w - Lw)_+||_{L_{d+1}(C_{T,R})} + \lambda||\varphi_\lambda||_{L_{d+1}(C_{\tau,\rho})}||w_+||_{C(\partial' C_{T,R})}$$

for any $\lambda \geqq N$ and $w \in W^{1,2}_{d+1}(C_{T,R})$.

We express the function w as $w_0 + w_1$, where $w_i \in W^{1,2}_{d+1}(C_{T,R})$, $w_0 = 0$, $w_1 = w$ on $\partial' C_{T,R}$, $Lw_0 - \lambda w_0 = Lw - \lambda w$, and $Lw_1 - \lambda w_1 = 0$ (a.e. $C_{T,R}$). By Lemma 5,

$$\lambda \| (w_0)_+ \|_{L_{d+1}(C_{T,R})} \leqq N \| (\lambda w - Lw)_+ \|_{L_{d+1}(C_{T,R})} \tag{8}$$

for $\lambda \geqq N$ where N is taken from Lemma 5. Moreover, if $\lambda \geqq K$ and $K_1 := K(\lambda - K)^{-1}$, we obtain $c - \lambda \leqq 0$ and $r + \text{tr}\, a + |b| \leqq K_1 |c - \lambda|$. Hence, applying Lemma 7 to $L - \lambda$ and w_1, we obtain

$$w_1 \leqq \varphi_\lambda \| w_+ \|_{C(\partial' C_{T,R})}.$$

It remains to combine this inequality with (8) and the relationships

$$w_+ = (w_0 + w_1)_+ \leqq (w_0)_+ + (w_1)_+.$$

This proves the lemma.

We now prove a basic result of this section.

9. THEOREM. Let $\varepsilon > 0$ and let F be an ε-nondegenerate L-type operator in a domain Q, F_0 an operator of L-type in Q, $u^n \in W^{1,2}_{d+1,loc}(Q) \cap C(\bar{Q})$, $n = 0, 1, 2, \ldots$, and u^n converges to u^0 in the Lesbegue measure on Q as $n \to \infty$. Then:

(a) if $\sup_n \| u^n_- \|_{C(\partial' Q)} < \infty$ and $\sup_n F[u^n] \in L_{d+1}(Q)$, then

$$\overline{\lim_{n\to\infty}}\, F_0[u^n] \geqq F_0[u^0] \quad \text{(a.e.}\, Q); \tag{9}$$

(b) if $\sup_n \| u^n_+ \|_{C(\partial' Q)} < \infty$ and $\inf_n F[u^n] \in L_{d+1}(Q)$, then

$$F_0[u^0] \geqq \underline{\lim_{n\to\infty}}\, F_0[u^n] \quad \text{(a.e.}\, Q).$$

Proof. By virtue of Lemma 4, it suffices to prove assertion (a). We first prove this assertion when $F_0 = F$. By Definition 3, for any n there exists an operator L^n of the form (1) such that

$$L^n(-u^n) = -L^n u^0 - F[u^n] + F[u^0].$$

Since $F[u^n]$ are uniformly bounded by a function from $L_{d+1}(Q)$, then from Theorem 3.9 it follows that the functions $(-u^n)$ are uniformly bounded from above on Q, i.e.,

$$\sup_n ||u_-^n||_{C(Q)} < \infty.$$

Note also that the inequality $\overline{\lim}\, F[u^n] \geq F[u^0]$ (a.e.Q) will be proved if it is proved almost everywhere in every cylinder contained in Q. Let $\overline{C}_{T,R} \subset Q$, $\tau \in [0,T)$, and $\rho \in [0,R)$. We put $f^n = F[u^n]$ and $f = \overline{\lim}\, f^n$ and follow the proof of Theorem 6. By lemma 8,

$$\begin{aligned}\lambda||(u - u^n)_+||_{L_{d+1}(C_{\tau,\rho})} \\ \leq N||(\lambda(u - u^n) - F[u] + f^n)_+||_{L_{d+1}(C_{T,R})} + \\ + \lambda||\varphi_\lambda||_{L_{d+1}(C_{\tau,\rho})} ||(u - u^n)_+||_{C(Q)} \end{aligned} \qquad (10)$$

with $\lambda \geq N$ and $u \in W^{1,2}_{d+1}(C_{T,R})$. Here $(u - u^n)_+$ are uniformly bounded and converges to $(u - u^0)$ with respect to measure. Thus, on the left of (10) we can pass to the limit by the Lebesgue dominated convergence theorem. In the first summand on the right we first replace f^n by $\sup\{f^n;\ n \geq n_0\}$, then let $n \to \infty$ and then $n_0 \to \infty$ both times using the Lebesgue dominated convergence theorem. It follows that

$$\begin{aligned}\lambda||(u - u^0)_+||_{L_{d+1}(C_{\tau,\rho})} \\ \leq N||(\lambda(u - u^0) - F[u] + f)_+||_{L_{d+1}(C_{T,R})} + \\ + \lambda||\varphi_\lambda||_{L_{d+1}(C_{\tau,\rho})}(||u||_{C(Q)} + N). \end{aligned} \qquad (11)$$

We substitute $u = u^0 + \lambda^{-1}w$, where $w \in W^{1,2}_{d+1}(Q)$, into this expression and let first $\lambda \to \infty$ and then $\tau \uparrow T$ and $\rho \uparrow R$. We can easily obtain from the explicit form of φ_λ (see Lemma 8) that for any $\tau < T$ and $\rho < R$

$$\lim_{\lambda\to\infty} \lambda||\varphi_\lambda||_{L_{d+1}(C_{\tau,\rho})} = 0.$$

Therefore, after passing to the limit the last summand in (11) disappears and we can complete the proof of this theorem for $F_0 = F$ following the proof of Theorem 6.

We now abandon the additional assumption $F_0 = F$. Selecting constants $m > 0$ and $\delta \in (0,1]$, we note that by Lemma 4, $m^0 : u \to m$, $\min(F_0[u], m)$, and $\delta F[u] + (1 - \delta) \min (F_0[u], m)$ are L-type operators in Q. Moreover, the latter is obviously $\varepsilon\delta$-nondegenerate. Furthermore,

$$\sup_n (\delta F[u^n] + (1 - \delta)(m \wedge F_0[u^n]))$$

$$\leq \delta \sup_n F[u^n] + m \in L_{d+1}(Q).$$

Thus it is possible to apply assertion (a), which was proved above for ε-nondegenerate operators, to the operator $\delta F + (1 - \delta)(m \wedge F_0)$. Hence,

$$\delta \sup_n F[u^n] + (1 - \delta)\left(m \wedge \overline{\lim_{n\to\infty}} F_0[u^n]\right)$$

$$\geq \overline{\lim_{n\to\infty}} (\delta F[u^n] + (1 - \delta)(m \wedge F_0[u^n]))$$

$$\geq \delta F[u^0] + (1 - \delta)(m \wedge F_0[u^0]).$$

As $\delta \downarrow 0$ and $m \to \infty$, this yields assertion (a) in its complete formulation. The proof is complete.

10. Remark. One can present examples demonstrating that the assumptions adopted to prove (9) do not imply $\sup \| u^n \|_{C(Q)} < \infty$, moreover, they do not imply that $u^n \to u^0$ in $L_{d+1}(Q)$.

11. COROLLARY. Let $\varepsilon > 0$ and let $B(\varepsilon)$ be the set of operators taken from the proof of Lemma 4, let $u^n \in W^{1,2}_{d+2,loc}(Q) \cap C(\bar{Q})$, $n \geq 0$, u^n converge to u^0 in measure in Q, and u^n be uniformly bounded on $\partial' Q$. Also, for any $n \geq 1$ let there exist an operator $L^n \in B(\varepsilon)$ such that $L^n u^n = 0$ (a.e.Q). Then there exists an operator $L^0 \in B(\varepsilon)$ such that $L^0 u^0 = 0$ (a.e.Q).

Indeed, for any n

$$\bigwedge_{B(\varepsilon)} Lu^n \leq L^n u^n = 0, \qquad \bigvee_{B(\varepsilon)} Lu^n \geq L^n u^n = 0.$$

on Q. As $n \to \infty$, from Lemma 4 and Theorem 9 it follows that

$\bigwedge_{B(\varepsilon)} Lu^0 \leqq 0 \leqq \bigvee_{B(\varepsilon)} Lu^0$.

Furthermore, it is easy to find a measurable $s \in [0, 1]$ such that the equation $F[u^0] = 0$ holds for $F := s \bigwedge_{B(\varepsilon)} L + (1 - s) \bigvee_{B(\varepsilon)} L$. Since F is an ε-nondegenerate L-type operator and $F[0] = 0$, there exists an operator $L^0 \in B(\varepsilon)$ such that $0 = F[u^0] - F[0] = L^0u^0$. This completes the proof.

This corollary relates to the so-called G-convergence of linear operators (see Zhikov and Sirazhudinov [1], Zhikov, Kozlov, and Oleinik [1] and Krylov [25]).

12. THEOREM. Let $\varepsilon > 0$ and let F be an ε-nondegenerate operator of L-type in Q, let $u^n \in W^{1,2}_{d+1,loc}(Q) \cap C(\bar{Q})$, $n = 0, 1, 2, \ldots$, let u^n converge to u^0 in measure in Q, $F[u^n] \to f$ in $L_{d+1}(Q)$, and let u^n be uniformly bounded on $\partial' Q$. Then $F[u^0] = f$ (a.e.Q).

Proof. We select a subsequence $n(k) \to \infty$ such that

$$||f - F[u^{n(k)}]||_{L_{d+1}(Q)} \leqq 2^{-k}.$$

From this subsequence we select another subsequence which yields convergence of $F[u^n]$ to f almost everywhere. Then we pass from the functional sequence u^n to the sequence v^n such that $F[v^n] \to f$ (a.e.Q), $v^n \to u^0$ in measure, and $||f - F[v^n]||_{L_{d+1}(Q)} \leqq 2^{-n}$.

From the latter relationship it follows that a series with the nth term $|F[v^{n+1}] - F[v^n]|$ converges in $L_{d+1}(Q)$ and, since $|F[v^n] - F[v^1]|$ does not exceed the sum of this series, we obtain

$$\sup_n |F[v^n]| \in L_{d+1}(Q).$$

Hence, by Theorem 9

$$f = \overline{\lim_{n\to\infty}} F[v^n] \geqq F[u^0] \geqq \underline{\lim}_{n\to\infty} F[v^n] = f \quad (\text{a.e.}Q).$$

This proves the theorem.

13. COROLLARY. Let $\varepsilon > 0$ and let F be an ε-nondegenerate operator of L type in Q, a set Γ belong to $W^{1,2}_{d+1,loc}(Q) \cap C(\bar{Q})$, Γ be a compactum

in $L_{d+1}(Q)$, and let the restriction of functions of Γ to $\partial'Q$ make up a bounded set in $C(\partial'Q)$. Then the set $H := \{F[u] : u \in \Gamma\}$ is closed in $L_{d+1}(Q)$.

Indeed, let $f^n \in H$ and $f^n \to f$ in $L_{d+1}(Q)$. Select a $u^n \in \Gamma$ such that $F[u^n] = f^n$. By hypothesis, there exists a $u^0 \in \Gamma$ such that $u^n \to u^0$ in $L_{d+1}(Q)$ in the sense of some subsequence. Hence, by Theorem 12, $F[u^0] = f$ and $f \in H$, which completes the proof.

14. COROLLARY. Let $u^n \in W^{1,2}_{d+1,loc}(Q) \cap C(\bar{Q})$ and let $L^n \in B(\varepsilon)$, $n = 1, 2, \ldots$, ($B(\varepsilon)$ was introduced in the proof of Lemma 4), let f be a nonnegative function on Q, and let $L^n u^n \leq -f$ (a.e.) for all n. Suppose that $u^n \to 0$ in the Lebesque measure and u^n are uniformly bounded on $\partial'Q$. Then $f = 0$ (a.e.).

Indeed, put $F[u] = \wedge\{Lu : L \in B(\varepsilon)\}$. Then $F[u^n] \leq -(f \wedge m)$ for any m. By Theorem 9(a) it follows that $F[0] \leq -(f \wedge m)$, $0 \leq -(f \wedge m)$, and $f = 0$ (a.e.).

This corollary will be substantially strengthened in Corollary 4.1.8.

We now extend Assertions 9 to 14 to nonlienar elliptic operators.

15. THEOREM. Let D be a bounded domain in E_d. In Definition 3, we replace r, $W^{1,2}_{d+1}(Q)$, $L_{d+1}(Q)$, and Q by 0, $W^2_d(D)$, $L_d(D)$, and D, respectively, abandon the condition $r \geq \varepsilon$, and regard the resulting definition as the definition of an ε-nondegenerate L-type operator in D. Then after replacing $d + 1$, $C_{T,R}$, $\partial'C_{T,R}$, $\partial'Q$, Q, $W^{1,2}_{d+1}$, and L_{d+1} by d, S_R, ∂S_R, ∂D, D, W^2_d, and L_d, respectively, Assertions 5, 6, and 9 — 14 remain in force.

To prove this theorem, it suffices almost literally to repeat the proofs of Assertions 5, 6, and 9 — 14 using Theorem 2.3 instead of Theorem 2.4, to change Lemma 7 in an obvious manner, and to put $\varphi_\lambda = (\cosh \mu|x|)(\cosh \mu R)^{-1}$ in Lemma 8.

3.6. Passage to the Limit for Nonlinear Operators in Classes of Convex Functions

It is convenient to apply the results of the previous section in further study of degenerate nonlinear equations when the derivatives u_x, u_t, and u_{xx} of a solution u are *a priori* known to be bounded. Unfortunately, such *a priori* boundedness is often unavailable for degenerate equations, although, as we shall see in Chapters 7 and 8, the matrix u_{xx} can be estimated from below (or only from above) in a rather broad class of problems. In other words, it is possible to prove that the function

$u + N|x|^2$ is convex in x for a constant N. In this connection, we shall consider functions of the form $u_1 + u_2$ where u_1 is convex in x and $u_2 \in W^{1,2}_{d+1}$. Note that, although this class of functions is broader than that considered in Section 5, nevertheless the class of operators under whose action we are going to pass to the limit is narrower than that considered in Section 5.

We fix constants K, T, R, and $\varepsilon > 0$. Let Ω be a set of indices and and let every $\omega \in \Omega$ correspond to real-valued functions $r(\omega)$, $a^{ij}(\omega)$, $b^i(\omega)$, $c(\omega)$, and $f(\omega)$, i, j = 1, ..., d which are Borel on $C_{T,R}$. Put $a(\omega) = (a^{ij}(\omega))$, $b(\omega) = (b^i(\omega))$ and assume that $a = a^*$, $r + \operatorname{tr} a + |b| + |c| \leq K$, $r \geq 0$, $(a\lambda, \lambda) \geq 0$ for all $\omega \in \Omega$, $(t, x) \in C_{T,R}$, and $\lambda \in E_d$. Also, assume that there exists a function $\bar{f} \in L_{d+1}(C_{T,R})$ such that $|f(\omega)| \leq \bar{f}$ on $C_{T,R}$ for all ω.

Let $W^{bv,+}(C_{T,R})$ be the set of all Borel bounded real-valued functions $u(t, x)$ on $C_{T,R}$ which are convex in x and whose variation with respect to t on [0, T] is integrable with respect to x on S_R. We discuss the properties of elements of $W^{bv,+}(C_{T,R})$. Let $\zeta \geq 0$, $\zeta \in C_0^\infty(E_{d+1})$, $\zeta(t, x) = 0$ for $|t|^2 + |x|^2 \geq 1$, $\int \zeta \, dt \, dx = 1$, and $\zeta(t, x) = \zeta(|t|^2 + |x|^2)$. We put $\zeta^\tau(t, x) = \tau^{-d-1}\zeta(\tau^{-1}t, \tau^{-1}x)$, where $\tau > 0$, and let

$$w^{(\tau)} := w * \zeta^\tau, \quad \mu^{(\tau)}(t, x) := \int \zeta^\tau(t - s, x - y)\mu(ds\, dy) \tag{1}$$

for a function w and a measure μ.

Also, let $\tau > 0$ be sufficiently small and Q^τ be a cylinder consisting of those points of $C_{T,R}$ which are further from the boundary of $C_{T,R}$ than 2τ. For a $\tau_0 > 0$, it is obvious that $Q^\tau \neq \phi$ if $\tau \in (0, \tau_0)$.

Consider a function $u \in W^{bv,+}(C_{T,R})$. It is clear that $u^{(\tau)}$ is infinitely differentiable on Q^τ with $\tau \in (0, \tau_0)$. Moreover, $u^{(\tau)}$ is convex in x on Q^τ, its variation with respect to t on $[2\tau, T - 2\tau]$ is integrable with respect to x over $S_{R-2\tau}$, and this integral does not exceed the integral of the variation of u on [0, T] over S_R.

The latter fact, along with the differentiability of $u^{(\tau)}$ with respect to t, implies that the measures with differentials $u_t^{(\tau)}(t, x)\, dt\, dx$ considered on Q^τ are uniformly bounded in variation. For $\tau \downarrow 0$ in the sense of weak measure convergence on every compactum of $C_{T,R}$, every accumulation point of such measures is a measure on C which is a generalized derivative of u with respect to t in the sense of the theory of generalized functions. Indeed, if μ is a weak accumulation point of these measures and $\varphi \in C_0^\infty(C_{T,R})$, then

$$\int \varphi_t u \, dt \, dx = \lim_{\tau \downarrow 0} \int \varphi_t u^{(\tau)} \, dt \, dx = -\lim_{\tau \downarrow 0} \int \varphi u_t^{(\tau)} \, dt \, dx$$

$$= -\int \varphi \mu (dt \, dx).$$

The established measure μ is uniquely determined by u. We denote it by u_t. It is obvious that the measure u_t has bounded variation on $C_{T,R}$.

We further repeat some argument from Section 1.2. Considering a one-dimensional cross-section of the graph of a smooth convex function, we straightforwardly obtain that the gradient of such a function can be estimated in magnitude in S_{R_1}, where $R_1 < R$, only in terms of R, R_1, and the oscillation of the convex function in question on S_R. It follows that $u_x^{(\tau)}$ is uniformly bounded on Q^{τ_1} for $\tau \in (0, \tau_1)$ with any $\tau_1 \in (0, \tau_0)$. Its weak limit, for instance in $L_2(Q^{\tau_1})$, obviously yields the Sobolev derivatives of the form u_x, which therefore exist on $C_{T,R}$ and are locally bounded.

We now consider the second derivatives. Since u^τ is convex in x on Q^τ, for any $l \in E_d$ on Q we obtain the inequality $u^{(\tau)}_{(l)(l)} \geqq 0$. Specifically, for any $\tau < \tau_1 < \tau_0$ and any nonnegative $\eta \in C_0^\infty(Q^\tau)$ which is equal to 1 on Q^{τ_1} we have

$$\int \eta_{(l)(l)} u^{(\tau)} \, dt \, dx = \int \eta u^{(\tau)}_{(l)(l)} \, dt \, dx \geqq \int_{Q^{\tau_1}} u^{(\tau)}_{(l)(l)} \, dt \, dx. \qquad (2)$$

It follows that the measures with differentials $u^{(\tau)}_{(l)(l)} \, dt \, dx$, as considered on Q^1 for $\tau < \tau_1$, are positive and uniformly bounded (in variation). It follows that there exists a measure of locally bounded variation in $C_{T,R}$ which is the weak limit of the measures $u^{(\tau)}_{(l)(l)} \, dt \, dx$ as $\tau \downarrow 0$ on every compactum in $C_{T,R}$. This measure is denoted by $u_{(l)(l)}$. From the first equation in (2) it follows that $u_{(l)(l)}$ is the second derivative of u along l in the sense of generalized functions. It is obvious that $u_{(l)(l)}$ is a quadratic form and the measures $u_{x^i x^j}$, which are coefficients of $l^i l^j$ in this form, are generalized derivatives of u with respect to $x^i x^j$. Thus, we have proved the following fact.

1. LEMMA. If $u \in W^{bv,+}(C_{T,R})$, then the Sobolev derivatives of the form u_x are locally bounded in $C_{T,R}$ and the generalized derivatives u_t and u_{xx} are measures of locally bounded variation in $C_{T,R}$. Moreover, $u_{(l)(l)} \geqq 0$ in $C_{T,R}$ for any $l \in E_d$.

For a function $u \in W^{1,2}_{d+1}(C_{T,R})$ it is also reasonable to consider u_x as a Sobolev derivative, while u_t and u_{xx} as measures (with differentials $u_t\,dt\,dx$ and $u_{xx}\,dt\,dx$, where u_t and u_{xx} are Sobolev derivatives). Therefore, in what follows, given $u \in \tilde{W}^{bv,+}(C_{T,R}) := W^{bv,+}(C_{T,R}) + W^{1,2}_{d+1}(C_{T,R})$, u_x will be regarded as Sobolev derivative and by u_t and u_{xx} will be meant measures which are respective derivatives in the sense of the theory of generalized functions.

Next follows a simple sufficient condition that a function belongs to the class $\tilde{W}^{bv,+}(C_{T,R})$. If $u \in C(\bar{C}_{T,R})$, Sobolev derivatives u_x and u_t are bounded, the generalized derivatives u_{xx} are measures, and $u_{x^ix^j}l^il^j \geqq -N_1|l|^2\Lambda$ on $C_{T,R}$ for all $l \in E_d$, where N_1 is a constant; Λ is the Lebesgue measure on $C_{T,R}$, then $u \in \tilde{W}^{bv,+}(C_{T,R})$ and, moreover, $u + (1/2)N_1|x|^2$ is convex in x on $C_{T,R}$. This fact readily follows from considering the function $u^{(\tau)}$ by taking into account that $(u^{(\tau)})_{x^ix^j} = (u_{x^ix^j})^{(\tau)}$.

Recall that, if μ is a measure of locally bounded variation and g is a function on $C_{T,R}$ which is locally summable with respect to μ, then $g\circ\mu$ is the indefinite integral of g with respect to $\mu(g\circ\mu(\Gamma) := \int_\Gamma g\mu(dt\,dx))$, $|\mu|$ is the variation of measure μ, $\mu^a(\mu^s)$ is the absolutely continuous (singular) part of the measure μ with respect to Λ, and $\mu^{(0)} := \mu^a(dt\,dx)/dt\,dx$. For every measure μ the values $\mu^{(0)}$ are assumed to be fixed at every point of the domain $C_{T,R}$. It is known (see Dunford and Schwartz [1]) that

$$\mu^{(0)} = \lim_{\tau\downarrow 0} \mu^{(\tau)} \quad (\text{a.e. } C_{T,R}). \tag{3}$$

(Remember that unless a special convention is used, "a.e." means "almost everywhere with respect to Lebesgue measure".) Thus the denotion $\mu^{(0)}$ is consistent with (1).

For $u \in \tilde{W}^{bv,+}(C_{T,R})$ we denote

$$Fu = \bigvee_{\omega\in\Omega}\Big[r(\omega)\circ u_t + a^{ij}(\omega)\circ u_{x^ix^j} + \Big(b^i(\omega)u_{x^i} + c(\omega)u + f(\omega)\Big)\circ\Lambda\Big], \tag{4}$$

where the supremum is naturally taken in the lattice of measures on

$C_{T,R}$. Since r, a, b, and c are bounded and $|f| \leqq \bar{f}$, the measures between the braces in (4) are uniformly bounded from above (and even in variation) by the measure

$$N\left(|u_t| + \Sigma|u_{x^i x^j}| + \left(\Sigma|u_{x^i}| + N + \bar{f}\right)\circ\Lambda\right),$$

where N is a constant. Therefore, (see Section 1.2) the expression of Fu is well defined and it is a measure on $C_{T,R}$ of locally bounded variation.

Some techniques of computing the suprema are discussed in Section 1.2. Here we pay particular attention to the rule that enables us to determine the aboslutely continuous and singular parts of a supremum in terms of the measure parts involved in the operation of taking the supremum. This fact implies the following result.

2. LEMMA. Let $u \in \tilde{W}^{bv,+}(C_{T,R})$ and let the Sobolev derivative of u with respect to t exist on $C_{T,R}$. Then the equation Fu = 0 holds in $C_{T,R}$ iff $a^{ij}(\omega)\cdot u_{x^i x^j} \ll \Lambda$ for all $\omega \in \Omega$ and

$$\bigvee_{\omega\in\Omega}\Big[a^{ij}(\omega)u^{(0)}_{x^i x^j} + r(\omega)u_t + b^i(\omega)u_{x^i} +$$

$$+ c(\omega)u + f(\omega)\Big] = 0 \text{ (a.e. } C_{T,R}). \tag{5}$$

Proof. Since $(Fu)^{(0)}$ is involved in the left-hand side of (5) and, obviously, $u^s_t = 0$,

$$(Fu)^s = \bigvee_{\omega\in\Omega}\left(r\circ u^s_t + a^{ij}\circ u^s_{x^i x^j}\right) = \bigvee_{\omega\in\Omega} a^{ij}\circ u^s_{x^i x^j}, \tag{6}$$

therefore, to prove the lemma, it suffices to show that

$$a^{ij}(\omega)\circ u^s_{x^i x^j} \leqq 0 \Rightarrow a^{ij}(\omega)\circ u_{x^i x^j} \ll \Lambda \Rightarrow a^{ij}(\omega)\circ u^s_{x^i x^j} = 0 \tag{7}$$

for every $\omega \in \Omega$. The latter implication is obvious. While proving the former, one can assume that $u \in W^{bv,+}(C_{T,R})$. In this case $a^{ij}(\omega)\circ u_{x^i x^j} \geqq 0$. Indeed, let $\mu = \Sigma|u_{x^i x^j}|$ and let $\bar{u}_{ij}(t, x) = u_{x^i x^j}(dt\,dx)/\mu(dt\,dx)$. Since the inequality $u_{(l)(l)} \geqq 0$ holds, $\bar{u}_{ij}\, l^i l^j \geqq 0$ almost everywhere in μ for every $l \in E_d$.

The function $\bar{u}_{ij} l^i l^j$ is continuous in l, therefore, there exists a set $\Gamma \subset C_{T,R}$ of μ-measure zero such that $\bar{u}_{ij} l^i l^j \geqq 0$ outside Γ for all $l \in E_d$. Then $a^{ij}\bar{u}_{ij} > 0$ outside Γ, since both matrices (a^{ij}) and $(\bar{u}_{ij})$ are symmetric and positive. Hence $a^{ij}(\omega) \circ u_{x^i x^j} \geqq 0$, too. From this inequality we obtain $a^{ij}(\omega) \circ u^s_{x^i x^j} \geqq 0$, which gives the first implication in (7). The lemma is proved.

Sometimes it is useful to apply the following fact. If the Sobolev derivatives u_t and u_{xx} exist, then $Fu \ll \Lambda$ and $(Fu)^{(0)}$ coincides with the left-hand side of (5) (where $u^{(0)}_{x^i x^j}$ is the Sobolev derivative). Also for instance, if Ω is a separable metric space, and r, a, b, c, and f are continuous in ω, then the supremum in the lattice of measurable functions with the Lebesgue measure on the left of (5) can be replaced by an ordinary supremum.

We now formulate the major result of this section.

3. THEOREM. Let $r \geqq \varepsilon$ and let $(a\lambda, \lambda) \geqq \varepsilon|\lambda|^2$ for all ω, t, x, λ. Also, let v, $u^n \in \tilde{W}^{bv,+}(C_{T,R}) \cap C(\bar{C}_{T,R})$, n = 1, 2, ..., let u^n uniformly converge to v on $C_{T,R}$ as $n \to \infty$, and let $Fu^n \geqq 0$ in $C_{T,R}$. Then, $(Fv)^a \geqq 0$ in $C_{T,R}$, and if the Sobolev derivative of v with respect to t exists on $C_{T,R}$, then $Fv \geqq 0$ in $C_{T,R}$.

This theorem will be proved after some preliminary results have been established. First of all, note that computing Fu^n and Fv, one can replace the supremum in (4) by the supremum taken with respect to an appropriate countable subset Ω, which possibly depends on n (see Section 1.2). Combining these subsets with respect to n, we obtain a countable subset Ω which can be substituted into (4) so that Fv and Fu^n do not change for any n. Therefore, we can assume that Ω is countable.

We further introduce the function

$$F(u_0, u_{ij}, u_i, u, t, x) := \sup_{\omega\in\Omega}\left[r(\omega)u_0 + a^{ij}(\omega)u_{ij} + b^i(\omega)u_i + c(\omega)u + f(\omega)\right]. \qquad (8)$$

Since Ω is countable, $F(u_0, u_{ij}, u_i, u, t, x)$ is measurable with respect to (t, x), and since the difference between suprema does not exceed the supremum of differences, this function satisfies the Lipschitz condition with respect to (u_0, u_{ij}, u_i, u) with a constant which does not depend on (t, x). For $u \in \tilde{W}^{bv,+}(C_{T,R})$ we put

$$F[u] = F\left(u_t^{(0)}, u_{x^ix^j}^{(0)}, u_{x^i}, u, t, x\right).$$

As mentioned above,

$$(Fu)^{(0)} = F[u] \quad (\text{a.e.} C_{T,R}). \tag{9}$$

We now reduce our consideration to the case when $r(\omega) \equiv 1$. Supposing the contrary and using the inequality $K \geqq r(\omega) \geqq \varepsilon$, we can easily show that for $u \in \tilde{W}^{bv,+}(C_{T,R})$ the inequality $Fu \geqq 0$ holds iff

$$\bigvee_{\omega\in\Omega}\left(u_t + r^{-1}(\omega)a^{ij}(\omega)\circ u_{x^ix^j} + \right.$$
$$\left. + r^{-1}(\omega)(b^i(\omega)u_{x^i} + c(\omega)u + f(\omega))\circ\Lambda\right) \geqq 0.$$

The same is true for the inequality $F[u] \leqq 0$. Consequently, without loss of generality we can assume that $r(\omega) \equiv 1$.

Finally, denote by $\mathbf{A}$ the set of all Borel functions on $C_{T,R}$ whose values lie in (a countable set) Ω. From (8), (9), and the major property of the supremum, we obtain the following lemma.

4. LEMMA. Let $u \in \tilde{W}^{bv,+}(C_{T,R})$ and let $\delta > 0$. Then there exists a function $\omega \in \mathbf{A}$ such that

$$u_t^{(0)} + a^{ij}(\omega)u_{x^ix^j}^{(0)} + b^i(\omega)u_{x^i} + c(\omega)u + f(\omega) \geqq (Fu)^{(0)} - \delta.$$
$$(\text{a.e.} C_{T,R})$$

An important role in the proof of Theorem 3 is played by the following lemma.

5. LEMMA. Let $u \in \tilde{W}^{bv,+}(C_{T,R})$, $w \in W_{d+1}^{1,2}(C_{T,R})$, and $g \in L_{d+1}(C_{T,R})$. Suppose that u is continuous in $\bar{C}_{T,R}$, $Fu \geqq g\circ\Lambda$ on $C_{T,R}$, $\gamma \in [0,T]$, and $\rho \in [0,R]$. Then there exists a constant N, which depends only on K, d, and ε and such that

$$\lambda||(u - w)_+||_{L_{d+1}(C_{\gamma,\rho})}$$
$$\leqq N||(\lambda(u - w) + F[w] - g)_+||_{L_{d+1}(C_{T,R})} +$$
$$+ \lambda||\varphi_\lambda||_{L_{d+1}(C_{\gamma,\rho})}||(u - w)_+||_{C(\partial' C_{T,R})}, \tag{10}$$

for $\lambda \geq N$, where φ_λ is taken from Lemma 5.8 with replacing K by $4\varepsilon^{-1}K^2d + K$.

Proof. We express u as $u_0 + u_1$, where $u_1 \in W^{1,2}_{d+1}(C_{T,R})$ and $u_0 - |x|^2 \in W^{bv,+}(C_{T,R})$. Denote

$$w_1 = w - u_1, \quad g_1 = g - F[w],$$

$$f_1(\omega) = f(\omega) + u^{(0)}_{1t} + a^{ij}(\omega)u^{(0)}_{1x^ix^j} + b^i(\omega)u_{1x^i} + c(\omega)u_1 - F[w]$$

and construct F_1 replacing $f(\omega)$ by $f_1(\omega)$ in (4). It is obvious that $u - w = u_0 - w_1$ and inequality (10) becomes

$$\lambda||(u_0 - w_1)_+||_{L_{d+1}(C_{\gamma,\rho})}$$

$$\leq N||(\lambda(u_0 - w_1) - g_1)_+||_{L_{d+1}(C_{T,R})} +$$

$$+ \lambda||\varphi_\lambda||_{L_{d+1}(C_{\gamma,\rho})}||(u_0 - w_1)_+||_{C(\partial' C_{T,R})}. \tag{11}$$

Note that $F_1\cdot = F(\circ + u_1) - Fw$ and $F_1u_0 \geq g^0_1\circ\Lambda$ with $F_1w_1 = 0$. Hence, inequality (11) is exactly of the form (10), if u, w, F and g are replaced by u_0, w_1, F_1, and g_1 respectively in the latter inequality. Here u_0, w_1, F_1, and g_1 satisfy the lemma conditions and, moreover, $F_1w_1 = 0$ and $u_0 - |x|^2 \in W^{bv,+}(C_{T,R})$.

Thus, to prove the lemma, we can assume without loss of generality that

$$Fw = 0 \text{ on } C_{T,R}, \quad u - |x|^2 \in W^{bv,+}(C_{T,R}).$$

Denote by Γ a Borel subset of $C_{T,R}$ such that $\Lambda(\Gamma) = 0$ and all measures u_t and u_{xx} are absolutely continuous on $C_{T,R}\setminus\Gamma$.

We now fix a $\delta > 0$, take the respective function $\omega \in \mathbf{A}$ from Lemma 4 (for brevity we shall write $a = a(\omega(t, x), t, x), \ldots, f = f(\omega(t, x), t, x))$, and put

$$\tilde{a} = a + K\chi_\Gamma I,$$

where I is a $d \times d$ unity matrix. Consider the inequalities

$$2K\delta^{ij}u_{x^ix^j} \geq \tilde{a}^{ij} \circ u_{x^ix^j} \geq \varepsilon\delta^{ij}u_{x^ix^j} \geq \varepsilon\Lambda. \tag{12}$$

Inequalities like those can easily be obtained from well-known properties of symmetric positive semi-definite matrices, if we pass from the measures $u_{x^ix^j}$ to their densities with respect to the measure $\Lambda + \Sigma|u_{x^ix^j}|$ and note that the convexity of $u - |x|^2$ in x implies the inequality $2|\xi|^2\Lambda \leqq \xi^i\xi^j u_{x^ix^j}$ for any $\xi \in E_d$. The latter inequality can be established, for example, by considering the Sobolev means. Moreover, the inequalities

$$2K\Delta u^{(\tau)} \geqq \tilde{a}^{ij}u^{(\tau)}_{x^ix^j} \geqq \varepsilon\Delta u^{(\tau)} \geqq \varepsilon. \tag{13}$$

hold on Q.

From (13) it follows that the following definition of the matrix a^τ on Q^τ makes sense:

$$a^\tau := \tilde{a}\left(\tilde{a}^{ij} \circ u_{x^ix^j}\right)^{(\tau)}\left(\tilde{a}^{pq}u^{(\tau)}_{x^px^q}\right)^{-1}.$$

By (12) and (13), we obtain $4\varepsilon^{-1}K^2I \geqq a^\tau \geqq \varepsilon^2(2K)^{-1}I$. Moreover, (see (3)), $a^\tau \to \tilde{a} = a$ (a.e.C) as $\tau \downarrow 0$.

If $\Psi \in W^{1,2}_{d+1}(Q^\tau)$, we put

$$L^\tau\psi = \psi_t + a^{ij\tau}\psi_{x^ix^j} + b^i\psi_{x^i} + c\psi,$$

where by Ψ_t and Ψ_{xx} the Sobolev derivatives are meant. It is obvious that on Q^τ

$$L^\tau u^{(\tau)} = \left[u_t + \tilde{a}^{ij} \circ u_{x^ix^j}\right]^{(\tau)} + b^i u^{(\tau)}_{x^i} + cu^{(\tau)}. \tag{14}$$

From the choice of $\omega \in \mathbf{A}$ and the inequality $Fu \geqq g \circ \Lambda$ it follows that

$$\begin{aligned} u_t + \tilde{a}^{ij} \circ u_{x^ix^j} &= \\ &= \left(u_t^{(0)} + a^{ij}u^{(0)}_{x^ix^j}\right) \circ \Lambda + \chi_\Gamma \circ \left(u_t + (a^{ij} + K\delta^{ij}) \circ u_{x^ix^j}\right) \\ &\geqq \left(g - \delta - b^i u_{x^i} - cu - f\right) \circ \Lambda + \chi_\Gamma \circ \left(u_t + K\delta^{ij}u_{x^ix^j}\right). \end{aligned} \tag{15}$$

The latter summand is denoted by ν. For any $\pi \in \Omega$ the inequality $K \geqq \operatorname{tr} a(\pi)$ implies

$$\nu \geqq \chi_\Gamma \circ \left(u_t + a^{ij}(\pi) \circ u_{x^i x^j}\right).$$

Since $\Lambda(\Gamma) = 0$, it follows that

$$\nu \geqq \chi_\Gamma \circ \left(u_t + a^{ij}(\pi) \circ u_{x^i x^j} + \left(b^i(\pi) u_{x^i} + c(\pi)u + f(\pi)\right) \circ \Lambda\right).$$

In other words, $\nu \geqq \chi_\Gamma \circ Fu$. Since $Fu \geqq g \circ \Lambda$, we obtain $\nu \geqq 0$.

Finally, from (14) and (15) we obtain $L^\tau u^{(\tau)} \geqq g - \delta - f + h^\tau$, where

$$h^\tau = g^{(\tau)} - g + f - f^{(\tau)} + b^i u^{(\tau)}_{x^i} - \left(b^i u_{x^i}\right)^{(\tau)} + cu^{(\tau)} - (cu)^{(\tau)}.$$

From well-known properties of the Sobolev mean it follows that $h^\tau \to 0$ in $L_{d+1}(Q^\sigma)$ as $\tau \downarrow 0$ for any $\sigma > 0$. On estimating $L^\tau u^{(\tau)}$, we now estimate $L^\tau w$. Taking into account that $Fw \leqq 0$, we obtain

$$L^\tau w \leqq -f + (a^{\tau ij} - a^{ij}) w_{x^i x^j} =: -f + h_1^\tau,$$

where the latter summand tends to zero in $L_{d+1}(Q^\sigma)$ as $\tau \downarrow 0$ for any $\sigma > 0$, since a^τ are uniformly bounded and $a^\tau \to a$ as $\tau \downarrow 0$ (a.e.$C_{T,R}$).

Thus, $L^\tau(u^{(\tau)} - w) \geqq g - \delta + h^\tau - h_1^\tau$ on Q^σ. From Lemma 5.8 applied to the cylinder Q^σ with $\tau \leqq \sigma$, the operator L^τ instead of F, and constants $4\varepsilon^{-1}K^2 d + K$ and $\varepsilon^2(2K)^{-1}$ instead of K and ε, it follows that

$$\lambda \| (u^{(\tau)} - w)_+ \|_{L_{d+1}(C_{\gamma,\rho} \cap Q^\sigma)}$$

$$\leqq N \left\| \left(\lambda(u^{(\tau)} - w) - g + \delta - h^\tau + h_1^\tau\right)_+ \right\|_{L_{d+1}(Q^\sigma)} +$$

$$+ \lambda \| \psi_\lambda^\sigma \|_{L_{d+1}\left(C_{\gamma,\rho} \cap Q^\sigma\right)} \| (u^{(\tau)} - w)_+ \|_{C(\partial' Q^\sigma)}$$

with $\lambda \geqq N$, where φ_λ^σ is defined as in Lemma 5.8 with replacing T, R, K by $T - 2\sigma$, $R - 2\sigma$, and $4\varepsilon^{-1}K^2 d + K$, respectively. Letting first $\tau \downarrow 0$, then $\sigma \downarrow 0$ and finally $\delta \downarrow 0$, we obtain the lemma assertion. The proof is complete.

The following lemma completes the preparatory work preceding the proof of Theorem 3.

6. LEMMA. Let a function $\Phi(\psi, t, x)$ be given on $E_{d_1} \times C_{T,R}$, which is measurable with respect to (t, x), satisfies the Lipschitz condition with respect to ψ with a constant independent of t, x, and such that $\Phi(0, \cdot, \cdot) \in \mathcal{L}_{d+1}(C_{T,R})$. Let two sets of functions: Ψ and $\{\psi^i, i = 1, 2, \dots\}$ into E_{d_1} be defined on $C_{T,R}$. Suppose that $\psi^i \in \Psi \subset \mathcal{L}_{d+1}(C_{T,R})$ and every function of Ψ can be uniformly on $C_{T,R}$ approximated by functions of $\{\psi^i, i \geq 1\}$ as accurately as desired. Then there exists a set $\Gamma \subset C_{T,R}$ such that $\Lambda(C_{T,R} \setminus \Gamma) = 0$ and for any $(t_0, x_0) \in \Gamma$ and $\psi \in \Psi$ we obtain

$$\lim_{\rho \downarrow 0} \Lambda^{-1/(d+1)}\left(C_{\rho^2, \rho}\right) ||\Phi(\psi, \cdot, \cdot) -$$

$$\Phi(\psi(t_0, x_0), t_0, x_0)||_{\mathcal{L}_{d+1}((t_0, x_0) + C_{\rho^2,\rho})} = 0. \tag{16}$$

When $\Psi = \{\psi^i, i \geq 1\}$, the existence of an appropriate set Γ follows from the fact that, for a summable function, almost every point is Lebesgue*) and that Ψ is countable while $\Phi(\psi, t, x) \in \mathcal{L}_{d+1}(C_{T,R})$ whenever $\psi \in \mathcal{L}_{d+1}(C_{T,R})$. In the general case, it suffices to note that, by hypothesis, $\Phi(\psi, t, x)$ can be uniformly approximated by a function of the form $\Phi(\psi^i, t, x)$, while the expression for ψ involved in the left-hand side of (16) differs little from the same expression for ψ^i.

7. <u>Proof of Theorem 3</u>. From (6) and the inequality $(a^{ij}(\omega) \circ v_{x^i x^j})^s \geq 0$ established in the proof of Lemma 2 it immediately follows that $(Fv)^s \geq 0$ if $v_t^s = 0$, i.e., if there exists a Sobolev derivative of v with respect to t. Therefore, we need only to prove the inequality $(Fv)^a \geq 0$ or, by virtue of (9), the inequality $F[v] \geq 0$ (a.e. $C_{T,R}$).

Consider a point $(t_0, x_0) \in C_{T,R}$ and apply Lemma 5 to u^n and $(t_0, x_0) + C_{\rho^2,\rho}$ instead of u and $C_{T,R}$ assuming that ρ is so small that $(t_0, x_0) + C_{\rho^2,\rho} \subset C_{T,R}$. We now let $n \to \infty$. Then for all $w \in W^{1,2}_{d+1}(C_{T,R})$ and sufficiently small ρ we obtain

$$\lambda||(v - w)_+||_{(t_0, x_0) + C_{r^2,r}}$$

---

*) See, however, Lemma 2 in Appendix 2.

$$\leq N\|(\lambda(v - w) + F[w])_+\|_{(t_0, x_0) + C_{\rho^2,\rho}} +$$
$$+ \lambda\|\varphi_{\lambda,\rho}\|_{(t_0, x_0)} +$$
$$+ C_{r^2,r}\|(v - w)_+\|_{C((t_0, x_0) + \partial' C_{\rho^2,\rho})}, \tag{17}$$

where $r = (1/2)\rho$, $\lambda \geq N$, and N is independent of λ and ρ,

$$\varphi_{\lambda,\rho}(t, x) = \frac{\cosh \mu|x - x_0|}{\cosh \mu\rho} + \exp((\frac{\lambda}{K_1} - 1)(t - t_0 - \rho^2)),$$

$K_1 = 4\varepsilon^{-1}K^2d + K$, μ is a positive root to the equation $\lambda - K_1 = K_1(\mu + \mu^2)$, and the subscripts L_{d+1} are omitted for the sake of simplicity.

Furthermore, we represent v as $u + v_1$, where $u \in W^{bv,+}(C_{T,R})$, $v_1 \in W^{1,2}_{d+1}(C_{T,R})$ and denote $\Phi[\cdot] = F[\cdot + v_1]$. From (17) it obviously follows that, for the same w, ρ, and λ,

$$\lambda\|(u - w)_+\|_{(t_0, x_0) + C_{r^2,r}}$$
$$\leq N\|(\lambda(u - w + \Phi[w])_+\|_{(t_0, x_0) + C_{\rho^2,\rho}} +$$
$$+ \lambda\|\varphi_{\lambda,\rho}\|_{(t_0, x_0)}$$
$$+ C_{r^2,r}\|(u - w)_+\|_{C((t_0, x_0) + \partial' C_{\rho^2,\rho})}. \tag{18}$$

We now denote by P (or P_0 respectively) the set of all polynomials of t, x (with rational coefficients) and, using

$$\Phi(\psi_0, \psi_{ij}, \psi_i, \Psi, t, x)$$
$$:= F(\psi_0 + v_{1t}, \psi_{ij} + v_{1x^ix^j}, \psi_i + v_{1x^i}, \psi + v_1, t, x),$$
$$\left\{\left(\psi_t, \psi_{x^ix^j}, \psi_{x^i}, \psi\right) : \psi \in P_0\right\},$$

$$\Psi := \left\{ \left(\psi_t, \psi_{x^i x^j}, \psi_{x^i}, \psi \right) : \psi \in P \right\},$$

we construct the corresponding set Γ on the basis of Lemma 6. According to the extension of the Aleksandrov–Buseman–Feller theorem proved in Appendix 2, one can assume that, for any $(t_0, x_0) \in \Gamma$,

$$u(t, x) - \Big[u(t_0, x_0) + u_t^{(0)}(t_0, x_0)(t - t_0) + \\ + u_{x^i}(t_0, x_0)\left(x^i - x_0^i\right) + \\ + \frac{1}{2} u^{(0)}_{x^i x^j}(t_0, x_0)\left(x^i - x_0^i\right)\left(x^j - x_0^j\right)\Big]$$

$$= o(|t - t_0| + |x - x_0|^2) \tag{19}$$

as $(t, x) \to (t_0, x_0)$. We now prove that, for any $(t_0, x_0) \in \Gamma$, the inequality $\Phi[u](t_0, x_0) \geqq 0$ holds. By virtue of the equation $\Phi[u] = F[v]$, this suffices to prove the theorem.

We fix a point $(t_0, x_0) \in \Gamma$, a constant $\xi \in (-\infty, \infty)$ and substitute the sum of $(-\xi/\lambda)$ and ψ for w in (18). Here ψ is the expression between the braces in (19). We next divide both sides of (18) by $\rho^{(d+2)/(d+1)}$ and let $\lambda \to \infty$ and $\rho \downarrow 0$ so that $\lambda\rho^2 \to \infty$ and $\lambda o(\rho^2) \to 0$ where $o(\rho^2)$ is taken from (19). It is clear that the maximum of $\varphi_{\lambda,\rho}$ with respect to $(t_0, x_0) + C_{r^2,r}$ approaches zero and the latter summand of (18) disappears after performing the operations mentioned. Moreover,

$$\left| \left(\lambda\left(u + \frac{\xi}{\lambda} - \psi\right) + \Phi\left[-\frac{\xi}{\lambda} + \psi\right]\right)_+ - (\xi + \Phi[\psi])_+ \right|$$

$$\leqq \lambda|u - \psi| + \left|\Phi\left[-\frac{\xi}{\lambda} + \psi\right] - \Phi[\psi]\right| \leqq \lambda o(\rho^2) + N\frac{1}{\lambda}.$$

Therefore, by virtue of the choice of the point (t_0, x_0), from (18) we obtain $\xi_+ \leqq N(\xi + \Phi[\psi](t_0, x_0))_+$, where N is independent of ξ. This inequality holds for all $\xi \in (-\infty, \infty)$, hence $\Phi[\psi](t_0, x_0) \geqq 0$ and since the equation $\Phi[\psi](t_0, x_0) = \Phi[u](t_0, x_0)$ is obvious, the theorem is proved.

3.7. The Uniqueness of a Solution and the Comparison Theorems for Non-Linear Operators in the Class of Convex Functions

In this section, we present some results on the solution uniqueness for degenerate non-linear equations in the class of functions which are convex in x. The basic result, namely Theorem 1, is taken from Krylov [14] and presented here without proof. Unfortunately, the author does not know a sufficiently short differential-equation interpretation of the probabilistic arguments evolved in that paper. This can partly be excused by the fact that the results of this section will be used only in Chapters 7 and 8, where in most cases one can, possibly with some loss of generality, use results of Section 4 instead.

We fix a domain $Q \subset E_{d+1}$. Let $W^{0,1}_{bv}(Q)$ be the set of all locally summable functions u on Q which have first Sobolev derivatives with respect to x bounded on Q, and such that their generalized derivatives (in the sense of the theory of generalized functions) of the form u_t and u_{xx} are measures of bounded variation on Q. We denote by $W^{0,1}_{bv,loc}(Q)$ the set of functions each of which coincides with a function of $W^{0,1}_{bv}(Q)$ on every bounded domain $Q_1 \subset \bar{Q}_1 \subset Q$. Let $W^{bv,+}_{loc}(Q)$ be the set of all funcitons on Q which belong to $W^{bv,+}(z + C_{\rho^2,\rho})$ (see Section 6) for any z and ρ such that $z + \bar{C}_{\rho^2,\rho} \subset Q$. We denote $\tilde{W}^{bv,+}_{loc}(Q) = W^{bv,+}_{loc}(Q) + W^{1,2}_{d+1,loc}(Q)$. It is clear that $\tilde{W}^{bv,+}_{loc}(Q) \subset W^{0,1}_{bv,loc}(Q)$.

Further, let Ω be a separable metric space and let $d_1 \geq 1$ be an integer. For all $\omega \in \Omega$ and $(t, x) \in Q$, let the following objects be defined: a $d \times d_1$-matrix $\sigma(\omega, t, x)$, a d-dimensional vector $b(\omega, t, x)$, and real-valued functions $r(\omega, t, x) \geq 0$, $c(\omega, t, x)$, and $f(\omega, t, x)$. It is assumed that all the functions are Borel with respect to (ω, t, x), continuous in ω, continuous in x uniformly with respect to ω for any t, and bounded on every set $\Omega \times Q_1$, where Q_1 is a bounded domain and $Q_1 \subset \bar{Q} \subset Q$. Moreover, we assume that on every such set there exist bounded Sobolev derivatives r_x, r_{xx}, σ_x, b_x, and a_{xx}, where $a = (1/2)\sigma\sigma^*$. Finally, suppose that c is bounded from above on $\Omega \times Q$.

For $u \in W^{0,1}_{bv,loc}(Q)$, we introduce the measure Fu on Q as in (6.4). In applying results of this section, one should bear in mind that, as noted in Lemma 6.2, if $u \in \tilde{W}^{bv,+}_{loc}(Q)$ and the Sobolev derivative u_t exists, then $Fu = 0$ on Q iff $a^{ij}(\omega) \circ u_{x^ix^j} \ll \Lambda$ for any $\omega \in \Omega$ and $F(u_t, u^{(0)}_{x^ix^j}, u_{x^i}, u, z) = 0$ in Q(a.e), where F is introduced in formula (6.8).

1. THEOREM. Let Q be a bounded domain, $r \equiv 1$, $u_1, u_2 \in C(\bar{Q})$, $u_1 \in \tilde{W}^{bv,+}_{loc}(Q)$, $u_2 \in W^{0,1}_{bv,loc}(Q)$, $Fu_1 \geqq 0 \geqq Fu_2$ on Q. Also, let $(s, x) \in Q$, $u_1 \leqq u_2$ on $\partial' Q(s, x)$. Then $u_1(s, x) \leqq u_2(s, x)$.

As we said above, we shall not prove this theorem. We merely discuss its assertion and deduce some corollaries which will be needed later. Theorem 1 obviously implies the following.

2. COROLLARY. If in addition to the assumptions of Theorem 1 we have $u_2 \in \tilde{W}^{bv,+}_{loc}(Q)$, $Fu_1 = Fu_2$, and $u_1 = u_2$ on $\partial' Q(s, x)$, then $u_1(s, x) = u_2(s, x)$.

In this corollary, the space $\tilde{W}^{bv,+}_{loc}(Q)$ cannot be replaced by $W^{0,1}_{bv,loc}(Q)$. This follows from the example.

3. <u>Example</u> (see Conway and Hopf [1]). Let $d = 1$, $\Omega = [-2, 2]$, $\sigma = c = 0$, $b = 2\omega$, $f = -\omega^2$, and $Q = C_{1,1}$. It is easy to see that the functions $u_1 := 0$, $u_2 := (1 - t - |x|)_+$ coincide on $\partial' Q$ and $Fu_1 = Fu_2 = 0$ in Q. Moreover, $u_2 \in W^{0,1}_{bv,loc}(Q)$ and $u_1(0, 0) = 0 \neq 1 = u_2(0, 0)$.

We now extend Theorem 1 to the case of arbitrary r. For $\omega \in \Omega$ and $u \in W^{0,1}_{bv,loc}(Q)$, we denote

$$L(\omega)u = r(\omega)\circ u_t + a^{ij}(\omega)\circ u_{x^ix^j} + \left(b^i(\omega)u_{x^i} + c(\omega)u\right)\circ\Lambda,$$

$$L_0(\omega)u = a^{ij}(\omega)\circ u_{x^ix^j} + \left(b^i(\omega)u_{x^i} + c(\omega)u\right)\circ\Lambda.$$

4. THEOREM. Let Q be a bounded domain and let $\varepsilon_0 > 0$, $\psi \geqq 0$ be a function,

$$u_1,\ (-\psi) \in \tilde{W}^{bv,+}_{loc}(Q), \quad u_2 \in W^{0,1}_{bv,loc}(Q),\ \psi,\ u_1,\ u_2 \in C(\bar{Q}),$$

$u_{1t} \geqq L(\omega)\psi + \varepsilon\psi_t$, $-u_{2t} \geqq L(\omega)\psi + \varepsilon\psi_t$ in Q for any $\omega \in \Omega$, $\varepsilon \in (0, \varepsilon_0)$ and $Fu_1 \geqq 0 \geqq Fu_2$ on Q. Also, let $(s, x) \in Q$ and $u_1 \leqq u_2$ on $\partial' Q(s, x)$. Then $u_1(s, x) \leqq u_2(s, x)$.

<u>Proof</u>. For $\varepsilon \in (0, \varepsilon_0)$ we obtain

$$0 \leqq \bigvee_{\omega\in\Omega}\left[L(\omega)(u_1 - \varepsilon\psi) + f(\omega)\circ\Lambda + \varepsilon(u_1 - \varepsilon\psi)_t\right],$$

$$0 \leqq \bigvee_{\omega\in\Omega}\left[(u_1 - \varepsilon\psi)_t + \frac{1}{r(\omega)+\varepsilon}\circ L_0(\omega)(u_1 - \varepsilon\psi) + \frac{1}{r(\omega)+\varepsilon}f(\omega)\circ\Lambda\right], \tag{1}$$

$$0 \geqq \bigvee_{\omega\in\Omega}\Big[(u_2 + \varepsilon\psi)_t + \frac{1}{r(\omega)+\varepsilon} \circ L_0(\omega)(u_2 + \varepsilon\psi) + \\ + \frac{1}{r(\omega)+\varepsilon} f(\omega) \circ \Lambda\Big]. \tag{2}$$

on Q.

By Theorem 1, from the latter two expressions we obtain $(u_1 - \varepsilon\psi)(s, x) \leqq (u_2 + \varepsilon\psi)(s, x)$. Since $\varepsilon \in (0, \varepsilon_0)$ is arbitrary, the proof is complete.

5. Remark. The necessity of using the auxiliary function ψ can be seen when $r = c = f = 0$, $a = 0$, and $b = 0$.

6. COROLLARY. Let Q, ε_0, and ψ satisfy the conditions of Theorem 4, let $u_1, u_2 \in C(\bar{Q}) \cap \tilde{W}^{bv,+}_{loc}(Q)$, let $-|u_{it}| \geqq L(\omega)\psi + \varepsilon\psi_t$, $i = 1, 2$, on Q for any $\omega \in \Omega$, and let $\varepsilon \in (0, \varepsilon_0)$ and $Fu_1 = Fu_2 = 0$ on Q. Also, let $(s, x) \in Q$ and $u_1 = u_2$ on $\partial' Q(s, x)$. Then $u_1(s, x) = u_2(s, x)$.

We now turn to the case of an unbounded domain.

7. THEOREM. Let Q be an unbounded domain and $\varepsilon_0 > 0$, let $\psi \geqq 0$ be a function continuous on $\bar{Q}$ such that $(-\psi) \in \tilde{W}^{bv,+}_{loc}(Q)$, let u_1 and u_2 be functions which are continuous on $\bar{Q}$ and such that $u_1 \in \tilde{W}^{bv,+}_{loc}(Q)$ and $u_2 \in W^{0,1}_{bv,loc}(Q)$, and let $u_{1t} \geqq L(\omega)\psi + \varepsilon\psi_t$, $-u_{2t} \geqq L(\omega)\psi + \varepsilon\psi_t$ on Q for any $\omega \in \Omega$ and $\varepsilon \in (0, \varepsilon_0)$, $Fu_1 \geqq 0 \geqq Fu$ on Q. Also, let $(s, x) \in Q$, let $u_1 \leqq u_2$ on $\partial' Q(s, x)$, and let

$$\lim \frac{|u_i(t, y)|}{\psi(t, y)} = 0 \quad \left(\frac{0}{0} = 0\right), \quad i = 1, 2, \tag{3}$$

where the limit is computed as $t + |y| \to \infty$ with $t \geqq s$, and $(t, y) \in Q$. (If there is no such set of (t, y), then the limit is assumed to be zero.) Then $u_1(s, x) \leqq u_2(s, x)$.

Proof. Consider an $\varepsilon \in (0, \varepsilon_0)$ and note that $u_1 - \varepsilon\psi \leqq u_2 + \varepsilon\psi$ on $Q(s, x) \setminus \{s < t < s + R, |y - x| < R\}$ for sufficiently large R. The inequality $u_1 - \varepsilon\psi \leqq u_2 + \varepsilon\psi$, therefore holds on a parabolic boundary of the domain $Q(s, x) \cap \{s < t < s + R, |y - x| < R\}$ for all sufficiently large R. Thus, by Theorem 1 applied to this domain, from (1) and (2) we obtain $(u_1 - \varepsilon\psi)(s + \delta, x) \leqq (u_2 + \varepsilon\psi)(s + \delta, x)$ for sufficiently small $\delta > 0$. It remains to let $\delta \downarrow 0$ and $\varepsilon \downarrow 0$. This proves the theorem.

8. COROLLARY. Let Q, ε_0, and ψ satisfy the conditions of Theorem 7, let functions u_1 and u_2 be continuous in $\bar{Q}$ and belong to $\tilde{W}^{bv,+}_{loc}(Q)$, let $-|u_{it}| \geq L(\omega)\psi + \varepsilon\psi_t$, $i = 1, 2$ on Q for any $\omega \in \Omega$ and $\varepsilon \in (0, \varepsilon_0)$, and let $Fu_1 = Fu_2 = 0$ on Q. Also, let $(s, x) \in Q$, let $u_1 = u_2$ on $\partial'Q(s, x)$, and let condition (3) hold. Then $u_1(s, x) = u_2(s, x)$.

We now formulate one more extension of Theorem 1 to unbounded domains.

9. THEOREM. Let Q be an unbounded domain and let $r \equiv 1$. Let $\psi \geq 0$ be a function which is continuous on Q and such that $(-\psi) \in \tilde{W}^{bv,+}_{loc}(Q)$ and $L(\omega)\psi \leq 0$ on Q for all $\omega \in \Omega$. Let functions u_1 and u_2 be continuous on $\bar{Q}$, $u_1 \in \tilde{W}^{bv,+}_{loc}(Q)$, and $u_2 \in W^{0,1}_{bv,loc}(Q)$, and let $Fu_1 \geq 0 \geq Fu_2$ on Q. Also, let $(s, x) \in Q$, let $u_1 \leq u_2$ on $\partial'Q(s, x)$, and let condition (3) hold. Then $u_1(s, x) \leq u_2(s, x)$.

This theorem can be proved by reiterating the proof of Theorem 7 and using the obvious inequality $(Fu_1 - \varepsilon\psi) \geq 0 \geq F(u_2 + \varepsilon\psi)$ which holds true for any $\varepsilon > 0$.

10. COROLLARY. If in addition to the assumptions of Theorem 9 we have $u_2 \in \tilde{W}^{bv,+}_{loc}(Q)$, $Fu_1 = Fu_2 = 0$ on Q, and $u_1 = u_2$ on $\partial'Q(s, x)$, then $u_1(s, x) = u_2(s, x)$.

11. Remark. We present a specific case when the function ψ satisfying the conditions of Theorem 9 can easily be found. Let $T > 0$, and let $Q = (0, T) \times E_d$, and let there exist constants K, $m \geq 0$ such that

$$\operatorname{tr} a + |b|^2 \leq K(1 + |x|)^2, \quad |u_1| \leq K(1 + |x|)^m$$

on $\Omega \times Q$.

The reader can easily check that in Theorem 9 one can set $\psi = (1 + |x|^2)^n \exp N(T - t)$ with sufficiently large constants N and n.

3.8. Comments

Sections 1 and 2. These sections are based on the papers Krylov [7,8]. Inequality (2.3) is called the Hadamard inequality.

Section 3. The results of this section are taken from Krylov [11] where they are proved by probabilistic methods. The proofs presented in this book seem to be original, although they use many elements of Krylov's book [10] (see Chapter 2). There is an example showing that the assertion of Lemma 2 is not true if ε is permitted to be strictly greater than R^{-1} in (1). The Aleksandrov estimates for parabolic equations with $|b| \in L_{d+1}(Q)$ can be found in recent papers of Nazarov and Ural'tseva [1], Krylov [26].

Section 4. This section is based on the results of Krylov [11]. A theorem like Theorem 2 was first proved in Krylov [3], and Examples 8 and 9 are taken from the same paper. In connection with the maximum principle for linear equations, it is worth mentioning the Aleksandrov serial research into the principle (see Aleksandrov [3]). Bony [1] proved Theorem 10 for $u \in W_p^2(D)$ with $p > d$.

Section 5. The significance of the theorems on passage to the limit in the theory of controlled diffusion processes has long been realized (see Krylov [3, 5, 10, 12]). Theorem 9 for elliptic operators is presented in Krylov [3] with somewhat different notation. Prior to Evans [1] such theorems were proved by probabilistic methods. (It should be said that the probabilistic proofs of such theorems and Theorem 6.3 in their idea are very similar to the well-known proof of the fact that the limit of harmonic functions is a harmonic function.) Using the theory of accretive operators for elliptic operators, Evans [1] proved a theorem which is similar to the corresponding counterpart of Theorem 12 under the assumption that u^n weakly converges to u in $W_p^2(D)$ and $p > d$ (cf. Remark 10). The proofs in this book are essentially close both to the Evans argument and to the probabilistic technique. The notion of positive operator is taken from Krasnosel'skii et al. [1]. A proposition like Corollary 14 was used as a basis for the theory of controlled diffusion processes (see Krylov [4]).

Section 6. This section presents alternate proofs of the author's results (see Krylov [13, 16]) under alleviated assumptions of Theorem 3 that u^n are uniformly bounded and converge to v almost everywhere rather than $u^n \to v$ uniformly on $C_{T,R}$. Moreover, the continuity condition on v and u^n can be omitted.

Section 7. In Kruzhkov [2], the methods of the theory of differential equations are used to prove the uniqueness of the solution to a first-order nonlinear equation in the class of convex functions. The author does not know whether his argument can be extended to the case of second-order equations. Lions [3] presents a result concerning elliptic equations in the whole space, which in a way extends Corollary 2. It is worth mentioning that some details of his argument (see Lions [2], 3.3) seem to be insufficiently justified. We should also mention Pragarauskas [2], where the uniqueness of a solution to a nonlinear integro-differential equation is examined by probabilistic methods.

Chapter 4

A PRIORI ESTIMATES IN C^{α} FOR SOLUTIONS OF LINEAR AND NONLINEAR EQUATIONS

The aim of this chapter is to prepare the derivation of estimates in $C^{2+\alpha}$ for solutions of nonlinear equations. We begin with the Krylov–Safonov estimates for linear equations, set out in Sections 1 and 2. We then study the properties of simultaneous linear inequalities which arise in differentiation of nonlinear equations along different directions. The last section prepares the proof of the Hölder condition for derivatives of a solution on the domain boundary.

4.1. Research into properties of two special functions

Let us fix constants K and $\varepsilon > 0$ and denote by A the set of all parabolic operators L of the form

$$L = \frac{\partial}{\partial t} + a^{ij}(t, x) \frac{\partial^2}{\partial x^i \partial x^j} + b^i(t, x) \frac{\partial}{\partial x^i} + c(t, x), \tag{1}$$

whose coefficients are defined on E_{d+1}, measurable with respect to (t, x), and satisfy the conditions

$$a^{ij} = a^{ji}, \quad \varepsilon|\lambda|^2 \leqq a^{rs}\lambda^r\lambda^s \leqq K|\lambda|^2, \quad |b^i| \leqq K,$$
$$0 \leqq -c \leqq K \tag{2}$$

for all $(t, x) \in E_{d+1}$, $\lambda \in E_d$, $i, j = 1, \ldots, d$.

Let A_0 be a subset of A consisting of all operators with infinitely differentiable coefficients. For $q \in [0, 1]$, we denote by $\mathbf{A}^m_q$ the set of all functions $u \in C^2(C_{1,1})$ each of which equals zero on $\partial' C_{1,1}$ and corresponds to an operator $L \in A_0$ such that $Lu \leqq 0$ on $C_{1,1}$ and $\Lambda(C_{1,1} \cap \{Lu \leqq -1\}) \geqq q\Lambda(C_{1,1})$. We now define one of the two functions that will be considered in this section, as

$$m(t, x, q) := \inf(u(t, x): u \in \mathbf{A}^m_q), \quad (t, x) \in C_{1,1},$$
$$q \in [0, 1]. \tag{3}$$

Note that if $u \in \mathbf{A}^m_q$, then $u \geqq 0$ by the maximum principle and therefore $m \geqq 0$. One more property, which shows that the infimum in (3) can be calculated with respect to a wider set with no change in m, is formulated in the following lemma.

1. LEMMA. Let $q \in [0,1)$, let $u \in W^{1,2}_{d+1}(C_{1,1})$ and $u \geqq 0$ on $\partial' C_{1,1}$, let $L \in A$ and $Lu \leqq 0$ on $C_{1,1}$(a.e.), and let $\Lambda(C_{1,1} \cap \{Lu \leqq -1\}) \geqq q\Lambda(C_{1,1})$. Then $u(t,x) \geqq m(t,x,q)$ on $\bar{C}_{1,1}$,

Proof. We first suppose that $L \in A_0$. Denote $\Gamma = C_{1,1} \cap \{Lu \leqq -1\}$ and select a sequence of closed sets $\Gamma_n \subset C_{1,1}$ such that $\Lambda(\Gamma_n) \geqq q\Lambda(C_{1,1})$ and $\chi_{\Gamma_n} \to \chi_{\Gamma}$ on $C_{1,1}$ (a.e). To construct Γ_n one can use, for example, the regularity of the Lebesgue measure and first construct closed sets $\Delta_n \subset \Gamma$ such that $\chi_{\Delta_n} \to \chi_{\Gamma}$ (a.e.), and if $\Lambda(\Gamma) > q\Lambda(C_{1,1})$, then for sufficiently large n, one can take $\Gamma_n = \Delta_n$. But if $\Lambda(\Gamma) = q\Lambda(C_{1,1})$, then $\Lambda(C_{1,1} \setminus \Gamma) = (1-q)\Lambda(C_{1,1}) > 0$, and for all sufficiently large n, one can find closed sets δ_n in $C_{1,1} \setminus \Gamma$ such that $\chi_{\delta_n} \to 0$ (a.e.) and $\Lambda(\delta_n) \geqq \Lambda(\Gamma \setminus \Delta_n)$. Furthermore, setting $\Gamma_n = \Delta_n \cup \delta_n$ for sufficiently large n, we obtain $\Lambda(\Gamma_n) = \Lambda(\Gamma) - \Lambda(\Gamma \setminus \Delta_n) + \Lambda(\delta_n) \geqq q\Lambda(C_{1,1})$.

On constructing Γ_n and noting that the characteristic function of any closed set in $C_{1,1}$ can be represented as the limit of a decreasing sequence of functions of the class $C_0^{\infty}(C_{1,1})$, we can easily find the sequence of $\zeta_n \in C_0^{\infty}(C_{1,1})$ such that $\Lambda\{\zeta_n \geqq 1\} \geqq q\Lambda(C_{1,1})$, $\zeta_n \to \chi_{\Gamma}$ (a.e.), and ζ_n are uniformly bounded.

Now, let v and v_n be $\overset{0}{W}{}^{1,2}_{d+1}(C_{1,1})$-solutions to the equations $Lv = -\chi_{\Gamma}$ and $Lv_n = -\zeta_n$, respectively. It is clear that $v_n \in C^{\infty}(\bar{C}_{1,1})$ and $v_n \in \mathbf{A}^m_q$. By the maximum principle and Theorem 3.3.9, we obtain $u \geqq v$ and $v - v_n \to 0$. Consequently,

$$u(t,x) \geqq v_n(t,x) + [v(t,x) - v_n(t,x)] \geqq m(t,x,q) + [v(t,x) - -v_n(t,x)].$$

Letting $n \to \infty$, we obtain the assertion of the lemma in the specific case under consideration.

In the general case, we take a sequence of operators $L_n \in A_0$ such that the coefficients in L_n converge almost everywhere to corresponding coefficients of the operator L. Let $Lu = f$, $L_n u = f_n$, and let u_n and w_n be $\overset{0}{W}{}^{1,2}_{d+1}(C_{1,1})$-solutions to the equations $L_n u_n = f_n$ and $L_n w_n = f - f_n$. It is obvious that $L_n(u+w_n) = L_n(u_n+w_n) = f$. Hence, by the maximum principle and the result established for the specific case above, we obtain $u + w_n \geqq u_n + w_n \geqq m(q)$. It remains to be said that $w_n \to 0$ by Theorem 3.3.9 since $f_n \to f$ in $\mathcal{L}_{d+1}(C_{1,1})$. The proof is complete.

A basic property of the function m is formulated in the following theorem.

2. THEOREM. Let $q \in (0,1]$ and $\kappa \in [0,1)$. Then there exists a constant $m_0 > 0$ which depends only on d, K, ε, q, and κ and such that $m(t,x,q) \geqq m_0$ for $|x| \leqq \kappa$ and $0 \leqq t \leqq \kappa q$.

To prove Theorem 2, we need some auxiliary propositions. We begin with the 'lemma on an oblique cylinder'.

3. LEMMA. Let $\kappa \in (0,1]$, and let Q lying in $C_{8,2}$ be a cylinder whose bases are balls in E_d of the same radius R such that one base lies in the hyperplane $\{(t,x) : t = t_1\}$ and the other in the hyperplane $\{(t,x) : t=t_2\}$ and $t_1 < t_2$. Suppose that the tangent of the angle beween the axis of Q and the t-axis does not exceed κ^{-1} and $\kappa \leqq R^2(t_2-t_1)^{-1} \leqq \kappa^{-1}$. Let $L \in A$, let $u \in W^{1,2}_{d+1}(Q)$, $u \geqq 0$, and $Lu \leqq 0$ (a.e.Q), and for a $\gamma \in [0,1]$ let $u(t_2,x) \geqq 1$ with x belonging to the ball of radius γR whose centre coincides with that of the base of Q lying in $\{t=t_2\}$. Then there exist constants $\delta > 0$ and $n \geqq 0$ which depend only on d, K, ε, and κ and such that $u(t_1,x) \geqq \delta\gamma^n$ for x belonging to the ball of radius $(1-\kappa)R$ whose centre coincides with that of the base of Q lying in the plane $\{t=t_1\}$.

Proof. If in Q we have

$$Lu = u_t + a^{ij}u_{x^ix^j} + b^iu_{x^i} + cu,$$

and $\bar{b} \in E_d$, then for $\tilde{u}(t,x) := u(t,x + \bar{b}t)$ we obtain

$$\tilde{L}\tilde{u}(t,x) = (Lu)(t,x + \bar{b}t) \text{ in } \tilde{Q} := \{(t,x) : (t,x + \bar{b}t) \in Q\},$$

where the operator $\tilde{L}$ is of the form (1) and its coefficients $\tilde{a}$, $\tilde{b}$, and $\tilde{c}$ are determined from the formulas

$$\tilde{a}^{ij}(t,x) = a^{ij}(t,x + \bar{b}t),$$

$$\tilde{b}^i(t,x) = b^i(t,x + \bar{b}t) - \bar{b}^i, \quad \tilde{c}(t,x) = c(t,x + \bar{b}t).$$

For an appropriate $\bar{b}$ the set $\tilde{Q}$ is a cylinder whose axis is parallel to the t-axis and the coefficients of $\tilde{L}$ satisfy conditions (2) if K is replaced by $K + \kappa^{-1}$. Thus, without loss of generality we can assume that the cylinder Q is right.

Furthermore, it is easy to see that, given an appropriately chosen $x_0 \in E_d$, for the function $\tilde{u}(t,x) := u(t_1 + t(t_2 - t_1), x_0 + Rx)$ in $C_{1,1}$ there exists an operator $\tilde{L}$ of the form (1) such that $\tilde{L}\tilde{u}(t,x) = (t_2 - t_1)(Lu)(t_1 + t(t_2 - t_1), x_0 + Rx)$ in $C_{1,1}$ and in view of the conditions $\kappa \leqq R^2(t_2 - t_1)^{-1} \leqq \kappa^{-1}$ and $R \leqq 2$ ($Q \subset C_{8,2}$), the coefficients of $\tilde{L}$ satisfy (2) with constants $\tilde{\varepsilon} > 0$ and $\tilde{K}$ which depends only on ε, K, and κ. It follows that we need to consider only the case $Q = C_{1,1}$.

In this case we have to prove that $u(0,x) \geqq \delta\gamma^n$ for $|x| \leqq 1 - \kappa$, where $\delta > 0$ and $n \geqq 0$ depend only on d, K, ε, and κ. We fix an x_0 such that $|x_0| \leqq 1 - \kappa$ and consider the oblique cylinder $\tilde{Q}$ with bases $x_0 + S_{\kappa}$ and S_{κ} lying in the plane $\{t = 0\}$ and $\{t = 1\}$, respectively. It is obvious that $u \geqq 0$, $Lu \leqq 0$ in $\tilde{Q}$, $u(1,x) \geqq 1$ for $x \in S_{\kappa\wedge\gamma}$, and we have to prove that $u \geqq \delta\gamma^n$ at the centre of the left base.

A transformation like that at the beginning of the proof 'straightens' the cylinder $\tilde{Q}$ and demonstrates that it suffices to check whether $u(0,0) \geqq \delta\gamma^n$ if $u \in W^{1,2}_{d+1}(C_{1,\kappa})$, $u \geqq 0$, $Lu \leqq 0$ on $C_{1,\kappa}$, where $L \in A$, $\gamma \in [0,1]$, $u(1,x) \geqq 1$ for $|x| < \kappa\wedge\gamma$, and $\delta > 0$ and $n \geqq 0$ depend only on d, K, ε, and κ. It is clear that the assumption $\gamma \leqq \frac{1}{2}\kappa$ implies no loss of generality.

Denote

$$\psi = \psi(t,x) = (\xi(1 - t) - |x|^2 + \gamma^2)^2(\xi(1 - t) + \gamma^2)^{-n},$$

$$Q_1 = \{(t,x) : 0 < t < 1, \ \xi(1 - t) - |x|^2 + \gamma^2 > 0\},$$

where $\xi = \kappa^2 - \gamma^2$ and the constant n will be defined later. We obtain $Q_1 \subset C_{1,\kappa}$, $\psi(t,x) = 0$ for $(t,x) \in \partial' Q_1$, $t < 1$ and $\psi(1,x) \leqq \gamma^{4-2n}$ if $(1,x) \in \partial' Q_1$. Thus,

$$\gamma^{2n-4}\psi \leqq u \text{ on } \partial' Q_1.$$

Furthermore, a simple calculation shows that on Q_1 we have

$$(\xi(1 - t) + \gamma^2)^n L\psi = n\xi(\xi(1 - t) + \gamma^2)^{-1}\varphi^2 + \qquad (4)$$
$$+ 8a^{ij}x^i x^j - [2\xi + 4 \operatorname{tr} a + 4b^i x^i - c\varphi]\varphi,$$

where $\varphi = \xi(1 - t) - |x|^2 + \gamma^2$. It is clear that $0 \leqq \varphi \leqq \xi + \gamma^2 = \kappa^2 \leqq 1$

and $|2\xi + 4\,\mathrm{tr}\, a + 4b^i x^i - c\varphi| \leqq N_0 := 2 + (4d + 5)K$ on Q_1. Therefore, from (4) it follows that $L\psi \geqq 0$ on the part of Q_1 where $8\varepsilon|x|^2 \geqq N_0\varphi(t,x)$. Everywhere on Q_1 where $8\varepsilon|x|^2 < N_0\varphi$, we obtain $8\varepsilon(\xi(1 - t) + \gamma^2 - \varphi) < N_0\varphi$, $8\varepsilon(\xi(1 - t) + \gamma^2) < (N_0 + 8\varepsilon)\varphi$, and

$$(\xi(1 - t) + \gamma^2)^{-1}\varphi^2 > \varphi 8\varepsilon(N_0 + 8\varepsilon)^{-1}.$$

Since $\xi \geqq (3/4)\kappa^2$, on the same part of Q_1 with $n = (4/3)\kappa^{-2}N_0(N_0 + 8\varepsilon)(8\varepsilon)^{-1}$ we also obtain from (4) that $L\psi \geqq 0$. Thus, $L\psi \geqq 0$ in Q_1. By the maximum principle, $\gamma^{2n-4}\psi \leqq u$ in $\bar{Q}_1$ and

$$u(0,0) \geqq \gamma^{2n-4}\varphi(0,0) = \gamma^{2n-4}\frac{\kappa^4}{\kappa^{2n}} = \kappa^{4-2n}\gamma^{2n-4}.$$

This proves the lemma.

4. LEMMA. Let $\kappa \in [0,1)$. Then there exist $q_0 \in (0,1)$ and $m_0 > 0$ which depend only on K, ε, d, and κ and such that $m(0,x,q) \geqq m_0$ for $|x| \leqq \kappa$ and $q \in [q_0,1]$.

Proof. Let $q \in [0,1]$ and let $u \in \mathfrak{A}^m_q$. Consider an operator $L \in A_0$ associated with q and u and denote $\Gamma = C_{1,1} \cap \{Lu \geqq -1\}$. Let v and w be $\overset{0}{W}{}^{1,2}_{d+1}(C_{1,1})$-solutions of the equations $Lv = -\chi_\Gamma$ and $Lw = -1$, respectively. By the maximum principle, the obvious inequality $L(w-v) \geqq Lu$ implies $u \geqq w-v$. By Theorem 3.3.9, we have

$$|v| \leqq N_1\Lambda^{1/(d+1)}(\Gamma) \leqq N_2(1 - q)^{1/(d+1)},$$

where the constants N_1 and N_2 depend only on d, K, and ε. We now estimate w from below.

Let $\varphi = 1 - |x|^2$ and let ψ be a $\overset{0}{W}{}^{1,2}_{d+1}(C_{1,1})$-solution of the equation $L\psi = L\varphi$.

From the relationships $L(\varphi-\psi) = 0$ and $0 \leqq \varphi - \psi \leqq 1$ on $\partial' C_{1,1}$ it follows, by the maximum principle, that $0 \leqq \varphi - \psi \leqq 1$ in $\bar{C}_{1,1}$. Hence, $L[(1-t)(\varphi-\psi)] = -(\varphi-\psi) + (1-t)L(\varphi-\psi) \geqq -1$ and, by the maximum principle, $(1-t)(\varphi-\psi) \leqq w$. Finally, $L(\varphi-\psi) = 0$ and $\varphi-\psi = \varphi \geqq \frac{1}{2}$ for $t = 1$ and $|x|^2 \leqq \frac{1}{2}$. By Lemma 3, there exists a $\delta > 0$ which depends only on d, K, ε, and κ and such that $(\varphi-\psi)(0,x) \geqq \delta$ whenever $|x| \leqq \kappa$. Thus,

$$u(0,x) \geqq \delta - N(1-q)^{1/(d+1)} \tag{5}$$

for $|x| \leqq \kappa$ and any $q \in [0,1]$. One can choose q_0 so that the right-hand side of (5) be greater than $\frac{1}{2}\delta$ for $q \geqq q_0$. It remains to take the infimum with respect to $u \in \mathbb{A}_q^m$ in (5). This proves the lemma.

The following lemma asserts that 'the permanent gain on the small scale yields a gain on the whole'.

5. LEMMA. Let $L \in A_0$, let Γ be an open subset of $C_{1,1}$, and let $f, g \in L_{d+1}(C_{1,1})$, f and g be nonpositive on $C_{1,1}$, and $g = 0$ outside Γ. Denote by u and v the $\overset{0}{W}{}^{1,2}_{d+1}(C_{1,1})$-solutions of the equations $Lu = f$ and $Lv = g$. Suppose that for any point $(t_0, x_0) \in \Gamma$ there exist numbers $\tau, \rho > 0$ and a point $(t_1, x_1) \in C_{1,1}$ such that $(t_0, x_0) \in (t_1, x_1) + \bar{C}_{\tau,\rho} \subset \bar{C}_{1,1}$ and u^1 and v^1, which are $\overset{0}{W}{}^{1,2}_{d+1}((t_1, x_1) + C_{\tau,\rho})$-solutions of the equations $Lu^1 = f$ and $Lv^1 = g$ on $(t_1, x_1) + C_{\tau,\rho}$, satisfy the inequalities $u^1(t_0, x_0) \geqq v^1(t_0, x_0)$ and $u^1(t_0, x_0) > 0$. Then $u \geqq v$ on $\bar{C}_{1,1}$.

Proof. If for fixed L, Γ, and f the hypothesis of the lemma is true for a function g, it is also true for $g\chi_{\Gamma^1}$, where Γ^1 is any measurable subset of $C_{1,1}$. This follows directly from the maximum principle. By Theorem 3.3.9, $\overset{0}{W}{}^{1,2}_{d+1}(C_{1,1})$-solutions of equations $Lv_n = g\chi_{\Gamma^n}$ converge to v if $\Gamma^n \uparrow \Gamma$. Therefore, to prove the lemma, it suffices to consider the only case when $g = 0$ outside a closed subset Γ^1 of Γ.

In this case it suffices to prove that $v \leqq (1+\delta)u$ on $\bar{C}_{1,1}$ for any $\delta > 0$. Suppose that this is not the case for a $\delta > 0$. Then $\gamma := \max(v - (1+\delta)u, \bar{C}_{1,1}) > 0$. By the maximum principle applied in the domain $C_{1,1} \setminus \Gamma^1$ it follows that this maximum is attained at a point $(t_0, x_0) \in \Gamma^1$.

We take appropriate $\tau, \rho > 0$ and (t_1, x_1) and note that $L(u-u^1) = L(v-v^1) = 0$ on $(t_1, x_1) + C_{\tau,\rho}$ and $v-v^1 = v \leqq \gamma + (1+\delta)u = \gamma + (1+\delta)(u-u^1)$ on $(t_1, x_1) + \partial' C_{\tau,\rho}$. By the maximum principle, $(L\gamma \leqq 0)$ $v-v^1 \leqq \gamma + (1+\delta)(u-u^1)$ at the point (t_0, x_0). Therefore, at this point

$$\gamma = v - (1 + \delta)u \leqq \gamma + v^1 - (1 + \delta)u^1 < \gamma + v^1 - u^1 \leqq \gamma.$$

We have come to the false inequality $\gamma < \gamma$, so the lemma is proved.

The following lemma on 'running ink spots' relates to measure theory and we shall prove it in Appendix 1. Before formulating it, we introduce some notions. Let Γ be a measurable subset of $C_{1,1}$ and let q,

η, $\zeta \in (0,1)$. We denote by $\mathbf{B}$ the collection of all sets Q of the form $(t_0, x_0) + C_{\rho^2,\rho}$ such that $Q \subset C_{1,1}$ and $\Lambda(Q \cap \Gamma) \geq q\Lambda(Q)$. If $Q = (t_0, x_0) + C_{\rho^2,\rho} \in \mathbf{B}$, then let

$$Q^1 = (t_0, x_0) - C_{\eta^{-1}\rho^2,\rho}, \quad Q^2 = (t_0 - \eta^{-1}\rho^2, x_0) + C_{\eta^{-1}\rho^2\zeta^2,\rho\zeta}.$$

Note that Q^1 is immediately adjacent to Q (on the left along the t-axis), the two cylinders have a common base, and along the t-axis Q^1 is longer than Q by a factor of η^{-1}. It is quite possible that Q^1 comes out of $C_{1,1}$. The cylinder Q^2 is obtained from Q^1 by compression centred at the centre of the left base of Q^1, the compression coefficient being ζ^{-2} for the t-axis and ζ^{-1} for the spatial axes. Finally, denote

$$\Gamma^2 = \bigcup_{Q\in\mathbf{B}} Q^2.$$

Since $Q \in \mathbf{B}$ and Q^2 are open cylinders, the set Γ^2 is open and measurable.

6. LEMMA. If $\Lambda(\Gamma) \leqq q\Lambda(C_{1,1})$, then

$$\Lambda(\Gamma^2) \geqq \left(1 - \frac{1-q}{3^{d+1}}\right)^{-1} (1+\eta)^{-1}\zeta^{d+2}\Lambda(\Gamma).$$

7. <u>Proof of Theorem</u> 2. We fix a $\kappa \in (0,1)$ and first prove that

$$\mu(q) := \inf\{m(0, x, q): |x| \leqq \kappa\} > 0 \tag{6}$$

for all $q \in (0,1]$. Since the sets $\mathbf{A}^m_q$ are obviously nested, $m(t, x, q)$ and $\mu(q)$ are increasing functions of q.

Let $\bar{q}$ be the supremum of those q's which satisfy the equation $\mu(q) = 0$. It is obvious that $\mu(q) = 0$ for $q \in [0,\bar{q})$ if $\bar{q} > 0$. We have to prove that $\bar{q} = 0$. Assume on the contrary that $\bar{q} > 0$. Since, by Lemma 4, $\mu(q_0^2) > 0$ for a $q_0 \in (0,1)$, one can choose numbers q_1 and q_2 as close to $\bar{q}$ as desired so that $0 < q_1 < \bar{q} < q_2 \leqq 1$, $\mu(q_1) = 0$ and $\mu(q_2) > 0$. In what follows we will show how q_1 and q_2 should be chosen to yield a contradiction with the equation $\mu(q_1) = 0$. Here we merely fix some q_1 and q_2 and some η, $\zeta \in (0,1)$.

Let $u \in \mathbf{A}^m_{q_1}$ and let $\Gamma = C_{1,1} \cap \{Lu \leqq -1\}$, where L is the operator associated with u. If $\Lambda(\Gamma) \geqq q_0\Lambda(C_{1,1})$, then

$$u(0, x) \geq m(0, x, q_0) \geq \mu(q_0) \geq \mu(q_0^2) > 0 \text{ for } |x| \leq \kappa \tag{7}$$

We show how $u(0, x)$ can be estimated from below if $\Lambda(\Gamma) < q_0\Lambda(C_{1,1})$, From chosen Γ, q_0, η, and ζ we construct a set Γ^2 in the same manner as prior to Lemma 6 and consider two cases:

(i) $\Lambda(\Gamma^2 \setminus C_{1,1}) \leq (q_2 - \bar{q})\Lambda(C_{1,1})$,

(ii) $\Lambda(\Gamma^2 \setminus C_{1,1}) > (q_2 - \bar{q})\Lambda(C_{1,1})$.

In the first case we denote by $\tilde{u}$ and $\tilde{v}$ the $\overset{0}{W}{}^{1,2}_{d+1}(C_{1,1})$-solutions of the equations $L\tilde{u} = -\chi_{\Gamma}$ and $L\tilde{v} = -\chi_{\Gamma^0}$, where $\Gamma^0 = \Gamma^2 \cap C_{1,1}$. By Lemma 6,

$$q_1\Lambda(C_{1,1}) \leq \Lambda(\Gamma) \leq \left(1 - \frac{1-q_0}{3^{d+1}}\right)(1 + \eta)\zeta^{-d-2}\Lambda(\Gamma^2).$$

Moreover, by hypothesis, $\Lambda(\Gamma^2) = \Lambda(\Gamma^2 \setminus C_{1,1}) + \Lambda(\Gamma^0) \leq (q_2 - \bar{q})\Lambda(C_{1,1}) + \Lambda(\Gamma^0)$. Therefore, if q_1, q_2, η, and ζ are such that

$$q_2 = q_1\left(1 - \frac{1-q_0}{3^{d+1}}\right)^{-1} (1 + \eta)^{-1}\zeta^{d+2} - (q_2 - \bar{q}), \tag{8}$$

then $q_2\Lambda(C_{1,1}) \leq \Lambda(\Gamma^0)$ and, by definition, $\tilde{v}(0, x) \geq m(0, x, q_2) \geq \mu(q_2) > 0$ for $|x| \leq \kappa$.

We have estimated $\tilde{v}$ from below. By the maximum principle, $u \geq \tilde{u}$ and it follows that, to estimate u from below, it suffices to estimate $\tilde{u}$ from below in terms of $\tilde{v}$. This will be done by use of Lemma 5 where f, g, and Γ are replaced by $(-N_0\chi_{\Gamma})$, $-\chi_{\Gamma^0}$, and Γ^0, respectively and N_0 will be defined later. We now check whether the conditions of Lemma 5 are satisfied.

If $(t_0, x_0) \in \Gamma^0$, then there exists a cylinder $Q \in \mathbf{B}$ such that $(t_0, x_0) \in Q^2$. Let $(t_1, x_1) + C_{\tau,\rho}$ be the cylinder $Q^3 = (Q^1 \cup Q) \cap C_{1,1}$ and let u^1 and v^1 be $\overset{0}{W}{}^{1,2}_{d+1}(Q^3)$-solutions of the equations $Lu^1 = -N_0\chi_{\Gamma}$ and $Lv^1 = -\chi_{\Gamma^0}$ on Q^3.

By the choice of Q, we obtain $q_0\Lambda(Q) \leq \Lambda(\Gamma \cap Q)$. This inequality is preserved under the elongation $x \to x\rho^{-1}$, $t \to t\rho^{-2}$, which transforms Q into a cylinder that is congruent to $C_{1,1}$. Under this transformation the equation $Lu^1 = -N_0\chi_{\Gamma}$ becomes, as can easily be seen, $\tilde{L}\tilde{u}^1 = -N_0\rho^2\chi_{\tilde{\Gamma}}$, where $\tilde{L} \in A_0$. It follows that if the points (t, x) lie in the left base of Q and $|x-x_1| \leqq \kappa\rho^2$, then the inequality $u^1(t, x) \geqq N_0\rho^2 m(0, x, q_0) \geqq \mu(q_0)N_0\rho^2$ holds. Since $(t_0, x_0) \in Q^2$, the distance from (t_0, x_0) to the left base of Q falls into the range $[\rho^2\eta^{-1}(1 - \zeta^2), \rho^2\eta^{-1}]$ and the distance from (t_0, x_0) to the cylindrical surface Q^3 lies in the interval $[\rho(1 - \zeta), \rho]$. By Lemma 3 on an oblique cylinder, $u^1(t_0, x_0) \geqq N_0\rho^2\delta\mu(q_0)$, where $\delta > 0$ and δ depends only on ε, η, ζ, κ, K, and d. On the other hand, $Lt \leqq 1$ if $t \geqq 0$ and, by the maximum principle, $v^1(t, x) \leqq (t_1 + \tau) - t$ on Q^3 and $v^1(t, x) \leqq (1 + \eta^{-1})\rho^2$. Thus, if we take $N_0 = \delta^{-1}\mu^{-1}(q_0)(1 + \eta^{-1})$, then $u^1(t_0, x_0) \geqq v^1(t_0, x_0)$.

By Lemma 5, $N_0\tilde{u} \geqq \tilde{v}$ on $\bar{C}_{1,1}$, hence summarizing the above argument we obtain

$$u(0, x) \geqq \tilde{u}(0, x) \geqq N_0^{-1}\tilde{v}(0, x) \geqq N_0^{-1}\mu(q_2) > 0 \tag{9}$$

for $|x| \leqq \kappa$.

We now consider the second case

$$\Lambda(\Gamma^2 \setminus C_{1,1}) > (q_2 - \bar{q})\Lambda(C_{1,1}).$$

In this case it is obvious that Γ^2 has a nonempty intersection with the set $\{(t, x) : t \leqq \bar{q} - q_2\}$. It follows that there exists a cylinder $Q = (t_0, x_0) + C_{\rho^2,\rho} \in \mathbf{B}$ such that Q^1 intersects this set. Then

$$\rho^2\eta^{-1} \geqq q_2 - \bar{q}, \quad \rho^2 \geqq \eta(q_2 - \bar{q}) =: \rho_0^2. \tag{10}$$

Moreover, $\Lambda(\Gamma \cap Q) \geqq q_0\Lambda(Q)$ and, for $\tilde{Q} = (t_0 + (1 - \alpha^2)\rho^2, x_0) + C_{\alpha^2\rho^2,\alpha\rho}$, where $\alpha = (1 + q_0)^{-1/(d+2)}$, we have

$$\begin{aligned} \Lambda(\Gamma \cap \tilde{Q}) &\geqq \Lambda(\Gamma \cap Q) - \Lambda(Q \setminus \tilde{Q}) \\ &\geqq q_0\Lambda(Q) - \Lambda(Q)(1 - \alpha^{d+2}) = q_0^2\Lambda(\tilde{Q}). \end{aligned}$$

Hence, if (t, x) lies in the left base of $\tilde{Q}$ and $|x - x_0| \leqq \kappa\alpha\rho$, then the inequality $u(t, x) \geqq \mu(q_0^2) > 0$ holds. By virtue of (10), this inequality holds for $|x - x_0| \leqq \kappa\alpha\rho_0$. Since the left base of $\tilde{Q}$ is separated from the plane $\{t = 0\}$ at a distance no less than $(1 - \alpha^2)\rho_0^2$, by Lemma 3 on an oblique cylinder, $u(0, x) \geqq \delta_1\mu(q_0^2)$ for $|x| \leqq \kappa$, where δ_1 depends only on K, ε, d, q_0, q_2, $\bar{q}$, η, and κ.

Thus, from (7), (9) and the latter argument we obtain

$$u(0, x) \geqq \mu(q_0) \wedge N_0^{-1}\mu(q_2) \wedge \delta_1\mu(q_0^2) > 0.$$

in every case whenever $|x| \leqq \kappa$.

Since this is true for any $u \in \mathbf{A}_{q_1}^m$, it follows that $\mu(q_1) > 0$ and this will give the proof of (6) for all $q \in (0, 1]$, if we can show that (8) and the inequalities $0 < q_1 < \bar{q} < q_2 \leqq 1$ can be furnished by appropriate choice of q_1, q_2, η, and ζ. To this end we take some η, $\zeta \in (0,1)$ so that the coefficient β of q_1 in (8) is strictly greater than 1. This can be done since $q_0 \in (0,1)$. We now take $q_1 < \bar{q}$ such that $1 \geqq q_1\beta > \bar{q}$ and determine q_2 from (8). Then $q_2 = \frac{1}{2}(\bar{q} + q_1\beta)$ and $\bar{q} < q_2 \leqq 1$.

To estimate $m(t, x, q)$ with $|x| \leqq \kappa$, $t \geqq 0$, and $t \leqq \kappa q$, note that if $u \in \mathbf{A}_q^m$ and $0 \leqq t_0 \leqq \kappa q$, then the function u, which is extended to be zero outside $C_{1,1}$, belongs to $\overset{0}{W}{}_{d+1}^{1,2}((t_0, 0) + C_{1,1})$ and

$$\begin{aligned}\Lambda(\{Lu \leqq -1\} \cap ((t_0, 0) + C_{1,1})) &\geqq \Lambda(\{Lu \leqq -1\} \cap C_{1,1}) - \\ &- \Lambda(C_{1,1} \setminus ((t_0, 0) + C_{1,1})) \\ &\geqq q\Lambda(C_{1,1}) - t_0\Lambda(C_{1,1}) \geqq (1 - \kappa)q\Lambda(C_{1,1}).\end{aligned}$$

Hence, by Lemma 1 and (6), we obtain $u(t_0, x) \geqq \mu((1 - \kappa)q) > 0$ for $|x| \leqq \kappa$ and $m(t_0, x, q) \geqq \mu((1 - \kappa)q) > 0$ for $0 \leqq t_0 \leqq \kappa q$ and $|x| \leqq \kappa$, which proves the theorem

8. <u>Corollary</u>. Let $L_n \in A$, let $u_n \in W_{d+1}^{1,2}(C_{1,1})$, and let $u_n \geqq 0$ on $C_{1,1}$. Also, let $L_n u_n = f_n \leqq 0$ on $C_{1,1}$(a.e.), $n = 1, 2, \ldots$, let $|x_0| < 1$, and let $u_n(0, x_0) \to 0$ as $n \to \infty$. Then $f_n \to 0$ in measure as $n \to \infty$.

Indeed, for every $\delta > 0$, Lemma 1 implies

$$\delta^{-1} u_n(0, x_0) \geqq m(0, x_0, \Lambda(\{f_n \leqq -\delta\})\Lambda^{-1}(C_{1,1})).$$

By Theorem 2, the right-hand side vanishes only if $\Lambda(\{f_n \leqq -\delta\}) \to 0$.

This corollary substantially strengthens Corollary 3.5.14.

Using the properties of the function m, we now examine a function $p(t, x, q)$ which is defined in the following manner. Let $q \in [0,1)$. We denote by $\mathbf{A}_q^p$ the set of all functions $u \in C^2(\bar{C}_{1,1})$ such that for each of them there exists an operator $L \in A_0$ satisfying the inequality $Lu \leq 0$ on $C_{1,1}$ and, in addition, $\Lambda(C_{1,1} \cap \{u \geq 1\}) \geq q\Lambda(C_{1,1})$ and $u = 0$ on $\partial' C_{1,1}$. Let

$$p(t, x, q) := \inf(u(t, x) : u \in \mathbf{A}_q^p),$$

$$(t, x) \in \bar{C}_{1,1}, \quad q \in [0,1).$$

The basic properties of p are accumulated in the following theorem.

9. THEOREM. If $(t, x) \in \bar{C}_{1,1}$ and $q \in [0,1)$, then $1 \geq p(t, x, q) \geq m(t, x, q)$. Specifically, for any $q \in (0,1)$ and $\kappa \in [0,1)$ there exists a constant $p_0 > 0$ which depends only on d, K, ε, q, and κ and such that $p(t, x, q) \geq p_0$ if $|x| \leq \kappa$ and $0 \leq t \leq \kappa q$. Moreover, for any $\kappa \in [0,1)$

$$\inf(p(t, x, q) : |x| \leq \kappa, \quad 0 \leq t \leq \kappa) \to 1 \tag{12}$$

as $q \uparrow 1$.

Proof. It is more convenient for us to prove the inequality $1 \geq p$ in the following theorem. The reader will see that it is possible to avoid the vicious circle. To prove the inequality $p \geq m$, we take a $u \in \mathbf{A}_q^p$ and denote $\Gamma = \bar{C}_{1,1} \cap \{u \geq 1\}$ and by v a $\overset{0}{W}{}^{1,2}_{d+1}(C_{1,1})$-solution of the equation $Lv = -\chi_\Gamma$, where L is the operator associated with u. Since $Lt \leq 1$, the maximum principle implies $v \leq 1 - t \leq 1$ on $C_{1,1}$. By the maximum principle applied to the domain $Q = C_{1,1} \setminus \Gamma$, it follows that $u \geq v$ on $\bar{Q}$ and clearly on Γ. Thus, $u \geq v$ on $\bar{C}_{1,1}$. By Lemma 1, $v \geq m(q)$ and $u \geq m(q)$, and, since the latter is true for any $u \in \mathbf{A}_q^p$, we otbain $p \geq m$.

We now prove (12). Let $\kappa \in [0,1)$ and $q \in (0,1)$, let (t_0, x_0) be a point such that $|x_0| \leq \kappa$ and $0 \leq t_0 \leq \kappa$, and let $u \in \mathbf{A}_q^p$ be a function and L be the operator associated with this function. Suppose that $u(t_0, x_0) < 1$ and consider the function $v(t, x) = 1 - u(t, x) - 2\sigma^{-1}(t - t_0) - 2\sigma^{-2}|x - x_0|^2$, where $\sigma = 1 - \kappa$.

Let $Q = \{(t, x) : t > t_0, v(t, x) > 0\} \cap C_{1,1}$. Since $u \geq 0$, we obtain that $v(t, x) < 0$ if $t \geq t_0 + \sigma$ or $|x - x_0| \geq \sigma$. It follows that $\partial' Q \subset C_{1,1}$ and $v = 0$ on $\partial' Q$. Moreover, $Lv \geq -N$, where N depends only on K, d, and κ. By Theorem 3.3.9, $v \leq N\Lambda^{1/(d+1)}(Q)$ on Q, where N depends only on K, d, κ, and ε. This inequality is also true on $\bar{Q}$, specifically, at the point (t_0, x_0). Considering that $Q \subset C_{1,1} \setminus \{u \geqq 1\}$, we conclude

$$1 - u(t_0, x_0) \leqq N\Lambda^{\frac{1}{d+1}}(Q),$$

$$u(t_0, x_0) \geqq 1 - N\Lambda^{\frac{1}{d+1}}(C_{1,1})(1 - q)^{\frac{1}{d+1}}.$$

The latter inequality is proved under the assumption that $u(t_0, x_0) < 1$. This inequality is obviously true as well if $u(t_0, x_0) \geqq 1$. It remains to take the infimum with respect to $u \in \mathbf{A}^p_q$, $|x| \leqq \kappa$, and $0 \leqq t_0 \leqq \kappa$, letting $q \uparrow 1$. This proves the theorem.

The following Theorem 10 and Lemma 11 make the function p more convenient for studying the properties of functions of $W^{1,2}_{d+1}$.

10. THEOREM. Let $\kappa \in (0,1]$, let $u \in W^{1,2}_{d+1}(C_{\kappa,1})$, let $L \in A$, let $Lu \leqq 0$ on $C_{\kappa,1}$ (a.e.), and let $u \geqq 0$ on $\partial' C_{\kappa,1}$. Also, let $q \in [0,1)$ and let $\Lambda(C_{\kappa,1} \cap \{u \geqq 1\}) \geqq q\Lambda(C_{\kappa,1})$. Then $u(t, x) \geqq p(t, x, \kappa q)$ on $\bar{C}_{\kappa,1}$.

Proof. Keeping the promise of the previous proof, we first of all note that the theorem assertion on being applied to $u \equiv 1$ yields the inequality $1 \geqq p$.

We further prove the theorem in the case when $L \in A_0$ and $u = 0$ on $\partial' C_{\kappa,1}$. Let $Lu = f$ and let $f_n \in C^{\infty}_0(C_{\kappa,1})$ be a sequence such that $f_n \leqq 0$ and $f_n \to f$ in $\mathcal{L}_{d+1}(C_{\kappa,1})$. Let u_n be a $C^2(C_{1,1})$-solution of the problem $Lu_n = f_n$ in $C_{1,1}$, $u_n = 0$ on $\partial' C_{1,1}$. It is clear that $u_n = 0$ in $C_{1,1} \setminus C_{\kappa,1}$. By Theorem 3.3.9 applied to the domain $C_{\kappa,1}$, we obtain $u_n \to u$ uniformly on $\bar{C}_{\kappa,1}$. Hence, for any $\delta \in (0,1)$ and all sufficiently large n the following inequalities

$$\Lambda(C_{1,1} \cap \{u_n \geqq \delta\}) = \Lambda(C_{\kappa,1} \cap \{u_n \geqq \delta\})$$

$$> \Lambda(C_{\kappa,1} \cap \{u \geqq 1\}) \geqq q\Lambda(C_{\kappa,1}) = \kappa q\Lambda(C_{1,1})$$

hold.

Hence, by definition, $u_n\delta^{-1} \geqq p(\kappa q)$ on $C_{1,1}$. Letting $n \to \infty$ and $\delta \uparrow 1$, we obtain the desired assertion.

We now abandon the assumption $u = 0$ on $\partial' C_{1,1}$. Let $L \in A_0$ and $u \geqq 0$ on $\partial' C_{1,1}$. For $\delta \in (0,1)$, we denote $\Gamma^{\delta} = \{(t, x) \in \bar{C}_{\kappa,1} : u(t, x) \geqq \delta\}$.

It is clear that if δ is sufficiently close to 1 and $u \geqq 1$ everywhere on $C_{\kappa,1}$, or if $u < 1$ at some points of $C_{\kappa,1}$, we obtain

$$\Lambda(\Gamma^{\delta}) > q\Lambda(C_{\kappa,1}). \tag{13}$$

For an appropriately fixed $\delta \in (0,1)$ with $n = 1, 2, \ldots$, we define v_n as $\overset{0}{W}{}^{1,2}_{d+1}(C_{\kappa,1})$-solutions of the equations $Lv_n + n\chi_{\Gamma\delta}\cdot(1 - v_n) = 0$ on $C_{\kappa,1}$. Since $L1 \leqq 0$, it follows that $L1 + n\chi_{\Gamma\delta}(1 - 1) \leqq 0$ and, by the maximum principle, $v_n \leqq 1$, $Lv_n \leqq 0$, $v_n \geqq 0$ on $C_{\kappa,1}$, and $Lv_n = 0$ on $C_{\kappa,1}\setminus\Gamma^{\delta}$. From the latter relationships and the maximum principle applied to the domain $C_{\kappa,1}\setminus\Gamma^{\delta}$, it follows that $v_n \leqq \delta^{-1}u$ on $C_{\kappa,1}\setminus\Gamma^{\delta}$ and, obviously, on Γ^{δ}. Hence, $v_n \leqq \delta^{-1}u$ on $\bar{C}_{\kappa,1}$.

From $L(\frac{1}{n}v_n) + \chi_{\Gamma\delta}(1 - v_n) = 0$, by Corollary 8 (or by Corollary 3.5.14), we obtain $v_n \to 1$ in measure on Γ^{δ}. Specifically, $\Lambda(C_{\kappa,1} \cap \{v_n \geqq \delta\}) \geqq q\Lambda(C_{\kappa,1})$ (see (13)) for all sufficiently large n. By the above argument, this yields $v_n\delta^{-1} \geqq p(\kappa q)$ on $\bar{C}_{1,1}$ for large n. Thus, $u \geqq \overline{\lim}\, \delta v_n \geqq \delta^2 p(\kappa q)$ and, letting $\delta \uparrow 1$, we conclude that $u \geqq p(\kappa q)$ on $\bar{C}_{\kappa,1}$.

We finally abandon the supposition that $L \in A_0$. Let us take a sequence of operators $L_n \in A_0$ such that the coefficients of L_n converge to the corresponding coefficients of L almost everywhere. Let w_n be a $\overset{0}{W}{}^{1,2}_{d+1}(C_{\kappa,1})$-solution of the equation $L_n w_n = (L_n - L)u$ in $C_{\kappa,1}$. Since $w_n \to 0$ uniformly on $\bar{C}_{\kappa,1}$, we obtain $\delta_n - w_n \geqq 0$, $L_n(u - w_n + \delta_n) \leqq L_n(u - w_n) = Lu \leqq 0$ on $C_{\kappa,1}$, and $u - w_n + \delta_n \geqq 1$ on $C_{\kappa,1} \cap \{u \geqq 1\}$ for a sequence $\delta_n \downarrow 0$. By construction, $L_n \in A_0$ and, therefore, by the above argument, $u - w_n + \delta_n \geqq p(\kappa q)$ on $\bar{C}_{\kappa,1}$. Letting $n \to \infty$, we complete the proof.

11. LEMMA. Let $\kappa \in (0,1]$, let $u \in W^{1,2}_{d+1}(C_{\kappa,1})$, let $L \in A$, let $Lu \geqq 0$ on $C_{\kappa,1}$(a.e.), let $\zeta \in (-\infty, \infty)$, let $q \in [0,1)$, and let $\Lambda(C_{\kappa,1} \cap \{u \leqq \zeta\}) \geqq q\Lambda(C_{\kappa,1})$. Also, let one of the following two conditions hold: (a) $c \equiv 0$ or (b) $\max(u, \bar{C}_{\kappa,1}) \geqq 0$. Then

$$u(t, x) \leqq p(t, x, \kappa q)\zeta + (1 - p(t, x, \kappa q))\max(u, \bar{C}_{\kappa,1}). \tag{14}$$

on $\bar{C}_{\kappa,1}$.

<u>Proof</u>. Denote the latter maximum in (14) by M. If $\zeta \geqq M$, then (14) is

clear. In case $\zeta < M$ we set $v = (m-u)(M-\zeta)^{-1}$ and note that, either (a) or (b) being assumed, the inequality $Lv \leqq 0$ is true on $C_{\kappa,1}$. Moreover, $\{u \leqq \zeta\} = \{v \geqq 1\}$ and $v \geqq 0$ on $C_{\kappa,1}$. By Theorem 10, we obtain $v \geqq p(\kappa q)$ on $\bar{C}_{\kappa,1}$. The latter is equivalent to (14). This proves the lemma.

We complete the section by extending the established results to elliptic operators. Let $\tilde{A}$ be the set of all elliptic operators of the form

$$L = a^{ij}(x)\frac{\partial^2}{\partial x^i \partial x^j} + b^i(x)\frac{\partial}{\partial x^i} + c(x),$$

whose coefficients are defined on E_d, measurable with respect to x, and satisfy condition (2) for all λ, $x \in E_d$, $i, j = 1, \ldots, d$. Let $\tilde{A}_0$ be a subset of A consisting of all operators with infinitely differentiable coefficients. We introduce sets $\tilde{\mathbf{A}}^m_q$ and $\tilde{\mathbf{A}}^p_q$ which differ from $\mathbf{A}^m_q$ and $\mathbf{A}^p_q$ in that $C_{1,1}$, $C^2(\bar{C}_{1,1})$, $\partial' C_{1,1}$, and A_0 are replaced by S_1, $C^2(\bar{S}_1)$, ∂S_1, and $\tilde{A}_0$, respectively (and Λ is now Lebesgue measure on E_d). Denote

$$m(x, q) = \inf(u(x) : u \in \tilde{\mathbf{A}}^m_q), \quad x \in \bar{S}_1, \quad q \in [0, 1],$$

$$p(x, q) = \inf(u(x) : u \in \tilde{\mathbf{A}}^p_q), \quad x \in \bar{S}_1, \quad q \in [0, 1).$$

12. THEOREM. (a) $m(x, q) \geqq m(0, x, q)$, $1 \geqq p(x, q) \geqq p(0, x, q)$.

(b) If $q \in [0,1)$, $u \in W^2_d(S_1)$, $u \geqq 0$ on ∂S_1, $L \in \tilde{A}$, $Lu \leqq 0$ on S_1 (a.e.) and $\Lambda(S_1 \cap \{Lu \leqq -1\}) \geqq q\Lambda(S_1)$, then $u(x) \geqq m(x, q)$ on $\bar{S}_1$.

(c) If $q \in [0,1)$, $u \in W^2_d(S_1)$, $u \geqq 0$ on ∂S_1, $L \in \tilde{A}$, $Lu \leqq 0$ on S_1(a.e.) and $\Lambda(S_1 \cap \{u \geqq 1\}) \geqq q\Lambda(S_1)$, then $u(x) \geqq p(x, q)$ on $\bar{S}_1$.

(d) If $q \in [0,1)$, $u \in W^2_d(S_1)$, $L \in \tilde{A}$, $Lu \geqq 0$ on S_1(a.e.), $\zeta \in (-\infty, \infty)$, $\Lambda(S_1 \cap \{u \leqq \zeta\}) \geqq q\Lambda(S_1)$, and one of the following two conditions holds: $c \equiv 0$ or $\max(u, \bar{S}_1) \geqq 0$, then

$$u(x) \leqq p(x, q)\zeta + (1 - p(x, q)) \max(u, \bar{S}_1).$$

on $\bar{S}_1$.

The inequalities $m(x, q) \geqq m(0, x, q)$ and $p(x, q) \geqq p(0, x, q)$ trivially follow from Lemma 1 and Theorem 10, if $C^2(\bar{S}_1)$ is regarded as a subset of $C^2(\bar{C}_{1,1})$ composed of functions which are constant in t. The proof of the remaining assertions is almost the same as the proofs of Lemmas 1 and 11 and Theorem 10. It only differs in that Theorem 3.3.11 is used in place of Theorem 3.3.9.

13. COROLLARY. Let $L_n \in \tilde{A}$, let $u_n \in W_d^2(S_1)$ and $u_n \geqq 0$ on S_1, let $L_n u_n = f_n \leqq 0$ on S_1 (a.e.), $n = 1, 2, \ldots$, let $x_0 \in S_1$ and let $u_n(x_0) \to 0$ as $n \to \infty$. Then $f_n \to 0$ in measure as $n \to \infty$.

This corollary can be proved in the same manner as Corollary 8. From Theorem 12 and (12) we easily obtain the following corollary.

14. COROLLARY. If $q \uparrow 1$, then $p(x, q) \uparrow 1$ uniformly on every compactum of S_1.

4.2. The Harnack Inequality and the Hölder Condition for Solutions of Linear Equations with Measurable Coefficients

We fix constants K, K_1, and $\varepsilon > 0$. Let

$$L = \frac{\partial}{\partial t} + a^{ij}(t, x) \frac{\partial^2}{\partial x^i \partial x^j} + b^i(t, x) \frac{\partial}{\partial x^i} + c(t, x) \tag{1}$$

be a parabolic operator whose coefficients are defined on E_{d+1}, measurable with respect to (t, x), and such that

$$a^{ij} = a^{ji}, \quad \varepsilon|\lambda|^2 \leqq a^{rs}\lambda^r\lambda^s \leqq K|\lambda|^2, \quad |b^i| \leqq K_1, \quad |c| \leqq K_1 \tag{2}$$

for all $(t, x) \in E_{d+1}$, $\lambda \in E_d$, $i, j = 1, \ldots, d$.

The Harnack inequality, which is considered first in this section, shows that, given a nonnegative solution to the equation $Lu = 0$ in a domain Q, which is greater than 1 at a point $(t_0, x_0) \in Q$, this solution cannot be arbitrarily small at a point $(t, x) \in Q$ which is sufficiently close to (t_0, x_0) but such that $t \leqq t_0 - \delta$ with $\delta > 0$. It should be said that the condition $t \leqq t_0 - \delta$ is essential. For example (see Moser [2]), the function $u(t, x) = \exp(\lambda x - \lambda^2 t)$ satisfies the equation $u_t + u_{xx} = 0$ on E_2 whereas $u(0, 0) = 1$ and $u(0, -\delta) \to 0$ as $\lambda \to \infty$ for any $\delta > 0$.

The Harnack inequality can be interpreted in terms of potential theory. Such an interpretation looks very much like a probabilistic one. Suppose that the coefficients of L and the domain Q are sufficiently regular, namely they are such that for every sufficiently smooth function φ defined on $\partial'Q$ the problem $Lu = 0$ in Q, with $u = \varphi$ on $\partial'Q$, has a sufficiently smooth solution. Then for any (t_0, x_0) we obtain the mapping $\varphi \to u(t_0, x_0)$. By the maximum principle, this mapping defined on a subset of $C(\overline{\partial'Q})$ is a linear bounded functional on this subset.

The above subset is dense in $C(\overline{\partial'Q})$ and, therefore, this functional can be continuously extended to the whole $C(\overline{\partial'Q})$. By the Riesz representation theorem,

$$u(t_0, x_0) = \int \varphi(t, x)\, \mu(t_0, x_0, dt\, dx),$$

where the measure μ is nucleated on $\overline{\partial' Q}$ and, by the maximum principle, it is nonnegative. In terms of μ the Harnack inequality means that $\mu(t_0, x_0) \leqq N\mu(t, x)$ if (t, x) is sufficiently close to (t_0, x_0) and such that $t \leqq t - \delta$ where $\delta > 0$ and N is a constant.

Also, note that if (t_0, x_0) is constructed from an operator L, while $\mu(t, x)$ is based on a different operator, then this inequality with a constant N which is independent of the smoothness of the operator coefficients cannot be true. The respective counterexample (and even a much stronger result) can easily be derived from Safonov [5].

We prove the Harnack inequality in the following form.

1. THEOREM. Let $\theta > 1$, let $R \leqq 2$, and let $u \in W^{1,2}_{d+1}(C_{\theta R^2, R})$ such that $u \geqq 0$ in $C_{\theta R^2, R}$ and $Lu = 0$ in $C_{\theta R^2, R}$ (a.e.). Then there exists a constant N which depends only on θ, ε, K, K_1, and d and such that $u(R^2, 0) \leqq Nu(0, x)$ whenever $|x| \leqq \frac{1}{2}R$. Moreover, if $(1-\theta)^{-1}$, ε^{-1}, K, and K_1 change within a bounded range, then N also changes within a bounded range. Finally, if $b \equiv 0$ and $c \equiv 0$, then our assertions hold true for $R > 2$ as well.

Proof. Denote $\tilde{u}(t, x) = u(\frac{1}{4}R^2 t, \frac{1}{2}Rx)$. It is easy to see that the function $\tilde{u}$ in $C_{4\theta, 2}$ satisfies the equation $\tilde{L}\tilde{u} = 0$, where $\tilde{L}$ is of the form (1) and its coefficients satisfy conditions (2). It clearly follows that one need only prove the theorem for $R = 2$. Using the appropriate transformation $t \to \tau(t)$, one can reduce the general case $\theta > 1$ to the case $\theta \geqq 2$. Since the best constant N is obviously nondecreasing in θ, it suffices to consider the case $\theta = 2$. If $\theta = R = 2$, the best admissible constant N is decreasing in ε and increasing in K and K_1. Therefore, the latter assertion of the theorem follows from the first one and the previous argument.

To prove the first assertion, we again assume that $\theta = R = 2$ and without loss of generality suppose that $u(4, 0) > 0$. Replacing $u(t, x)$ by $e^{K_1 t} u(t, x)$, we can see that $c \leqq 0$ can also be assumed. Finally, selecting, if necessary, the greatest constant among K and K_1, one can easily limit consideration to the case when condition (1.2) holds. For $\kappa = 1/5$ we determine the constants δ and n from Lemma 1.3 on an oblique cylinder and from Theorem 1.9 find $q \in (0,1)$ such that

$$1 > p(0, 0, (1 - q))\frac{1}{2} + (1 - p(0, 0, (1 - q)))2^n. \tag{3}$$

Furthermore, for $r \in [0,1)$ we put

$$\nu(r) = u(4,\ 0)(1 - r)^{-n},\quad n(r) = \max\left\{u,\ (4,\ 0) + \bar{C}_{r^2,r}\right\}.$$

Let r_0 be the greatest root of the equation $n(r) = \nu(r)$. Since $n(0) = \nu(0)$ and $\nu(r) \to \infty$ as $r \uparrow 1$, while u is continuous and bounded in $C_{8,2}$, it follows that the definition of r_0 is meaningful and $r_0 < 1$. Let $(t_1,\ x_1) \in (4,\ 0) + \bar{C}_{r_0^2,r_0}$ and $\nu(r_0) = u(t_1,\ x_1)$. We introduce the cylinder

$$Q := \left\{(t,\ x) : 0 < t - t_1 < \frac{(1 - r_0)^2}{4},\quad |x - x_1| < \frac{1 - r_0}{2}\right\}.$$

It is easy to show that $\bar{Q} \subset (4,\ 0) + \bar{C}_{r_1^2,r_1}$, where $r_1 = (1 + r_0)/2$ and, by the definition of r_0, it therefore follows that

$$u < \nu(r_1) = u(4,\ 0)\left(\frac{1 - r_0}{2}\right)^{-n} = 2^n \nu(r_0).$$

on $\bar{Q}$.

We assert that hence, by the choice of q from (3),

$$\Lambda\left(Q \cap \left\{u > \frac{\nu(r_0)}{2}\right\}\right) \geqq q\ \Lambda\ (Q). \tag{4}$$

Indeed, assume the contrary and perform the elongation of the cylinder Q to the dimensions of $C_{1,1}$ as at the beginning of the proof. We then shift the elongated Q so that it coincides with $C_{1,1}$. As a result the function u is transformed to a function $\tilde{u}$ defined on $C_{1,1}$ and satisfying the equation $\tilde{L}\tilde{u} = 0$ on $C_{1,1}$ such that the operator $\tilde{L}$ is of the form (1) and its coefficients satisfy conditions (1.2). It is clear that under our assumption

$$\Lambda(C_{1,1} \cap \{\tilde{u} > (1/2)\nu(r_0)\}) < q\Lambda(C_{1,1})$$

and

$$\Lambda(C_{1,1} \cap \{\tilde{u} \leqq (1/2)\nu(r_0)\}) > (1 - q)\Lambda(C_{1,1}).$$

By Lemma 1.11 with $\zeta = (1/2)\nu(r_0)$,

$$\nu(r_0) = u(t_1,\ x_1) = \tilde{u}(0,\ 0)$$

$$\le p(0, 0, (1-q))\tfrac{1}{2}\nu(r_0) + (1-p(0, 0, (1-q)))\max(\tilde{u}, \bar{C}_{1,1})$$

$$= p(0, 0, (1-q))\tfrac{1}{2}\nu(r_0) + (1-p(0, 0, (1-q)))\max(\tilde{u}, \bar{Q})$$

$$\le p(0, 0, (1-q))\tfrac{1}{2}\nu(r_0) + (1-p(0, 0, (1-q)))2^n\nu(r_0).$$

By virtue of (3), this is impossible, hence (4) is proved. We now use the conversion of Q and u to $C_{1,1}$ and $\tilde{u}$ along with (4) and by Theorems 1.10 and 1.9 conclude that $u(t_1, x) \ge (1/2)p_0(r_0)$ if $|x - x_1| \le (1/4)(1 - r_0)$, where $p_0 > 0$ and p_0 depends only on d, ε, K, and q. Hence, by Lemma 1.3 with $|x| \le 1$,

$$u(0, x) \ge \tfrac{1}{2}p_0\nu(r_0)\delta\left(\frac{1 - r_0}{4}\right)^n = 2^{-2n-1}p_0\delta u(4, 0).$$

This proves the theorem.

We now estimate the Hölder norm of a solution to the equation $Lu = f$. We need two lemmas, but before formulating them let us recall that if $z_1 = (t_1, x_1)$ and $z_2 = (t_2, x_2)$, we set

$$\rho(z_1, z_2) = |x_1 - x_2| + |t_1 - t_2|^{1/2}. \tag{5}$$

The value $\rho(z_1, z_2)$ is called the parabolic distance between z_1 and z_2.

2. LEMMA. Let $r \in (0, K_1^{-1} \wedge 1]$, let $u \in W^{1,2}_{d+1}(C_{4r^2,2r})$, let $Lu = 0$, and let $c = 0$ on $C_{4r^2,2r}$ (a.e.). Then there exist constants $N < \infty$ and $\alpha \in (0, 1)$ which depend only on d, K, and ε and such that, for any $z_1, z_2 \in \bar{C}_{r^2,r}$

$$|u(z_1) - u(z_2)| \le Nr^{-\alpha}\rho^{\alpha}(z_1, z_2)\sup(|u|, C_{4r^2,2r}). \tag{6}$$

If, in addition, $b = 0$, then (6) is also true for $r > K_1^{-1} \wedge 1$.

Proof. Using the transformation $(t, x) \to (r^{-2}t, r^{-1}x)$, we can easily see that it suffices to prove both assertions of the lemma in the case $r = 1$, $K_1 \le 1$. If we further replace K by $K \vee 1$, our consideration will reduce to the case when condition (1.2) holds (and $c \equiv 0$). Furthermore, for $r = 1$ we use the argument from De Giorgi [1], applying, when necessary, Lemma 1.11 and Theorem 1.9.

For $R \in (0, 2]$, we define $w(R)$ as the oscillation of u on $C_{R^2,R}$, i.e. the difference between the least upper bound $M(R)$ and the greatest lower bound $m(R)$ of the function on $C_{R^2,R}$. If $R \leqq 1/2$ and

$$\Lambda\left(C_{4R^2,2R} \cap \left\{u \leqq \frac{1}{2}(m(R) + M(R))\right\}\right) \geqq \frac{1}{2}\Lambda(C_{4R^2,2R}), \tag{7}$$

then, by Lemma 1.11,

$$u(t, x) \leqq p\left(4^{-1}R^{-2}t,\ 2^{-1}R^{-1}x,\ \frac{1}{2}\right)\frac{1}{2}(m(R) + M(R)) +$$

$$+ \left(1 - p\left(4^{-1}R^{-2}t,\ 2^{-1}R^{-1}x,\ \frac{1}{2}\right)\right)M(2R).$$

on $C_{4R^2,2R}$.

We calculate the supremum with respect to $C_{R^2,R}$ and note that $(1/2)(m(R) + M(R)) \leqq M(2R)$ and, by Theorem 1.9,

$$p_0 := \inf\left(p\left(t,\ x,\ \frac{1}{2}\right):\quad 0 \leqq t \leqq \frac{1}{4},\quad |x| \leqq \frac{1}{2}\right) > 0,$$

where p_0 depends only on d, K, and ε. We then obtain

$$M(R) \leqq p_0\,\frac{1}{2}(m(R) + M(R)) + (1 - p_0)M(2R).$$

Adding this inequality to the obvious inequality $(p_0 - 1)m(R) \leqq (p_0 - 1)m(2R)$, we obtain

$$\left(1 - \frac{1}{2}\,p_0\right)w(R) \leqq (1 - p_0)w(2R), \quad w(R) \leqq \delta w(2R), \tag{8}$$

where $\delta < 1$, δ depends only on d, K, and ε, and it can be assumed that $\delta > 1/2$.

Inequalities (8) have been proved under the assumption that inequality (7) holds. If the inequality opposite to (7) is true, then it is easily seen that the function $(-u)$ satisfies an inequality similar to (7) and considering the case at hand as above, we again obtain (8).

Thus, $w(R) \leqq \delta w(2R)$ for all $R \leqq 1/2$. Iterating on this inequality, we obtain $w(R) \leqq \delta^2 w(4R)$ for $R \leqq 1/4$, ..., $w(R) \leqq \delta^n w(2^n R)$ with $R \leqq 2^{-n}$. If $R \leqq 1/2$ and $n := [-\log_2 R]$, we obtain $R \leqq 2^{-n}$ and, therefore,

$$w(R) \leqq \delta^n w(2^n R) \leqq \delta^{-1}R^{\alpha}w(1) \leqq 2\delta^{-1}R^{\alpha}\sup(|u|,\ C_{1,1}),$$

where $\alpha = -\log_2\delta \in (0,1)$. We have estimated the oscillation of u in any cylinder of the form $C_{R^2,R}$ with $R \leqq 1/2$. The oscillation of u in a cylinder of the form $(t, x) + C_{R^2,R}$ with $R \leq 1/2$ and $(t, x) \in \bar{C}_{1,1}$ can be estimated in the same manner. It should only be noted that in this case $(t, x) + C_{1,1} \subset C_{4,2}$.

Now let $z_1, z_2 \in C_{1,1}$ and $R := \rho(z_1, z_2) \leq 1/2$. We express z_1 and z_2 as (t_1, x_1) and (t_2, x_2) and set $t = t_1 \wedge t_2$ and $x = (1/2)(x_1 + x_2)$. Then, as can easily be seen, $z_i \in (t, x) + \bar{C}_{R^2,R}$, $i = 1, 2$ and, therefore,

$$|u(z_1) - u(z_2)| \leq 2\delta^{-1}R^{\alpha}\sup(|u|, (t, x) + C_{1,1})$$
$$\leq 2\delta^{-1}\rho^{\alpha}(z_1, z_2)\sup(|u|, C_{4,2}).$$

It remains to consider the case when $\rho(z_1, z_2) > 1/2$. Then it is clear that

$$|u(z_1) - u(z_2)| \leq 2\sup(|u|, C_{4,2})$$
$$\leq 2^{1+\alpha}\rho^{\alpha}(z_1, z_2)\sup(|u|, C_{4,2}).$$

Thus, we can always set $N = 2^{1+\alpha} + 2\delta^{-1}$ in (6) with $r = 1$, which proves the lemma.

3. COROLLARY (two-sided Liouville theorem). If $c \equiv 0$, $b \equiv 0$, $u \in W^{1,2}_{d+1}(C_{r^2,r})$ for all $r > 0$, $Lu = 0$ almost everywhere in $\{t > 0\} \times E_d$, and u is bounded on this set, then $u = \text{const}$.

Indeed, from (6) with $r \to \infty$ it follows that $u = \text{const}$.

We take α from Lemma 2 and abandon the condition $Lu = 0$.

4. LEMMA. Let $r \in (0, K_1^{-1} \wedge 1]$, let $u \in W^{1,2}_{d+1}(C_{4r^2,2r})$, and let $c \equiv 0$. Then there exists a constant N which depends only on d, K, and ε and such that

$$|u(z_1) - u(z_2)| \leq N\Big(r^{-\alpha}\rho^{\alpha}(z_1, z_2)\sup_{C_{4r^2,2r}}|u| + r^{d/(d+1)}||Lu||_{L_{d+1}(C_{4r^2,2r})}\Big). \tag{9}$$

for $z_1, z_2 \in \bar{C}_{r^2,r}$. If, in addition, $b \equiv 0$, then (9) is also true for $r > K_1^{-1} \wedge 1$.

Proof. Similarity transformation shows that it suffices to prove (9) when $r = 1$ and $K_1 \le 1$. Moreover, it is easy to perform a limiting process in (9) passing from operators L with smooth coefficients to operators with measurable coefficients. We therefore assume that $r = 1$ and the coefficients of L are infinitely differentiable.

We define u_1 and u_2 as $W^{1,2}_{d+1}(C_{4,2})$-solutions of the problems: $Lu_1 = 0$, $Lu_2 = Lu$ in $C_{4,2}$(a.e.), $u_1 = u$, $u_2 = 0$ on $\partial' C_{4,2}$. It is clear that $u = u_1 + u_2$. We apply Lemma 2 to the function u_1 taking note that, by the maximum principle, $\sup(|u_1|, C_{4,2}) \le \sup(|u|, \partial' C_{4,2}) \le \sup(|u|, C_{4,2})$. Then it only remains to show that

$$|u_2| \le N||Lu||_{\mathcal{L}_{d+1}(C_{4,2})} = N||Lu_2||_{\mathcal{L}_{d+1}(C_{4,2})}.$$

in $\bar{C}_{1,1}$.

Even in $\bar{C}_{4,2}$ this estimate immediately follows from Theorem 3.3.9. The lemma is proved.

In the following Theorems 5 to 7 it is assumed that α is the constant in Lemma 2, $\beta = (\alpha d)[d + \alpha(d+1)]^{-1}$. Q is a domain of E_{d+1}, and

$$\rho(z) = \inf(\rho(z, z') : z' = (t', x') \in \partial Q, t' > t), \quad \inf\phi = \infty,$$

$$\tilde{\rho}(z) = 1 \wedge \rho(z) \tag{10}$$

for $z = (t, x) \in Q$. The value $\rho(z)$ is called the parabolic distance from z to the boundary of Q.

5. THEOREM. Let $u \in W^{1,2}_{d+1}(Q)$, let $z_1, z_2 \in Q$, $c \equiv 0$, and $b \equiv 0$, and let $4\rho(z_1, z_2) \le \rho(z_1) \wedge \rho(z_2)$. Then there exists a constant N which depends only on d, K, and ε and such that for any finite $\rho \in [4\rho(z_1, z_2), \rho(z_1) \wedge \rho(z_2)]$

$$|u(z_1) - u(z_2)| \le N\rho^{-\beta}\left(\sup_Q |u| + \rho^{d/(d+1)}||Lu||_{\mathcal{L}_{d+1}(Q)}\right)\rho^{\beta}(z_1, z_2). \tag{11}$$

Proof. Let $z_1 = (t_1, x_1)$ and $z_2 = (t_2, x_2)$. Without loss of generality we assume that $t_1 \le t_2$. Denote

$$\xi = \frac{\alpha(d + 1)}{d + \alpha(d + 1)}, \quad r = 4^{\xi-1}\rho^{1-\xi}\rho^{\xi}(z_1, z_2).$$

By the conditions on ρ, it is not difficult to show that

$$2r \leqq (1/2)\rho(z_1) \wedge \rho(z_2), \ \rho(z_1, z_2) \leqq r.$$

It follows that $z_2 \in z_1 + \bar{C}_{r^2,r}$ and $z_1 + C_{4r^2,2r} \subset Q$. Hence, using a shift, we can apply Lemma 4. It then remains to note that, by the choice of r,

$$r^{-\alpha}\rho^{\alpha}(z_1, z_2) \sup_{z_1+C_{4r^2,2r}} |u| + r^{\frac{d}{d+1}} ||Lu||_{L_{d+1}(z_1 + C_{4r^2,2r})}$$

$$= \left(4^{\beta}\rho^{-\beta} \sup_{z_1+C_{4r^2,2r}} |u| + \right. \tag{12}$$

$$\left. + 4^{\beta - \frac{d}{d+1}} \rho^{\frac{d}{d+1} - \beta} ||Lu||_{L_{d+1}(z_1+C_{4r^2,2r})}\right) \rho^{\beta}(z_1, z_2).$$

The theorem is proved.

6. THEOREM. Let $u \in W^{1,2}_{d+1}(Q)$. Then there exists a constant N which depends only on d, K, ε, and K_1 and such that inequality (11) is true for $\rho = \tilde{\rho}(z_1) \wedge \tilde{\rho}(z_2)$.

Proof. Since on the right of (11) there is the norm of u in $C(Q)$, one can assume that $4\rho(z_1, z_2) \leqq \rho := \tilde{\rho}(z_1) \wedge \tilde{\rho}(z_2)$. In this case we use the same argument as in the previous proof, taking note that here $\rho \leqq 1$ and $r \leqq 1/4$. If, in addition, $r \geqq K_1^{-1}$, then $4\rho^{-1}\rho(z_1, z_2) \geqq K_1^{-1/\xi}$ and inequality (11) is obviously true with an appropriate N. We therefore assume that $r \leqq K_1^{-1}$. It is now clear that, as in the proof of Theorem 5, one can use Lemma 4 to estimate $|u(z_1) - u(z_2)|$ in terms of the constant $N(d, K, \varepsilon)$ multiplied by the right-hand side of (12) where Lu should be replaced by Lu - cu. It then remains with the fact that on $z_1 + C_{4r^2,2r}$ the norm of u in L_{d+1} is estimated in terms of $\sup(|u|, Q)$ (since $r \leqq 1/4$). The theorem is proved.

The reader can find versions of Theorem 6 in Remark 3.4. The folloiwng theorem is an immediate consequence of the previous one.

7. THEOREM. Let Q' be a domain such that $Q' \subset Q$ and let $u \in W^{1,2}_{d+1}(Q)$.

Denote $\rho = \inf(\rho(z) : z \in Q')$ and suppose that $\rho > 0$. Then for a constant N which depends only on d, K, ε, E_1, and ρ

$$|u(z_1) - u(z_2)| \leqq N\left(\sup_Q |u| + ||Lu||_{L_{d+1}(Q)}\right)\rho^{\beta}(z_1, z_2) \tag{13}$$

for all $z_1, z_2 \in Q$.

Indeed, if $4\rho(z_1, z_2) \leqq \rho \wedge 1$, then in (13) we can set $N = N_1 = N_2(\rho \wedge 1)^{-\beta}$ where N_2 is the constant involved in (11) in the case of Theorem 6. But if $4\rho(z_1, z_2) \geqq \rho \wedge 1$, we can take $N_3 = 2\cdot 4^{\beta}(\rho \wedge 1)^{-\beta}$. In either case $N = N_1 + N_3$ is suitable for (13).

We now extend the above results to elliptic equations. Consider the operator

$$\tilde{L} = a^{ij}(x)\frac{\partial^2}{\partial x^i \partial x^i} + b^i(x)\frac{\partial}{\partial x^i} + c(x), \tag{14}$$

whose coefficients are defined on E_d, measurable with respect to x and satisfy conditions (2) for all $x, \lambda \in E_d$, $i, j = 1, \ldots, d$, whereas $c \leqq 0$.

8. THEOREM. Let $R \in (0,2]$ and let $u \in W_d^2(S_R)$ be such that $u \geqq 0$ in S_R and $\tilde{L}u = 0$ in S_R (a.e.). Then there exists a constant N which depends only on d, K, ε, and K_1 and such that $u(0) \leqq Nu(x)$ with $|x| \leqq (1/2)R$. Moreover, if ε^{-1}, K, and K_1 vary in a bounded range, the N also varies in a bounded range. Finally, if $b \equiv 0$ and $c \equiv 0$, then our assertions are also true for $R > 2$.

Proof. If $u \in C^2(\bar{S}_R)$, then our assertion follows from Theorem 1. To show this, it suffices to put $\theta = 2$ and $L = \frac{\partial}{\partial t} + \tilde{L}$ in Theorem 1. In the general case, we take sequences of functions $\varphi_n \in C^3(\bar{S}_R)$ and operators $\tilde{L}_n$ of the form (14) with smooth coefficients such that $\varphi_n \geqq 0$ in S_R, $\varphi_n \to u$ uniformly on ∂S_R, coefficients of $\tilde{L}_n$ satisfy conditions (2) and converge to the corresponding coefficients of $\tilde{L}$ almost everywhere on S_R. Let u_n be a $C^2(\bar{S}_R)$-solution of the equation $\tilde{L}_n u_n = 0$ with the boundary condition $u_n = \varphi_n$ on ∂S_R. Then $\tilde{L}_n(u_n - u) = (\tilde{L} - \tilde{L}_n)u$ and by Theorem 3.3.11, $u_n \to u$ uniformly on $\bar{S}_R$. Since the theorem is proved for u_n, it is also true for u. This completes the proof.

9. COROLLARY (one-sided Liouville's theorem). If $b \equiv 0$, $c \equiv 0$, $u \in W_d^2(S_R)$ for all R, $u \geqq 0$ in E_d, $\tilde{L}u = 0$ on E_d (a.e.), then $u = \text{const}$.

Indeed, replacing u by $u - \inf(u, E_d)$, we can assume that $\inf(u, E_d) = 0$. Let $x_n \in E$ be a sequence such that $u(x_n) \to 0$. By Theorem 8, we obtain $0 \leqq u(0) \leqq Nu(x_n)$, where N is independent of n. Hence, $u(0) = 0$ and, since the origin can be changed, $u(x) = 0$ for all x.

10. THEOREM. Let $r \in (0, K_1^{-1} \wedge 1]$, let $u \in W_d^2(S_{2r})$, and let $c \equiv 0$. Then there exist constants N and $\alpha > 0$ which depend only on d, K, and ε and such that

$$|u(x_1) - u(x_2)| \leqq N\left(r^{-\alpha}|x_1 - x_2|^{\alpha}\sup_{S_{2r}}|u| + r||\tilde{L}u||_{\mathcal{L}_d(S_{2r})}\right). \quad (15)$$

whenever $x_1, x_2 \in S_r$. If, in addition, $b \equiv 0$, then (15) also holds for $r > K_1^{-1} \wedge 1$.

Similarity transformation reduces the proof of (15) to the case $r = 1$ and $K_1 \leqq 1$. If $r = 1$, $u \in C^2(\bar{S}_{2r})$, and $\tilde{L}u = 0$, inequality (15) follows from Lemma 2, in which we set $z_1 = (0, x_1)$, $z_2 = (0, x_2)$, and $L = \frac{\partial}{\partial t} + \tilde{L}$. Following the proof of Theorem 8, we then come to the case $u \in W_d^2(S_{2r})$ and $\tilde{L}u = 0$. The general case is treated as in Lemma 4, where, when needed, Theorem 3.3.9 is replaced by Theorem 3.3.11.|

11. THEOREM. Let D be a domain in E_d, let $x_1, x_2 \in D$, let $2|x_1 - x_2| \leqq d(x_1, \partial D)$, let $b \equiv 0$ and $c \equiv 0$, and let $u \in W_d^2(D)$. Then for all finite $\rho \in [2|x_1 - x_2|, d(x_1, \partial D)]$,

$$|u(x_1) - u(x_2)| \leqq N\rho^{-\sigma}|x_1 - x_2|^{\sigma}\left(\sup_D|u| + \rho||\tilde{L}u||_{\mathcal{L}_d(D)}\right),$$

where $\sigma = \alpha/(1 + \alpha)$ and the constant N depends only on d, K, ε, and K_1.

To prove this theorem, it suffices to put

$$r = 2^{\sigma-1}\rho^{1-\sigma}|x_1 - x_2|^{\sigma},$$

to note that $2|x_1 - x_2| \leqq 2r \leqq \rho \leqq d(x_1, \partial D)$, and to apply Theorem 10 to the balls $x_1 + S_r$ and $x_1 + S_{2r}$.

Recalling the derivation of Theorem 6 from Lemma 4, we obtain the following theorem from Theorem 10.

12. THEOREM. Let D be a domain of E_d, let $x_1, x_2 \in D$, and let $u \in W_d^2(D)$. Then

$$|u(x_1) - u(x_2)| \leq N[1 \wedge d(x_1, \partial D)]^{-\sigma}|x_1 - x_2|^{\sigma}\Big\{\sup_D |u| + $$

$$+ ||\tilde{L}u||_{\mathcal{L}_d(D)}\Big\},$$

where N depends only on d, K, ε, and K_1.

4.3. The Hölder Condition for Solutions of Simultaneous Linear Inequalities

To prove estimates in $C^{2+\alpha}$ of solutions of nonlinear equations, we shall differentiate these equations with respect to independent variables and derive some inequalities for the partial derivatives. The system of inequalities will further yield the Hölder condition for each of the derivatives under consideration.

We fix constants $K \geqq \varepsilon > 0$, K_1 and $K_2 \geqq 0$. Let $L_1, \ldots, L_n$, L be differential operators of the form (2.1) whose coefficients satisfy conditions (2.2). We now take a domain $Q \subset E_{d+1}$ and, using notation (2.5) and (2.10) for points of Q, formulate the basic result of this section for parabolic inequalities.

1. THEOREM. Let $u^1, \ldots, u^n \in W^{1,2}_{d+1,loc}(Q) \cap C(Q)$, let $f_1, \ldots, f_n \in C^2_{loc}(E_1)$, let $f_i' > 0$ and $f_i'' \leqq 0$ on E_1, and let $L_i u^i \geqq -K_2$ in Q (a.e.) with $i = 1, \ldots, n$. Summation over i is assumed neither here nor in the proof. Suppose that, given constants $\alpha \in (0,2]$, $\nu \in [0,\alpha]$, and $\delta > 0$,

$$K_2\tilde{\rho}^{\nu-\alpha}\rho^{\alpha}(z_1, z_2) \geqq \delta \sum_{i=1}^{n} [f_i(u^i(z_1)) - f_i(u^i(z_2))]_+ -$$

$$- \sum_{i=1}^{n} [f_i(u^i(z_1)) - f_i(u^i(z_2))]_- \tag{1}$$

for any $z_1, z_2 \in Q$ such that $4\rho(z_1, z_2) \leqq \tilde{\rho} := \tilde{\rho}(z_1) \wedge \tilde{\rho}(z_2)$.

Then for any $z_1, z_2 \in Q$

$$\sum_{i=1}^{n} |u^i(z_1) - u^i(z_2)| \leqq \tilde{\rho}^{-\beta}\rho^{\beta}(z_1, z_2)N\Big(K_2\tilde{\rho}^{\nu} + \sum_{i=1}^{n} \sup_Q |u^i|\Big), \tag{2}$$

where $\beta = \beta_0 \wedge \alpha$, $\beta_0 \in (0,1)$, β_0 depends only on n, d, K, ε, and δ, the constant N depends on the same parameters plus K_1 and α if $f_i(r) = r$ for all i, r, but if $f_i(r) \not\equiv r$, N also depends on K_2, the bounds M_i of $|u^i|$ on Q, and the maximum values of f_i', $(f_i')^{-1}$, $|f_i''|$ on $[-M_i, M_i]$.

The major assertion of the theorem relates to the case $f_i(r) \equiv r$. This specific case can easily be extended to the general case if β is permitted to depend on min f_i' and max f_i'. This dependence is, however, undesirable since, if it is the case in our considerations, the Hölder exponent for the derivatives of a solution to a nonlinear equation would depend on the estimates of their maxima. In addition to the results established in Section 1, the proof of this theorem requires the following lemma.

2. LEMMA. Let $N_1, N_2 \geqq 0$, $\alpha > 0$, $\mu, \gamma \in (0,1)$, and $\rho_0 > 0$ be constants. Let, for any $\rho \in [0,\rho_0]$, a set $A(\rho) \subset \{1, \dots, n\}$ (possibly empty) be defined and let nonnegative nondecreasing functions in ρ, $g_i(\rho)$ and $w_i(\rho)$, $i = 1, \dots, n$, which do not exceed μ^{-1}, be defined on $[0, 2\rho_0]$. Finally, for $j = 1, \dots, n$, $\rho \in [0,\rho_0]$, and $k \in A(\rho)$ (if there is such a k), let the inequalities

$$\mu w_j(\rho) \leqq g_j(\rho) \leqq \mu^{-1} w_j(\rho), \quad g_k(\rho) w_k(2\rho)$$

$$\leqq w_k(\rho) g_k(2\rho)(1 + \mu^{-1} w_k(2\rho)), \tag{3}$$

$$w_k(\rho) \leqq \gamma w_k(2\rho) + N_2 \rho^\alpha \rho_0^{-\alpha}, \tag{4}$$

$$\sum_{i \notin A(\rho)} g_i(\rho) \leqq N_1 \sum_{i \in A(\rho)} g_i(2\rho) + N_2 \rho^\alpha \rho_0^{-\alpha}, \tag{5}$$

be true, where the sum over an empty set of indices is assumed to be zero. Then for $\rho \in [0, 2\rho_0]$

$$\sum_{i=1}^{n} w_i(\rho) \leqq N \rho^\beta \rho_0^{-\beta} \Big(\sum_{i=1}^{n} w_i(2\rho_0) + N_2 \Big), \tag{6}$$

where $\beta = \beta_0 \wedge \alpha$, β_0 depends only on γ, N_1 and $\beta_0 \in (0,1)$, the constant N depends only on α, γ, N_1, n, N_2, and μ, and if $g_i = w_i$ for all i and ρ, then N depends only on α, γ, N_1, and n.

Proof. We first suppose that $g_i \equiv w_i$. Consider a small $\kappa = \kappa(\gamma, N_1) > 0$ such that

$$0 < (N_1 + 1) \frac{\kappa(1 + \kappa)}{1 - \gamma(1 + \kappa)} \leqq \frac{1}{2}, \tag{7}$$

and prove that for any $\rho \in [0,\rho_0]$ one of the two following inequalities is true:

$$(1 + \kappa)\sum_i w_i(\rho) \leqq \sum_i w_i(2\rho),$$

$$\sum_i w_i(\rho) \leqq \left(2 + \frac{n}{\kappa}\right)N_2\rho^{\alpha}\rho_0^{-\alpha}. \tag{8}$$

We fix a $\rho \in [0, \rho_0]$ and assume that the inverse of the former inequality holds. Then

$$\sum_i w_i(2\rho) \leqq (1 + \kappa)\sum_i w_i(\rho).$$

By condition (4), we obtain

$$\sum_i w_i(\rho) = \sum_{i \notin A(\rho)} + \sum_{i \in A(\rho)}$$

$$\leqq \sum_{i \notin A(\rho)} w_i(2\rho) + \gamma \sum_{i \in A(\rho)} w_i(2\rho) + nN_2\rho^{\alpha}\rho_0^{-\alpha}.$$

Combining these inequalities, we obtain

$$\sum_{i \in A(\rho)} w_i(2\rho) + \sum_{i \notin A(\rho)} w_i(2\rho) \leqq (1 + \kappa) \sum_{i \notin A(\rho)} w_i(2\rho) +$$

$$+ (1 + \kappa)\gamma \sum_{i \in A(\rho)} w_i(2\rho) + (1 + \kappa)nN_2\rho^{\alpha}\rho_0^{-\alpha},$$

$$(1 - \gamma - \kappa\gamma) \sum_{i \in A(\rho)} w_i(2\rho) \leqq \kappa \sum_{i \notin A(\rho)} w_i(2\rho) + (1 + \kappa)nN_2\rho^{\alpha}\rho_0^{-\alpha}$$

$$\leqq \kappa(1 + \kappa)\sum_i w_i(\rho) + (1 + \kappa)nN_2\rho^{\alpha}\rho_0^{-\alpha},$$

which together with inequality (5) yield

$$\sum w_i(\rho) \leqq \sum_{i \in A(\rho)} w_i(2\rho) + \sum_{i \notin A(\rho)} w_i(\rho)$$

$$\leqq N_2\rho^{\alpha}\rho_0^{-\alpha} + (1 + N_1) \sum_{i \in A(\rho)} w_i(2\rho)$$

$$\leqq N_2\rho^{\alpha}\rho_0^{-\alpha} + (1 + N_1) \frac{\kappa(1 + \kappa)}{1 - \gamma(1 + \kappa)} \sum_i w_i(\rho) +$$

$$+ (1 + N_1) \frac{\kappa(1 + \kappa)}{1 - \gamma(1 + \kappa)} \frac{n}{\kappa} N_2\rho^{\alpha}\rho_0^{-\alpha}.$$

Collecting similar terms in the inequality and using (7), we obtain the latter inequality in (8). We have thus proved (8). We now put

$\beta_0 = \log_2(1 + \kappa)$ and $\beta = \beta_0 \wedge \alpha$, and

$$\psi(\rho) = \max\left(\sum_i w_i(\rho),\ \left(2 + \frac{n}{\kappa}\right)N_2\rho^{\alpha}\rho_0^{-\alpha}\right).$$

If the latter inequality in (8) is true, then

$$\psi(\rho) = 2^{-\alpha}\left(2 + \frac{n}{\kappa}\right)N_2(2\rho)^{\alpha}\rho_0^{-\alpha} \leqq 2^{-\alpha}\psi(2\rho) \leqq 2^{-\beta}\psi(2\rho).$$

If the latter inequality in (8) is not true, then the former one holds and

$$\psi(\rho) = \sum_i w_i(\rho) \leqq (1 + \kappa)^{-1} \sum_i w_i(2\rho) \leqq (1 + \kappa)^{-1}\psi(2\rho) \leqq 2^{-\beta}\psi(2\rho).$$

Therefore, $\psi(\rho) \leqq 2^{-\beta}\psi(2\rho)$ for all $\rho \in [0, \rho_0]$. Hence, (see the discussion following formula (2.8)) we immediately obtain $\psi(\rho) \leqq N(\beta)\rho^{\beta}\rho_0^{-\beta}\psi(2\rho_0)$ with $\rho \in [0, \rho_0]$. Since this inequality is obviously true for $\rho \in [\rho_0, 2\rho_0]$ and $N = 1$, while the maximum of two nonnegative numbers does not exceed their sum, we have proved the lemma for $g_i \equiv w_i$.

We now consider the general case and note that, by (3) and (5),

$$\sum_{i \notin A(\rho)} w_i(\rho) \leq N_1' \sum_{i \in A(\rho)} w_i(2\rho) + N_2'\rho^{\alpha}\rho_0^{-\alpha},$$

where $N_1' = N_1\mu^{-2}$ and $N_2' = N_2\mu^{-1}$. By what we have proved, $w_i(\rho) \leqq M\rho^{\sigma}\rho_0^{-\sigma}$ for $\rho \in [0, 2\rho_0]$, where $\sigma > 0$ and σ and M depend only on α, γ, N_1, N_2, μ, and n. Let $\rho_1 = \xi\rho_0$, where the constant $\xi = \xi(\alpha, \gamma, N_1, N_2, \mu, n)$ is determined from the formula

$$\xi = \frac{1}{2} \wedge \left\{\frac{1}{2}[(\gamma^{-1/2} - 1)\mu M^{-1}]^{1/\sigma}\right\}.$$

It is clear that $\xi > 0$ and $w_i(2\rho)\mu^{-1} + 1 \leqq \gamma^{-1/2}$ with $\rho \leqq \rho_1$. Therefore, by virtue of inequalities (3) and (4) with $\rho \leqq \rho_1$, $k \in A(\rho)$, and $g_k(\rho) > 0$, we obtain

$$\begin{aligned} g_k(\rho) &= \frac{g_k(\rho)}{w_k(\rho)} w_k(\rho) \leqq \gamma^{-1/2} \frac{g_k(2\rho)}{w_k(2\rho)} w_k(\rho) \\ &\leqq \gamma^{1/2} g_k(2\rho) + \gamma^{-1/2}\mu^{-1}N_2\rho^{\alpha}\rho_0^{-\alpha} \\ &= \gamma^{1/2} g_k(2\rho) + (N_2\gamma^{-1/2}\mu^{-1}\xi^{\alpha})\rho^{\alpha}\rho_1^{-\alpha}. \end{aligned}$$

Here the inequality between the extreme terms is also clear for $g_k(\rho) = 0$.

By what we have proved and in view of (5), it follows that, for $\rho \in [0, 2\rho_1]$,

$$\sum_i w_i(\rho) \leqq \mu^{-1} \sum_i g_i(\rho) \leqq N\rho^\beta \rho_1^{-\beta}\Big(\sum_i g_i(2\rho_1) + N_2\gamma^{-1/2}\mu^{-1}\xi^\alpha\Big)$$
$$\leqq N\rho^\beta \rho_0^{-\beta}\Big(\sum_i w_i(2\rho_0) + N_2\Big),$$

where the constants N and β depend on the initial constants exactly as it is said in the statement. It remains to note that, for $\rho \in [2\xi\rho_0, 2\rho_0]$, inequality (6) is obviously true with $N = (2\xi)^{-\beta}$. This proves the lemma.

3. The proof of Theorem 1. Let $c_i = L_i 1$. Then $(L_i - c_i)u^i \geqq -K_2 - K_1 \times \Sigma \sup|u^i| =: -K_2'$. If K_2 in (1) is replaced by K_2', then this inequality remains valid, and if (2) can be proved with K_2' rather than K_2, then, recalling that $\tilde{\rho}^\nu \leqq 1$ and $K_2 \leqq K_2(1 + K_1)$, we obtain (2) in its original form. Thus, we can assume that $c_i = 0$ for all i.

We further fix a point $z_0 = (t_0, x_0) \in Q$ and denote

$$\rho_0 = \tilde{\rho}(z_0)28^{-1}(1 + K_1)^{-1} (\varepsilon \wedge 1),$$

$$m_i(\rho) = \inf(u^i, z_0 + C_{\rho^2,\rho}), \quad M_i(\rho) = \sup(u^i, z_0 + C_{\rho^2,\rho}),$$

$$w_i(\rho) = M_i(\rho) - m_i(\rho), \quad g_i(\rho) = f_i(M_i(\rho)) - f_i(m_i(\rho)),$$

$$Q(\rho) = (z_0 + C_{4\rho^2, 2\rho}) \cap \{t > t_0 + \rho^2\}, \quad \xi = \frac{1}{2(n + \delta)}\delta,$$

$$\Gamma_i(\rho) = Q(\rho) \cap \{z : f_i(u^i(z)) < (1 - \xi)f_i(M_i(\rho)) + \xi f_i(m_i(\rho))\},$$

$$A(\rho) = \Big\{i : \Lambda(\Gamma_i(\rho)) \geqq \frac{1}{n}\Lambda(Q(\rho))\Big\},$$

$$B(\rho) = \Big\{i : \Lambda(\Gamma_i(\rho)) < \frac{1}{n}\Lambda(Q(\rho))\Big\}.$$

Note that $28\rho_0 \leqq \tilde{\rho}(z_0)$ and m_i, M_i, w_i, and g_i are defined with $\rho \in [0, (1/2)\tilde{\rho}(z_0)]$, while remaining objects will be considered only if $\rho \in [0, 2\rho_0]$ (when $Q(\rho)$ evidently lies in Q). It is obvious that the union of all $\Gamma_i(\rho)$ with $i \in B(\rho)$ does not make up the whole $Q(\rho)$. Therefore, there exists a point $z(\rho)$ belonging to $Q(\rho)$ which does not belong to $\Gamma_i(\rho)$ for any $i \in B(\rho)$:

$$z(\rho) \in Q(\rho), \quad f_i(u^i(z(\rho))) \geqq (1-\xi)f_i(M_i(\rho)) + \\ + \xi f_i(m_i(\rho)) \quad \forall i \in B(\rho). \tag{9}$$

We complete the list of objects which will be needed by introducing points $z_1^r(\rho) \in z_0 + C_{\rho^2,\rho}$ $(\rho \leqq 2\rho_0)$ such that $u^i(z_i^r(\rho)) \to m_i(\rho)$ as $r \to \infty$.

We now show that

$$\sum_{i\in B(\rho)} g_i(\rho) \leqq \frac{2n}{\delta} \sum_{i\in A(\rho)} g_i(2\rho) + (NK_2\rho_0^\nu)\rho^\alpha\rho_0^{-\alpha}. \tag{10}$$

if $\rho \leqq \rho_0$ and $N = N(n, \delta, \alpha)$.

Consider a $\rho \leqq \rho_0$ without loss of generality assume that $B(\rho) \neq \emptyset$, and apply inequality (1) to the points $z(\rho)$ and $z_j^r(\rho)$ with $j \in B(\rho)$. Both points lie in $z_0 + C_{4\rho^2,2\rho}$ and $\rho \leqq \rho_0$, and their parabolic distances to $\partial' Q$ are, therefore, no smaller than $24\rho_0$, while the parabolic distance between them is no greater than $6\rho \leqq (1/4)24\rho_0$. Taking into account that $\nu \leqq \alpha$, we obtain from (1)

$$(NK_2\rho_0^\nu)\rho^\alpha\rho_0^{-\alpha} \geqq \delta \sum_i [f_i(u^i(z(\rho))) - f_i(u^i(z_j^r(\rho)))]_+ - \\ - \sum_i [f_i(u^i(z(\rho))) - f_i(u^i(z_j^r(\rho)))]_- . \tag{11}$$

In the first sum, we omit all summands except for that corresponding to $i = j$ and, using (9), we estimate this summand as

$$f_j(u^j(z(\rho))) - f_j(u^j(z_j^r(\rho)))$$

$$\geqq (1-\xi)f_j(M_j(\rho)) + \xi f_j(m_j(\rho)) - f_j(u(z_j^r(\rho))) \to (1-\xi)g_j(\rho).$$

In the second sum, we obtain

$$f_i(u^i(z(\rho))) - f_i(u^i(z_j^r(\rho)))$$

$$\geqq (1-\xi)f_i(M_i(\rho)) + \xi f_i(m_i(\rho)) - f_i(M_i(\rho)) = -\xi g_i(\rho)$$

with $i \in B(\rho)$.

When $i \in A(\rho)$, we roughly estimate respective expressions and their negative parts from below in terms of $(-g_i(2\rho))$ and from above in terms of $g_i(2\rho)$, respectively. As a result, we obtain from (11)

$$(NK_2\rho_0^\nu)\rho^\alpha\rho_0^{-\alpha} \geqq \delta(1-\xi)g_j(\rho) - \xi \sum_{i\in B(\rho)} g_i(\rho) - \sum_{i\in A(\rho)} g_i(2\rho).$$

Summing all these inequalities over $j \in B(\rho)$ and using the relationship $\delta(1-\xi) - n\xi = (1/2)\sigma$, we obtain (10).

We further fix a $\rho \leqq \rho_0$, assume that $A(\rho) \neq \emptyset$, and prove that

$$w_k(\rho) \leqq \gamma w_k(2\rho) + NK_2\rho^2 \leqq \gamma w_k(2\rho) + (NK_2\rho_0^\nu)\rho^\alpha\rho_0^{-\alpha}, \tag{12}$$

for all $k \in A(\rho)$, where $\gamma \in (0,1)$, γ depends only on d, K, ε, n, and δ, while the constant N depends only on the same parameters and K_1. To prove this fact, we set

$$v(t, x) = \varepsilon^{-1}K_2(1+K_1)(4\rho^2 - |x-x_0|^2)$$

and note that for every k ($c_k = 0$)

$$L_k u^k \geqq -K_2,$$

$$K_2^{-1}(1+K_1)^{-1}\, L_k v \leqq -2d + 4\varepsilon^{-1}\rho K_1 d \leqq -2 + \tilde{\rho}(z_0) \leqq -1, \tag{13}$$

$$L_k(u^k - v) \geqq 0.$$

at $z_0 + C_{4\rho^2,2\rho}$.

We now perform the coordinate transformation $(t, x) \to (4^{-1}\rho^{-2}(t - t_0), (2\rho)^{-1}(x - x_0))$, which converts $z_0 + C_{4\rho^2,2\rho}$ to $C_{1,1}$, and denote images produced by this transformation by the use of the tilde. Inequality (13) is converted to $\tilde{L}_k(\tilde{u}^k - \tilde{v}) \geqq 0$ on $C_{1,1}$, where the operator $\tilde{L}_k$ is of the form (2.1), $\tilde{c}_k := \tilde{L}_k 1 = 0$, and since $2\rho \leqq (1+K_1)^{-1}$ the coefficients $\tilde{b}_k^j$ of u_{x^j} in the expression of $\tilde{L}_k u$ satisfy the inequality $|\tilde{b}_k^j| \leqq \tilde{K}_1$, where $\tilde{K}_1 = 1$. Also, we fix $k \in A(\rho)$, set

$$\zeta = (1-\xi)M_k(\rho) + \xi m_k(\rho)$$

and note that, by the concavity of the increasing function f_k, we obtain

$$f_k(u^k) < (1-\xi)f_k(M_k(\rho)) + \xi f_k(m_k(\rho)) \leqq f_k(\zeta)$$

on $\Gamma_k(\rho)$ and, therefore, $u^k < \zeta$. Hence it is clear that, for $t \in [0, 1/4]$,

$$\Lambda[((t, 0) + C_{1-t,1}) \cap \{\tilde{u}^k \leqq \zeta\}] \geqq (2\rho)^{-d-2}\Lambda(\Gamma_k(\rho))$$

$$\geqq (2\rho)^{-d-2}\,\frac{1}{n}\Lambda(Q(\rho)) = \frac{3}{4n(1-t)}\,\Lambda((t, 0) + C_{1-t,1}).$$

Since $\tilde{v} \geqq 0$ on $C_{1,1}$, these inequalities remain valid if $\tilde{u}^k$ is replaced by $\tilde{u}^k - \tilde{v}$. From the above discussion concerning $\tilde{L}_k$ and $\tilde{u}^k - \tilde{v}$, by Lemma 1.11 (with $\kappa = 1 - t$), we obtain

$$\tilde{u}^k(t, x) - \tilde{v}(t, x) \leqq p\left(0, x, \frac{3}{4n}\right)\zeta + \left(1 - p\left(0, x, \frac{3}{4n}\right)\right)M_k(2\rho)$$

for all $t \in (0, 1/4]$ and $|x| \leqq 1$.

Here $\zeta \leqq M_k(2\rho)$ and the latter inequality is only strengthened if p is replaced by its estimate. By Theorem 1.9 with $t \in (0, 1/4]$ and $|x| \leqq 1/2$,

$$\tilde{u}^k(t, x) \leqq p_0\zeta + (1 - p_0)M_k(2\rho) + \tilde{v}(t, x),$$

where $p_0 \in (0,1)$ and p_0 depends only on d, K, ε, n (and $\tilde{K}_1$ which is equal to 1). Calculating the least upper bounds, we obtain

$$M_k(\rho) \leqq p_0\zeta + (1 - p_o)M_k(2\rho) + 4\varepsilon^{-1}(1 + K_1)K_2\rho^2$$

$$= p_0M_k(\rho) - \xi p_0w_k(\rho) + (1 - p_0)M_k(2\rho) + NK_2\rho^2,$$

$$(1 - p_0)M_k(\rho) + \xi p_0w_k(\rho) \leqq (1 - p_0)M_k(2\rho) + NK_2\rho^2.$$

To complete the proof of (12), we should add the latter inequality to the obvious inequality $-(1 - p_0)m_k(\rho) \leqq -(1 - p_0)m_k(2\rho)$, divide the result by $1 - p_0 + \xi p_0$, and use the inequalities $\alpha \leqq 2$ and $\nu \leqq \alpha$.

We now complete the proof of the theorem in the case when $f_i(r) = r$ for all i, r. Then $g_i = w_i$ and from (10) and (12), by Lemma 2, it follows that

$$\sum_i w_i(\rho) \leqq N\rho^{\beta}\rho_0^{-\beta}\left(K_2\rho_0^{\nu} + \sum_i \sup_Q|u^i|\right), \tag{14}$$

with $\rho \leqq \rho_0$, where β and N depend on the initial parameters as it is stipulated in the statement of the theorem in the case $f_i(r) \equiv r$. The same inequality with $N = 2$ holds for all $\rho \geqq \rho_0$ such that the left-hand side of (14) makes sense. Recalling the definition of ρ_0, we obtain from (14)

$$\sum_i w_i(\rho) \leq N\rho^\beta(\tilde{\rho}(z_0))^{-\beta}\Big((K_2(\tilde{\rho}(z_0))^\nu + \sum_i \sup_Q |u^i|\Big). \tag{15}$$

Finally, if $z_1 = (t_1, x_1)$, $z_2 = (t_2, x_2) \in Q$, $4\rho(z_1, z_2) \leq \tilde{\rho}(z_1) \wedge \tilde{\rho}(z_2)$, and for the sake of definiteness $t_1 \leq t_2$, then, putting $z_0 = z_1$, $\rho = \rho(z_1, z_2)$ and taking note that

$$z_1, z_2 \in z_0 + \bar{C}_{\rho^2,\rho} \subset \bar{Q}, \quad \tilde{\rho}(z_1) - \tilde{\rho}(z_2) \leq \rho,$$

$$\tilde{\rho}(z_1) \leq \frac{5}{4}\tilde{\rho}(z_2), \quad \tilde{\rho}(z_1) \wedge \tilde{\rho}(z_2) \leq \tilde{\rho}(z_0) \leq \frac{5}{4}(\tilde{\rho}(z_1) \wedge \tilde{\rho}(z_2)),$$

we can readily obtain (2) from (15). But if $4\rho(z_1, z_2) > \tilde{\rho}$, then (2) with $N = 8$ is clear.

In the case of arbitrary f_i the assertion of the theorem, by virtue of (10) and (12), also follows from Lemma 2, if we prove inequalities (3) with μ which depends only on the estimates of f_i', $(f_i')^{-1}$, and $|f_i''|$ from above. The first inequality in (3) is clear and the second one becomes clear if $g_k(\rho) = 0$. But if $g_k(\rho) \neq 0$, then

$$\frac{g_k(\rho)}{w_k(\rho)} = f_k'(u^k(z_k^1)) = f_k'(u^k(z_k^1)) - f_k'(u^k(z_k^2))$$

$$+ \frac{g_k(2\rho)}{w_k(2\rho)} \leq \frac{g_k(2\rho)}{w_k(2\rho)}(1 + \mu^{-1}w_k(2\rho)),$$

by virtue of the Lagrange theorem for some $z_k^1, z_k^2 \in z_0 + \bar{C}_{4\rho^2,2\rho}$ with μ^{-1} being the maximum of $|f_i''|(f_i')^{-1}$ over the respective values of r. This proves the theorem.

4. <u>Remark</u>. From Theorem 1 we obtain a result which is similar to Theorem 2.6. Namely let $u^1 \in W^{1,2}_{d+1,loc}(Q)$ and let $|Lu^1| \leq K_2$ in Q. Let us take $u^2 = -u^1$. Then for every $z_1, z_2 \in Q$

$$\sum_{i=1}^2 [u^i(z_1) - u^i(z_2)]_+ - \sum_{i=1}^2 [u^i(z_1) - u^i(z_2)]_-$$

$$= \sum_{i=1}^2 (u^i(z_1) - u^i(z_2)) = 0.$$

Therefore, one can take $f_i(r) = r$ and $2\delta = \alpha = \nu = 2$ and obtain

$$|u^1(z_1) - u^1(z_2)| \leq \tilde{\rho}^{-\beta}\rho^\beta(z_1, z_2)N(K_2\tilde{\rho}^2 + \sup(|u|, Q))$$

from (2).

Here we require that $|Lu^1| \leqq K_2$. In fact, the condition $L_i u^i \geqq -K_2$ of Theorem 1 can be replaced by the condition $L_i u^i \geqq -f^i$, $||f^i||_{L_{d+1}(Q)} \leqq K_2$ and a more complete version of Theorem 2.6 can then be deduced from this modification of Theorem 1. In what follows we shall not need the modification, so its formulation and proof is left to the reader.

From Theorem 1 it also follows that inequality (2) holds true with $n = 1$ and $u^1 := u$ if $u, v \in W^{1,2}_{d+1,loc}(Q)$, $|Lu| \leqq Lv + K_2$ in Q, and $|v(z_1) - v(z_2)| \leqq K_2 \tilde{\rho}^{\nu-\alpha}\rho^{\alpha}(z_1, z_2)$ with $4\rho(z_1, z_2) \leqq \tilde{\rho} := \tilde{\rho}(z_1) \wedge \tilde{\rho}(z_2)$.

Indeed, if $u^1 = v - u$, $u^2 = v + u$, we obtain $Lu^i \geqq -K_2$ and the right-hand side of (1) is equal to $2(v(z_1) - v(z_2))$ for $f_i(r) = r$ and $\delta = 1$. This version of Theorem 1 is essentially stronger than Theorem 2.6 as can be seen from Theorems 3.2.3 and 3.2.4.

Condition (1) with zero on the left is satisfied for a $\delta > 0$ if, roughly speaking, at every point and for every direction such that there exists at least one function u^i increasing at this point along this direction there also exists a function u^j decreasing at this point along this direction. The reader will find the realization of this idea in the following theorem.

5. THEOREM. Let $\tau \in (0,1)$, let $u^1, \ldots, u^n \in W^{1,2}_{d+1,loc}(Q)$, and let $g^1, \ldots, g^d$ be some functions on Q. Suppose that $|u^j| \leqq K_2$ and

$$|Lu^j - u^j_{x^k}g^k| \leqq K_2(1 + (u^i_x, u^i_x))^{\tau}, \quad g^k g^k \leqq K_2(1 + (u^i_x, u^i_x))$$

on Q (a.e.) for all $j = 1, \ldots, n$ (the sum is taken over i and k). Then there exist constants $\beta \in (0,1)$ and $N < \infty$ such that

$$\sum_{i=1}^{n} |u^j(z_1) - u^j(z_2)| \leqq N\rho^{\beta}(z_1, z_2)(\tilde{\rho}(z_1) \wedge \tilde{\rho}(z_2))^{-\beta} \tag{16}$$

for $z_1, z_2 \in Q$. Moreover, β depends only on d, K, ε, n, while N depends on the same parameters and τ, K_1, and K_2.

<u>Proof</u>. Let $Lu = a^{ij}u_{x^ix^j} + b^iu_{x^i} + cu + u_t$ and $L_0u = a^{ij}u_{x^ix^j} + u_t$. Then

$$|L_0u^j - u_{x^k}g^k| \leqq |Lu^j - u^j_{x^k}g^k| + dK_1(|u^j_x| + |u^j|)$$
$$\leqq N(K_1, K_2, \tau)(1 + (u^i_x, u^i_x))^{\tau\vee 1/2}.$$

Thus, varying K_2, we can assume that $b^i = c = 0$. So, let $b^i = c = 0$. We fix a constant $\gamma > 0$, whose choice will be defined later, and denote

$$v(p, t, x) = \exp\left\{\gamma \sum_{j=1}^{n} (u^j(t, x) - p^j)^2\right\}$$

for $(t, x) \in Q$ and $p = (p^1, \ldots, p^n) \in E_n$. The embedding theorems imply that $v \in W^{1,2}_{d+1,loc}(Q)$ for any p. Straightforward calculations yield

$$2^{-1}\gamma^{-1}v^{-1}Lv = 2\gamma a^{ij}\left[(u^k - p^k)u^k_{x^i}\right]\left[(u^r - p^r)u^r_{x^j}\right] +$$

$$+ a^{ij}u^r_{x^i}u^r_{x^j} + (u^k - p^k)Lu^k \quad (a.e.Q). \tag{17}$$

We estimate the first two terms on the right from below using inequality $a^{ij}\lambda^i\lambda^j \geq \varepsilon|\lambda|^2$ and replace the latter term by

$$(u^k - p^k)(Lu^k - u^k_{x^r}g^r) + \left[(u^k - p^k)u^k_{x^r}\right]g^r. \tag{18}$$

Further estimates will be derived under the assumption that $|p^k| \leq 2K_2$ with $k = 1, \ldots, n$. Then, by Young's inequality, expression (18) is estimated from below as

$$-3dK_2^2(1 + (u^i_x, u^i_x))^\tau - \frac{\varepsilon g^r g^r}{2(K_2 + 1)} -$$

$$- \frac{K_2 + 1}{2\varepsilon}(u^k - p^k)(u^r - p^r)\, u^k_{x^i}u^r_{x^i}$$

$$\geq -\varepsilon\left(u^i_x, u^i_x\right) - \frac{K_2 - 1}{2\varepsilon}(u^k - p^k)(u^r - p^r)\, u^k_{x^i}u^r_{x^i} -$$

$$- N(d, \tau, \varepsilon, K_2).$$

From the above discussion of equation (17) it now follows that $Lv \geq -N$ for $\gamma = (K_2 + 1)(2\varepsilon)^{-2}$ almost everywhere in Q whenever $|p^k| \leq 2K_2$ for all k, where $N = N(d, \tau, \varepsilon, K_2)$.

We now make use of the idea preceding the statement of the theorem. Let $m = 2^n$ and let $p_1, \ldots, p_m$ be all vertices of the cube $\{p \in E_n : |p^i| \leq 2K_2, i = 1, \ldots, n\}$, let $v^r(t, x) := v(p_r, t, x)$, $r = 1, \ldots, m$. It is clear that if $|u^i_1|, |u^i_2| \leq K_2$ for $i = 1, \ldots, n$, and the motion from the point $(u^1_1, \ldots, u^n_1)$ to the point $(u^1_2, \ldots, u^n_2)$ increases the distance to a vertex p_i, then the distance to a different vertex decreases. More precisely, we fix $z_1, z_2 \in Q$, put $\bar{p}^i = 2K_2 \cdot \operatorname{sgn}(u^i(z_2) - u^i(z_1))$, $i = 1, \ldots, n$, and note that $\bar{p} = (\bar{p}^1, \ldots, \bar{p}^n) \in \{p_i\}$ and

$$\begin{aligned}
&\frac{1}{3}(\ln v^r(z_1) - \ln v^r(z_2)) \\
&\quad = \frac{\gamma}{3} \sum_{i=1}^{n} \Big[(u^i(z_1) - u^i(z_2))(u^i(z_1) + u^i(z_2)) + \\
&\qquad + 2p_r^i(u^i(z_2) - u^i(z_1)) \Big] \\
&\quad \leqq \frac{\gamma}{3} 6K_2 \sum_{i=1}^{n} |u^i(z_2) - u^i(z_1)| \\
&\quad = \gamma \sum_{i=1}^{n} \Big[2\bar{p}^i(u^i(z_2) - u^i(z_1)) - 2K_2 |u^i(z_2) - u^i(z_1)| \Big] \\
&\quad \leqq \ln v^s(z_2) - \ln v^s(z_1),
\end{aligned} \tag{19}$$

where s is such that $p_s = \bar{p}$. Hence

$$\begin{aligned}
\frac{1}{3} \sum_{r=1}^{m} (\ln v^r(z_1) - \ln v^r(z_2))_+ &\leqq m(\ln v^s(z_1) - \ln v^s(z_2))_- \\
&\leqq m \sum_{r=1}^{m} (\ln v^r(z_1) - \ln v^r(z_2))_- .
\end{aligned}$$

By Theorem 1, every expression $|v^r(z_1) - v^r(z_2)|$ is less than or equal to the right-hand side of (16). This is also true for $|\ln v^r(z_1) - \ln v^r(z_2)|$ and, by virtue of (19), for the left-hand side of (16). This proves the theorem.

6. COROLLARY. Let $u \in W^{1,2}_{d+1,loc}(Q)$, let $|u| \leqq K_2$, and let $|Lu| \leqq K_2(1 + |u_x|^2)$ (a.e.Q). Then there exist constants $\beta \in (0,1)$ and $N < \infty$ such that

$$|u(z_1) - u(z_2)| \leqq N\rho^\beta(z_1, z_2)(\tilde{\rho}(z_1) \wedge \tilde{\rho}(z_2))^{-\beta}$$

with $z_1, z_2 \in Q$. Moreover, β depends only on d, K, ε, while N depends only on d, K, ε, K_1, and K_2.

Indeed, $Lu = g(1 + |u_x|^2)$, where $|g| \leqq K_2$ and for $g^k := gu_{x^k}$ it is clear that $Lu - u_{x^k} g^k = g$ and $g^k g^k \leqq K_2^2 |u_x|^2$.

We now extend the results of this section to inequalities with elliptic operators. Below in this section we assume that D is a domain in E_d, the coefficients $L, L_1, \ldots, L_n$ do not depend on t, and $\tilde{L} = L - \frac{\partial}{\partial t}$, $\tilde{L}_i = L_i - \frac{\partial}{\partial t}$.

7. THEOREM. Let $u^1, \ldots, u^n \in W^2_{d,loc}(D)$, let $f_1, \ldots, f_n \in C^2_{loc}(E_1)$, let $f'_i > 0$ and $f''_i \leq 0$ on E_1, and let $\tilde{L}_i u^i \geq -K_2$ in D (a.e.) for every $i = 1, \ldots, n$. Suppose that

$$K_2\rho^{\nu-\alpha}|x_1 - x_2|^\alpha \geq \delta \sum_{i=1}^n [f_i(u^i(x_1)) - f_i(u^i(x_2))]_+ - \\ - \sum_{i=1}^n [f_i(u^i(x_1)) - f_i(u^i(x_2))]_-$$

for constants $\alpha \in (0, 2]$, $\nu \in [0, \alpha]$ and $\delta > 0$ and for any $x_1, x_2 \in D$ such that $4|x_1 - x_2| \leq \rho := d(x_1, \partial D) \wedge d(x_2, \partial D) \wedge 1$. Then, for all $x_1, x_2 \in D$,

$$\sum_{i=1}^n |u^i(x_1) - u^i(x_2)| \leq \rho^{-\beta}|x_1 - x_2|^\beta N\Big(K_2\rho^\nu + \sum_{i=1}^n \sup_D|u^i|\Big),$$

where $\beta = \beta_0 \wedge \alpha$, $\beta_0 \in (0, 1)$, β_0 depends only on n, d, K, ε, δ and the constant N depends only on the same parameters plus K_1 if $f_i(r) = r$ for all i, r, and if $f_i(r) \not\equiv r$, then N also depends on K_2, the constants M_i (which are not necessarily least possible), which are bounds on $|u^i|$ on D, and the maximum values of f'_i, $(f'_i)^{-1}$, and $|f''_i|$ on $[-M_i M_i]$.

This theorem can be proved by almost repeating the proof of Theorem 1 replacing $C_{\rho^2,\rho}$ there by S_ρ, omitting t, and using Theorem 1.12 instead of Lemma 1.11. The proof can be simplified if S_ρ, rather than $S_{2\rho}$, is taken as $Q(\rho)$. In this case a weaker assertion can be used in place of Lemma 2, where $g_i(2\rho)$ is replaced by $g_i(\rho)$ on the right of (5).

8. THEOREM. Let $\tau \in (0, 1)$, $u^1, \ldots, u^n \in W^2_{d,loc}(D)$ and let $g^1, \ldots, g^d$ be some functions on D. Suppose that $|u^j| \leq K_2$ and

$$|\tilde{L}u^j - u^j_{x^k}g^k| \leq K_2(1 + (u^i_x, u^i_x))^\tau, \quad g^k g^k \leq K_2(1 + (u^i_x, u^i_x)),$$

on D(a.e.) for all $j = 1, \ldots, n$. Then there exist constants $\beta \in (0, 1)$ and $N < \infty$ such that

$$\sum_{j=1}^{n} |u^j(x_1) - u^j(x_2)|$$

$$\leq N|x_1 - x_2|^{\beta}(d(x_1, \partial D) \wedge d(x_2, \partial D) \wedge 1)^{-\beta} \tag{20}$$

with x_1, $x_2 \in D$. Moreover, β depends only on d, K, ε, and n, while N depends only on d, K, ε, n, τ, K_1, and K_2.

The proof of this theorem is identical to the proof of Theorem 5.

The analogue of Corollary 6 can be proved by the above argument.

9. COROLLARY. Let $u \in W_d^2(D)$, let $|u| \leq K_2$, and let $|\tilde{L}u| \leq K_2(1 + |u_x|^2)$ (a.e.D). Then there exist constants $\beta \in (0, 1)$ and $N < \infty$, such that inequality (20) is true with n = 1 and $u^1 = u$ whenever x_1, $x_2 \in D$. Moreover, β depends only on d, K and ε, while N depends only on d, K, ε, K_1, and K_2.

4.4. The Hölder Condition for Solutions of Elliptic and Parabolic Equations near the Boundary

When proving the solvability of nonlinear equations, we first construct a priori interior estimates of a solution in $C^{2+\alpha}$, then boundary estimates in $C^{2+\alpha}$, and then 'paste' these estimates together. To paste them, we need some additional results which it is relevant to prove at this point and which are applied, first of all, to estimating the Hölder constant for solutions of linear equations near the boundary under the assumption that boundary data satisfy a Hölder condition.

Consider some constants R, T, K, K_1, K_2 and $\varepsilon > 0$. Let

$$L = \frac{\partial}{\partial t} + a^{ij}(t, x)\frac{\partial^2}{\partial x^i \partial x^j} + b^i(t, x)\frac{\partial}{\partial x^i} + c(t, x) \tag{1}$$

be a parabolic operator whose coefficients are defined on E_{d+1}, measurable in (t, x), and such that

$$a^{ij} = a^{ji}, \quad \varepsilon|\lambda|^2 \leq a^{rs}\lambda^r\lambda^s \leq K|\lambda|^2, \quad |b^i| \leq K_1, \quad |c| \leq K_2 \tag{2}$$

for all $(t, x) \in E_{d+1}$, $\lambda \in E_d$, and i, j = 1, ..., d. Recall that

$$Q_{T,R} = \{(t, x) : 0 < t < T, \quad |x^i| < R, \quad i = 1, \dots, d\},$$

$$Q^+_{T,R} = Q_{T,R} + (0, R, 0, \dots, 0).$$

We need some barrier functions. Let $\varphi(t)$ and p(t) be nonnegative functions defined on $[0, \infty)$ such that φ is continuous on $[0, \infty)$ φ' is piecewise continuous and nonnegative on $(0, \infty)$, and p is measurable. Also, let $q \geqq 0$ and $\nu > 0$ be some constants. For $i = 1, 2, \ldots, d$, we denote

$$\xi^i_{(\pm)}(T - t, x) = \exp[q^2Kt + q(K_1t \pm x^i - R)], \quad t \geqq 0,$$

$$\eta^i_{(\pm)}(T - t, x) = \int_0^\infty \varphi'(y) \exp[p^2(y)Kt + p(y)(K_1t \pm x^i - R)dy, \quad t \geqq 0,$$

$$\zeta(T - t, x) = \frac{1}{\sqrt{2\pi t}} \int_{-\infty}^{\tilde{x}^1} e^{-\frac{(y + K_3t)^2}{2t}} dy - \frac{e^{-2K_3\tilde{x}^1}}{\sqrt{2\pi t}} \int_{-\infty}^{-\tilde{x}^1} e^{-\frac{(y + K_3t)^2}{2t}} dy,$$

$$\tilde{x}^1 = \frac{1}{\sqrt{2\varepsilon}} x^1, \quad K_3 = \frac{1}{\sqrt{2\varepsilon}} K_1, \quad t \in (0, T], \quad x^1 \geqq 0,$$

$$v^\nu(s, x) = \frac{1}{\sqrt{4\pi\nu}} s^{-\frac{3}{2}} x^1 e^{-\frac{(x^1)^2}{4\nu s}}, \quad s > 0,$$

$$\tau(t, x) = \int_0^\infty \varphi(t + s)v^\varepsilon(s, x)\, ds, \quad x^1 > 0, \quad t \geqq 0,$$

$$\tau(t, x) = \varphi(t), \quad x^1 = 0, \quad t \geqq 0,$$

$$\sigma^i(x) = \int_0^\infty \varphi((x^i)^2 + s)v^\nu(s, x)\, ds, \quad x^1 > 0,$$

$$\sigma^i(x) = \varphi((x^1)^2), \quad x^1 = 0.$$

The assertions of the following lemma concerning one or another property of the introduced functions without specifying a set of arguments always mean that the property in question holds for all possible arguments of the function. The only exception is the assertions concerning derivatives, which are meant to hold on the interior of the respective domains of definition. In addition, we leave it for the reader to check that the functions, which are described there, are continuous (everywhere on the domain of definition) under the conditions of assertions (a) to (g) and belong to the class C^2_{loc} on the interior of the domains of their definition.

1. LEMMA. Suppose that $c \leqq 0$. Then the following assertions hold true:

(a) $\xi^i_{(\pm)}(T, x) = 1$ for $x^i = \pm R$ and $L\xi^i_{(\pm)} \leqq 0$.

(b) If φ' is a function with compact support, and p is bounded, then $\eta^i_{(\pm)}(T, x) \geqq \varphi(|x^i|) - \varphi(0)$ for $\pm x^i \geqq 0$ and $L\eta^i_{(\pm)} \leqq 0$.

(c) $\zeta(T-, x) = 1$ for $x^1 > 0$, $L\zeta \leqq 0$, $\zeta \geqq 0$, $\zeta(t, x) \leqq N(\varepsilon)x^1(K_1 + (T - t)^{-1/2})$.

(d) If $\alpha \in (0, 1)$, $\varphi(t) = t^{\alpha/2}$, and $b^1 \leqq 0$, then $\tau_t \geqq 0$, $\tau(0, x) \leqq N(x^1)^\alpha$, $t^{\alpha/2} + (x^1)^\alpha \leqq N\tau(t, x)$, where $N = N(\alpha, \varepsilon)$, and $L\tau \leqq 0$.

(e) If $i = 2, \ldots, d$, $\alpha \in (0, 1)$, $\varphi(t) = t^{\alpha/2}$, and $b^1 \leqq 0$, then

$$N^{-1}(|x^i| + |x^1|)^\alpha \leqq \sigma^i(x) \leqq N(|x^i| + |x^1|)^\alpha,$$

where $N = N(\alpha, \nu)$, moreover, there exists a $\nu = \nu(\alpha, \varepsilon, K, K_1) > 0$ such that $L\sigma^i \leqq 0$ for $|x^i| \leqq 2$, $x^1 > 0$.

(f) If $\alpha > 1$, $\varphi(t) = t^{\alpha/2}$ for $t \leqq T + 1$, $\varphi(t) = (T + 1)^{\alpha/2}$ for $t \geqq T + 1$, and $b^1 \leqq 0$, then

$$N\tau(T, x) \geqq x^1 \text{ for } x^1 \in [0, 6],\ \tau(0, x)$$
$$\leqq Nx^1(T + 1)^{\frac{\alpha}{2} - \frac{1}{2}}, \quad \tau(t, x) \geqq \varphi(t), \quad L\tau \leqq 0$$

for $N = N(\alpha, \varepsilon)$.

(g) If $i = 2, \ldots, d$, $\alpha \in (1, 2)$, $\varphi(t) = t^{\alpha/2}$ for $t \leqq 10$, $\varphi(t) = (11)^{\alpha/2}$ with $t \geqq 11$, φ'' is bounded for $t \geqq 9$, $\varphi'' \leqq 0$, and $b^1 \leqq 0$, then

$$\sigma^i(x^1, 0, \ldots, 0) \leqq N(\nu, \varphi)x^1;$$

moreover, there exist constants ν and $\delta > 0$ depending only on ε, K, K_1, and φ such that $L\sigma^i \leqq 0$ for $|x^i| \leqq 3$, $x^1 \leqq \delta$.

Proof. Inequalities $L\xi^i_{(\pm)} \leqq 0$ and $L\eta^i_{(\pm)} \leqq 0$ can be checked by simple calculations taking into account that p, q, and φ' are positive. The equality $\xi^i_{(\pm)}(T, x) = 1$ with $x^i = \pm R$ is clear, while the inequality

$$\eta^i_{(\pm)}(T, x) \geqq \varphi(|x^i|) - \varphi(0)$$

with $\pm x^i \geqq 0$ follows from

$$\eta^i_{(\pm)}(T, x) = \int_0^\infty \varphi'(y) \exp p(y) (\pm x^i - y)\, dy$$

$$\geqq \int_0^{\pm x^i} \varphi'(y)\, dy = \varphi(x^i) - \varphi(0).$$

To prove (c), we denote

$$u(t, x^1) = \frac{1}{\sqrt{2\pi t}} \int_{-\infty}^{x^1} e^{-(y + K_3 t)^2/(2t)}\, dy.$$

It is well known that $u_t = (1/2)u_{x^1x^1} + K_3 u_{x^1}$ with $t > 0$, It obviously follows that the function

$$w(t, x^1) := u(t, x^1) - u(t, -x^1) \exp(-2K_3 x^1)$$

satisfies the same equation. Moreover, the formulas for w_{x^1} and $w_{x^1x^1}$ directly imply that $w_{x^1} \geqq 0$ and $w_{x^1x^1} \leqq 0$ for $x^1 \gneqq 0$. From these properties of the derivatives of w and the relationship $\zeta(t, x) = w(T - t, \tilde{x}^1)$ it follows that $L\zeta \leqq 0$.

The inequality $\zeta \geqq 0$ (for $x^1 \geqq 0$) is clear, the equality $\zeta(T-, x) = 1$ with $x^1 > 0$ follows from the well-known equalities $u(0+, x^1) = 1$ for $x^1 > 0$ and $u(0+, x^1) = 0$ for $x^1 < 0$. The inequality $\zeta(t, x) \leqq Nx^1(K_1 + (T - t)^{-1/2})$ follows from the equality $\zeta(t, x) = 0$ for $x^1 = 0$, and the inequality $w_{x^1} \leqq 2(K_3 + (2\pi t)^{-1/2})$ for $x^1 \geqq 0$, which in turn straightforwardly follows from the explicit expression for w_{x^1}.

Also, note that, if $K_1 = 0$, we can obtain another expression for ζ:

$$\zeta(t, x) = \int_{T-t}^{\infty} v^{\varepsilon}(s, x)\, ds, \quad x^1 > 0. \tag{3}$$

It follows, for example, from the fact that $\zeta_t + \varepsilon\zeta_{x^1x^1} = 0$ in $(0, T) \times \{x^1 > 0\}$, the right-hand side of (3) satisfies the same equation, both functions are bounded and their limit values coincide on the parabolic boundary of $(0, T) \times \{x^1 > 0\}$ (everwhere except for the 'corner').

To prove (d), we first note that the condition $\alpha \in (0, 1)$ provides the convergence of the integral defining τ. After the substitution $s = (x^1)^2 r$, we can see that $\tau(0, x) = N(x^1)^{\alpha}$ and, since

$$\varphi(t + s) \geqq 2^{\frac{\alpha}{2} - 1}\left(t^{\frac{\alpha}{2}} + s^{\frac{\alpha}{2}}\right),$$

we obtain $N\tau(t, x) \geqq t^{\alpha/2} + |x^1|^\alpha$. Furthermore, the function v^ε satisfies the equation $v^\varepsilon_s = \varepsilon v^\varepsilon_{x^1x^1}$ for $s > 0$. Therefore,

$$\tau_{x^1x^1}(t, x) = \int_0^\infty \varphi(t + s) v^\varepsilon_{x^1x^1}(s, x)\, ds$$

$$= \varepsilon^{-1} \int_0^\infty \varphi(t + s) v^\varepsilon_s(s, x)\, ds$$

$$= -\varepsilon^{-1} \int_0^\infty \varphi_t(t + s) v^\varepsilon(s, x)\, ds$$

$$= -\varepsilon^{-1} \tau_t(t, x) \leqq 0, \tag{4}$$

the latter inequality is true since $\varphi_t \geqq 0$. The following formulas are also true:

$$v^\varepsilon_{x^1}(s, x) = -2\left(\frac{s}{x^1} v^\varepsilon(s, x)\right)_s, \tag{5}$$

$$\tau_{x^1}(t, x) = -2 \int_0^\infty \varphi(t + s)\left(\frac{s}{x^1} v^\varepsilon(s, x)\right)_s ds$$

$$= 2 \int_0^\infty \varphi_t(t + s) \frac{s}{x^1} v^\varepsilon(s, x)\, ds \geqq 0, \tag{6}$$

where integration by parts is possible in view of the condition $\alpha \in (0, 1)$. Hence, recalling that $b^1 \leqq 0$, we conclude that

$$L\tau = \tau_t + a^{11}\tau_{x^1x^1} + b^1\tau_{x^1} + c\tau \leqq \tau_t + \varepsilon\tau_{x^1x^1} = 0.$$

The same argument proves the inequality $L\tau \leqq 0$ in assertion (f). One only needs to note that the convergence of integrals and the possibility of integration by parts are ensured by the boundedness of φ.

We have proved (d). Let us now complete the proof of (f). The inequality $\tau(t, x) \geqq \varphi(t)$ follows from the relationships $\tau(t, x) = \varphi(t)$, $x^1 = 0$, and $\tau_{x^1} \geqq 0$. It follows from the inequalities $\tau_t \geqq 0$ and $\tau_{x^1x^1} \leqq 0$ that τ increases in t, is convex in x^1, and

$$(x^1)^{-1}\, \tau(T, x) \geqq 6^{-1}\tau(T, 6, 0, \ldots, 0)$$

$$\geqq 6^{-1}\tau(0, 6, 0, \ldots, 0) \tag{7}$$

for $x^1 \in (0, 6]$. Moreover, τ as a function of the parameter T obviously increases. Therefore, the latter expression in (7) decreases if τ in it is found from T = 0. On performing such a substitution, we see that $(x^1)^{-1}\tau(t, x) \geq N^{-1}(\alpha, \varepsilon)$.

To prove the remaining inequality in (f), we note that

$$\tau(0, x) = \int_0^{T+1} s^{\alpha/2} v^{\varepsilon}(s, x)\, ds + (T + 1)^{\alpha/2} \int_{T+1}^{\infty} v^{\varepsilon}(\varepsilon, x)\, ds.$$

Here the second integral is similar to the right-hand side of (3) and it follows from the properties of ζ(with $K_1 = 0$) that it does not exceed $x^1 N(\varepsilon)(T + 1)^{-1/2}$. The former integral can be estimated from above by removing the exponential function involved in the expression of v^{ε}. Then there remains a convergent integral ($\alpha > 1$) which is equal to

$$N(\alpha)(T + 1)^{\frac{\alpha}{2} - \frac{1}{2}} x^1.$$

We now prove assertion (e). Let us fix an $i = 2, \ldots, d$ and stipulate that in what follows there is no summation over the repeated superscript i. By analogy with (4) to (6), we have

$$\sigma^i_{x^1x^1} = -\nu^{-1} \int_0^{\infty} \varphi_s((x^i)^2 + s) v^{\nu}(s, x)\, ds \leqq 0, \tag{8}$$

$\sigma^i_{x^1} \geqq 0$. From the inequality in (8) and following (6) we also find

$$\sigma^i_{x^i}(x) = 2x^i \int_0^{\infty} \varphi_s((x^i)^2 + s) v^{\nu}(s, x)\, ds = -2\nu x^i \sigma^i_{x^1x^1}(x),$$

$$\sigma^i_{x^ix^1}(x) = -\,4x^i \int_0^{\infty} \varphi_s((x^i)^2 + s)\left(\frac{s}{x^1} v^{\nu}(s, x)\right)_s ds =$$

$$= 4\,\frac{x^i}{x^1} \int_0^{\infty} \varphi_{ss}((x^i)^2 + s) s v^{\nu}(s, x)\, ds,$$

$$\sigma^i_{x^ix^i}(x) = -2\nu\sigma^i_{x^1x^1} + 4(x^i)^2 \int_0^{\infty} \varphi_{ss}((x^i)^2 + s) v^{\nu}(s, x)\, ds.$$

So far the explicit expression φ has been used only to provide the convergence of the integral defining σ and the possibility of integration by parts. We now make the substitution $s = (x^1)^2(4\nu)^{-1}r$ and $s = (x^i)^2 r$ in the integral expression defining $\sigma^i_{x^1x^1}$ and $\sigma^i_{x^ix^1}$. We then obtain

$$\sigma^i_{x^1x^1}(x) = -2\alpha(4\nu)^{\frac{1-\alpha}{2}}(x^1)^{\alpha-2}I_1(x).$$

$$\sigma^i_{x^ix^1}(x) = \alpha(\alpha-2)(4\nu)^{\frac{1-\alpha}{2}}(x^1)^{\alpha-2}(4\nu)^{\frac{1}{2}}I_2(x),$$

where

$$I_1(x) = \frac{1}{2\sqrt{\pi\nu}}\int_0^\infty (z^2+r)^{\frac{\alpha}{2}-1}r^{-\frac{3}{2}}e^{-\frac{1}{r}}\,dr, \quad z = \frac{x^i}{x^1}(4\nu)^{\frac{1}{2}},$$

$$I_2(x) = \frac{z|z|^{\alpha-3}}{2\sqrt{\pi\nu}}\int_0^\infty (1+r)^{\frac{\alpha}{2}-2}r^{-\frac{1}{2}}e^{-\frac{1}{z^2r}}\,dr.$$

From the inequalities $z^2 \leqq z^2 + r \leqq z^2(1+r)$ which hold true for $|z| \geqq 1$ it follows that $I_1(x)$ is of the order $|z|^{\alpha-2}$ as $|z| \to \infty (\alpha < 3)$. Moreover, it is obvious that $I_2(x)$ is of the order $z|z|^{\alpha-3}$. Therefore, the ratio I_2/I_1 is bounded at infinity and since it is continuous, we have $|I_2| \leqq N_1I_1$, where $N_1 = N_1(\alpha)$. It is important that the constant N_1 is independent of ν. Omitting the obviously negative summand ($\varphi'' \leqq 0$, $\alpha < 2$) in the expression for $\sigma^i_{x^ix^i}$, we sum the results of the above computation:

$$\sigma^i_{x^1} \geqq 0, \quad \sigma^i_{x^i} = -2\nu x^i\sigma_{x^1x^1}, \quad \sigma^i_{x^1x^1} \leqq 0,$$

$$\sigma^i_{x^ix^i} \leqq -2\nu\sigma^i_{x^1x^1}, \quad \left|\sigma^i_{x^ix^1}\right| \leqq -\nu^{1/2}N_2(\alpha)\sigma^i_{x^1x^1}. \tag{9}$$

Hence we conclude

$$L\sigma^i \leqq \left(\varepsilon - 2K\nu - 2N_2K\nu^{1/2} - 2K_1\nu|x^i|\right)\sigma^i_{x^1x^1}.$$

The latter is negative for $|x^i| \leqq 2$ if $\nu = \nu(\alpha, \varepsilon, K, K_1)$ is sufficiently small. Finally, the first assertion in (e) is merely an implication of the formula

$$\sigma^i(x) = \frac{1}{2\sqrt{\pi\nu}}\int_0^\infty \varphi((x^i)^2 + (x^1)^2s)s^{-\frac{3}{2}}e^{-\frac{1}{4\nu s}}\,ds$$

and the inequalities $a + bs \leqq (a+b)(1+s)$ for all $s, a, b \geqq 0$, $a + bs \geqq a + b$ with $s \geqq 1$ and $a, b \geqq 0$.

We now come to the last assertion (g). As above, it is easy to obtain all relationships of (9) for σ^i (under the conditions of (g)) except for the latter one. Using the above calculations with $|x|^i \leq 3$ and $0 < x^1 \leq 2\sqrt{\nu}$ we obtain

$$-2\sqrt{\pi\nu}\sigma_{x^1x^1}(x) = \frac{2}{\sqrt{\nu}}\int_0^\infty \varphi_s\left((x^i)^2 + \frac{(x^1)^2}{4\nu}r\right)r^{-\frac{3}{2}}e^{-\frac{1}{r}}\,dr$$

$$\geq \frac{2}{\sqrt{\nu}}\int_0^1 \varphi_s\left((x^i)^2 + \frac{(x^1)^2}{4\nu}r\right)r^{-\frac{3}{2}}e^{-\frac{1}{r}}\,dr$$

$$= (4\nu)^{\frac{1-\alpha}{2}}2\alpha(x^1)^{\alpha-2}\int_0^1 (z^2+r)^{\frac{\alpha}{2}-1}r^{-\frac{3}{2}}e^{-\frac{1}{r}}\,dr,$$

$$2\sqrt{\pi\nu}\left|\sigma^i_{x^ix^1}(x)\right|$$

$$= 4|x^i|^2\int_0^\infty |\varphi_{ss}((x^i)^2 + (x^i)^2 r)|r^{-\frac{1}{2}}e^{-\frac{1}{z^2r}}\,dr$$

$$\leq N(\varphi)\alpha(2-\alpha)4|x^i|^2\int_0^\infty ((x^i)^2 + (x^i)^2 r)^{\frac{\alpha}{2}-2}r^{-\frac{1}{2}}e^{-\frac{1}{z^2r}}\,dr$$

$$= N(\varphi)\alpha(2-\alpha)(4\nu)^{\frac{1-\alpha}{2}}(x^1)^{\alpha-2}(4\nu)^{\frac{1}{2}}|z|^{\alpha-2}\int_0^\infty (1+r)^{\frac{\alpha}{2}-2}$$

$$r^{-\frac{1}{2}}e^{-\frac{1}{z^2r}}\,dr.$$

Hence, just as above, comparing I_2 and I_1, we conclude that the latter relationship in (9) is also true (after replacing $N_2(\alpha)$ by $N(\varphi)$), but this time for $|x^i| \leq 3$ and $0 < x^1 \leq 2\sqrt{\nu}$. This obviously confirms the statement of (g) concerning $L\sigma^i$.

Finally,

$$\sigma^i(x^1, 0, \ldots, 0)$$

$$= \int_0^\infty \varphi(s)v^\nu(s, x)\,ds \leq Nx^1\left(\int_0^{10} s^{\frac{\alpha-3}{2}}\,ds + \int_{10}^\infty s^{-\frac{3}{2}}\,ds\right) = Nx^1.$$

This proves the lemma.

Using the constructed auxiliary functions, we now estimate the difference between a solution to the equation and the boundary data. The established estimates are considered separately for the three different cases: near $\partial_t Q_{T,R}$ but far from $\partial_x Q_{T,R}$, near $\partial_x Q_{T,R}$ but far from $\partial_t Q_{T,R}$, and near $\partial_{tx} Q_{T,R}$.

2. LEMMA. Let $c = 0$, $\alpha > 0$, $u \in W^{1,2}_{d+1,loc}(Q_{T,R}) \cap C(\bar{Q}_{T,R})$, let M, M_α, and M_2 be some nonnegative constants, let $u \leqq M + u(T, 0)$ on $\partial' Q_{T,R}$, let $u(T, x) - u(T, 0) \leqq M_\alpha |x|^\alpha$ on $\partial_t Q_{T,R}$, and let $Lu \geqq -M_2$ in $Q_{T,R}$ (a.e.). Then, for $K_1 t \leqq (1/2)R$,

$$u(T - t, 0) - u(T, 0)$$
$$\leqq \gamma^{-1}\left[\left(t^{\frac{\alpha}{2}} + (K_1 t)^\alpha\right)M_\alpha + M_2 t + Me^{-\gamma \frac{R^2}{t}}\right],$$

where $\gamma = \gamma(K, K_1, \alpha, d) > 0$.

Proof. Let $\varphi(t) = t^\alpha$ for $t \in [0, R]$, while for $t > R$, we define φ so that it will be nondecreasing continuously differentiable and $\varphi'(t)$ will vanish for large values of t. For a nonnegative measurable bounded function $p(y)$ and a number $q \geqq 0$, we further define functions $\xi^i_{(\pm)}$ and $\eta^i_{(\pm)}$ and put

$$w(t, x) = M_2(T - t) +$$
$$+ M \sum_{i=1}^{d} (\xi^i_{(+)} + \xi^i_{(-)})(t, x) +$$
$$+ M_\alpha \sum_{i=1}^{d} (\eta^i_{(+)} + \eta^i_{(-)})(t, x).$$

Moreover, without loss of generality we assume that $u(T, 0) = 0$. From Lemma 1(a) and (b) and the maximum principle it follows that $u \leqq Nw$ in $\bar{Q}_{T,R}$, where $N = N(\alpha, d)$. Specifically, for $t \in [0, T]$ we obtain $u(T - t, 0) \leqq Nw(T - t, 0)$ for any choice of p and q and any continuation of φ. We fix t such that $K_1 t \leqq R$ and put $q = (2Kt)^{-1}(R - K_1 t)$, $p(y) = 0$ for $y \leqq K_1 t$, and $p(y) = (2Kt)^{-1}\cdot(y - K_1 t)$ for $y \geqq K_1 t$. Then

$$N^{-1}u(T - t, 0) \leqq w(T - t, 0)$$
$$= M_2 t + 2dMe^{-\frac{(R - K_1 t)^2}{4Kt}} +$$

$$+ 2dM_\alpha\left(\varphi(K_1t) + \int_{K_1t}^{\infty} \varphi'(y)e^{-\frac{(y - K_1t)^2}{4Kt}} dy\right).$$

The inequality between the extreme terms remains valid for any of the above continuations of t^α with $t > R$. Passing to the limit, we replace the infinite upper limit of integration by R. Then it remains for us to note that $\varphi(K_1t) = (K_1t)^\alpha$ and

$$(K_1t)^\alpha + \alpha \int_{K_1t}^{R} y^{\alpha-1}e^{-\frac{(y - K_1t)^2}{4Kt}} dy \leqq (K_1t)^\alpha +$$

$$+ \alpha \int_{K_1t}^{\infty} y^{\alpha-1}e^{-\frac{(y - K_1t)^2}{4Kt}} dy =$$

$$= \frac{1}{2Kt}\int_0^\infty y(y + K_1t)^\alpha e^{-\frac{y^2}{4Kt}} dy,$$

and using the substitution $y = z\sqrt{t}$, we can see that

$$\frac{1}{t}\int_0^\infty (y^{\alpha+1} + (K_1t)^\alpha y)e^{-\frac{y^2}{4Kt}} dy = N_1t^{\frac{\alpha}{2}} + N_2(K_1t)^\alpha.$$

which proves the lemma.

3. LEMMA. Let $u \in W^{1,2}_{d+1,loc}(Q^+_{T,R}) \cap C(\bar{Q}^+_{T,R})$, let $R \leqq 1$, $c = 0$, and $\alpha \in (0, 1)$, let M, M_α, and M_2 be nonnegative constants, let $t_0 \in [0, T)$, let $u \leqq M + u(t_0, 0)$ in $Q_{T,R}$, let $Lu \geqq -M_2$ in $Q^+_{T,R}$ (a.e.), and let $u(t, x) \leqq u(t_0, 0) + M_\alpha(|x|^\alpha + |t - t_0|^{\alpha/2})$ for $(t, x) \in \partial' Q^+_{T,R}$ and $x^1 = 0$, $t \in [t_0, T]$. Then

$$\gamma(u(t_0, x) - u(t_0, 0))$$

$$\lesssim (M_\alpha + MR^{-\alpha})(x^1)^\alpha + [M(R^{-1} + K_1 + (T - t_0)^{-\frac{1}{2}}) +$$

$$+ RM_2]x^1, \tag{10}$$

for $0 \leqq x^1 \leqq 2R$ and $x = (x^1, 0, \dots, 0)$, where $\gamma = \gamma(\alpha, \varepsilon, K, K_1, d) > 0$.

Proof. The similarity transformation $(t, x) \to (R^{-2}t, R^{-1}x)$ reduces the case $R \leqq 1$ to the case $R = 1$. Making, if needed, a shift along the t-axis, we can assume that $t_0 = 0$. Finally, the assumption $c = 0$ allows us to assume that $u(t_0, 0) = 0$. Thus, let $R = 1$, $t_0 = 0$, and $u(0, 0) = 0$.

We take $\delta = \delta(\varepsilon, K, K_1) > 0$ and $\lambda = \lambda(\varepsilon, K, K_1) > 0$ such that the inequalities

$$Lv \leqq 0, \quad b^1 v + 2a^{11} v_{x^1} \leqq 0, \quad v \geqq 1,$$

hold for $0 < x^1 < \delta$ where $v(x) = 3 - \exp(\lambda x^1)$. The reader can easily check that such a choice is possible. If $x^1 \geqq \delta$, then γ in (10) can obviously be taken equal to δ. Therefore, one needs to consider only the case $x^1 \leqq \delta$.

For any smooth function ψ, we put

$$\tilde{L}\psi := \frac{1}{v} L(v\psi) =: \psi_t + a^{ij}\psi_{x^i x^j} + \tilde{b}^i \psi_{x^i} + \tilde{c}\psi, \quad 0 < x^1 \leqq \delta.$$

It is clear that $\tilde{c} = v^{-1} Lv \leqq 0$, $\tilde{b}^1 = v^{-1}(b^1 v + 2a^{11} v_{x^1}) \leqq 0$, and $\tilde{b}^i = b^i$ for $i \geqq 2$. Moreover, for $\tilde{u} := v^{-1}$ we have $\tilde{L}\tilde{u} = v^{-1} Lu \geqq -M_2$ in $Q^+_{T,R} \cap \{x^1 < \delta\}$ and in this domain u also satisfies other conditions of the lemma with the same constants. Also, note that if inequality (10) with $x^1 \leqq \delta$ is proved to be valid for $\tilde{u}$ with a γ, then it is also valid for u with $(1/3)\gamma$ instead of γ.

Thus, considering $x^1 \leqq \delta$, we need to prove the lemma for the case when $c \leqq 0$ (rather than $=0$) and $b^1 \leqq 0$. We now note that $Lx^1(2\delta - x^1) \leqq -\varepsilon$ for $x^1 \leqq \delta$, and for $\omega > 0$, we put

$$w(t, x) = M\zeta(t, x^1 + \omega, 0, \dots, 0) + M\delta^{-2}x^1(2\delta - x^1) +$$

$$+ (M + M_\alpha) \sum_{i=2}^{d} \sigma^i(x) + M_\alpha \tau(t, x) +$$

$$+ M_2 \varepsilon^{-1} x^1 (2\delta - x^1), \tag{11}$$

where σ^i and τ are constructed using $\varphi = t^{\alpha/2}$ (and for $d = 1$, we put $\Sigma\sigma^i = 0$).

From the above properties of the functions σ^i, τ, ζ it follows that the inequality $u \leq Nw$ holds on the parabolic boundary of the domain $Q^{+}_{T,R} \cap \{x^1 < \delta\}$ with a constant $N(\alpha, \varepsilon, K, K_1, d)$. By virtue of the maximum principle, Lemma 1(c) - (e) implies that $u \leq Nw$ on the closure of this domain. Specifically, $u(0, x) \leq Nw(0, x)$ for $x = (x^1, 0, \ldots, 0)$ and $0 < x^1 \leq \delta$. It remains to let $\omega \downarrow 0$ and to make use of the estimates of Lemma 1(c) to (e). This proves the lemma.

4. LEMMA. Let $c = 0$. Let $u \in W^{1,2}_{d+1,loc}(Q^{+}_{T,R}) \cap C(\bar{Q}^{+}_{T,R})$, let M, M_α, and M_2 be nonnegative constants, and let $|u| \leq M$ in $Q^{+}_{T,R}$. Also, let $Lu \geq -M_2$ in $Q^{+}_{T,R}$ (a.e.).

(a) If $\alpha > 0$, $0 < x^1_0 \leq R$, $x_0 = (x^1_0, 0, \ldots, 0)$, and $u(t, x) - u(T, x_0) \leq M_\alpha(|x - x_0|^\alpha + (T - t)^{\alpha/2})$ on $\partial' Q^{+}_{T,R} \cap (\{t = T\} \cup \{x_1 = 0\})$, then for $\gamma = \gamma(K, K_1, \alpha, d) > 0$ and $K_1 t \leq (1/2)R$ we obtain

$$u(T - t, x_0) - u(T, x_0) \leq \gamma^{-1}\left[\left(t^{\frac{\alpha}{2}} + (K_1 t)^\alpha\right)M_\alpha + M_2 t + Me^{-\nu\frac{R^2}{t}}\right].$$

(b) If $t_0 \in [0, T)$, $R \leq 1$, $\alpha \in (0, 1)$, and $u(t, x) - u(t_0, 0) \leq M_\alpha(|t - t_0|^{\alpha/2} + |x|^\alpha)$ on $\partial' Q^{+}_{T,R} \cap (\{t = T\} \cup \{x_1 = 0\})$, then for $\gamma = \gamma(\varepsilon, K, K_1, \alpha, d) > 0$ and $x = (x^1, 0, \ldots, 0)$ we obtain

$$u(t_0, x) - u(t_0, 0) \leq \gamma^{-1}[(M_\alpha + MR^{-\alpha})(x^1)^\alpha + (MR^{-1} + M^2R)x^1].$$

The proof of this lemma is nearly identical to the proofs of Lemmas 2 and 3. It is only needed that, using the proof of Lemma 2 to prove assertion (a), one should replace w by the function

$$M_2(T - t) + M\left[\xi^1_{(+)}(t, x - x_0) + \sum_{i=2}^{d} (\xi^i_{(+)}(t, x) + \xi^i_{(-)}(t, x))\right] +$$

$$+ M_\alpha\left[(T - t)^{\alpha/2} + \sum_{i=1}^{d} (\eta^i_{(+)}(t, x - x_0) + \eta^i_{(-)}(t, x - x_0))\right],$$

while when proving assertion (b), one should replace w involved in the proof of Lemma 3 by the function

$$M_\alpha \tau + (M\alpha + M)\sum_{i=2}^{d} \sigma^i + (M\delta^{-2} + M_2\varepsilon^{-1})x^1(2\delta - x^1).$$

We now prove the major result of this section.

5. THEOREM. Let $u \in W^{1,2}_{d+1,loc}(Q^+_{T,3}) \cap C(\bar{Q}^+_{T,3})$, let M and M_2 be nonnegative constants, let $|u| \le M$ in $Q^+_{T,3}$, let $|Lu| \le M_2$ in $Q^+_{T,3}$ (a.e.), and let κ be a constant such that $\kappa \in (0, 1]$. Then there exists a constant $\alpha_0 \in (0, 1)$ depending only on d, K, and ε such that for any $\alpha \in (0, \alpha_0]$ and $N = N(\alpha, \varepsilon, d, K, K_1, K_2)$:

(a) if $z_1, z_2 \in Q^+_{T,1} \cap \{x^1 > \kappa\}$, then

$$|u(z_1) - u(z_2)| \le \rho^\alpha(z_1, z_2)(M_\alpha + M_2 + M\kappa^{-\alpha})N, \tag{12}$$

where M_α is the Hölder constant of order α for the function u on the face $t = T$ of the parallelepiped $\bar{Q}^+_{T,3}$;

(b) if M_α is the Hölder constant of order α for the function u on the face $x^1 = 0$ of the parallelepiped $\bar{Q}^+_{T,3}$, then inequality (12) is true for any $z_1, z_2 \in Q^+_{T,1} \cap \{t < T - \kappa^2\}$;

(c) if M_α is the Hölder constant of order α for the function u on the union of the faces $t = T$ and $x^1 = 0$ of the parallelepiped $\bar{Q}^+_{T,3}$, then inequality (12) is true for any $z_1, z_2 \in Q^+_{T,1}$ with $\kappa = 1$.

Proof. Since $|Lu - cu| \le M_2 + K_2 M$ and $K_2 M + M\kappa^{-\alpha} \le M(K_2 + 1)\kappa^{-\alpha}$, we can and will assume without loss of generality that $c \equiv 0$. Furthermore, let $\rho(z)$ be the parabolic distance from $z = (t, x)$ to the parabolic boundary of $Q^+_{T,3}$. It is clear that $\rho(z) = (T - t)^{1/2} \wedge x^1$ for $z \in Q^+_{T,1}$. We also stipulate that N denotes various constants depending only on d, ε, α, K, and K_1. We first prove assertion (a). It is clear that it suffices to prove it for $2K_1\kappa \le 1$ and $\rho(z_1, z_2) \le (1/5)\kappa$. Let $z_1 = (t_1, x_1)$ and $z_2 = (t_2, x_2)$.

For definiteness, assume that $\rho(z_1) \ge \rho(z_2)$, denote $\bar{\rho} = \rho(z_1)$ $\underline{\rho} = \rho(z_2)$, $\rho = \rho(z_1, z_2)$, and consider the case

$$\underline{\rho} \le \frac{1}{5}\kappa, \quad \underline{\rho} \le 4\rho. \tag{13}$$

Using the fact that the parabolic distance satisfies the triangle inequality, we obtain $\bar{\rho} - \underline{\rho} \le \rho$. Therefore, it follows from (13) that $\bar{\rho} \le 5\rho \le \kappa$ and $\rho^2(z_i) = T - t_i$ for $i = 1, 2$. Moreover, $(0, x_i) + Q_{T,\kappa} \subset Q^+_{T,3}$. Let α_0 be equal to the cosntant β from Section 2. Assuming that

$\alpha \in (0, \alpha_0]$, we apply Lemma 2 to the function in the parallelepiped $(0, x_i) + Q_{T,\kappa}$. It should be taken into account that

$$\exp\left(-\gamma\frac{\kappa^2}{T - t_i}\right) \leqq N(\kappa^{-2}(T - t_i))^{\alpha/2} \leqq N\kappa^{-\alpha}\bar{\rho}^{\alpha},$$

$$K_1(T - t_i) \leqq K_1\bar{\rho}^2 \leqq K_1\kappa^2 \leqq \frac{1}{2}\kappa \leqq 1.$$

We also take advantage of the fact that Lemma 2 can be applied to the function $(-u)$. Then we obtain

$$|u(z_i) - u(T, x_i)| \leqq N\bar{\rho}^{\alpha}(M_\alpha + M_2 + M\kappa^{-\alpha})$$

$$\leqq N\rho^{\alpha}(M_\alpha + M_2 + M\kappa^{-\alpha}).$$

Adding this to the available inequality $|u(T, x_1) - u(T, x_2)| \leqq M_\alpha|x_1 - x_2|^\alpha \leqq M_\alpha\rho^\alpha$, we obtain inequality (12).

Having completed the case (13), we pass to the case when

$$\underline{\rho} \leqq \frac{1}{5}\kappa, \quad \underline{\rho} \geqq 4\rho. \tag{15}$$

We make use of the following argument communicated by M.V. Safonov. Since $\bar{\rho} \leqq \underline{\rho} + \rho$, it follows from (15) that $\bar{\rho} \leqq 5/4\underline{\rho}$ and $2\bar{\rho} \leqq (1/2)\kappa$. Therefore, the domain $Q := z_1 + Q_{\bar{\rho}^2, 2\bar{\rho}}$ which is adjacent to the right base of $Q^+_{T,3}$ along the t-axis, lies in $Q^+_{T,3}$, and for any point $(t, x) \in Q$ we obtain $x^1 > (1/2)\kappa$. Moreover $\rho \leqq (1/4)\underline{\rho} \leqq (1/4)\bar{\rho}$ and the parabolic distances from the points z_1 and z_2 to the parabolic boundary of Q remain unchanged, i.e., $\bar{\rho}$ and $\underline{\rho}$.

Furthermore, let the function u attain its maximum and minimum values over $\bar{Q}$ at the points $\tilde{z}_1$ and $\tilde{z}_2 \in \bar{Q}$, respectively. We now apply Theorem 2.6 to $u(z) - u(\tilde{z}_2)$ in the domain Q, recalling that $c \equiv 0$. Then

$$|u(z_1) - u(z_2)| \leqq N\left(\frac{\rho}{\underline{\rho}}\right)^{\alpha_0}(|u(\tilde{z}_1) - u(\tilde{z}_2)| + \bar{\rho}^2 M_2)$$

$$\leqq N\left(\frac{\rho}{\underline{\rho}}\right)^{\alpha}(|u(\tilde{z}_1) - u(\tilde{z}_2)| + \bar{\rho}^2 M_2). \tag{16}$$

To estimate the difference $|u(\tilde{z}_1) - u(\tilde{z}_2)|$, we use the same argument that was employed in the case (13). Namely, we apply Lemma 2 to the function u in the parallelepiped $(t_1, \tilde{x}_1) + Q_{\bar{\rho}^2, 1/2\kappa}$ (lying in $Q^+_{T,3}$

with $\tilde{z}_i = (\tilde{t}_i, \tilde{x}_i))$, taking note that $K_1\bar{\rho}^2 \leqq K_1(1/16)\kappa^2 \leqq (1/4)\kappa$. Similarly to (14),

$$|u(\tilde{z}_i) - u(T, \tilde{x}_i)| \leqq N\bar{\rho}^{\alpha}(M_\alpha + M_2 + M\kappa^{-\alpha}).$$

The latter inequality together with inequality (16) and the inequalities $|u(T, \tilde{x}_1) - u(T, \tilde{x}_2)| \leqq M_\alpha\bar{\rho}^{\alpha}4^{\alpha}$ and $\bar{\rho} \leqq (5/4)\underline{\rho}$ again yield (12).

To complete the proof of assertion (a), it remains to note that if $\underline{\rho} \geqq (1/5)\kappa$, then inequality (12) immediately follows from Theorem 2.6, applied to the function u in the domain $Q^+_{T,3} \cap (z_1 + Q_{1,3})$ and from the inequalities $\rho \leqq (1/5)\kappa$ and $\alpha_0 \leqq d(d+1)^{-1}$.

Proving assertion (b), we follow the notation introduced and assume that $\rho(z_1) \geqq \rho(z_2)$ and $\rho \leqq (1/5)\kappa$. We first consider the case (13). Since in this case we have $\bar{\rho} \leqq \kappa$, $\rho(z_i) = x_i^1 < \kappa$. By Lemma 3 applied to $(0, 0, x_i^2, \ldots, x_i) + Q^+_{T,1}$, we obtain

$$\begin{aligned}|u(z_i) &- u(t_i, 0, x_i^2, \ldots, x_i^d)| \\ &\leqq N(x_i^1)^{\alpha}(M_\alpha + M + M_2) + NM\kappa^{-1}x_i^1 \\ &\leqq N(x_i^1)^{\alpha}(M_\alpha + M_2 + M\kappa^{-\alpha}).\end{aligned} \tag{17}$$

Here $x_i^1 \leq 5\rho$ and $|u(t_1, 0, x_1^2, \ldots, x_1^d) - u(t_2, 0, x_2^2, \ldots, x_2^d)| \leqq M_\alpha\rho^\alpha$, which obviously yields (12).

We next consider the case (15). In this case we have $\bar{\rho} \leqq (5/4)\underline{\rho}$, $2\bar{\rho} \leqq (1/2)\kappa$, and $\rho \leqq (1/4)\bar{\rho}$. Denote

$$Q_1 = \left[((t_1, x_1) + Q_{\bar{\rho}^2, 2\bar{\rho}}) \cup ((t_2, x_1) + Q_{\bar{\rho}^2, 2\bar{\rho}})\right] \cap Q^+_{T,3}.$$

Since $|t_1 - t_2| \leqq \rho^2 < \bar{\rho}^2 < (1/16)\kappa^2$, the parabolic distances from the points z_1 and z_2 to the parabolic boundary of Q_1 remain equal to $\bar{\rho}$ and $\underline{\rho}$. Let the function u attain its maximum and minimum values on $\bar{Q}_1$ at the points $\tilde{z}_1$ and $\tilde{z}_2 \in \bar{Q}_1$, respectively. Like (16), by Theorem 2.6 applied to Q_1, we obtain

$$|u(z_1) - u(z_2)| \leqq N\left(\frac{\rho}{\underline{\rho}}\right)^{\alpha}(|u(\tilde{z}_1) - u(\tilde{z}_2)| + \bar{\rho}^2M_2).$$

If $\tilde{z}_i = (\tilde{t}_i, \tilde{x}_1)$, then $T - \tilde{t}_i \geqq \kappa^2 - \bar{\rho}^2 \geqq (15/16)\kappa^2$ and $\tilde{x}_i^1 \leqq 4\bar{\rho}$.

Therefore, like (17), we have

$$|u(\tilde{z}_i) - u(\tilde{t}_i, 0, \tilde{x}_i^2, \ldots, \tilde{x}_i^d)| \leqq N\bar{\rho}^{\alpha}(M_\alpha + M_2 + M\kappa^{-\alpha})$$

and inequality (12) can further be obtained in a standard way using the inequality $\bar{\rho} \leqq (5/4)\rho$.

Finally, when $\rho \geqq (1/5)\kappa$, to prove assertion (b), it suffices to apply Theorem 2.6 in the domain

$$Q^+_{T,3} \cap [(z_1 + Q_{1,3}) \cup (z_2 + Q_{1,3})].$$

We now prove the latter assertion (c). Put $\kappa = (2K_1)^{-1} \wedge 1$. It clearly suffices to prove inequality (12) with this κ. Let, first, (13) and the inequality $\rho \leqq (1/5)\kappa$ be true. Just as above, using Lemma 4 similarly to (14) and (17), we obtain

$$|u(z_i) - u(T, x_i)|$$
$$\leqq N(T - t_i)^{\alpha/2}(M_\alpha + M_2 + M) \text{ for } \rho^2(z_i) = T - t_i,$$
$$|u(z_i) - u(t_i, 0, x_i^2, \ldots, x_i^d)| \leqq N(x_i^1)^{\alpha}(M_\alpha + M_2 + M).$$

Since $\rho(z_i) = x_i^1 \wedge (T - t_i)^{1/2}$, we can see that the absolute value of the difference between $u(z_i)$ and the value of u at the point which is the projection of z_i on the plane $t = T$ or the plane $x^1 = 0$ is no greater than $N\rho^{\alpha}(z_i)(M_\alpha + M_2 + M)$. Moreover, recall that $\rho(z_i) \leqq 5\rho$ by virtue of (13). The magnitude of the difference between the values of u at the projections of z_i on $t = T$ or the projections of z_i on $x^1 = 0$ does not exceed $M_\alpha\rho^\alpha$ by hypothesis.

Moreover, if, for example, $\rho^2(z_1) = T - t_1$ and $\rho(z_2) = x_2^1$, then, by the triangle inequality for the parabolic distance, the parabolic distance between the projections of z_1 on $t = T$ and z_2 on $x^1 = 0$ does not exceed $\rho(z_1) + \rho(z_2) + \rho \leqq 10\rho$. Then the magnitude of the difference between the values of u at these projections does not exceed $M_\alpha 10^\alpha\rho^\alpha$ by hypothesis. Thus, (12) can easily be obtained by expressing $u(z_1) - u(z_2)$ as a sum of differences between the values of u at the points z and their appropriate projections on the plane $t = T$ or $x^1 = 0$.

Considering the case (15), we put

$$Q_2 = \Big\{(t, x) : |t - t_1| < \bar{\rho}^2, |x^i - x_1^i| <$$

$$< \bar{\rho},\ i = 1, \ldots, d\Big\} \cap Q^+_{T,3}.$$

By Theorem 2.6 applied in the domain Q_2, we have

$$|u(z_1) - u(z_2)| \le N(\rho/\underline{\rho})^\alpha(|u(\tilde{z}_1) - u(\tilde{z}_2)| + \bar{\rho}^{-2}M_2), \tag{18}$$

where $\tilde{z}_1$ and $\tilde{z}_2 \in \bar{Q}_2$. Furthermore, it is clear that Q_2 is adjacent to the face $x^1 = 0$ or the face $t = T$ of the parellelepiped $\bar{Q}^+_{T,3}$. In the former case, to estimate $|u(z_1) - u(z_2)|$, we apply Lemma 4(b) with $R = 1$, in the latter case we use Lemma 4(a) and note that the parabolic distance between any two points of Q_2 does not exceed $N\bar{\rho}$ and, moreover, $\bar{\rho} \le (5/4)\underline{\rho}$ and $2K_1\bar{\rho}^{-2} \le (1/8)K_1\kappa^2 \le 1/2$. Then in both cases we can easily obtain the inequality $|u(\tilde{z}_1) - u(\tilde{z}_2)| \le N\underline{\rho}^\alpha(M_\alpha + M_2 + M)$. This together with (18) yield (12).

Finally, if $\rho \ge (1/5)\kappa$, then inequality (12) is clear with $N = 2.5^\alpha$, and if $\underline{\rho} > (1/5)\kappa$, then it follows from assertions (a) and (b). This proves the theorem.

We now extend the statement of the above theorem to the case of elliptic operators.

6. THEOREM. Let the coefficients of L be independent of t, let $\tilde{L} = L - \frac{\partial}{\partial t}$, let $u \in W^2_{d,loc}(D^+_3) \cap C(\bar{D}^+_3)$, let M, M_α, and M_2 be nonnegative constants, let $|u| \le M$ in D^+_3, let $|\tilde{L}u| \le M_2$ (a.e.), let $\alpha \in (0, \alpha_0]$, where α_0 is the constant from Theorem 5, and let $|u(x_1) - u(x_2)| \le M_\alpha|x_1 - x_2|^\alpha$ for $x_1, x_2 \in \bar{D}^+_3$, $x^1_1 = x^1_2 = 0$. Then

$$|u(x_1) - u(x_2)| \le N|x_1 - x_2|^\alpha(M_\alpha + M_2 + M)$$

for $x_1, x_2 \in D^+_1$, where $N = N(\alpha, \varepsilon, d, K, K_1, K_2)$.

If $u \in W^2_{d+1,loc}(D^+_3)$, then the assertion of this theorem follows from assertion (b) of Theorem 5 with $T = 2$, $\kappa = 1$, $u(t, x) = u(x)$, $z_1 = (0, x_1)$ and $z_2 = (0, x_2)$. In the general case, the reader can easily deduce this assertion, regarding appropriate solutions to the equations $\tilde{L}_n u_n = \tilde{L}u$, for which $\tilde{L}_n(u_n - u) = (\tilde{L} - \tilde{L}_n)u$, and applying the estimates established in Section 3.3 which guarantee that u_n converges to u.

Theorem 6 can be proved in a different way by essentially repeating the proof of assertion (b) of Theorem 5, where Theorem 2.12 is used in place of Theorem 2.6 and Lemma 3 is replaced by the following lemma.

7. LEMMA. Let $R \leqq 1$, let $c \equiv 0$, let $u \in W^2_{d,loc}(D^+_R) \cap C(\overline{D^+_R})$, let $\alpha \in (0, 1)$, let M, M_α, and M_2 be nonnegative constants, let $u \leqq M + u(0)$ in D^+_R, let $\tilde{L}u \geqq -M_2$ in D^+_R (a.e.), and let $u(x) \leqq u(0) + M_\alpha |x|^\alpha$ for $x^1 = 0$ and $x \in \partial D^+_R$. Then, if $0 \leqq x^1 \leqq 2R$ and $x = (x^1, 0, \ldots, 0)$, we obtain

$$\gamma(u(x) - u(0)) \leqq (M_\alpha + MR^{-\alpha})(x^1)^\alpha + (MR^{-1} + M_2 R)x^1,$$

where $\gamma = \gamma(\alpha, \varepsilon, K, K_1, d) > 0$.

In turn, the proof of this lemma can easily be obtained from the proof of Lemma 3 by considering the auxiliary function $w(x) = w(0, x)$, where $w(t, x)$ is defined by formula (11) where the term with ζ is omitted.

4.5. The Hölder Condition for Solutions of Simultaneous Linear Inequalities near the Boundary and Estimation of their Normal Derivatives.

This section has the same aim as Section 3, except that the estimates of Section 3 will be applied to estimating the Hölder constant of a derivative on the interior of a domain, while the estimates of this section will be used for estimating it on the closure of the domain. In addition, this section is devoted to estimating the magnitude of a derivative only on the boundary.

Let $K \geqq \varepsilon > 0$, T, K_1, $K_2 \geqq 0$ be some constants and let $L_1, \ldots, L_n$, L be differential operators of the form (2.1) whose coefficients satisfy conditions (2.2). We will use the notation $Q_{T,R}$ and $Q^+_{T,R}$ from the beginning of the previous section. Remark 3.4 shows that the following theorem essentially covers Theorem 4.5 (although its proof, as we shall see, is in turn covered by the proof of Theorem 4.5).

1. THEOREM. Let $u^1, \ldots, u^n \in W^{1,2}_{d+1,loc}(Q^+_{T,3}) \cap C(\overline{Q^+_{T,3}})$, let $f_1, \ldots, f_n \in C^2_{loc}(E_1)$, let $f'_i > 0$ and $f''_i \leqq 0$ on E_1, let M and M_2 be nonnegative constants, let $|u^i| \leqq M$ in $Q^+_{T,3}$, let $L_i u^i \geqq -M_2$ (again there is no summation over i) in $Q^+_{T,3}$ (a.e.) for $i = 1, \ldots, n$. Also, suppose that for the constants $\delta > 0$, α, $\kappa \in (0, 1]$ and any z_1, $z_2 \in Q^+_{T,3}$ we have

$$M_2\rho^\alpha(z_1, z_2) \geqq \delta \sum_{i=1}^n [f_i(u^i(z_1)) - f_i(u^i(z_2))]_+ -$$

$$- \sum_{i=1}^n [f_i(u^i(z_1)) - f_i(u^i(z_2))]_- . \tag{1}$$

Then there exists a constant $\beta_0 \in (0, 1)$ depending only on n, d, K, ε and δ such that for any $\beta \in (0, \beta_0 \wedge \alpha]$:

(a) if $z_1, z_2 \in Q^+_{T,1} \cap \{x^1 > \kappa\}$, then

$$\sum_{i=1}^{n} |u^i(z_1) - u^i(z_2)| \leq \rho^\beta(z_1, z_2)N(M_\beta + M_2 + M\kappa^{-\beta}), \tag{2}$$

where M_β is the greatest Hölder constant of order β for the functions u^i on the face $t = T$ of the parallelepiped $\bar{Q}^+_{T,3}$;

(b) if M_β is the greatest Hölder constant of order β for the functions u^i on the face $x^1 = 0$ of the parallelepiped $\bar{Q}^+_{T,3}$, then inequality (2) is true for any $z_1, z_2 \in Q^+_{T,1} \cap \{t < T - \kappa^2\}$;

(c) if M_β is the greatest Hölder constant of order β for the functions u^i on the union of the faces $t = T$ and $x^1 = 0$ of the parallelepiped $\bar{Q}^+_{T,3}$, the inequality (2) is true for any $z_1, z_2 \in Q^+_{T,1}$ with $\kappa = 1$. Furthermore, if $f_i(r) = r$ for all i and r, then N depends only on n, d, K, ε, α, δ, K_1, and β, while in the general case, it depends only on the above parameters along with M_2, M, and the maxima of f_i', $(f_i')^{-1}$, and $|f_i''|$ on $[-M, M]$.

<u>Proof</u>. Let β_0 be the constant from Theorem 3.1. We follow the proof of Theorem 4.5. As in this proof, we first reduce our consideration to the case when $L_i 1 = 0$ for $i = 1, \ldots, n$. Then we prove (a), assuming that

$$2K_1\kappa \leq 1, \quad \rho(z_1, z_2) \leq (1/5)\kappa,$$

where

$$z_1 = (t_1, x_1), \quad z_2 = (t_2, x_2) \in Q^+_{T,1} \cap \{x^1 > \kappa\}.$$

Suppose that $\rho(z_1) \geq \rho(z_2)$, denote $\bar{\rho} = \rho(z_1)$, $\underline{\rho} = \rho(z_2)$, $\rho = \rho(z_1, z_2)$, and consider the case (4.13). In this case, by Lemma 4.2 with $\beta \in (0, \beta_0 \wedge \alpha]$,

$$u^i(z_j) - u^i(T, x_j) \leq N\rho^\beta(M_\beta + M_2 + M\kappa^{-\beta})$$

for all $i = 1, \ldots, n$, $j = 1, 2$. Hence we obtain an obvious inequality for the positive part of $u^i(z_j) - u^i(T, x_j)$, while the inequality for the negative part is obtained from the estimate of the positive part and inequality (1), where z_1 and z_2 are replaced by (T, x_j) and z_j. We then recall that $\beta \leqq \alpha$ and make use of the inequality $N^{-1} \leqq f_i' \leqq N$. Thus we obtain

$$|u^i(z_j) - u^i(T, x_j)| \leqq N\rho^\beta(M_\beta + M_2 + M\kappa^{-\beta}).$$

This inequality is very like (4.14) and, as in Section 4, it yields (2).

We now consider the case (4.15). Let the function u^i attain its maximum and minimum values on $\bar{Q}$ at the points $\tilde{z}_{1i}$ and $\tilde{z}_{2i} \in \bar{Q}$ $(Q := z_1 + Q_{\bar{\rho}^2, 2\bar{\rho}})$. We apply Theorem 3.1 with $\nu = \alpha$ to the functions $u^i - u^i(\tilde{z}_{2i})$. It follows that

$$\sum_{i=1}^n |u^i(z_1) - u^i(z_2)| \leqq N\left(\frac{\rho}{\bar{\rho}}\right)^\beta \left[M_2\bar{\rho}^\alpha + \sum_{i=1}^n |u^i(\tilde{z}_{1i}) - u^i(\tilde{z}_{2i})|\right].$$

Estimation of the latter sum can be performed using the arguments which were employed in the case (4.13) (see the proof of Theorem 4.5). Finally, in the last case when $\underline{\rho} \geqq (1/5)\kappa$, inequality (2) immediately follows from Theorem 3.1 with $\nu = \alpha$.

Assertions (b) and (c) can be proved very similarly to the proof of Theorem 4.5. As compared to the proof of Theorem 4.5 there are only two differences; first, we cannot apply Lemmas 4.2 to 4.4 to $(-u^i)$ and therefore we use inequality (1) in addition and second, we use Theorem 3.1 with $\nu = \alpha$ in place of Theorem 2.6. This proves the theorem.

By essentially repeating the proof of Theorem 3.5, we obtain the following result from Theorem 1.

2. THEOREM. Let $\tau \in (0, 1)$, let $u^1, \ldots, u^n \in W^{1,2}_{d+1,loc}(Q^+_{T,3}) \cap C(\bar{Q}^{+}_{T,3})$, let M_β be a nonnegative constant, let $\kappa \in (0, 1]$, and let $g^1 \ldots, g^d$ be functions on $Q^+_{T,3}$. Suppose that $|u^j| \leqq K_2$ and

$$|Lu^j - u^j_{x^k} g^k| \leqq K_2\left(1 + \left(u^i_x, u^i_x\right)\right)^\tau, \quad g^k g^k \leqq K_2\left(1 + \left(u^i_x, u^i_x\right)\right)$$

on $Q^+_{T,3}$ (a.e.). Then there exists a constant $\beta_0 \in (0, 1)$ depending only on d, K, ε, and n such that for any $\beta \in (0, \beta_0]$ and $N = N(d, K, K_1, K_2, \varepsilon, \tau, n, \beta, M_\beta, \kappa)$:

(a) if the Hölder constants or order β for the functions u^i on the face $t = T$ of the parallelepiped $\bar{Q}^+_{T,3}$ do not exceed M_β, then

$$\sum_{i=1}^{n} |u^i(z_1) - u^i(z_2)| \leq N\rho^{\beta}(z_1, z_2) \tag{3}$$

for $z_1, z_2 \in Q^+_{T,1} \cap \{x^1 > \kappa\}$;

(b) if the Hölder constants of order β for the functions u^i on the face $x^1 = 0$ of the parallelepiped $\bar{Q}^+_{T,3}$ do not exceed M_β, then inequality (3) holds true for any $z_1, z_2 \in Q^+_{T,1} \cap \{t < T - \kappa^2\}$;

(c) if the Hölder constants of order β for the functions u^i on the union of the faces $t = T$ and $x^1 = 0$ of the parallelepiped $\bar{Q}^+_{T,3}$ do not exceed M_β, then inequality (3) holds true for all $z_1, z_2 \in Q^+_{T,1}$ (and $\kappa = 1$ in the expression of N).

Just as Corollary 3.6 follows from Theorem 3.5, Theorem 2 implies the following.

3. COROLLARY. Let $u \in W^{1,2}_{d+1,loc}(Q^+_{T,3}) \cap C(\bar{Q}^+_{T,3})$, let $|u| \leq K_2$, let $|Lu| \leq K_2(1 + |u_x|^2)$ (a.e.), let M_β be a nonnegative constant, and let $\kappa \in (0, 1]$. We put $n = 1$, $\tau = 1/2$; $u^1 = u$, $(L_1 = L)$. Then all assertions of Theorem 2 are true.

For our purpose, it is convenient to prove the estimate of the normal derivative of u^j immediately after Theorem 2.

4. THEOREM. Let $\alpha, \beta \in (0, 1)$, let $\kappa \in (0, 1]$, and let $u^1, \ldots, u^n \in W^{1,2}_{d+1,loc}(Q^+_{T,3}) \cap C(\bar{Q}^+_{T,3})$. Also, let $g^1, \ldots, g^d$ be functions on $Q^+_{T,3}$. Suppose that

$$|Lu^j - u^j_{x^k} g^k| \leq K_2\left(\frac{1}{x^1} + (u^i_x, u^i_x)\right)^{\alpha},$$

$$g^k g^k \leq K_2((x^1)^{-\alpha} + (u^i_x, u^i_x))$$

and $|u^j| \leq K_2$ on $Q^+_{T,3}$ (a.e.). Also, let

$$|u^j(t_1, x_1) - u^j(t_2, x_1)| \leq K_2|t_1 - t_2|^{(1+\beta)/2},$$

$$\left|u^j_{x^i}(t_1, x_1) - u^j_{x^i}(t_1, x_2)\right| \leq K_2|x_1 - x_2|^{\beta} \tag{4}$$

for $j = 1, \ldots, n$, $i = 2, \ldots, d$, $(t_1, x_1), (t_2, x_2) \in \Sigma_{T,3}$ ($u^j_{x^i}$ are supposed to exist on $\Sigma_{T,3}$) and $u^j_{x^1}$ exists on $\Sigma_{T,1}$. Then

(a) $|u^j_{x^1}| \leq \kappa^{-1}N$ on $\Sigma_{T-\kappa^2,1}$ for $j = 1, \ldots, n$;

(b) if, in addition,

$$|u^j(T, x) - u^j(T, 0, x^2, \ldots, x^d)| \leq K_2 x^1$$

for $(T, x) \in \partial_t Q^+_{T,3}$, then $|u^j_{x^1}| \leq N$ on $\Sigma_{T,1}$ with $j = 1, \ldots, n$.

In both cases $N = N(\varepsilon, d, K, K_1, K_2, \beta, \alpha, n)$.

Proof. As in the proof of Theorem 3.5, we can assume that the coefficients b^i and c of the operator L are zero. Furthermore, it follows from the inequality $|u^j| \leq K_2$ that $|u^j_{x^i}| \leq K_2$ for all $j = 1, \ldots, n$, $i = 2, \ldots, d$ and $t \in [0, T)$ at least for one point $x \in \Sigma_3$. It then follows from the second condition of (4) that $|u^j_{x^i}| \leq N(\beta, K_2)$ on $\Sigma_{T,3}$. In what follows we will not specially stress the dependence of constants on n, ε, d, K, K_1, K_2, β, and α. We fix a point $(t_0, x_0) \in \Sigma_{T,1}$ and put

$$\tilde{u}^j(t, x) = u^j(t, x) - u^j(t_0, x_0) - \sum_{i=2}^{d} u^j_{x^i}(t_0, x_0)(x^i - x^i_0).$$

It is clear that $\tilde{u}^j$ satisfies the conditions of the theorem with the same functions g^k if K_2 is replaced with a constant N_1 and α is replaced by $\alpha \vee (1/2)$. Therefore, following the proof of Theorem 3.5, we can choose $\gamma > 0$ such that the function

$$v(p, t, x) := \exp\left\{\gamma \sum_{i=1}^{n} (\tilde{u}^j(t, x) - p^j)^2\right\} - \exp \gamma|p|^2$$

satisfies the inequality $Lv \geq -N_2(x^1)^{-\alpha}$ in $Q^+_{T,3}$ for all $p = (p^1, \ldots, p^n) \in E_n$ such that $|p^j| \leq N_1$. Also, note that, by virtue of (4),

$$\begin{aligned}|\tilde{u}^j(t, x)| \leq{} & |u^j(t, x) - u^j(t_0, x)| + \\ & + |u^j(t_0, x) - u^j(t_0, x_0) - \\ & - \sum_{i=2}^{d} u^j_{x^i}(t_0, x_0)(x^i - x^i_0)| \leq\end{aligned}$$

$$\le N\Big(|t - t_0|^{(1+\beta)/2} + \sum_{i=2}^{d} |x^i - x_0^i|^{1+\beta}\Big)$$

for $(t, x) \in \Sigma_{T,3}$. It follows that

$$|v(p, t, x)| \le N_3\Big(|t - t_0|^{(1+\beta)/2} + \sum_{i=2}^{d} |x^i - x_0^i|^{1+\beta}\Big). \tag{5}$$

for the p's under consideration and the same (t, x).

Using the established properties of v and the maximum principle, we further estimate $(x^1)^{-1} v(t_0, x)$ for $x^2 = x_0^2, \ldots, x^d = x_0^d$. Assume that $T - t_0 \le 1$, since otherwise we can appropriately cut $Q^+_{T,3}$. We can obviously assume that $t_0 = 0$.

We now choose a sufficiently large m and a sufficiently small $\delta \in (0, 1)$ such that the function $w(x) = 1 - |x - \tilde{x}_0|^{-m} - (x^1)^{2-\alpha}$ with $\tilde{x}_0 = (-1, x_0^2, \ldots, x_0^d)$ satisfies the inequality $Lw \le -\delta(x^1)^{-\alpha}$ in the domain $G := Q^+_{T,3} \cap \{|x - \tilde{x}_0| < 1 + \delta\}$. The formula

$$Lw = m|\tilde{x}|^{-m-2}(\delta^{ij}a^{ij} - (m + 2)|\tilde{x}|^{-2}a^{ij}\tilde{x}^i\tilde{x}^j) -$$

$$- (2 - \alpha)(1 - \alpha)a^{11}(x^1)^{-\alpha} \text{ with } \tilde{x} = x - \tilde{x}_0$$

shows that it is possible to choose such m and δ. Reducing if needed δ, we can assume that $w \ge 0$ on G and $w(x) \ge \delta^2$ with $|\tilde{x}| = 1 + \delta$.

The function w will play the role of a majorant for v on the part of $\partial' G$ which lies on $\{(t, x) : |\tilde{x}| = 1 + \delta\}$. We further reduce δ, if necessary, so that it suits Lemma 4.1(g), where α is replaced by $1 + \beta$ and according to the sum in (5) construct the barriers dominating v on the other parts of $\partial' G$. For $\partial_t G$ we take ζ as introduced at the beginning of Section 4 (setting $K_1 = 0$ in its definition). For $|t|^{(1+\beta)/2}$ we use the function τ from Lemma 4.1(f)), setting α equal to $1 + \beta$. For other summands of (5) we take σ^i from Lemma 4.1(g). Finally, we fix $\omega > 0$ and denote $M = \delta^{-2}\exp(\gamma 4nN_1^2)$,

$$h(t, x) = (N_2 + 1)Mw(x) + M\zeta(t, x^1 + \omega, 0, \ldots, 0) +$$

$$+ N_3\tau(t, x) + N_3 \sum_{i=2}^{d} \sigma^i(t, x^1, x^2 - x_0^2, \ldots, x^d - x_0^d). \tag{6}$$

By Lemma 4.1, we have $Lh \le -N_2(x^1)^{-\alpha} \le Lv$ in G. Moreover, $v \le h$ on $\partial' G$. By the maximum principle, $v \le h$ in G. In particular,

$$v(p, t_0, x) \leq \lim_{\omega\downarrow 0} h(t, x)$$
$$\leq Nx^1 \left[\frac{1}{x^1}(1 - |1 + x^1|^{-m}) + T - t_0)^{-1/2}\right] \tag{7}$$

for $x = (x^1, x_0^2, \ldots, x_0^d)$ and $x^1 \in (0, \delta)$ (see Lemma 4.1). This obviously implies the estimate of the derivative $v_{x^1}(p, t_0, x_0)$ and hence the estimate

$$-\sum_{j=1}^{n} p^j u^j_{x^1}(t_0, x_0) \leq N(1 + (T - t_0)^{-1/2}).$$

To prove assertion (a), it remains to recall the stipulated arbitrariness permitted in the choice of p. To prove (b), it suffices to omit the summand $M\zeta$ in (6), to note that

$$|\tilde{u}^j(T, x)| \leq |\tilde{u}^j(T, 0, x^1, \ldots, x^d)| + K_2 x^1$$
$$\leq N\left[(T - t_0)^{(1+\beta)/2} + x^1 + \sum_{i=2}^{d} |x^i - x_0^i|^{1+\beta}\right]$$

for $(T, x) \in \partial_t Q^+_{T,3}$ and to increase the coefficient N_3 of τ in (6) according to Lemma 4.1(f), which implies that $N\tau(T, x) \geq T^{(1+\beta)/2} + x^1$ for $x^1 \leq 6$. This proves the theorem.

5. Remark. In fact our arguments imply a stronger result, namely $\Sigma|u^j(t, x) - u^j(t, 0, x^2, \ldots, x^d)|$ does not exceed $N\kappa^{-1}x^1$ for $(t, x) \in Q^+_{T-\kappa^2,1}$, and if the hypothesis of (b) is true, this sum does not exceed Nx^1 in $Q^+_{T,1}$. It is not required here that $u^j_{x^1}$ exist on $\Sigma_{T,1}$.

To prove this result, we make use of the following consequence of (7):

$$\sum_{j=1}^{n} (\tilde{u}^j(t_0, x) - p^j)^2 - |p|^2 \leq Nx^1(1 + (T - t_0)^{-1/2}).$$

Without loss of generality assuming that $N_1 \geq 1$, we remove the brackets and substitute $p^j = -\operatorname{sgn} \tilde{u}^j(t_0, x)$. We then obtain

$$\sum_{j=1}^{n} |\tilde{u}^j(t_0, x)| \leq Nx^1(1 + (T - t_0)^{-1/2}) \quad .$$

and thus our remark concerning assertion (a) is true. Assertion (b) can be considered in a similar manner.

6. COROLLARY (cf. Corollary 3.6). If $|u| \leq K_2$, $|Lu| \leq K_2(1 + |u_x|^2)$ (a.e. $Q^+_{T,3}$), $u \in W^{1,2}_{d+1,loc}(Q^+_{T,3}) \cap C(\bar{Q}^+_{T,3})$, u_x exist on $\Sigma_{T,3}$, and inequalities (4) hold true for $n = 1$ and $u^1 = u$, then the assertions of Theorem 4 are also true for $n = 1$ and $u^1 = u$.

7. Remark. Assertions 1 to 6 remain valid if $Q^+_{T,3}$, $Q^+_{T,1}$, $\Sigma_{T,3}$, $\Sigma_{T,1}$ and $\Sigma_{T-\kappa^2,1}$ are replaced by $(0, T) \times D^+_{\vec{R}_1}$, $(0, T) \times D^+_{\vec{R}_2}$, $[0, T) \times \Sigma_{\vec{R}_1}$, $[0, T) \times \Sigma_{\vec{R}_2}$, and $[0, T - \kappa^2) \times \Sigma_{\vec{R}_1}$, where for $\vec{R} = (R^1, \ldots, R^d)$

$$D^+_{\vec{R}} = \left\{x \in E_d\colon 0 < x^1 < 2R^1,\ |x^i| < R^i,\ i = 2, \ldots, d\right\},$$

$$\Sigma_{\vec{R}} = \{x^1 = 0\} \cap \partial D^+_{\vec{R}},$$

$R^i_1 > R^i_2 > 0$. Here the constants N should be allowed to depend on $R^i_1 - R^i_2$, while β_0 should *remain unchanged.*

If $3 \geq R^1_1 = \ldots = R^d_1 \geq 3R^1_2, \ldots, 3R^d_2$, then the validity of the remark can be established using the self-similarity transformation $(t, x) \to (9(R^1_1)^{-2}t, 3(R^1_1)^{-1}x)$ which *preserves* the constants ε and K in condition (2.2) and converts $D^+_{\vec{R}}$ to D^+_3.

In case when the R^i_1 are distinct, and the linear transformation of $D^+_{\vec{R}_1}$ to D^+_3 variously acts on different coordinates, it changes the constants ε and K in (2.2) and β_0. In this case $D^+_{\vec{R}_2}$ should be covered with cubes of the form $x + D^+_R$ (the quantity is immaterial) such that $x + D^+_{3R} \subset D^+_{\vec{R}_1}$. A suitable R can be chosen so that it will be smaller than 1 and depend only on $R^i_1 - R^i_2$. It then remains to convert $(0, T) \times (x + D^+_R)$ to $(0, T) \times (y + D^+_1)$ using the self-similarity transformation and to note that, to estimate the Hölder exponent or the normal derivative, it suffices to consider only points from $(0, T) \times (x + D^+_R)$.

We now formulate similar results concerning elliptic equations. It seems unnecessary to explain how their proofs can be obtained from the previous discussion. As usually, we assume that the coefficients of L_i do not depend on t and put $\tilde{L}_i = L_i - \frac{\partial}{\partial t}$, $\tilde{L} = L - \frac{\partial}{\partial t}$.

8. THEOREM. Let $u^1, \ldots, u^n \in W^2_{d,loc}(D_3^+) \cap C(\overline{D_3^+})$, $f_1, \ldots, f_n \in C^2_{loc}(E_1)$, let $f_i' > 0$ and $f_i'' \leqq 0$ on E_1, let M and M_2 be nonnegative constants, and let $|u^i| \leqq M$ in D_3^+ and $\tilde{L}_i u^i \geqq -M_2$ in D_3^+(a.e.) for i = 1, ..., n. Suppose that

$$M_2|x_1 - x_2|^{\alpha} \geqq \sigma \sum_{i=1}^{n} [f_i(u^i(x_1)) - f_i(u^i(x_2))]_+ -$$

$$- \sum_{i=1}^{n} [f_i(u_i(x_1)) - f_i(u^i(x_2))]$$

with constants $\delta > 0$ and $\alpha \in (0, 1)$ and for any $x_1, x_2 \in D_3^+$.

Then for any $x^1, x^2 \in D_1^+$

$$\sum_{i=1}^{n} |u^i(x_1) - u^i(x_2)| \leqq |x_1 - x_2|^{\beta} N(M_\beta + M_2 + M),$$

where β is any number from the interval $(0, \beta_0 \wedge \alpha]$, $\beta_0 \in (0, 1)$, $N \geqq 0$; β_0 and N depend on the initial constants as specified in Theorem 1, and finally M_β is the greatest Hölder constant of order β for the functions u^i on the face $x^1 = 0$ of the cube $\overline{D_3^+}$.

9. THEOREM. Let $\tau \in (0, 1)$, let $u^1, \ldots, u^n \in W^2_{d,loc}(D_3^+) \cap C(\overline{D_3^+})$, let M_β be a nonnegative constant, and let $g^1, \ldots, g^d$ be functions on D_3^+. Suppose that $|u^j| \leqq K_2$,

$$|Lu^j - u^j_{x^k} g^k| \leqq K_2(1 + (u^i_x, u^i_x))^{\tau}, \quad g^k g^k \leqq K_2(1 + (u^i_x, u^i_x))$$

in D_3^+ (a.e.). Then there exists a constant $\beta_0 \in (0, 1)$, which depends only on d, K, ε, and n and such that if $\beta \in (0, \beta_0]$ and the Hölder constants of order β for the functions u^i on the face $x^1 = 0$ of the cube $\overline{D_3^+}$ do not exceed M_β, then for any $x_1, x_2 \in D_1^+$

$$\sum_{i=1}^{n} |u^i(x_1) - u^i(x_2)| \leqq N|x_1 - x_2|^{\beta},$$

where N depends only on d, K, ε, n, K_1, K_2, τ, M_β, and β.

10. COROLLARY. Let $u \in W^2_{d,loc}(D_3^+) \cap C(\overline{D_3^+})$, let $|u| \leq K_2$, and let $|\tilde{L}u| \leq K_2(1 + |u_x|^2)$ in D_3^+ (a.e.). We put $n = 1$, $\tau = 1/2$, and $u^1 = u$. Then the assertion of Theorem 9 is true.

11. THEOREM. Let α, $\beta \in (0, 1)$, let $u^1, \ldots, u^n \in W^2_{d,loc}(D_3^+) \cap C(\bar{D}_3^+)$, and let $g^1, \ldots, g^d$ be functions on D_3^+. Suppose that $|u^j| \leq K_2$ and

$$|\tilde{L}u^j - u^j_{x^k} g^k| \leq K_2\left(\frac{1}{x^1} + (u^i_x, u^i_x)\right)^{\alpha},$$

$$\bar{g}^k g^k \leq K_2((x^1)^{-\alpha} + (u^i_x, u^i_x))$$

on D_3^+ (a.e.). Also, let the norm of u^j in $C^{1+\beta}(\Sigma_3)$ be no greater than K_2 and let $u^j_{x^1}$ exist on Σ_1. Then $|u^j_{x^1}| \leq N(\varepsilon, d, K, K_1, K_2, \beta, \alpha, n)$ on Σ_1 for all j.

12. COROLLARY. In addition to the hypothesis of Corollary 10, suppose that $\beta \in (0, 1)$, the norm of u in $C^{1+\beta}(\Sigma_3)$ does not exceed K_2, and u_{x^1} exists on Σ_1. Then $|u_{x^1}| \leq N(\varepsilon, d, K, K_1, K_2, \beta)$ on Σ_1.

4.6. The Hölder Condition for Solutions of Some Degenerate Linear Equations on the Boundary

The proof of an estimate in $C^{2+\alpha}$ in a closed domain for the solutions of nonlinear equations is based on the interior estimate and the estimate of the normal derivative in $C^{1+\alpha}$ on the boundary of the domain. In the theory of linear equations, this derivative is estimated by the method of 'freezing' the coefficients using the explicit integral representation of a solution. For obvious reasons, in the theory of nonlinear equations we have to use an appropriate equation for the normal derivative rather than the integral representations.

Following the author's paper [21], we construct an equation which is used to examine the normal derivative by substituting the solution expressed in the form $x^1 v$ if a part of the domain boundary lies in the plane $x^1 = 0$ and the zero boundary condition is specified on this part. It is clear that v coincides with the normal derivative of a solution on this part and we can write an equation for v based on the initial equation. This equation is degenerate for $x^1 = 0$ and possesses some pecularities.

The major difficulties involved in such equations can be surmounted using linear equations and linear inequalities with the same peculiarities. It is specifically degenerate linear inequalities that constitute the subject of this section. Also, note that by the above argument we shall deal with the Hölder condition for their solutions exclusively on the boundary.

Consider some constants $K \geqq \varepsilon > 0$, R, $T > 0$, K_1, and $K_2 \geqq 0$, and let

$$L = x^1 \frac{\partial}{\partial t} + x^1 a^{ij}(t, x) \frac{\partial^2}{\partial x^i \partial x^j} + b^i(t, x) \frac{\partial}{\partial x^i} + c(t, x) \tag{1}$$

be an operator whose measurable coefficients are defined in $Q^+_{T,3}$ and such that

$$a^{ij} = a^{ji}, \quad \varepsilon|\lambda|^2 \leqq a^{rs}\lambda^r\lambda^s \leqq K|\lambda|^2, \quad |c| \leqq K_1 \text{ in } Q^+_{T,3}, \tag{2}$$

$$|b^i| \leqq K \text{ with } (t, x) \in Q^+_{T,3} \cap \{x^1 \leqq (K_1 + 1)^{-1}\}, \tag{3}$$

$$2a^{11} - b^1 + cx^1 \leqq 0 \text{ in } Q^+_{T,3} \tag{4}$$

for all $\lambda \in E_d$, $i, j = 1, \ldots, d$.

Note that somewhat strange condition (3) is dictated by the wish to restrict the set of parameters involved in the results to be established (see Section 5.1), and which affect the Hölder exponent, to the parameters d, K, and ε. So far we have managed to avoid the dependence of the Hölder exponent on estimates of $|b^i|$ by reducing them by the self-similarity transformation, but as for operators (1) such a method is of no use since it can provide only a change of variables in the coefficients a and b.

It should be said that, of course, the results of this section could be formulated in the same way as the results of Sections 3 and 5. However, the statements of Sections 3 and 5 have already been used for studying quasi-linear inequalities and will be used later in a much more important case to estimate the Hölder constant for the second derivatives, where they are essential. At the same time we will need comparatively little information on the operators (1). Therefore, we do not intend to obtain results in the most general form, thereby gaining in simplicity.

We prove two estimates for the Hölder constant for the solution of the inequality $|Lu| \leqq Lv + K_2$, where v is a known function. One of these estimates is interior with respect to t, another is valid up to the right (along the t-axis) base of $Q^+_{T,R}$.

1. THEOREM. Let $R \leqq 3 \wedge \sqrt{T}$, let u and v belong to $C^2_{loc}(Q^+_{R^2,R})$, let u and v be bounded on $Q^+_{R^2,R}$, let $\alpha \in (0, 1)$, and let $v \in C^{\alpha}(Q^+_{R^2,R})$ and

$$|Lu| \leq Lv + K_2 \text{ on } Q^+_{R^2,R}. \tag{5}$$

Then u can be continuously extended to $Q^+_{R^2,R} \cup \Sigma_{R^2,R}$ and there exist constants $\alpha_0 = \alpha_0\ (d, K, \varepsilon) \in (0, 1)$ and $N = N(\alpha, d, K, K_1, \varepsilon)$ such that

$$|u(z_1) - u(z_2)| \leq Nr^\beta R^{-\beta}\Big(RK_2 + R^\alpha|[v]|_{C^\alpha(Q^+_{R^2,R})} + ||u||_{C(Q^+_{R^2,R})} + ||v||_{C(Q^+_{R^2,R})}\Big) \tag{6}$$

for any $r \leq R$, $z_1, z_2 \in Q^+_{r^2,r}$ and $\beta = \alpha_0 \wedge \alpha$.

To prove this theorem, we need a lemma which, in essence, shows that the face $x^1 = 0$ of the parallelepiped $\overline{Q^+_{R^2,R}}$ should not be regarded as a boundary one for the operator L. The latter is not surprising, since condition (4) means essentially that $L(x^1)^{-1} \leq 0$ in $Q^+_{T,3}$ and hence from the standpoint of potential theory this face is the polar set for the operator L.

2. LEMMA. Let R, $\gamma \in (0, 1]$, let $\gamma \leq T$, let $Q = (0, \gamma) \times (0, 3) \times (-R, R)^{d-1}$, let $c \geq 0$, $b^1 \leq 0$, $|b^i| \leq K$ on Q, let $\psi \in C^2_{loc}(Q) \cap C(\bar{Q} \setminus \{x^1 = 0\})$, let $L\psi \leq 0$ in Q, let $\psi \geq 0$ on $\partial'Q \setminus \{x_1 = 0\}$, and let $\psi \geq 1$ on the face $x^1 = 3$ of the parallelepiped $\bar{Q}$. Then there exists a constant $\delta > 0$ which depends only on R, γ, d, K, K_1, and ε and such that $\psi(0, x) \geq \delta$ for $0 < x^1 \leq 3$ and $x^2 = \dots = x^d = 0$. Moreover, $\psi \geq 0$ in Q.

<u>Proof</u>. Consider a $\xi > 0$ and a function $w(t, x) = \exp(\xi t)(\psi(t, x) + \xi(x^1)^{-1}$. This function attains its minimum over $\bar{Q} \setminus \{x^1 = 0\}$ at a point $(t_0, x_0) \in \bar{Q}$. Obviously, $x_0^1 > 0$. If $(t_0, x_0) \in \partial'Q$, then $w(t_0, x_0) \geq 0$. But if $(t_0, x_0) \notin \partial'Q$ and $w(t_0, x_0) \leq 0$, then, by virtue of well-known properties of the derivative of a function at its extremum points, we have $Lw \geq 0$ at the point (t_0, x_0). Computing Lw at the point (t_0, x_0), we obtain $x^1\xi w + \exp(\xi t)L(\psi + \xi(x^1)^{-1}) \geq 0$. Here $L\psi \leq 0$ and $L(x^1)^{-1} \leq 0$. Therefore, again $w(t_0, x_0) \geq 0$. Thus, $w(t_0, x_0) \geq 0$,

$w \geqq 0$ on Q and $\psi + \xi(x^1)^{-1} \geqq 0$ on Q, and since $\xi > 0$ is chosen arbitrarily, we have $\psi \geqq 0$ on Q.

We now estimate $\psi(0, x)$. We first estimate $\psi(t, x)$ for values of x^1 close to 3. To this end we take γ, K, $t^{1/4}$ as T, K_1, and φ and use them to construct the functions ζ and σ^i as at the beginning of Section 4. Denote $\tilde{\zeta}(t, x) = \zeta^1(t, \tilde{x})$, $\tilde{\sigma}^i(x) = \sigma^i(\tilde{x})$, where $\tilde{x} = (3, 0, \ldots, 0) - x$. By Lemma 4.1 (here we use the condition $b^1 \geqq 0$), we have $(x)^{-1} L\tilde{\zeta} \leqq 0$ and $L\tilde{\sigma}^i \leqq 0$ in $Q \setminus \{x^1 \leqq 1\}$. Moreover, from Lemma 4.1 it readily follows that for any $\delta_1 > 0$, the function

$$w_1(t, x) := N\left(\tilde{\zeta}(t, x) + \sum_{i=2}^{d} \tilde{\sigma}^i(x)\right)$$

with a constant $N(\delta_1, d, K, K_1, \varepsilon, R, \gamma)$ exceeds 1 everywhere on the parabolic boundary of the parallelepiped $Q \setminus \{x^1 \leqq 3 - \delta_1\}$, except for the face $x^1 = 3$ where it is nonnegative.

We now choose the constants δ_1 such that $L(x^1 - 3)(3 - \delta_1 - x^1) \leqq \varepsilon$ in $Q \setminus \{x^1 \leqq 3 - \delta_1\}$ and put

$$w_2(t, d) = 1 - w_1(t, x) - K_1\varepsilon^{-1}(x^1 - 3)(3 - \delta_1 - x^1).$$

Then, by the maximum principle, $w_2 \leqq \psi$ in $Q \setminus \{x^1 \leqq 3 - \delta_1\}$. Since $w_2(0, x) \to 1$ as $x = (x^1, 0, \ldots, 0)$ and $x^1 \uparrow 3$, we have $\psi(0, x^1, 0, \ldots, 0) \geqq 1/2$ for $x^1 \in [3 - \delta_2, 3]$ with $\delta_2 > 0$. Moreover, $w_2(3/4)\gamma, x) \to 1$ as $x = (x^1, 0, \ldots, 0)$ and $x^1 \uparrow 3$, thus it follows that there exists a ball S in E_d, whose centre lies on the x^1-axis below but sufficiently close to 3, such that $w_2(3/4)\gamma, x) \geqq 1/2$ and $\psi((3/4)\gamma, x) \geqq 1/2)$ for $x \in S$.

By Lemma 1.3 on an oblique cylinder, the established properties of ψ imply that $\psi(t, x^1, 0, \ldots, 0) \geqq \delta_3$ for $t \in [0, 1/2\gamma]$, $x^1 \in [\eta, 3]$, where the constant $\eta \in (0, 3]$, we can choose from the condition

$$K_1\varepsilon^{-1}\eta + 2\eta^2(\varepsilon\gamma)^{-1} + 2(d - 1)\exp\left(\mu - \frac{\nu}{\eta}R\right) \leqq \frac{1}{2}, \tag{7}$$

where $\mu = 1/2\varepsilon K^{-1}$ and $\nu = 1/2\mu^2$. Considering $x^1 \in [0, \eta]$, we compare ψ with the function w_3 which is defined as

$$w_3 = 1 - \sum_{i=2}^{d} (w_3^{(+i)} + w_3^{(-i)} - w_4 - w_5,$$

$$w_3^{(\pm i)} = \exp[q(\eta - x^1) + p(\pm x^i - R)], \quad p = \nu/\eta, \quad q = \mu/\eta,$$

$$w_4(t, x) = \eta\varepsilon^{-1}(\eta - x^1)\left(\tfrac{1}{2}\gamma - t\right)^{-1}, \quad w_5 = K_1\varepsilon^{-1}(\eta - x^1).$$

We put $Q_1 = Q \cap \{x^1 < \eta,\ w_4(t, x) < 1,\ t < (1/2)\gamma\}$. It is clear that Q_1 is a polyhedron such that $w_3 \leqq 0$ on $\partial' Q_1$ except for the face $x^1 = \eta$, where $w_3 \leqq 1$, and the face $x^1 = 0$, which is absolutely immaterial as we could see at the outset of the proof. Moreover, by virtue of (2), (4) and the inequality $K_1\eta \leqq \varepsilon$, we have $b^1 \geqq 2a^{11} + cx^1 \geqq \varepsilon$ in Q_1 and

$$\left(w_3^{(\pm i)}\right)^{-1} Lw_3^{(\pm i)} = x^1[a^{11}q^2 \mp 2a^{1i}pq + a^{ii}p^2] -$$

$$- b^1q \pm b^ip \leqq \eta K(p^2 + q^2) + Kp - \varepsilon q$$

$$= \eta^{-1}\left[K\nu^2 + K\nu - \tfrac{1}{4}\varepsilon^2 K^{-1}\right]$$

$$< \eta^{-1}\left[2K\nu - \tfrac{1}{4}\varepsilon^2 K^{-1}\right] = 0, \tag{8}$$

$$Lw_4 = \eta\left(\tfrac{1}{2}\gamma - t\right)^{-1}\left[x^1(\eta - x^1)\varepsilon^{-1}\left(\tfrac{1}{2}\gamma - t\right)^{-1} - \varepsilon^{-1}b^1\right]$$

$$\leqq \eta\left(\tfrac{1}{2}\gamma - t\right)^{-1}[w_4(t, x) - 1] < 0,$$

$$Lw_5 = -K_1\varepsilon^{-1}b^1 + cK_1\varepsilon^{-1}(\eta - x^1) \leqq -K_1 \leqq c.$$

Hence $Lw_3 > 0$ in Q_1 and, by the maximum principle $w_3 \leqq \delta_3^{-1}\psi$ in Q_1. Specifically, for $x^1 \in (0, \eta]$

$$\delta_3^{-1}\psi(0, x^1, 0, \ldots, 0)$$

$$\geqq w_3(0, x^1, 0, \ldots, 0)$$

$$= 1 - 2(d - 1)\exp[q(\eta - x^1) - pR] - \eta(\eta - x^1)2(\varepsilon\gamma)^{-1} -$$

$$- K_1\varepsilon^{-1}(\eta - x^1)$$

$$\geqq 1 - 2(d - 1)\exp(q\eta - pR) - \eta^2 2(\varepsilon\gamma)^{-1} - K_1\varepsilon^{-1}\eta.$$

The latter is greater than 1/2 by virute of (7). It remains to note that all the constants involved could be chosen depending only on R, γ, d, K, K_1, and ε. This proves the lemma.

3. Proof of Theorem 1. Since $|Lu - c_+u| \leqq |Lu| + K_1|u|$, a similar inequality holds for Lv and condition (4) remains valid if c is replaced by $(-c_-)$, we can, therefore, assume that $c \leqq 0$.

Also, note that for $R \geqq r \geqq \varepsilon 2^{-1}(K_1 + 1)^{-1}$ inequality (6) is clear since on the right there is the norm of u in $C(Q^+_{R^2,R})$, but if $R \geqq \varepsilon 2^{-1}(K_1 + 1)^{-1} \geqq r$, inequality (6) follows from the inequality which results from (6) if R is replaced by $\varepsilon 2^{-1}(K_1 + 1)^{-1}$. Thus, when proving (6), we can assume that

$$r \leqq R \leqq \varepsilon 2^{-1}(K_1 + 1)^{-1}. \tag{9}$$

Denote

$$M = ||u||_{C(Q^+_{R^2,R})} + ||v||_{C(Q^+_{R^2,R})}, \quad M_\alpha = |[v]|_{C^\alpha(Q^+_{R^2,R})},$$

$$H_1 = \varepsilon^{-1}(K_1M + K_2), \quad H_2 = M_\alpha + R^{1-\alpha}(M + K_2),$$

$$u^1 = v + u + H_1x^1, \quad u^2 = v - u + H_1x^1,$$

$$m^i(r) = \inf(u^i, Q^+_{r^2,r}), \quad M^i(r) = \sup(u^i, Q^+_{r^2,r}),$$

$$w^i(r) = M^i(r) - m^i(r), \quad \theta^i(r) = \frac{3}{4}M^i(r) + \frac{1}{4}m^i(r),$$

$$i = 1, 2.$$

It is clear that for any $z_1, z_2 \in Q^+_{r^2,r}$, $z_i = (t_i, x_i)$,

$$[u^1(z_1) - u^1(z_2)] + [u^2(z_1) - u^2(z_2)]$$

$$= 2[(v(z_1) - v(z_2)) + (x^1_1 - x^1_2)H_1] \leqq Nr^\alpha H_2 \tag{10}$$

(we do not stress the dependence of N on α, d, K, K_1, ε). Moreover, in $Q^+_{R^2,R}$ (see (9))

$$L(u^i - x^1H_1) \geqq -K_2,$$

$$b^1 \geqq 2a^{11} + x^1c \geqq 2\varepsilon - 2K_1R \geqq \varepsilon,$$

$$x^1u^i_t + x^1a^{kr}u^i_{x^kx^r} + b^ku^i_{x^k}$$

$$\geqq b^1 H_1 - K_2 - c(u^i - x^1 H_1) \geqq \varepsilon H_1 - K_2 - K_1 M = 0. \tag{11}$$

We now fix $r \leqq (1/2)R$ and write $i \in A(r)$ if

$$\Lambda\left[Q^+_{4r^2,2r} \cap \{u^i \leqq \theta^i(r)\}\right] \geqq \frac{1}{2}\Lambda(Q^+_{4r^2,2r}). \tag{12}$$

Let, first, $A(r) \neq \emptyset$, $i \in A(r)$. We put $\tilde{u}^i(t, x) = u^i(r^2t, rx)$. Then, by virtue of (11), we have

$$\tilde{u}^i_t + \tilde{a}^{ks}\tilde{u}^i_{x^k x^s} + (x^1)^{-1}\tilde{b}^k\tilde{u}^i_{x^k} \geqq 0 \tag{13}$$

in $Q^+_{4,2}$, where $\tilde{a}^{ks} = a^{ks}(r^2t, rx)$, $\tilde{b}^k = b^k(r^2t, rx)$. When $(t, x) \in Q^+_{4,2}$, it is obvious that $(r^2t, rx) \in Q^+_{4r^2,2r} \subset Q^+_{T,3}$ and

$$rx^1 \leqq 4r \leqq 2R \leqq \varepsilon(K_1 + 1)^{-1}.$$

Therefore $|\tilde{b}^k| \leqq K$ on $Q^+_{4,2}$.

Furthermore, it follows from (12) that the set $\{\tilde{u}^i \leqq \theta^i(r)\}$ takes up no less than one fifth of the volume of the parallelepiped $\tilde{Q} := (3/2, 2, 0, \ldots, 0) + Q_{5/2,2}$. We now wish to apply Lemma 1.11 to inequality (13) in Q. Here we meet with two difficulties. First, the coefficients $(x^1)^{-1}\tilde{b}^k$ are unbounded in $\tilde{Q}$ and, second, $\tilde{Q}$ is not a circular cylinder. The first difficulty can easily be surmounted if we consider $\tilde{Q} \cap \{x^1 > 10^{-1}\}$ in place of $\tilde{Q}$. As for the second problem, we can find an absolute constant $\gamma > 0$ such that the share which the set $\{\tilde{u}^i \leqq \theta^i(r)\}$ occupies in the volume of a cylinder congruent to $C_{\gamma^2,\gamma}$ and lying in $\tilde{Q} \cap \{x^1 > 10^{-1}\}$, is no less than γ. Then, by Lemma 1.11 and Theorem 1.9, the inequalities

$$\tilde{u}^i \leqq p_0\theta^i(r) + (1 - p_0)M^i(2r),$$

$$M^i(2r) - \tilde{u}^i \geqq p_0(M^i(2r) - \theta^i(r)),$$

where $p_0 \in (0, 1)$ and p_0 depends only on d, K, and ε, hold on the disk which concentrically lies in the left base of this cylinder along the t-axis and whose radius is $(1/2)\gamma$. By Lemma 1.3 on an oblique cylinder, for a constant $\delta_1 = \delta_1(d, K, \varepsilon) > 0$ with $t \in (0, 5/4)$, $x^1 = 3$, $|x^i| \leqq 3/2$, with $i \geqq 2$, we obtain

$$M^i(2r) - \tilde{u}^i(t, x) \geqq \delta_1 p_0(M^i(2r) - \theta^i(r)). \tag{14}$$

Since $M^i(2r) - \tilde{u}^i(t, x) \geq 0$ in $Q^+_{4,2}$ and $c \leq 0$, the operator

$$x^1 \frac{\partial}{\partial t} + x^1 \tilde{a}^{ks} \frac{\partial^2}{\partial x^k \partial x^s} + \tilde{b}^k \frac{\partial}{\partial x^k} + r\tilde{c},$$

with $\tilde{c} = c(r^2 t, rx)$ as applied to the left-hand side of (14) yields a negative expression on $Q^+_{4,2}$. Hence, by Lemma 2 applied to shifts of the parallelepiped $(0, 1/4) \times (0, 3) \times (-1/2, 1/2)^{d-1}$ in the plane $x^1 = 0$, it follows that inequality (14) is true for all $(t, x) \in Q^+_{1,1}$, except that it may be necessary to replace $\delta_1 p_0$ by $\delta_2 = \delta_2(d, K, \varepsilon) \in (0, 1)$. Thus,

$$M^i(2r) - M^i(r) \geq \delta_2\left(M^i(2r) - \left(\frac{3}{4} M^i(r) + \frac{1}{4} m^i(r)\right)\right),$$

$$\left(1 - \frac{3}{4}\delta_2\right)M^i(r) - \frac{1}{4}\delta_2 m^i(r) \leq (1 - \delta_2)M^i(2r).$$

Adding the latter to the obvious inequality $(\delta_2 - 1)m^i(r) \leq (\delta_2 - 1)m^i(2r)$, we conclude

$$w^i(r) \leq \delta_2 w^i(2r), \quad i \in A(r), \tag{15}$$

where $\delta_3 = (1 - \delta_2)\left(1 - \frac{3}{4}\delta_2\right)^{-1} < 1$.

We now consider the case when $A(r) = \phi$. Then for $i = 1,2$ the inequality opposite to (12) is true and there exists at least one point $z_0 \in Q^+_{4r^2,2r}$ such that

$$u^i(z_0) > \frac{3}{4} M^i(r) + \frac{1}{4} m^i(r), \quad i = 1, 2. \tag{16}$$

We substitute $z_1 = z_0$ into (10) and take a point of $Q^+_{r^2,r}$, where u^1 almost attains its minimum, as the point z_2. Then, of course, $u^2(z_2) \leq M^2(r)$ and from (10) it follows that

$$\frac{3}{4} w^1(r) - \frac{1}{4} w^2(r) \leq Nr^{\alpha} H_2. \tag{17}$$

The superscripts 1 and 2 can be interchanged and, on doing this and adding the resulting inequality to (17), we can see that

$$w^1(r) + w^2(r) \leq Nr^{\alpha} H_2.$$

Finally, if for example $1 \notin A(r)$, $2 \in A(r)$, then we can find a point z_0 satisfying (16) for $i = 1$ and obtain the inequality $(3/4)w^1(r) \leq w^2(2r) + Nr^\alpha H_2$ as it was done with (17). It follows that

$$\frac{3}{4} \sum_{i \notin A(r)} w^i(r) \leq \sum_{i \in A(r)} w^i(2r) + Nr^\alpha H_2. \tag{18}$$

From (15) and (18) using Lemma 3.2, we obtain

$$w^1(r) + w^2(r) \leq Nr^\beta R^{-\beta}(M + R^\alpha H_2)$$

for $r \leq R$, where $\beta = \alpha_0 \wedge \alpha$ and α_0 depends only on d, K, and ε. To complete the proof of (6), it remains to note that if $z_1, z_2 \in Q^+_{r^2,r}$, then

$$2|u(z_1) - u(z_2)| = |[u^1(z_1) - u^1(z_2)] - [u^2(z_1) - u^2(z_2)]|$$
$$\leq w^1(r) + w^2(r).$$

The fact that u can be continuously extended to $Q^+_{R^2,R} \cup \Sigma_{R^2,R}$ readily follows from (6). Indeed, let $t \in [0, R^2)$, $x^1 = 0$, $|x^i| < R$ for $i \geq 2$, and $(t_1, x_1), (t_2, x_2) \in Q^+_{R^2,R}$, $(t_1, x_1), (t_2, x_2) \to (t, x)$. Suppose, for definiteness, that $t_1 \leq t_2$ and take a constant $R_0 > 0$ such that the inclusion $(t_1, x) + Q^+_{R_0^2,R_0} \subset Q^+_{R^2,R}$ holds throughout the whole process of varying (t_1, x_1). Then applying (6) to $(t_1, x) + Q^+_{R_0^2,R_0}$ rather than $Q^+_{R^2,R}$, we obtain $|u(t_1, x_1) - u(t_2, x_2)| \to 0$. The latter, together with Cauchy's criterion, proves our assertion. The theorem is completely proved.

4. THEOREM. Let $\alpha \in (0, 1)$, let $v \in C^\alpha(Q^+_{T,3})$, let $u, v \in C^2_{loc}(Q^+_{T,3}) \cap C(Q^+_{T,3} \cup \Sigma_{T,3})$, let $|Lu| \leq Lv + K_2$ on $Q^+_{T,3}$, and let $\kappa \in (0, 1]$ be a constant such that $T > \kappa^2$. Then for any $z_1, z_2 \in \Sigma_{T-\kappa^2,1}$, $\beta \in (0, \alpha_0 \wedge \alpha]$, where α_0 is taken from Theorem 1, the inequality

$$|u(z_1) - u(z_2)|$$
$$\leq N\kappa^{-\beta}\rho^\beta(z_1, z_2)\Big(K_2 + ||u||_{C(Q^+_{T,3})} + ||v||_{C^\alpha(Q^+_{T,3})}\Big), \tag{19}$$

holds with $N = N(d, K, \varepsilon, K_1, \alpha)$.

To derive this theorem from Theorem 1, it suffices to note that when $\rho(z_1, z_2) \geqq \kappa$, inequality (19) is clear with $N = 2$, while when $\rho(z_1, z_1) < \kappa$, reducing β in (19) does not violate the inequality and for $z_1 = (t_1, x_1)$. $z_2 = (t_2, x_2)$, $z_0 = (t_1 \wedge t_2, x_1)$, and $\rho = \rho(z_1, z_2)$ we obtain

$$z_0 + Q^+_{\rho^2,\rho} \subset z_0 + Q^+_{\kappa^2,\kappa} \subset Q^+_{T,3}, \quad z_i \in z_0 + \overline{Q^+_{\rho^2,\rho}}.$$

To study the behaviour of $|u(z_1) - u(z_2)|$ for $x^1_i = 0$ and t_i near T, we shall need two lemmas.

5. **LEMMA.** Let $c \leqq 0$, $b' \leqq \varepsilon$, let $|b^i| \leqq K$ in $Q^+_{T,3}$, let ξ be a strictly positive constant, let φ be a function defined on $[0, \infty)$ such that $\varphi \geqq 0$, φ' is continuous on $[0, \infty)$, $\varphi' \geqq 0$, and $\varphi' = 0$ near zero and infinity. Denote

$$\mu = \frac{1}{2}\varepsilon K^{-1}, \quad \nu = \frac{1}{2}\mu^2, \quad \psi(T - t, x) = \xi \cosh \xi x^1 \exp(2K\xi^2 t),$$

$$\eta^i_{(\pm)}(t, x) = \psi(t, x) \int_{x^1}^{\infty} \frac{\sinh \xi r}{\cosh^2 \xi r} e^{\mu\left(1 - \frac{x^1}{r}\right)}\Big[\varphi(x^i_{(\pm)}) + \int_{x^i_{\pm}}^{\infty} \varphi'(s) e^{\frac{\nu}{r}(\pm x^i - s)}\, ds\Big] dr,$$

$$\zeta(t, x) = \psi(t, x) \int_{x^1}^{\infty} \varphi(r) \frac{\sinh \xi r}{\cosh^2 \xi r}\, dr.$$

Then the functions $\eta^i_{(\pm)}$ and ζ are twice continuously differentiable with respect to (t, x) for $x^1 > 0$ and $L\eta^i_{(\pm)} \leqq 0$, $L\zeta \leqq 0$, $\eta^i_{(\pm)} \geqq \varphi(x^i_{\pm})$, and $\zeta \geqq \varphi(x^1)$ for $i = 2, \ldots, d$ in $Q^+_{T,3}$.

Proof. We first note that $a^{11}, b^1 \leqq K$, $\tanh \zeta x^1 \leqq \zeta x^1$, and

$$\psi^{-1} L\psi - c = -x^1 2K\xi^2 + x^1 a^{11}\xi^2 + \xi b^1 \mathrm{th}\, \xi x^1 \leqq 0. \tag{20}$$

Hence, using the formula

$$L(\psi_1\psi_2) = \psi_1(L\psi_2 - c\psi_2) + \psi_2 L\psi_1 + 2x^1 a^{ij}\psi_{1x^i}\psi_{2x^j} \tag{21}$$

and the inequalities $b^1 > 0$ and $\varphi' \geqq 0$, we obtain

$$\psi^{-1}L\zeta \leqq b^1\varphi(x^1)\,\frac{\sinh \xi x^1}{\cosh^2 \xi x^1} - x^1a^{11}\left(\varphi(x^1)\,\frac{\sinh \xi x^1}{\cosh^2 \xi x^1}\right)_{x^1} - 2x^1a^{11}\xi\varphi(x^1)\,\frac{\sinh^2 \xi x^1}{\cosh^3 \xi x^1}$$

$$\leqq - x^1a^{11}\xi\varphi(x^1)\left[\left(\frac{\sinh \xi x^1}{\cosh^2 \xi x^1}\right)_{x^1} + 2\xi\,\frac{\sinh^2 \xi x^1}{\cosh^3 \xi x^1}\right]$$

$$= -x^1a^{11}\varphi(x^1)\xi\,\frac{1}{\cosh \xi x^1} \leqq 0 \qquad (22)$$

To prove the inequalities $L\eta^i_{(\pm)} \leqq 0$, we denote

$$\chi(r, x^i) = \varphi(x^i_\pm) + \int_{x^i_+}^{\infty} \varphi'(s)\,\exp\left[\frac{\nu}{r}\,(\pm x^i - s)\right] ds,$$

$$\kappa(x^1, x^i) = \int_{x^1}^{1} \frac{\sin \xi r}{\cosh^2 \xi r}\,\chi(r, x^i)\,\exp\left[\mu\left(1 - \frac{x^1}{r}\right)\right] dr.$$

Note that in view of our hypothesis, the functions χ and κ are sufficiently smooth and differentiating them, we can bring the sign of a derivative into the integral if $x^1 > 0$. For example,

$$\chi_r(r, x^i) = -\frac{\nu}{r^2}\int_{x^i_\pm}^{\infty} \varphi'(s)\,(\pm x^i - s)\,\exp\left[\frac{\nu}{r}\,(\pm x^i - s)\right] ds \geqq 0,$$

$$\chi_{x^i}(r, x^i) = \frac{\nu}{r}\int_{x^i_\pm}^{\infty} \varphi'(s)\,\exp\left[\frac{\nu}{r}\,(\pm x^i - s)\right] ds, \quad \left|\chi_{x^i}\right| \leqq \frac{\nu}{r}\,\chi,$$

$$\chi_{x^ix^i}(r, x^i) = -\frac{\nu}{r}\,\varphi'(x_\pm) + \frac{\nu^2}{r^2}\int_{x^i_\pm}^{\infty} \varphi'(s)\,\exp\left[\frac{\nu}{r}\,(\pm x^i - s)\right] ds$$

$$\leqq \frac{\nu^2}{r^2}\,\chi(r, x^i). \qquad (23)$$

Furthermore, directly using formula (21), we can see that (with no summation over i)

$$\begin{aligned}\psi^{-1}L(\psi\kappa) &= \kappa\psi^{-1}L\psi + \\ &+ x^1\left[2\xi \tanh \xi x^1\left(a^{11}\kappa_{x^1} + a^{1i}\kappa_{x^i}\right) - \right.\\ &\left. - a^{11}\left(\frac{\sinh \xi x^1}{\cosh^2 \xi x^1}\right)_{x^1} \chi\right] - x^1 a^{11}\frac{\sinh \xi x^1}{\cosh^2 \xi x^1}\chi_r + \\ &+ \frac{\sinh \xi x^1}{\cosh^2 \xi x^1}\left[a^{11}\mu\chi - b^1\chi - 2x^1 a\chi_{x^i}\right] + \\ &+ \int_{x^i}^{\infty} \frac{\sinh \xi r}{\cosh^2 \xi r}\left[x^1 a^{11}\frac{\mu^2}{r^2}\chi - 2x^1 a^{1i}\frac{\mu}{r}\chi_{x^i} + \right.\\ &\left. + x^1 a^{ii}\chi_{x^i x^i} + b^i\chi_{x^i} - b^i\frac{\mu}{r}\chi\right]\exp \mu\left(1 - \frac{x^1}{r}\right) dr\end{aligned} \tag{24}$$

where the arguments of the functions ψ, κ, and χ are omitted for brevity. This together with (20) and (23), implies that, to prove the inequality $L\eta^i_{(\mp)} \leqq 0$, it suffices to show that every pair of braces in (24) contains a negative function. The calculations in (23) provide the following bound for the expression between the last pair of braces in (24), where $r \geqq x^1$,

$$\frac{1}{r}\chi\left[a^{11}\mu^2 + 2|a^{1i}|\mu\nu + a^{ii}\nu^2 + K\nu - \varepsilon\mu\right].$$

This expression is negative owing to the specific choice of μ and ν (cf. (8)). The expression between the second pair of braces in (24) does not exceed

$$\chi[K\mu - \varepsilon + 2K\nu] = \chi\left[\varepsilon^2(4K)^{-1} - \frac{1}{2}\varepsilon\right] < 0.$$

To estimate the expression between the first pair of braces, we make use of the calculations in (22). Then we obtain that this expression is equal to

$$-a^{11}\left(\frac{\sinh \xi x^1}{\cosh^2 \xi x^1}\right)_{x^1}\chi - a^{11}\, 2\xi \tanh \xi x^1 \frac{\sinh \xi x^1}{\cosh^2 \xi x^1}\chi +$$

$$+ 2\xi \tanh \xi x^1 \int_{x^1}^{\infty} \frac{\sinh \xi r}{\cosh^2 \xi r} \left[a^{1i}\chi_{x^i} - a^{11} \frac{\mu}{r} \chi\right] \exp \mu\left(1 - \frac{x^1}{r}\right) dr$$

$$\leqq 2\xi \tanh \xi x^1 \int_{x^1}^{\infty} \frac{\sinh \xi r}{\cosh^2 \xi r} \frac{1}{r} \chi[K\nu - \varepsilon\mu] \exp \mu\left(1 - \frac{x^1}{r}\right) dr.$$

The latter term is negative since $K\nu - \varepsilon\mu = -\frac{3}{2}\mu^2 K < 0$. This asserts the inequality $L\eta^i_{(\pm)} \leqq 0$ for $x^i > 0$, $i \geqq 2$.

Finally, it is obvious that, for $t \leqq T$ and $x^1 \geqq 0$,

$$\zeta(t, x) \geqq \varphi(x^1)\psi(t, x) \int_{x^1}^{\infty} \frac{\sinh \xi r}{\cosh^2 \xi r} dr =$$

$$= \varphi(x^1)\exp(2K\xi^2(T - t)) \geqq \varphi(x^1),$$

$$\eta^i_{(\pm)}(t, x) \geqq \varphi(x^i_\pm)\psi(t, x) \int_{x^1}^{\infty} \frac{\sinh \xi r}{\cosh^2 \xi r} dr \geqq \varphi(x^i_\pm).$$

This proves the lemma.

6. LEMMA. Let $R \in (0, 3]$, let $u \in C^2_{loc}(Q^+_{T,R}) \cap C(\overline{Q^+_{T,R}} \setminus \{x^1 = 0\})$ be a funciton defined at the point $(T, 0)$, let M and M_α be some positive constants, let $|u|$, $|u(T, 0)| \leqq M$, let $\alpha \in (0, 1)$, and let

$$u(T, x) - u(T, 0) \leqq M_\alpha |x|^\alpha$$

for $x^1 > 0$. Also, let $Lu \geqq -K_2$, let $c \leqq 0$ in $Q^+_{T,R}$. Then

$$u(t, x) - u(T, 0) \leqq N(|T - t|^{\alpha/2} +$$

$$+ |x|^\alpha)[R^{-\alpha}M + M_\alpha + K_2 R^{1-\alpha}]$$

on $Q^+_{T,R}$, where $N = N(d, K, K_1, \varepsilon, \alpha)$.

Proof. We have $L(u - u(T, 0)) \geqq -K_2 - u(T, 0)c \geqq -K_2 - KM$, so it suffices to prove the lemma in the case $u(T, 0) = 0$.

As in our proof of Theorem 1, our further consideration is reduced to the case $R \leqq (1/4)\varepsilon(K_1 + 1)^{-1}$. In this case (we assume that $\varepsilon \leqq 1$) $|b^i| \leqq K$, $b^1 \geqq \varepsilon$, $R|c| \leqq (1/4)\varepsilon$ in $Q^+_{T,R}$, and the function $\tilde{u}(t, x) :=$

$u(9^{-1}R^2t, 3^{-1}Rx) + K_2R\varepsilon^{-1}x^1$ in $Q^+_{S,3}$, where $S = 9R^{-2}T$, satisfies the inequalities

$$x^1\tilde{u}_t + x^1\tilde{a}^{ij}\tilde{u}_{x^ix^j} + \tilde{b}^i\tilde{u}_{x^i} + \frac{1}{3}R\tilde{c}\tilde{u}$$

$$\geq -\frac{1}{3}RK_2 + K_2R - \frac{1}{3}\cdot\frac{1}{4}\varepsilon 6K_2R\varepsilon^{-1} = K_2R\frac{1}{6} \geq 0, \tag{25}$$

where $\tilde{a}^{ij}(t, x) = a^{ij}(9^{-1}R^2t, 3^{-1}Rx)$, etc. Moreover, $\tilde{u}(S, x) \leq (M_\alpha 3^{-\alpha}R^\alpha + K_2 6^{1-\alpha}R\varepsilon^{-1})|x|^\alpha$ with $x^1 > 0$. Finally, the coefficients of the operator on the left of (25) satisfy conditions (2) and (4), where K_1 is replaced with ε, and the inequalities $|\tilde{b}^i| \leq K$ and $\tilde{b}^1 \geq \varepsilon$ on $Q^+_{S,3}$.

In turn, it follows that the proof of the lemma reduces to the case when $K_2 = u(T, 0) = 0$, $R = 3$, $|b^i| \leq K$, and $b^1 \geq \varepsilon$ in $Q^+_{T,3}$, so we focus on this case. We take $\xi > 0$. Let the function φ satisfy the conditions of Lemma 5 and $\varphi(r) \geq r^\alpha$ for $r \in [0, 6]$. Using the functions ζ and $\eta^i_{(\pm)}$ introduced in Lemma 5, we construct the function

$$w = \zeta + \sum_{i=2}^{d} (\eta^i_{(\pm)} + \eta^i_{(-)}).$$

From Lemma 5 and the maximum principle it follows that there exists a constant $N = N(\alpha)$ such that

$$u \leq Nw[M_\alpha + M]$$

in $Q^+_{T,3}$.

Using an appropriate limiting process *in this inequality*, we can show that it remains valid if $\varphi(r)$ is replaced by r^α for constructing η^i, ζ, and w. Also, note that constructing the function w, we choose $\xi > 0$ arbitrarily. It is clear that the proof of the lemma now reduces to the inequality

$$\inf_{\xi>0} w(t, x) \leq N(|T - t|^{\alpha/2} + |x|^\alpha) \tag{26}$$

in $Q^+_{T,3}$ with $N = N(d, \varepsilon, K, \alpha)$.

To prove (26), we fix a $(t, x) \in Q^+_{T,3}$, denote $\rho = |T - t|^{1/2} + |x|$ and put $\xi = \rho^{-1}$. For such a ξ and $\varphi(r) = r^\alpha$, using the change of variables $\xi r \to r$, we easily obtain

$$\zeta(t, x) \leq N\int_0^\infty r^\alpha \frac{\sinh \xi r}{\cosh^2 \xi r}\,dr = N\rho^\alpha \int_0^\infty r^\alpha \frac{\sinh r}{\cosh^2 r}\,dr. \tag{27}$$

Since the calculations in (23) show that for $\varphi(s) = s^\alpha$ and $\pm x^i > 0$,

$$\left(\pm \int_{x_\pm^i} \varphi'(s) \exp \frac{\nu}{r} (\pm x^i - s)\, ds\right)_{x^i}$$

$$= -\varphi'(x_\pm^i) + \frac{\nu}{r} \int_{x_\pm^i}^\infty \varphi'(s) \exp \frac{\nu}{r} (\pm x^i - s)\, ds$$

$$\leq \varphi'(x_\pm^i)\left[\frac{\nu}{r} \int_{x_\pm^i}^\infty \exp \frac{\nu}{r} (+x^i - s)\, ds - 1\right] \leq 0,$$

we conclude that

$$\int_{x_\pm^i}^\infty \varphi'(s) \exp \frac{\nu}{r} (\pm x^i - s)\, ds \leq \int_0^\infty \varphi'(s)\, e^{-\frac{\nu}{r} s}\, ds = Nr^\alpha.$$

This and (27) imply

$$\eta^i_{(\pm)}(t, x) \leq N\varphi(x_\pm^i)\xi \int_0^\infty \frac{\sinh \xi r}{\cosh^2 \xi r}\, dr + N\rho^\alpha \leq N\rho^\alpha.$$

which together with (27) assert (26). The proof is complete.

7. THEOREM. Let $u, v \in C^2_{loc}(Q^+_{T,3}) \cap C(\overline{Q^+_{T,3}} \setminus \{t = T, x^1 = 0\})$, let $\alpha \in (0, 1)$, and let

$$v \in C^\alpha(Q^+_{T,3}), \quad |Lu| \leq Lv + K_2 \text{ on } Q^+_{T,3},$$

Also, let $\beta = \alpha_0 \wedge \alpha$ where α_0 is the constant in Theorem 1, let $\alpha_1 \in (0, \beta]$, let M_{α_1} be a constant, and let u be defined in $\overline{Q^+_{T,3}}$ such that

$$|u(T, x_1) - u(T, x_2)| \leq M_{\alpha_1} |x_1 - x_2|^{\alpha_1} \tag{28}$$

for $x, x_2 \in \overline{D^+_3}$, $x_1^1 = 0$, $x_2^1 > 0$. Then u is continuous in $\overline{Q^+_{T,1}}$ and

$$|u(z_1) - u(z_2)| \leq N\rho^{\alpha_1}(z_1, z_2)\left[M_{\alpha_1} + K_2 + ||u||_{C(Q^+_{T,3})} + ||v||_{C^\alpha(Q^+_{T,3})}\right] \tag{29}$$

for any $z_1, z_2 \in \bar{\Sigma}_{T,1}$, where $N = N(\varepsilon, d, K, K_1, \alpha, \alpha_1)$.

Proof. First, we note that, following the proof of Theorem 1, we can assume that $c \leqq 0$. Denote

$$u^1 = v + u, \quad u^2 = v - u,$$

$$M = ||u||_{C(Q^+_{T,3})} + ||v||_{C^{\alpha}(Q^+_{T,3})}$$

and, using Lemma 6, let us prove that u is continuous in $\overline{Q^{\mp}_{T,1}}$, i.e., on $\overline{Q^{\mp}_{T,1}} \cap \{t = T, x^1 = 0\}$. Without loss of generality, we can assume that the function v is continuous in $\overline{Q^{\mp}_{T,3}}$.

By hypothesis, $Lu^i \geqq -K_2$ in $Q^+_{T,3}$ and, for any point $x_0 \in \Sigma_2$,

$$|u^i(T, x) - u^i(T, x_0)| \leqq N(M_{\alpha_1} + M)|x - x_0|^{\alpha_1}$$

for $x^1 > 0$, $x \in x_0 + D^+_1$, the constant N, here and in what follows, depending only on ε, d, K, K_1, α, α_1. Hence, by Lemma 6,

$$u^i(t, x) - u^i(T, x_0) \leqq N(|T - t|^{\alpha_1/2} + |x - x_0|^{\alpha_1})(M + M_{\alpha_1} + K_2),$$

$$|u(t, x) - u(T, x_0)| \leqq N(|T - t|^{\alpha_1/2} + |x - x_0|^{\alpha_1})(M + M_{\alpha_1} + K_2) \tag{30}$$

for $(t, x) \in (0, x_0) + Q^+_{T,1}$. The latter inequality obviously implies that u is continuous in $\overline{Q^{\mp}_{T,2}}$.

We further use an argument which is similar to the one used in the proof of Theorem 4.5. Let $z_1, z_2 \in \Sigma_{T,1}$ and let, for definiteness, $z_i = (t_i, x_i)$, $t_1 \leqq t_2$. Furthermore, we denote $\rho = \rho(z_1, z_2)$ and separately consider three cases: (i) $\rho \geqq 1$, (ii) $\rho \geqq 1$, $T - t_2 \leqq \rho^2$, and (iii) $\rho \leqq 1$, $T - t_2 > \rho^2$.

In the first case, (29) is clear with N = 2. In the second case, we have $t_2 - t_1 \leqq \rho^2$, $T - t_1 \leqq 2\rho^2$ and (29) readily follows from (28) and (30) for $x_0 = x_1, x_2$.

In the third case, by Theorem 4, we can assume that $T - t_2 \leqq 1/2$. Then $R^2 := T - t_1 \leqq \rho^2 + T - t_2 \leqq 2(T - t_2) \leqq 1$, $R^2 \geqq T - t_2 \geqq \rho^2$ and both points z_1 and z_2 lie in the closure of $Q := z_1 + Q^+_{R^2,R} \subset Q^+_{T,3}$. Let $\tilde{z}_1$ be a point of Q and let Theorem 1 be applied to Q, $u - u(\tilde{z}_1)$, and $v - v(\tilde{z}_1)$ instead of $Q^+_{R^2,R}$, u, and v. It is clear that K_2 should be replaced by $K_2 + MK_1$ in order to preserve condition (5). Taking note that β in (6) can obviously be replaced by any $\alpha_1 \in (0, \beta]$, we obtain

$$|u(z_1) - u(z_2)| \leqq N\left(\frac{\rho}{R}\right)^{\alpha_1} [RK_2 + R^\alpha M + ||u - u(\tilde{z}_1)||_{C(Q)} + ||v - v(\tilde{z}_1)||_{C(Q)}]. \quad (31)$$

Here the latter norm does not exceed $2^\alpha R^\alpha M$, since v satisfies the Hölder condition. To estimate the norm of $u - u(\tilde{z}_1)$, it suffices to estimate $|u(\tilde{z}_1) - u(\tilde{z}_2)|$ for any $\tilde{z}_2 \in Q$. If (t, x) in (30) is replaced by $\tilde{z}_1$, $\tilde{z}_2$ and $x_0 = x_1$, $z_0 = (T, x_1)$, then we obtain $|u(\tilde{z}_i) - u(z_0)| \leqq NR^{\alpha_1}(M + M_{\alpha_1} + K_2)$. Therefore, $|u(\tilde{z}_1) - u(\tilde{z}_2)| \leqq NR^{\alpha_1}(M + M_{\alpha_1} + K_2)$, all summands between the braces in (31) are estimated by the same expression, and $|u(z_1) - u(z_2)| \leqq N\rho^{\alpha_1}(M + M_{\alpha_1} + K_2)$. Thus, we obtain (29) in the third case which asserts the theorem.

Finally, let us formulate a result which is similar to Theorems 1 and 4 for elliptic equations. Theorem 8 immediately follows from Theorems 1 and 4, if we take functions u and v independent of t and put $T = 2\kappa - 2$.

8. THEOREM. Let the coefficients of L be independent of t, let $\tilde{L} = L - x^1 \frac{\partial}{\partial t}$, let $u, v \in C^2_{loc}(D^+_3) \cap C(D^+_3)$, let $\alpha \in (0, 1)$, let $v \in C^\alpha(D^+_3)$, and let $|\tilde{L}u| \leqq \tilde{L}v + K_2$ on D^+_3. Then u can be continuously extended to $\bar{D}^+_1$ and for any $\beta \in (0, \alpha_0 \wedge \alpha]$, where α_0 is taken from Theorem 1, and for any $x_1, x_2 \in \Sigma_1$ we obtain

$$|u(x_1) - u(x_2)| \leqq N|x_1 - x_2|^\beta \Big(K_2 + ||u||_{C(D^+_3)} + ||v||_{C^\alpha(D^+_3)}\Big),$$

where $N = N(\varepsilon, d, K, K_1, \alpha)$ and $u(x_i)$ is the value of the extended function u at the point x_i.

9. Remark. Theorems 4 and 7 remain valid if we replace $Q^+_{T,3}$, $Q^+_{T,1}$ $\Sigma_{T,1}$, and $\Sigma_{T-\kappa^2,1}$ involved in the conditions at the beginning of this section and in their formulations by Q^+_{T,R_1}, Q^+_{T,R_2}, Σ_{T,R_2}, and $\Sigma_{T-\kappa^2,R_2}$, where $0 < R_2 < R_1$, and if N (but not α_0) can also depend on R_1 and R_2. The same is true for Theorem 8.

Indeed, if $R_2 \leqq (1/3)R_1 \leqq 1$, then, using the familiar similarity transformation, we can convert Q^+_{T,R_1} to $Q^+_{S,3}$ so that the image of Q^+_{T,R_2} under such a transformation will belong to $Q^+_{S,1}$. After applying this transformation it remains to use Theorems 4 and 7 and perform the inverse transformation. But if $R_2 \geqq (1/3)R_1$ or $R_1 > 3$, then estimating $|u(z_1) - u(z_2)|$ for $z_1, z_2 \in \Sigma_{T,R_2}$, we can always assume that $\rho(z_1, z_2) \leqq (1/3)\rho$, where $\rho = (R_1 - R_2) \wedge 1$, in which case z_1 and z_2 lie in a set of the form $(0, x_0) + \Sigma_{T,(1/3)\rho}$ and $(0, x_0) + \Sigma_{T,\rho} \subset \Sigma_{T,R_1}$. In this case we therefore deal with the sets $(0, x_0) + Q^+_{T,(1/3)\rho}$, $(0, x_0) + Q^+_{T,\rho} \subset Q^+_{T,R_1}$ which, as explained above, allows us to apply Theorems 4 and 7.

4.7. Comments

Sections 1 and 2. The results of these sections as they are presented here have been established by Krylov and Safonov [1, 2], and Safonov [4]. Their proofs are based on the combination of some methods of De Giorgi [1] and Landis [1] with Aleksandrov's estimates. The validity of Theorem 1.2 was reported in Krylov and Safonov [1, §8]. Lemma 1.11 resembles the so-called lemmas on 'increasing' which, prior to the Krylov and Safonov results, were proved (like the Hölder condition) on the condition that the eigenvalues of the matrix (a^{ij}) are not too much spread (see Landis [1]). There are also some other methods of estimating the Hölder constant for nondivergence form equations with measurable coefficients, see Novruzov [1, 2], Mamedov [1], Trudinger [1]. Krylov [21] studied the Hölder condition for solutions to some degenerate equations. For divergence form equations, which are beyond the scope of this book, estimates for the Hölder constant were obtained in De Giorgi [1], Nash [1]. These results were further developed by different authors (see Friedman [1]) and became a basis for the theory of quasi-linear equations (see Ladyzhenskaya and Ural'tseva [1], Ladyzhenskaya, Solonnikov, and Ural'tseva [1]). The Harnack inequality for divergence form parabolic equations was proved by Moser [2].

Section 3. Theorem 1 is a formalization of the method for estimating the Hölder constant for the second derivatives of solutions to nonlinear equations. This method was suggested by Krylov [19]. The method of using an auxiliary function in Theorem 5 is taken from the book of Ladyzhenskaya, Solonnikov, and Ural'tseva [1]. Corollaries 6 and 9 are known from Ladyzhenskaya and Ural'tseva [2–4], Novruzov [1], Mamedov [1], and Trudinger [1]. In Theorems 5 and 8, one cannot use $\tau = 1$, as can be seen in the Heinz example $n = 2$, $d = 1$, $u^i_t + u^i_{xx} = u^i\,(u^1_x)^2 + (u^2_x)^2$, $i = 1, 2$, $u^1 = \cos mx$, $u^2 = \sin mx$.

Section 4. Unfortunately, we should have assigned more room for the proof of Lemma 1. Some barriers constructed in Lemma 1 seem to be original, although the author believes that they may have been known previously. The idea of constructing these barriers is based on the Burkholder–Davis–Gundy inequalities, which are well known in the theory of martingales. Lemmas 2 to 4 extend results of Kamynin [1], and Kamynin and Khimchenko [1]. A version of Theorem 5 can be found in Ladyzhenskaya and Ural'tseva [4], a version of Theorem 6 is presented in Safonov [4].

Section 5. Corollary 3 is proved in Ladyzhenskaya and Ural'tseva [4]. The form of Theorem 4 calls for the barriers introduced in Section 4. This theorem is associated with 'straightening' the boundary if the boundary and the boundary data are not sufficiently regular (see Lemmas 1.1.1 and 1.1.2). In the 'regular' case this theorem is well known (see Ladyzhenskaya, Solonnikov and Ural'tseva [1, Chapter 7]).

Section 6. This section presents new results (see, however, Krylov [24]), although degenerate operators like (1) have previously been used by the author [21] in his research into the boundary properties of the derivatives of solutions and the proof of Theorem 1 is very similar to the corresponding argument of that paper. The estimates established in this section are more accurate than the estimates obtained in the paper just referred to. The major gain is obtained by use of more precise barriers in Lemma 5. It should be said that the idea of constructing these barriers is based on regarding operator (1) as a parabolic one when the x^1-axis plays the role of the time axis. If viewed from the standpoint of probability theory, the use of function w_4 involved in the proof of Lemma 2 is similar to that of Chebyshev's inequality.

Chapter 5

A-PRIORI ESTIMATES IN $C^{2+\alpha}$ FOR SOLUTIONS OF NONLINEAR EQUATIONS

In this chapter we construct the a-priori estimates which will allow us to prove the solvability of nonlinear equations in the classes of smooth functions. The proof will be given in the next chapter. In Section 1, we shall first estimate the derivatives of a solution and the Hölder constants for the derivatives only on the boundary. To this end we shall use the results of Section 4.6. Using other results of Chapter 4, we shall further prove $C^{2+\alpha}$ estimates both for the interior of a domain and up to the boundary. In Section 4, special attention is paid to the proof of the estimate of the solution norm in $C^2(Q)$, when the first-order matching condition can be violated. It should be said that the estimates up to the boundary look rather cumbersome and we should advise the reader to pay attention only to the interior estimates on the first reading.

5.1. The Boundedness of and the Hölder Condition for the Derivatives of Solutions on the Boundary

We fix constants K, K_1, T, $\varepsilon > 0$, α, $\kappa \in (0, 1]$, $T > \kappa^2$ and recall that

$$Q^+_{T,R} = \{(t, x) \in E_{d+1} : 0 < t < T,\ 0 < x^1 < 2R,$$

$$|x^i| < R,\ i = 2, \ldots, d\},$$

$$\Sigma_{T,R} = \{(t, x) \in E_{d+1} : 0 \leqq t < T,\ x^1 = 0,$$

$$|x^i| < R,\ i = 2, \ldots, d\},$$

$$\partial_t Q^+_{T,R} = \{(T, x) \in E_{d+1} : 0 \leqq x^1 \leqq 2R,$$

$$|x^i| \leqq R,\ i = 2, \ldots, d\}.$$

Let a Borel function $F(u_{ij}, u_i, u, t, x)$ be defined for all $(t, x) \in Q^+_{T,3}$ and real-valued u_{ij}, u_i, $(i, j = 1, \ldots, d)$, and u. Throughout this section, a function $u \in C^2_{loc}(Q^+_{T,3}) \cap C(\bar{Q}^+_{T,3})$ is considered such that

$$u_t(t, x) + F\left(u_{x^ix^j}(t, x), u_{x^i}(t, x), u(t, x), t, x\right) = 0 \quad (1)$$

in $Q^+_{T,3}$.

For the sake of convenience, we put together all the conditions on the function F, various combinations of which will be used in different assertions.

1. The differentiability with respect to (u_{ij}). For every $(t, x) \in Q^+_{T,3}$, u_i, and u, the function F is once continuously differentiable with respect to (u_{ij}).
2. The differentiability with respect to (u_{ij}, u_i, u, x). For every $t \in (0, T)$ and $x^1 \in (0, 6)$, the function F is once continuously differentiable with respect to $(u_{ij}, u_i, u, x^2, \ldots, x^d)$.

The following conditions 3-6 are assumed to hold for every $(t, x) \in Q^+_{T,3}$ on substituting the arguments $(u_{x^ix^j}(t, x), u_{x^i}(t, x), u(t, x), t, x)$ into F and its derivatives.

3. The operator F is elliptic. For $\lambda \in E_d$,

$$\varepsilon|\lambda|^2 \leqq F_{u_{ij}}\lambda^i\lambda^j \leqq K|\lambda|^2.$$

4. Equation (1) is almost linear:

$$|F - F_{u_{ij}}u_{x^ix^j}| \leqq K_1((x^1)^{\alpha-1} + |u_x|).$$

5. The 'coefficients' of equation (1) and their derivatives are linearly bounded along the tangent directions. For $j = 2, \ldots, d$, $i = 1, \ldots, d$,

$$|F_u| + |F_{x^j}| \leqq K_1((x^1)^{\alpha-1} + \|u_{xx}\|),$$

$$|F_{u_i}| \leqq K_1((x^1)^{(\alpha-1)/2} + \|u_{xx}\|).$$

6. The 'coefficients' of equation (1) and their derivatives are bounded along the tangent directions. For $j = 2, \ldots, d$, $i = 1, \ldots, d$,

$$|F_{x^j}| \leqq K_1((x^1)^{\alpha-1} + \|u_{xx}\|), \quad |F_{u_i}| + |F_u| \leqq K_1.$$

It is clear that the formulations of conditions 3 — 6 arose from the similarity between equation (1) and the linear equation. Thus, note that the results of this section are, of course, applicable to linear equations. For example, Theorem 8 for the solutions of linear equations with measurable coefficients provides an estimate of a Hölder constant on $\Sigma_{T,1}$ for the function u_{x^1}.

Let us clarify the connection between conditions 3 — 6 and those conditions which will be encountered in the sections to come. There we shall basically assume that condition 3 holds and

$$|F - F_{u_{ij}} u_{x^i x^j}| \leqq K_1 (1 + |u_x|^2), \tag{2}$$

$$|F_u| + |F_{u_i}|(1 + |u_x|) + |F_{x^j}|(1 + |u_x|)^{-1}$$
$$\leqq K_1 (1 + |u_x|^2 + ||u_{xx}||). \tag{3}$$

Of course, the latter conditions do not necessarily imply conditions 4 — 6. Therefore, it should be said that the results of the present section, which involve conditions 4 and 5, will be used later only after establishing an estimate of $|u_x|$ in a domain, i.e., when the stronger condition $|F - F_{u_{ij}} u_{x^i x^j}| \leqq N$ is valid in place of (2), while condition 5 holds with a constant N instead of K_1 by virtue of (3). Condition 6 is used to estimate a Hölder constant for u_{xx} on $\Sigma_{T,1}$. The respective result will be used only after estimating $|u_x|$ and $||u_{xx}||$ in a domain, when inequality (3) ensures that condition 6 holds.

7. LEMMA. Let conditions 1, 3, and 4 hold. Then there exist measurable functions b^i, f on $Q^+_{T,3}$ such that $|b^i| \leqq K_1$, $|f| \leqq K_1 (x^1)^{\alpha-1}$, and

$$u_t + a^{ij} u_{x^i x^j} + b^i u_{x^i} + f = 0 \tag{4}$$

in $Q^+_{T,3}$, where $a^{ij} = (1/2)(F_{u_{ij}} + F_{u_{ji}})$. Moreover, $\varepsilon|\lambda|^2 \leqq a^{ij}\lambda^i\lambda^j \leqq K|\lambda|^2$ in $Q^+_{T,3}$ for all $\lambda \in E_d$.

Proof. The latter assertion is clear. To prove the former one, it suffices to put

$$g = \left(F - F_{u_{ij}} u_{x^i x^j}\right)((x^1)^{\alpha-1} + |u_x|)^{-1},$$

$$f = g(x^1)^{\alpha-1}, \quad b^i = g u_{x^i} |u_x|^{-1},$$

where $0 \circ 0^{-1} = 0$, and take note that

$$0 = u_t + F = u_t + a^{ij}u_{x^ix^j} + g((x^1)^{\alpha-1} + |u_x|)$$

$$= u_t + a^{ij}u_{x^ix^j} + b^iu_{x^i} + f$$

which proves the lemma.

8. THEOREM. Let conditions 1, 3, and 4 hold. Let $|u| \leq K_1$ in $Q^+_{T,3}$, let u have derivatives with respect to $x^2, \ldots, x^d$ on $\Sigma_{T,3}$ which are continuous on $\Sigma_{T,3}$, and for $i = 2, \ldots, d$, $z_1 = (t_1, x_1)$, $z_2 = (t_2, x_2) \in \Sigma_{T,3}$, let the inequalities

$$|u(t_1, x_1) - u(t_2, x_1)| \leqq K_1|t_1 - t_2|^{(1+\alpha)/2},$$

$$|u_{x^i}(z_1) - u_{x^i}(z_2)| \leqq K_1\rho^\alpha(z_1, z_2)$$

hold.

Then u_{x^1} exists on $\Sigma_{T,1}$ and for any $\alpha_1 \in (0, \alpha_0 \wedge \alpha]$, where α_0 is the constant in Theorem 4.6.1, we have:

(a) $$|u_{x^1}(z_1) - u_{x^1}(z_2)| \leqq N\kappa^{-\alpha_1-1}\rho^{\alpha_1}(z_1, z_2), \tag{5}$$

where $z_1, z_2 \in \Sigma_{T-\kappa^2,1}$, $N = N(d, K, K_1, \varepsilon, \alpha, \alpha_1)$;

(b) if, in addition, u_{x^1} exists on $\partial_t Q^+_{T,3}$ and $|u_{x^1}|$ and a Hölder constant of order α_1 for the function u_{x^1} on $\partial_t Q^+_{T,3}$ do not exceed K_1, then inequality (5) with $\kappa = 1$ holds for all $z_1, z_2 \in \bar{\Sigma}_{T,1}$.

Proof. By virtue of Lemma 7, the assertions of this theorem essentially relate to linear equation (4). By Remark 4.5.5, (4) implies that

$$|u(t, x) - u(t, 0, x^2, \ldots, x^d)| \leq Nx^1\kappa^{-1} \text{ in } Q^+_{S,2}, \tag{6}$$

where (and in what follows) we do not stress the dependence of N on d, K, K_1, ε, α, α_1 and $S = T - (1/2)\kappa^2$. By Lemma 1.1.5, we extend $u(t, x)$ from $\Sigma_{T,3}$ to $Q^+_{T,2}$ and obtain a function φ such that

$$|\varphi_t(t, x)|, \quad ||\varphi_{xx}|| \le N(x^1)^{\alpha-1}, \quad |\varphi_x| \le N, \tag{7}$$

in $Q^+_{T,2}$ and for $z_1, z_2 \in \overline{Q^+_{T,2}}$

$$|\varphi_{x^1}(z_1) - \varphi_{x^1}(z_2)| \le N\rho^{\alpha}(z_1, z_2). \tag{8}$$

Denote

$$v = u - \varphi, \quad w = (x^1)^{1+\alpha}, \quad L = \frac{\partial}{\partial t} + a^{ij}\frac{\partial^2}{\partial x^i \partial x^j} + b^i \frac{\partial}{\partial x^i}$$

and note that in $Q^+_{T,2}$, by virtue of (4) and (7)

$$|Lv| \le (K_1 + N)(x^1)^{\alpha-1} \le N_1 Lw + N.$$

Moreover, from (6) and (7) it follows that $|v(t, x)| \le Nx^1\kappa^{-1}$ in $Q^+_{S,2}$. Therefore, on this set, the function $\tilde{v}(t, x) := (x^1)^{-1}\, v(t, x)$ is bounded and together with $\tilde{w}(t, x) = N_1(x^1)^{-1}\, w(t, x) = N_1(x^1)^{\alpha}$ satisfies the inequality

$$|\tilde{L}\tilde{v}| \le \tilde{L}\tilde{w} + N,$$

where

$$\tilde{L}h := L(x^1 h),$$

$$\tilde{L} = x^1 \frac{\partial}{\partial t} + x^1 a^{ij}\frac{\partial^2}{\partial x^i \partial x^j} + \tilde{b}^i \frac{\partial}{\partial x^i} + c,$$

$$\tilde{b}^i = x^1 b^i + 2a^{i1}, \quad c = b^1.$$

It is obvious that $2a^{11} - \tilde{b}^1 + cx^1 = 0$ and the coefficients of the operator $\tilde{L}$ satisfy conditions (4.6.2) to (4.6.4) with the constant 3K instead of K. By Theorem 4.6.1, the function $\tilde{v}$ can be extended to a function which is continuous on $\Sigma_{S,2}$ and clearly equal to $u_{x^1} - \varphi_{x^1}$ on this set. By Theorem 4.6.4 applied to the functions $\tilde{v}$, $\tilde{w}$ with $Q^+_{T,2}$ in place of $Q^+_{S,3}$ (cf. Remark 4.6.9), for $\alpha_1 \in (0, \alpha_0 \wedge \alpha]$ and $z_1, z_2 \in \Sigma_{T-\kappa^2,1}$, we obtain

$$|u_{x^1}(z_1) - u_{x^1}(z_2) - \varphi_{x^1}(z_1) + \varphi_{x^1}(z_2)|$$
$$\le N\kappa^{-\alpha_1}\rho^{\alpha_1}(z_1, z_2)(1 + \sup(|\tilde{v}|, Q^+_{S,2})). \tag{9}$$

This along with (8) and the derived estimate $|\tilde{v}| \leqq N\kappa^{-1}$ in $Q^+_{S,2}$ complete the proof of (5).

We now prove assertion (b). In particular, we now have the inequality

$$|u(T, x) - u(T, 0, x^2, \ldots, x^d| \leqq K_1 x^1,$$

hence, by Remark 4.5.5 and Theorem 4.5.4, we obtain

$$|u(t, x) - u(t, 0, x^2, \ldots, x^d)| \leqq Nx^1 \text{ in } Q^+_{T,2}.$$

Hence by virtue of (7) we obtain $|v| \leqq Nx^1$ and $|\tilde{v}| \geqq N$ in $Q^+_{T,2}$. Finally, from the obvious formula

$$\tilde{v}(T, x) = \int_0^1 v_{x^1}(T, rx^1, x^2, \ldots, x^d)\, dr,$$

which allows us to define $\tilde{v}(T, x)$ with $x^1 = 0$, inequlaity (8), and our hypotheses we obtain

$$|\tilde{v}(T, x_1) - \tilde{v}(T, x_2)| \leqq N|x_1 - x_2|^{\alpha}$$

for x_1, $x_2 \in \overline{D^+_2}$. By Theorem 4.6.7, inequality (9) holds true for any z_1, $z_2 \in \bar{\Sigma}_{T,1}$, if κ and $Q^+_{S,2}$ are replaced by 1 and $Q^+_{T,2}$. On using (8) once more, the theorem is proved.

In the following theorem we estimate $||u_{xx}||$ on $\Sigma_{T,1}$

9. THEOREM. Let conditions 2, 3, and 5 hold. Let $|u|$ and $|u_x|$ be no greater than K_1 in $Q^+_{T,3}$, let $u_{x^i x^j}$ be continuous in $Q^+_{T,1} \cup \Sigma_{T,1}$ for all i, j. Let

$$u_{x^j} \in C^2_{loc}(Q^+_{T,3}) \cap C(\overline{Q^+_{T,3}}),$$

$$j = 2, \ldots, d, \quad ||u||_{C^{2+\alpha}(\Sigma_{T,3})} \leqq K_1. \tag{10}$$

Suppose, finally, that either

(a) $|u_{x^1 x^1}| \leqq K_1$ in $Q^+_{T,3}$, or

(b) condition 4 with $\alpha = 1$ and $|u_t| \leqq K_1$ holds in $Q^+_{T,3}$. Then $||u_{xx}|| \leqq \kappa^{-1} N(d, K, K_1, \varepsilon, \alpha)$ on $\Sigma_{T-\kappa^2,1}$. If, in addition,

$$|u_{x^j}(T, x) - u_{x^j}(T, 0, x^2, \ldots, x^d)| \leqq K_1 x^1, \quad j \geqq 2 \tag{11}$$

for $(T, x) \in \partial_t Q^+_{T,3}$, then $||u_{xx}|| \leqq N(d, K, K_1, \varepsilon, \alpha)$ on $\Sigma_{T,1}$.

<u>Proof</u>. Denote $u^j = u_{x^j}$ for $j = 2, \ldots, d$. By differentiation of (1) with respect to x^j and setting

$$a^{rs} = \frac{1}{2}\left(F_{u_{rs}} + F_{u_{sr}}\right), \quad g^k = -F_{u_k}, \quad Lv = v_t + a^{rs}v_{x^r x^s},$$

we obtain

$$Lu^j - u^j_{x^k} g^k + F_u u^j + F_{x^j} = 0.$$

By hypothesis,

$$|F_u||u^j| + |F_{x^j}| \leqq N\left[(x^1)^{-1} + \sum_{j=2,\ k=1}^{d} \left(u^j_{x^k}\right)^2 + \right.$$
$$\left. + |u_{x^1x^1}|^2\right]^{\frac{1}{2\sqrt{(1-\alpha)}}},$$

$$g^k g^k \leqq N\left[(x^1)^{\alpha-1} + \sum_{j=2,\ k=1}^{d} \left(u^j_{x^k}\right)^2 + |u_{x^1x^1}|^2\right], \tag{12}$$

in $Q^+_{T,3}$, where $N = N(d, K_1)$. If condition 4 holds with $\alpha = 1$ and $|u_t| \leqq K_1$, then from (4) it follows that

$$|a^{rs}u_{x^r x^s}| \leqq N, \quad |a^{11}u_{x^1x^1}|$$
$$\leqq N + 2\sum_{j=2,\ k=1}^{d} |a^{jk}u_{x^j x^k}|,$$

$$|u_{x^1x^1}| \leqq N(d, K_1, \varepsilon) + 2K\varepsilon^{-1} \sum_{j=2,\ k=1}^{d} |u^i_{x^k}|, \tag{13}$$

and the summands $|u_{x^1x^1}|^2$ in (12) can be omitted by increasing the constant N. This can obviously be done in the case $|u_{x^1x^1}| \leqq K_1$ too. Now, the desired estimates of the derivatives $u_{x^j x^1}$ with $j \geq 2$ immediately follow from Lemma 1.1.4 and Theorem 4.5.4. The estimates of $u_{x^i x^j}$ with $i, j \geq 2$ are presented in (10). The estimate of $u_{x^1 x^1}$ on $\Sigma_{T,1}$ is available if condition (a) holds, otherwise, using the continuity, it can be derived from the latter inequality in (13), which is proved in $Q^+_{T,3}$, and the already proved estimates of $u_{x^i x^j}$ with $i \circ j \geq 2$. The theorem is proved.

10. THEOREM. Let conditions 2, 3, and 6 hold. Let $|u|$ and $|u_x|$ be no greater than K_1 in $Q^+_{T,3}$, and let conditions (10) hold.

Suppose that either (a) $|u_{x^1 x^1}| \leqq K_1$ in $Q^+_{T,3}$, or (b) condition 4 holds and $|u_t| \leqq K_1$ in $Q^+_{T,3}$. Then $\frac{\partial}{\partial x^1} u_{x^j}$ exists on $\Sigma_{T,1}$ for $j \geqq 2$. Moreover, there exists a constant $\alpha_0 = \alpha_0\ (d, K, \varepsilon) \in (0, 1)$ such that, for any $\alpha_1 \in (0, \alpha_0 \wedge \alpha]$

(i) we have

$$\left|\frac{\partial}{\partial x^1} u_{x^j}(z_1) - \frac{\partial}{\partial x^1} u_{x^j}(z_2)\right|$$

$$\leqq N\kappa^{-\alpha_1 - 1} \rho^{\alpha_1}(z_1, z_2), \quad j \geqq 2, \tag{14}$$

where $z_1, z_2 \in \Sigma_{T-\kappa^2,1}$, $N = N(d, K, K_1, \varepsilon, \alpha, \alpha_1)$;

(ii) if, in addition, there exist $u_{x^j x^1}$ for $j \geqq 2$ on $\partial_t Q^+_{T,3}$, and they themselves and their Hölder constants of order α_1 on $\partial_t Q^+_{T,3}$ do not exceed in magnitude K_1, then inequality (14) with $\kappa = 1$ holds for all $z_1, z_2 \in \bar{\Sigma}_{T,1}$.

Proof. By differentiation of (1) with respect to x^j with $j \geqq 2$, the function

$$v(t, x^1, \ldots, x^d, x^{d+2}, \ldots, x^{2d}) :=$$

$$:= \sum_{j=2}^{d} x^{d+j} u_{x^j}(t, x^1, \ldots, x^d)$$

on the set

$$\tilde{Q}^+_{T,3} := Q^+_{T,3} \times \{|x^{d+j}| \leq 3(K_1 + 1)^{-1}, \quad j \geq 2\}$$

satisfies the equation

$$v_t + a^{rs}v_{x^rx^s} + b^rv_{x^r} + cv + \sum_{j=2}^{d} x^{d+j}F_{x^j} = 0, \tag{15}$$

where $a^{rs} = (1/2)\left(F_{u_{rs}} + F_{u_{sr}}\right), b^r = F_{u_r}$, and $c = F_u$. By condition 6, we obtain $|c|$, $|b^r| \leq K_1$ and (cf. Lemma 7), for functions σ^{rs} and f such that $|\sigma^{rs}| \leq 3d$ and $|f| \leq 3d(x^1)^{\alpha-1}$, the equality

$$\sum_{j=2}^{d} x^{d+j}F_{x^j} = f + \sum_{r,s=1}^{d} \sigma^{rs}u_{x^rx^s}$$

$$= f + \sum_{i=2,\ s=1}^{d} \sigma^{rs}v_{x^{d+r}x^s} +$$

$$+ \sum_{r=1,\ s=2}^{d} \sigma^{rs}v_{x^rx^{d+s}} + \sigma^{11}u_{x^1x^1} \tag{16}$$

holds on $\tilde{Q}^+_{T,3}$.

If $|u_{x^1x^1}| \leq K_1$ in $Q^+_{T,2}$, then the latter summand is combined with f. But if condition 4 holds and $|u_t| \leq K_1$ in $Q^+_{T,3}$, then we can write (cf. (13))

$$u_{x^1x^1} = h + \sum_{r=1,\ s=2}^{d} \tau^{rs}v_{x^rx^{d+s}},$$

where $|\tau^{rs}| \leq N(d, K, \varepsilon)$, and $|h| \leq N(d, K_1, \varepsilon)(x^1)^{\alpha-1}$.

In any case, from (15) and (16) we obtain on $Q^+_{T,3}$

$$v_t + \sum_{r,s=1}^{d} a^{rs}v_{x^rx^s} + \sum_{r=2,\ s=1}^{d} \tilde{a}^{rs}v_{x^{d+r}x^s} +$$

$$+ N_1 \sum_{r=2}^{d} v_{x^{d+r}x^{d+r}} + \sum_{r=1}^{d} b^rv_{x^r} + cv + g = 0, \tag{17}$$

where N_1 is any constant, $|g| \leq N(d, K, K_1, \varepsilon)(x^1)^{\alpha-1}$, $|\tilde{a}^{rs}| \leq N(d, K,$

ε). It is clear that

$$\sum_{r,s=1}^{d} a^{rs}\lambda^r\lambda^s + \sum_{r=2,\ s=1}^{d} \tilde{a}^{rs}\lambda^{d+r}\lambda^s + N_1 \sum_{r=2}^{d} (\lambda^{d+r})^2$$

$$\geq \frac{1}{2}\varepsilon\left(\sum_{i=1}^{d}(\lambda^i)^2 + \sum_{i=d+2}^{2d}(\lambda^i)^2\right)$$

for an $N_1 = N_1(d, K, \varepsilon)$ and all $\lambda^r \in (-\infty, \infty)$. Using this constant N_1, we regard (17) as an equation in v. Of course, its coefficients depend on the function u, but this function is fixed so that the coefficients of equation (17) can be regarded as functions only of $t, x^1, \ldots, x^d, x^{d+2}, \ldots, x^{2d}$. Let us apply Theorem 8 to (17) while defining an obvious new 'nonlinear' function F and taking note that by the above argument, $|cv + g| \leq N(d, K, K_1, \varepsilon)(x^1)^{\alpha-1}$.

The applicability of the theorem, in the case when the bases of the parallelepiped are not cubes, may be put in question. However, we believe that Remark 4.5.7 has already answered this question. Theorem 8 proves the existence of v_{x^1} for $t \in [0, T)$, $x^1 = 0$, $|x^i| < 1$, $i = 2, \ldots, d$, $|x^j| < (K_1 + 1)^{-1}$, $j \geq d + 2$. Selecting a single nonzero x^j for $j \geq d + 2$, we obtain the existence of $\frac{\partial}{\partial x^i} u_{x^j}$ on $\Sigma_{T,1}$ with $j \geq 2$. Applying this technique to the assertion of Theorem 8 concerning the module of the difference with v_{x^1}, we can derive the remaining assertions of this theorem. This completes the proof.

Under some additional assumptions on F, we can estimate the Hölder constant on $\Sigma_{T,1}$ not only for the mixed derivatives, but also for the derivative $u_{x^1x^1}$.

11. THOEREM. Let

$$P(\tau) = \Big\{(u_{ij}, u_i, u, t, x) : \sum_{i,j}|u_{ij}| + \sum_i |u_i| + |u| < \tau,\quad u_{ij} = u_{ji},\quad (t, x) \in Q^+_{T,3}\Big\},$$

let $M(\tau)$ be a finite-valued nonnegative function on $[0, \infty)$ such that $M(\tau) \geq 1$. Suppose that condition 2 holds; on every $P(\tau)$

$$\varepsilon|\lambda|^2 \leqq F_{u_{rs}}(u_{ij}, u_i, u, t, x)\lambda^r\lambda^s \leqq K|\lambda|^2, \quad \lambda \in E_d, \tag{18}$$

the first derivatives of F with respect to (u_{ij}, u_i, u) do not exceed $M(\tau)$ in magnitude, for $(u_{ij}, u_i, u, z_r) \in P(\tau)$, $r = 1, 2$, we have

$$|F(u_{ij}, u_i, u, z_1) - F(u_{ij}, u_i, u, z_2)|$$
$$\leqq M(\tau)\rho^\alpha(z_1, z_2). \tag{19}$$

Also, let condition 4 with $\alpha = 1$ hold, condition 6 hold, let $u \in C^2_{loc}(Q^+_{T,3} \cup \Sigma_{T,3})$, $|u|$, $|u_x|$, $|u_t| \leqq K_1$ in $Q^+_{T,3}$, and let conditions (10) hold. Then there exists a constant $\alpha_0 = \alpha_0(d, K, \varepsilon) \in (0, 1)$ such that for any $\alpha_1 \in (0, \alpha_0 \wedge \alpha]$:

(a) for $z_1, z_2 \in \Sigma_{T-\kappa^2,1}$ with $N = N(d, K, K_1, \varepsilon, \kappa, \alpha, \alpha_1, M)$, the inequality

$$||u_{xx}(z_1) - u_{xx}(z_2)|| \leqq N\rho^{\alpha_1}(z_1, z_2) \tag{20}$$

holds;

(b) if, in addition, u_{x^1} and $u_{x^jx^1}$ exist on ${}_tQ^+_{T,3}$ for $j \geqq 2$, they themselves and their Hölder constants of order α_1 do not exceed K_1 in magnitude on $\partial_tQ^+_{T,3}$, then inequality (20) holds for all $z_1, z_2 \in \Sigma_{T,1}$ with $N = N(d, K, K_1, \varepsilon, \alpha, \alpha_1, M)$.

<u>Proof</u>. By virtue of Theorem 10, it suffices to estimate $|u_{x^1x^1}(z_1) - u_{x^1x^1}(z_2)|$. Let first $z_1, z_2 \in Q^+_{T,3}$, $z_i = (t_i, x_i)$, and $x^1_1 = x^1_2$. We subtract the value on the left in (1) at the point z_2 from this value at the point z_1 and write the result as

$$0 = u_t(z_1) - u_t(z_2) + \Big[F\Big(u_{x^ix^j}(z_1), u_{x^i}(z_1), u(z_1), z_1\Big) -$$
$$- F\Big(u_{x^ix^j}(z_2), u_{x^i}(z_2), u(z_2), z_1\Big)\Big] +$$
$$+ \Big[F\Big(u_{x^ix^j}(z_2), u_{x^i}(z_2), u(z_2), z_1\Big) -$$
$$- F\Big(u_{x^ix^j}(z_2), u_{x^i}(z_2), u(z_2), z_2\Big)\Big].$$

Estimating the expression in the second pair of square brackets, we use (19), whereas to transform the first pair, we apply the Hadamard formula. Then, if τ is such that $\left(u_{x^ix^j}(z_r),\ u_{x^i}(z_r),\ u(z_r),\ z_r\right) \in P(\tau)$, $r = 1, 2$, we obviously have

$$\left|u_t(z_1) - u_t(z_2) + a^{ij}\left(u_{x^ix^j}(z_1) - u_{x^ix^j}(z_2)\right)\right|$$

$$\leq N(d)M(\tau)\left(\rho^{\alpha}(z_1, z_2) + \sum_{i=1}^{d}\left|u_{x^i}(z_1) - u_{x^i}(z_2)\right| + \right.$$

$$\left. + \left|u(z_1) - u(z_2)\right|\right),$$

where $a^{ij} = a^{ji}$, and $\varepsilon|\lambda|^2 \leq a^{ij}\lambda^i\lambda^j \leq K|\lambda|^2$ (see (18)). We let $x_1^1 = x_2^1 \downarrow 0$ and make use of condition (10). Then, for $z_r \in \Sigma_{T,1}$ and τ related to z_r as above, we obtain

$$\left|u_{x^1x^1}(z_1) - u_{x^1x^1}(z_2)\right|$$

$$\leq NM(\tau)\rho^{\alpha}(z_1, z_2) +$$

$$+ N\sum_{i=2}^{d}\left|u_{x^jx^1}(z_1) - u_{x^jx^1}(z_2)\right| +$$

$$+ NM(\tau)\left|u_{x^1}(z_1) - u_{x^1}(z_2)\right|, \tag{21}$$

where $N = N(d, K, K_1, \varepsilon)$. If $z_1, z_2 \in \Sigma_{T-\kappa^2,1}$, then, by Theorem 9, we can take $\tau \leq \kappa^{-1}N(d, K, K_1, \varepsilon, \alpha)$ and estimate the right-hand side of (21) as desired using Theorems 8 and 10. This proves (a).

Under conditions (b), inequality (11) obviously holds and, using Theorem 9, we can take $\tau \leq N(d, K, K_1, \varepsilon, \alpha)$ for $z_1, z_2 \in \Sigma_{T,1}$. It then remains to apply Theorems 8 and 10. This proves the theorem.

12. Remark. The assertion (b) of Theorem 8 and assertion (ii) of Theorem 10 involve $\bar{\Sigma}_{T,1}$, while assertion (b) of Theorem 11 involves only $\Sigma_{T,1}$. This is a substantial distinction since even the equation $u_t + u_{x^1x^1} = 0$ in $(0, T) \times (0, 6)$ with the condition $u(t, 0) = 0$ would be associated with the discontinuity of the derivative $u_{x^1x^1}(t, 0)$ at the point $t = T$, if the matching condition $u_{x^1x^1}(T, 0) = 0$ does not hold.

13. Remark. All results of this section can be applied to the elliptic equations

$$F\left(u_{x^i x^j}(x),\ u_{x^i}(x),\ u(x),\ x\right) = 0 \text{ in } D_3^+. \tag{22}$$

To this end one should add $u_t\,(= 0)$ to the left-hand side of (22) and set $T = 2$, $\kappa = 1$ throughout the section as well as

$$Q_{T,3}^+ = (0,\ 2) \times D_3^+,\quad z_k = (0,\ x_k).$$

5.2. The Estimates of the First Derivatives with Respect to x

In this section, we prove two types of estimates: the interior estimates and the estimates up to the boundary. The method which is used here for $|u_x|$ is due to S.N. Bernstein and consists of considering the points where an appropriate combination of the function and its derivatives attains its maximum. When considering $|u_x|$ we apply a version of this method which is called the *monitored maximum method*.

Ladyzhenskaya and Ural'tseva [1] show how to apply the already established estimate of the Hölder constant of a solution to estimating the absolute value of the solution gradient under natural assumptions on the growth of the coefficients of quasi-linear equations relative to the derivatives of the solution. Once such a possibility is established, it can be implemented in different ways and our approach differs from that discussed by Ladyzhenskaya and Ural'tseva.

Except for the estimate of $|u_x|$, this section deals with an estimate of the Hölder constant for u_x.

We discuss some constructions of this section considering a simple nonlinear elliptic equation 'with constant coefficients'

$$F\left(u_{x^i x^j}(x),\ u_{x^i}(x),\ u(x)\right) = 0. \tag{1}$$

Let F and u be sufficiently smooth functions, let the matrix $(F_{u_{ij}})$ be bounded and positive definite, and let $F_u \leqq 0$. By differentiation of (1) with respect to x^k for $u^k = u_{x^k}$, we obtain the equation

$$F_{u_{ij}} u^k_{x^i x^j} + F_{u_i} u^k_{x^i} + F_u u^k = 0. \tag{2}$$

By the maximum principle, the greatest value of $|u^k|$ is attained on the boundary of the domain wherein equation (1) is considered. Therefore, the interior estimate of $|u_x|$ reduces to the boundary estimate.

The situation becomes more complicated when F also depends on x and the left-hand side of (2) involves F_{x^k}. If $|F_{x^k}| \leq N(1 + |u_x| + \|u_{xx}\|)$ and equation (1) can be written as

$$F_{u_{ij}} u_{x^i x^j} + F_{u_i} u_{x^i} + F_u u + f = 0, \tag{3}$$

where f is a bounded function, then we can apply Bernstein's method and following the case of linear equations, consider $v := |u_x|^2 + \tau u^2$, then consider the points where this function attains its maximum, and choose τ such that v could be estimated at every interior point where v attains its maximum. The latter is possible in view of (2), (3), and the fact that at an interior point where v attains its maximum, its first derivatives equal zero, while the matrix of second derivatives is negative. It remains to estimate v on the boundary.

Slight modification of this technique enables us to treat the case when F_{x^k} and f grow with respect to u_x no faster than $|u_x|^{2-\alpha}$ with $\alpha > 0$. In this case we can replace τ by $\tau(1 + \sup|u_x|^{\beta})$ for a $\beta > 0$. The case $\alpha = 2$ cannot be handled by this modification. In this case Ladyzhenskaya and Ural'tseva [1] use an auxiliary function of the form $\varphi(u) \quad |u_x|^2$, whereas we make use of the same function v, but choose τ according to a relationship between τ and sup v so that τ is involved in an equation which enables us to minotor the contribution of $|u_x|$ and $|u|$ into the maximum of v.

Note that estimating $|u_x|$ (but not $|u_t|$), we prevent F and its derivatives from growing with respect to u_{xx} faster than linearly and always assume (see (19) and condition (8) below) that the equation under consideration can be rewritten as (3), where f grows with respect to u_x no faster than $|u_x|^2$.

Once the estimate of $|u_x|$ is established, the rate of growth of F and its derivatives with respect to u_x is immaterial and the interior estimate of the Hölder constant of u_x can be obtained by use of Theorem 4.3.5 from relationships like (2).

We fix constants $K \geq \varepsilon > 0$, T, K_1, K_2, $\varepsilon_1 > 0$, $\kappa \in (0, 1]$, a domain $D \subset E_d$ and denote $Q = (0, T) \times D$. Let Γ be a (possibly empty) closed subset of $\overline{\partial' Q}$ and let $Q_1 \subset Q$ be a domain. Consider one more domain $Q_0 \subset Q_1$ and suppose that every its point has a neighbourhood (in E_{d+1}) such that the parabolic distance from any point of this neighbourhood to $\partial' Q_1 \setminus \Gamma$ is greater than κ. Also, assume that $Q_0 \neq \emptyset$. Since the distance to the empty set is assumed to be equal to infinity, we can take $Q_o = Q$ if $\Gamma = \overline{\partial' Q}$ and $Q_1 = Q$.

Let $F(u_{ij}, u_i, u, t, x)$ be a Borel function defined for $(t, x) \in Q_1$ and real-valued u_{ij}, u_i $(i, j = 1, \ldots, d)$, and u. Suppose that for every $t \in (0, T)$ the function F is continuously differentiable with respect to (u_{ij}, u_i, u, x) on $E_{d^2+d+1} \times \{x : (t, x) \in Q_1\}$. Finally, let $u \in C^2_{loc}(Q_1) \cap C(\bar{Q}_1)$ be a function such that

$$u_t(t, x) + F\left(u_{x^ix^j}(t, x), u_{x^i}(t, x), u(t, x), t, x\right) = 0 \tag{4}$$

in Q_1. These assumptions are supposed to be satisfied throughout this section.

The following lemma relates to properties of some auxiliary domains.

1. LEMMA. (a) Let Q' be the set of all those points of Q_1, each of which has a neighbourhoodl (in E_{d+1}) such that the parabolic distance from any point of this neighbourhood to $\partial'Q_1 \setminus \Gamma$ is greater than $(1/2)\kappa$. Then Q' is a domain, $Q_0 \subset Q'$, $\Gamma \cap \partial'Q' \subset \Gamma \cap \partial'Q_1$ and for every point of Q_0 there exists a neighbourhood such that the parabolic distance from any point of this neighbourhood to $\partial'Q' \setminus \Gamma$ is greater than $(1/2)\kappa$.

(b) If $z_0 \in Q_0$, $\rho \in (0, \frac{1}{2}\kappa]$, and $G = G(z_0, \rho)$ is the connected component of $(z_0 + C_{\rho^2\rho}) \cap Q$ whose closure contains z_0, then

$$G \subset Q_1, \quad \bar{G} \setminus \Gamma \subset Q_1, \quad \partial'G \cap \partial'Q = \Gamma \cap \partial'G \subset \Gamma \cap \partial'Q_1. \tag{5}$$

Proof. (a) The set Q' is open by definition, the inclusion $Q_0 \subset Q'$ is clear, the inclusion $\Gamma \cap \partial'Q' \subset \Gamma \cap \partial'Q_1$ immediately follows from the fact that $Q' \subset Q_1$ and $\Gamma \cap Q_1 = \emptyset$. To prove the latter assertion of (a), we take a point $z_0 \in Q_0$ and a spherical neighbourhood U of z_0 (in E_{d+1}) such that the parabolic distance from any point of this neighbourhood to $\partial'Q_1 \setminus \Gamma$ is greater than κ. We denote by U_1 the ball centred at the point z_0 whose radius is half as great as the radius of U. It is clear that, for a $\delta > 0$ and any point $z \in U_1$ we obtain $\rho(z, \partial'Q_1 \setminus \Gamma) \geq \kappa + \delta$. Assertion (a) will be proved, if we manage to show that $\rho(z, \partial'Q' \setminus \Gamma) \geq (1/2)\kappa$ for all $z \in U_1$.

We take $z_1 = (t_1, x_1) \in U_1$ and $z_2 = (t_2, x_2) \in \partial'Q' \setminus \Gamma$ such that $t_2 > t_1$ (if there are no such z_2, then $\rho(z_1, \partial'Q' \setminus \Gamma) = \infty > (1/2)\kappa$). In the case when $z_2 \notin Q_1$, the inclusions $Q' \subset Q_1$ and $z_2 \in \partial'Q'$ and the definition of the parabolic boundary imply that $z_2 \in \partial'Q_1$, and since

$z_2 \notin \Gamma$, we obtain $z_2 \in \partial'Q_1 \setminus \Gamma$ and $\rho(z_1, z_2) \geq \kappa + \delta \geq (1/2)\kappa$. But if $z_2 \in Q_1$, then by virtue of the condition $z_2 \notin Q'$, every neighbourhood of the point z_2 contains a point z such that $\rho(z, \partial'Q_1 \setminus \Gamma) \leq (1/2)\kappa$. We choose a $z = (t, x)$ such that $t > t_1$ and $\rho(z_1, z_2) \geq \rho(z_1, z) - \delta$. By the triangle inequality, we obtain

$$\frac{1}{2}\kappa + \rho(z_1, z_2) + \delta$$

$$\geq \rho(z, \partial'Q_1 \setminus \Gamma) + \rho(z_1, z)$$

$$\geq \rho(z_1, \partial'Q_1 \setminus \Gamma) \geq \kappa + \delta.$$

Hence $\rho(z_1, z_2) \geq (1/2)\kappa$. Thus, $\rho(z_1, z_2) \geq (1/2)\kappa$ for all $z_2 \in \partial'Q' \setminus \Gamma$ such that $t_2 > t_1$ and it follows that

$$\rho(z_1, \partial'Q' \setminus \Gamma) \geq \frac{1}{2}\kappa.$$

(b) The third relationship in (5) follows from the first two. Indeed,

$$\partial'G \cap \partial'Q = [\partial'G \cap (\Gamma \cap \partial'Q)] \cup [\partial'G \cap (\partial'Q \setminus \Gamma)],$$

$$\partial'G \cap (\partial'Q \setminus \Gamma) = (\partial'G \setminus \Gamma) \cap \partial'Q \subset Q_1 \cap \partial'Q = \emptyset,$$

$$\partial'G \cap (\Gamma \cap \partial'Q) \subset \Gamma \cap \partial'G, \tag{6}$$

and if $z \in \Gamma \cap \partial'G \subset \partial'G \cap \partial Q$, then $z \notin Q$, $z \notin Q_1$ and $z \in \partial'G$.From the definition of the parabolic boundary it follows that $z \in \partial'Q$, $z \in \partial'Q_1$, we can put the equality sign into (6), and $\Gamma \cap \partial'G \subset \Gamma \cap \partial'Q_1$.

We now prove the first relation in (5). Consider a $z_1 = (t_1, x_1) \in G$ and assume that $z_1 \notin Q_1$. The set G is a cylinder whose base is connected. Moreover, $t_1 > t_0$. Therefore, there exists a continuous curve (t, x_t) on $[t_0, t_1]$ which lies in G for $t > t_0$ and such that $x_{t_0} = x_0$ and $x_{t_1} = x_1$. The origin of this curve lies in $Q_1(Q_0)$, while the end point does not. Therefore, there exists a point z_2 of this curve such that $z_2 \in \partial Q_1$ and t has the minimum value. By the definition of the parabolic boundary, we obtain $z_2 \in \partial'Q_1$. Furthermore, from the inclusion $z_2 \in G$ we obtain two implications: $\rho(z_0, z_2) < \kappa$ and $z_2 \notin \Gamma$. Finally, $z_2 \in \partial'Q_1 \setminus \Gamma$ and $\rho(z_0, \partial'Q_1 \setminus \Gamma) < \kappa$, which is impossible.

It remains for us to show that $\partial G \setminus \Gamma \subset Q_1$. Assume the contrary. Then there exists a point $z_1 = (t_1, x_1)$ such that

$$z_1 \in \partial G, \quad z_1 \notin \Gamma, \quad z_1 \notin Q_1.$$

We select a sequence $z_n = (t_n, x_n) \in G$ such that $z_n \to z_1$ and, without loss of generality, assume that $t_n \neq t_1$. Also, we choose a small $\delta_1 > 0$ (cf. the argument in the proof of assertion (a)) such that the inequality

$$\rho(z_0', \partial' Q_1 \setminus \Gamma) \geqq \kappa + 2\delta_1 \tag{7}$$

holds for $z_0' := (t_0 - \delta_1^2, x_0)$. Finally, we put $z_n' = (t_1 - |t_1 - t_n|, x_n)$.

It is clear that $\rho(z_0', z) \leqq \kappa + \delta_1$ for all $z \in z_0 + C_{\rho^2,\rho}$ and $t_1 - |t_1 - t_n| \geqq t_o - |t_1 - t_n| > t_0 - \delta_1^2$ for sufficiently large n's, i.e., only those which attract our attention. Connecting z_0' with z_n' by a continuous curve and using (7) as above, we derive a contradiciton to the assumption that z_n' does not lie in Q_1. Hence, $z_n' \in Q_1$.

Furthermore, $z_1 \notin Q_1$, $z_n' \in Q_1$, and $t_1 - |t_1 - t_n| < t_1$. Therefore, on the line segment connecting the points z_n' and z_1 there exists a point $z_n'' \in \partial' Q_1$. Since $z_n' \to z_1$, we obtain $z_n'' \to z_1$ and the *closedness* of Γ implies that $z_n'' \notin \Gamma$ and $z_n'' \in \partial' Q_1 \setminus \Gamma$ for all sufficiently large n. Finally,

$$\kappa + \delta_1 \geqq \rho(z_0', z_1) = \lim_{n\to\infty} \rho(z_0', z_n'')$$

$$\geqq \rho(z_0', \partial' Q_1 \setminus \Gamma) \geqq \kappa + 2\delta_1.$$

We have derived a contradiction which proves the lemma.

In the following lemma, conditions (8) — (10) are satisfied if, for example, $F = a^{ij}(u_k, u, t, x)u_{ij} + a(u_k, u, t, x)$ holds under conditions (3.2) — (3.6) with m = 2 of Lydyzhenskaya, Solonnikov, and Ural'tseva [1], Chapter 6. This indicates that conditions (8) — (10) are quite reasonable.

2. LEMMA. For every point $z \in Q_1$ where the inequalities $|u(z)| \leqq K_1$ and $|u_x(z)| \geqq K_2$ hold (if such a point exists at all), let the inequalities

$$|F| + \left|F_{u_r}\right| |u_x| + \left|F_{x^r}\right| |u_x|^{-1} \leqq K(|u_x|^2 + ||u_{xx}||), \tag{8}$$

$$F_u \leqq K(|u_x|^2 + ||u_{xx}||), \tag{9}$$

$$\varepsilon|\lambda|^2 \leqq F_{u_{ij}} \lambda^i\lambda^j \leqq K|\lambda|^2, \tag{10}$$

also hold for arbitrary $\lambda \in E_d$, $r = 1, \ldots, d$, where for the sake of simplicity the arguments $(u_{x^ix^j}, u_{x^i}, u, z)$ of F and its derivatives and z of u and its derivatives are omitted. Let $\alpha \in (0, 1)$, let $|u|$ and the Hölder constant of order α for the function u on Q_1 do not exceed K_1, let $u_x \in C^2_{loc}(Q_1) \cap C(\bar{Q}_1)$, and let $|u_x| \leqq K_2$ on $\Gamma \cap \partial' Q_1$. Then

$$|u_x| \leqq N(d, \varepsilon, K, K_1, K_2, \alpha, \kappa)$$

in Q_0.

Proof. We fix $z_0 = (t_0, x_0) \in Q_0$ and take constants $\rho \in (0,(1/2)\kappa]$ and $\delta > 0$ and a nonnegative function $\zeta \in C^2(z_0 + \bar{C}_{\rho^2,\rho})$ which is zero on $z_0 + \partial' C_{\rho^2,\rho}$ and such that $\zeta \leqq 1$, $\zeta(z_0) = 1$, and

$$|\zeta_x|^2 \leqq N\rho^{-2}\zeta, \quad ||\zeta_{xx}|| \leqq N\rho^{-2}, \quad |\zeta_t| \leqq N\rho^{-2} \tag{11}$$

in $z_0 + C_{\rho^2,\rho}$, where $N = N(d)$. In order to construct such a function, it suffices to take any function $\eta \in C^\infty(\bar{C}_{1,1})$ such that $\eta(0) = 1$, $0 \leqq \eta \leqq 1$ on $C_{1,1}$, and $\eta = 0$ on $\partial' C_{1,1}$ and to put $\zeta(t, x) = \eta^2(\rho^{-2}(t - t_0), \rho^{-1}(x - x_0))$. In what follows the numbers ρ and δ will be specially chosen on establishing some preliminary estimates.

We denote by N various constants depending only on d, ε, K, and K_1, but independent of ρ, δ, α and assume that δ is associated with ρ as

$$(2\rho)^{2\alpha} K_1^2 \leq \frac{1}{2}\delta. \tag{12}$$

Also, we consider the domain G from Lemma 1(b). It follows from (5) and (12) that $(u(z) - u(z_0))^2 \leqq (1/2)\delta$ on G and the function

$$M(\tau) := \max[\zeta(z)|u_x(z)|^2 + \tau(u(z) - u_0)^2 : z \in \bar{G}], \tag{13}$$

where $u_0 = u(z_0)$, as a function of real τ, grows slower than $(1/2)\delta\tau$ as $\tau \to \infty$.

It is clear that $M(\tau) \geqq 0$, the function $M(\tau)$ is convex as the supremum of linear in τ functions, and the function $M(\tau)$ is continuous as a finite-valued convex function. It follows that there exists a (unique) number $\tau \in [0, \infty)$ such that $M(\tau) = \delta\tau$. We fix this τ and denote by z_1 a point of $\bar{G}$ where the supremum of (13) is attained. Then we obtain the following relationship which enables us to 'monitor' the maximum

$$\zeta(z_1)|u_x(z_1)|^2 + \tau(u(z_1) - u_0)^2 = M(\tau) = \delta\tau. \tag{14}$$

We further consider two separate cases:

(a) $\zeta(z_1)|u_x(z_1)|^2 > K_2^2$ and

(b) $\zeta(z_1)|u_x(z_1)|^2 \leqq K_2^2$.

Note that $\zeta|u_x|^2 \leqq K_2^2$ on $\partial'G$ and $\overline{\partial'G}$. Indeed, on the part of $\partial'G$ which does not belong to $z_0 + C_{\rho^2,\rho}$ we obtain $\zeta = 0$ by construction, whereas the part of $\partial'G$ which (possibly) does not belong to Q, necessarily lies on $\partial'Q$, on $\partial'Q_1 \cap \Gamma$ in view of (5), and on this part we have $|u_x|^2 \leqq K_2^2$. Therefore, in case (a), we obtain $z_1 \in \bar{G} \setminus \overline{\partial'G}$, $z_1 + C_{\sigma^2,\sigma} \subset G$ for a sufficiently small $\sigma > 0$ (cf. the proof of Lemma 3.3.6), $z_1 + \bar{C}_{\sigma^2,\sigma} \subset Q_1$ by virtue of (5), and at the point z_1 we can differentiate (4). Furthermore, at this point, the derivative with respect to t and the matrix of second derivatives with respect to x of the function $\zeta|u_x|^2 + \tau(u - u_0)^2$ are negative, while the first derivatives with respect to x are zero. Thus, at the point z_1, we obtain

$$\frac{1}{2}\zeta_t|u_x|^2 + \zeta u_{x^k t} u_{x^k} + \tau u_t(u - u_0) \leqq 0, \tag{15}$$

$$\frac{1}{2}\zeta_{x^i}|u_x|^2 + \zeta u_{x^k x^i} u_{x^k} + \tau u_{x^i}(u - u_0) = 0, \tag{16}$$

$$\Big(\zeta u_{x^k x^i x^j} u_{x^k} + \tau u_{x^i x^j}(u - u_0) + \zeta u_{x^k x^i} u_{x^k x^j} + \tau u_{x^i} u_{x^j} + \zeta_{x^j} u_{x^k x^i} u_{x^k} + \zeta_{x^i} u_{x^k x^j} u_{x^k} + \frac{1}{2}\zeta_{x^i x^j}|u_x|^2\Big) \leqq 0, \text{(1}$$

$$0 = u_{tx^k} + F_{u_{ij}} u_{x^i x^j x^k} + F_{u_i} u_{x^i x^k} + F_u u_{x^k} + F_{x^k}, \tag{18}$$

$$0 = u_t + F_{u_{ij}} u_{x^i x^j} + F_{u_j} u_{x^i} + f, \tag{19}$$

where $f := F - F_{u_{ij}}u_{x^ix^j} - F_{u_i}u_{x^i}$. Let us multiply (18) and (19) by ζu_{x^k} and $\tau(u - u_0)$ respectively, add together the results, and apply (15) and (16). Moreover, denote

$$a^{ij} = \frac{1}{2}\Big(F_{u_{ij}} + F_{u_{ji}}\Big),$$

$$g = f\tau(u - u_0) + \Big(F_u\zeta - \frac{1}{2}F_{u_i}\zeta_{x^i} - \frac{1}{2}\zeta_t - \frac{1}{2}a^{ij}\zeta_{x^ix^j}\Big)|u_x|^2 +$$

$$+ F_{x^k}\zeta u_{x^k} - 2a^{ij}\zeta_{x^i}u_{x^kx^j}u_{x^k}.$$

Then, at the point z_1, we obtain

$$0 \leqq a^{ij}\Big(\zeta u_{x^ix^jx^k}u_{x^k} + \tau u_{x^ix^j}(u - u_0) +$$

$$+ 2\zeta_{x^i}u_{x^kx^j}u_{x^k} + \frac{1}{2}\zeta_{x^ix^j}|u_x|^2\Big) + g$$

$$\leqq -a^{ij}\Big(u_{x^kx^i}u_{x^kx^j}\zeta + \tau u_{x^i}u_{x^i}\Big) + g$$

$$\leqq -\varepsilon\zeta||u_{xx}||^2 - \tau\varepsilon|u_x|^2 + g, \tag{20}$$

where the second inequlaity follows from (17) and the fact that the trace of the product of two nonnegative symmetric matrices is nonnegative, whereas the latter inequality is true by assumption (10).

We further estimate g from above. From our assumptions (8) — (10), (11) and the Young inequality it can easily be obtained that

$$g \leqq \varepsilon\zeta||u_{xx}||^2 + N\zeta^{-1}\tau^2(u - u_0)^2 + N|u_x|^2\tau|u - u_0| +$$

$$+ N\zeta|u_x|^4 + N\rho^{-2}|u_x|^2 + N\zeta^{1/2}|u_x|^3\rho^{-1}$$

$$\leqq \varepsilon\zeta||u_{xx}||^2 + N\zeta^{-1}\tau^2(u - u_0)^2 + N\zeta|u_x|^4 + N\rho^{-2}|u_x|^2.$$

This along with (20) and (14) yields

$$0 \leqq -\varepsilon\delta\tau^2 + N_1\tau^2(u - u_0)^2 + N_2\zeta^2|u_x|^4 + N_3\rho^{-2}\zeta|u_x|^2$$

$$\leqq -\varepsilon\delta\tau^2 + N_1\tau^2(u - u_0)^2 + N_2\delta^2\tau^2 + N_3\rho^{-2}\delta\tau. \tag{21}$$

We should stress that the constants N_1, N_2, and N_3 do not depend on ρ and δ. Assume that ρ and δ are chosen to satisfy not only the conditions $\rho \in (0,(1/2)\kappa]$ and (12), but also

$$N_2\delta \leqq \frac{1}{4}\varepsilon, \quad N_1K_1^2(2\rho)^{2\alpha} \leqq \frac{1}{4}\varepsilon\delta.$$

It is clear that appropriate ρ and δ can be chosen so that they depend only on K, K_1, d, ε, κ. Then from (21) we obtain $\tau \leqq 2N_3\varepsilon^{-1}\rho^{-2}$ and $\delta\tau \leqq N = 2N_3\delta\varepsilon^{-1}\rho^{-2}$, while from (14) it follows that $\zeta|u_x|^2 + \tau(u - u_0)^2 \leqq N$ in G, $|u_x(z_0)|^2 \leqq N$.

We have considered the case (a): $\zeta(z_1)|u_x(z_1)|^2 > K_2^2$. If $\zeta(z_1)|u_x(z_1)|^2 \leqq K_2^2$, then (with the ρ and δ fixed above) from (12) and (14) it follows that $K_2^2 + (1/2)\delta\tau \geqq \delta\tau$, $\delta\tau \leqq 2K_2^2$, and $|u_x(z_0)| \leqq 2K_2^2$. Since z_0 is an arbitrary point of Q_0, the lemma is proved.

Lemma 2 points out the significance of preliminary estimation of u in C^α. The preliminary estimate will be proved in the following lemma. In this lemma, introducing functions φ_1 and φ_2, we are able to abandon the excessively strong assumption that u satisfies Hölder's condition everywhere in $\Gamma \cap \partial' Q_1$.

3. LEMMA. Let the domain D be defined as $\{x : \psi(x) > 0\}$ such that $\psi \in C^2(E_d)$, $||\psi||_{C^2(E_d)} \leqq K_1$, and $|\psi_x| \geqq \varepsilon_1$ on ∂D (if $\partial D \neq \emptyset$). Suppose that we use the abbreviated notation introduced in Lemma 2 and for all $\lambda \in E_d$ inequalities (10) and

$$|F - F_{u_{ij}}u_{x^ix^j}| \leqq K_1(1 + |u_x|^2) \tag{22}$$

hold true on Q_1. Finally, let $\alpha \in (0, 1)$, let φ_1 and φ_2 be functions of $C^\alpha(\bar{Q})$, let their norms not exceed K_1 in this space, let $|u| \leqq K_1$ in Q_1 and $u = \varphi_1$ on $\Gamma \cap \partial' Q_1 \cap \partial_x Q$, and let $u = \varphi_2$ on $\Gamma \cap \partial' Q_1 \cap \partial_t Q$. Then for any $\alpha_1 \in (0, \alpha_0 \wedge \alpha]$, z_0, $z_1 \in Q_0$, we obtain

$$|u(z_0) - u(z_1)| \le N\rho^{\alpha_1}(z_0, z_1), \tag{23}$$

where $N = N(d, K, K_1, \varepsilon, \varepsilon_1, \kappa, \alpha, \alpha_1)$, $\alpha_0 = \alpha_0(d, K, \varepsilon) \in (0, 1)$.

Proof. We take the constant $\rho_0 = \rho_0(d, K_1\varepsilon_1^{-1}) > 0$ from Lemma 1.1.2 and note that (23) fails to be obvious only if

$$d \cdot \rho(z_0, z_1) \le \left(\frac{1}{4}\rho_0\right) \wedge \left(\frac{1}{30d}\kappa\right) =: \rho_1. \tag{24}$$

Therefore, without loss of generality we can assume that $z_0 = (t_0, x_0)$, $z_1 = (t_1, x_1) \in Q_0$, $t_0 \le t_1$, $\rho(z_0, z_1) \le \rho_1 d^{-1}$. Consider two separate cases: $d(x_0, \partial D) \ge 3\rho_1$ and $d(x_0, \partial D) < 3\rho_i$.

First let $d(x_0, \partial D) \ge 3\rho_1$. We take $G = G(z_0, 3\rho_1)$ from Lemma 1(b) and note that in the case under consideration

$$G = \left(z_0 + C_{9\rho_1^2, 3\rho_1}\right) \cap (t < T). \tag{25}$$

Furthermore, for $a^{ij} =(1/2)\left(F_{u_{ij}} + F_{u_{ji}}\right)$ we have (see (4), (22))

$$|u_t + a^{ij}u_{x^ix^j}| \le K_1(1 + |u_x|^2) \tag{26}$$

in Q_1 and G. This along with Corollary 4.3.6 imply that if $t_0 + 9\rho_1^2 < T$, then inequality (23) holds true for all $\alpha_1 \in (0, \beta_1]$, where $\beta_1 = \beta_1(d, K, \varepsilon) \in (0, 1)$. But if $t_0 + 9\rho_1^2 \ge T$, then by Lemma 1(b) and the hypothesis of this lemma, the function u on $\partial_t G$ satisfies the Hölder condition of order α with a constant which does not exceed K_1. We would like to apply Corollary 4.5.3 and Remark 4.5.7 to (26). To this purpose, it suffices to note that z_0, z_1 lie in the closure of the first of the domains in the following chain of inclusions:

$$z_0 + Q_{T-t_0,\, d^{-1}\rho_1} \subset z_0 + Q_{T-t_0,\, d^{-1}3\rho_1} \subset G. \tag{27}$$

By Corollary 4.5.3 (Theorem 4.5.2) and Remark 4.5.7, inequality (23) is true for all $\alpha_1 \in (0, \beta_2 \wedge \alpha]$, where $\beta_2 = \beta_2(d, K, \varepsilon) \in (0, 1)$.

Let us consider the second case when $d(x_0, \partial D) < 3\rho_1$. Let $x_2 \in \partial D$ and $|x_2 - x_0| < 3\rho_1$. It is obvious that

$$|x_2 - x_0|,\ |x_2 - x_1| < 4\rho_1,\quad x_2 + S_{12d\rho_1} \subset x_0 + S_{(1/2)\kappa}. \tag{28}$$

Moreover, $4\rho_1 \leqq \rho_0$ and therefore we can straighten the part of the boundary of D which lies in $x_2 + S_{12d\rho_1}$ using Lemma 1.1.2. According to this lemma, there exists a mapping f such that

$$f : \bar{D}_3^+ \to (x_2 + S_{12d\rho_1}) \cap \bar{D}, \quad f(D_1^+) \supset (x_2 + S_{4\rho_1}) \cap D,$$

$$f(\Sigma_3) \subset \partial D, \tag{29}$$

the inverse mapping f^{-1} is defined on $f(\overline{D_3^+})$; f and f^{-1} are twice continuously differentiable,

$$\frac{1}{2}|y_1 - y_2|4\rho_1 \leqq |f(y_1) - f(y_2)| \leqq 2|y_1 - y_2|4\rho_1 \tag{30}$$

for $y_1, y_2 \in \overline{D_3^+}$, the norms of the second partial derivatives of f (respectively, f^{-1}) do not exceed $N(d, K_1, \varepsilon_1)\rho_1^2$ (respectively, $N(d, K_1, \varepsilon_1)\rho_1^{-1}$).

We denote

$$G = G\left(z_0, \frac{1}{2}\kappa\right), \quad S = 16^{-1}\rho_1^{-2}\left[(T - t_0) \wedge \frac{\kappa^2}{4}\right], \tag{31}$$

where $16\rho_1^2 S$ is the length of the cylinder G along the t-axis, and using the formula

$$h(t, x) = (t_0 + 16\rho_1^2 t, f(x)), \tag{32}$$

we construct a mapping from $Q_{S,3}^+$ into a connected (f is continuous) domain which is contained (see (28) and (29)) in G. It is clear that $h(\overline{Q_{S,3}^+}) \subset G$ and $\bar{G} \subset \bar{Q}_1$ by Lemma 1. Therefore, functions and operators can naturally be brought from $h(Q_{S,3}^+)$ to $Q_{S,3}^+$ using the mapping h. We put

$$L = \frac{\partial}{\partial t} + a^{ij}\frac{\partial^2}{\partial x^i \partial x^j}, \quad \tilde{u} = u(h), \quad \tilde{L}v = 16\rho_1^2(L(v(h^{-1})))(h) \tag{33}$$

Note that the coefficient of $\frac{\partial}{\partial t}$ in the operator $\tilde{L}$ equals unity and in view of (30) the eigenvalues of the matrix composed of the coefficients of the second derivatives with respect to space variables fall within the segment $[4^{-1}\varepsilon, 4K]$. There are two possibilities: $\kappa^2 > 4(T - t_0)$ and $\kappa^2 \leqq 4(T - t_0)$. If $\kappa^2 > 4(T - t_0)$, then by Lemma 1, we obtain

$$h(\partial_t Q_{S,3}^+) \subset \partial_t G \cap \partial_t Q \subset \Gamma \cap \partial' Q_1 \cap \partial_t Q, \tag{34}$$

$$h(\Sigma_{S,3} \setminus \{t = 0\}) \subset \partial' G \cap \partial_x Q \subset \Gamma \cap \partial' Q_1 \cap \partial_x Q, \tag{35}$$

and therefore, the function $\tilde{u}$ coincides with $\varphi_1(h)$ on $\Sigma_{S,3} \setminus \{t = 0\}$ and hence, by virtue of continuity, it is also true on $\Sigma_{S,3}$, the function $\tilde{u}$ coincides with $\varphi_2(h)$ on $\partial_t Q^+_{S,3}$, by hypothesis and in view of the continuity of $\tilde{u}$ on $\Sigma_{S,3} \cup \partial_t Q^+_{S,3}$, the function $\tilde{u}$ satisfies the Hölder condition with the exponent d and the constnat $2K_1$. Moreover,

$$|\tilde{L}\tilde{u}| \leq 4K_1(1 + |\tilde{u}_x|^2) \tag{36}$$

in $Q^+_{S,3}$. Thus, by Corollary 4.5.3 (and Theorem 4.5.2(c)), we obtain $|\tilde{u}(\tilde{z}_0) - \tilde{u}(\tilde{z}_1)| \leq N\rho^{\alpha_1}(\tilde{z}_0, \tilde{z}_1)$ for $\alpha_1 \in (0, \beta_3 \wedge \alpha]$, where

$$\beta_3 = \beta_3(d, K, \varepsilon) \in (0, 1), \ \tilde{z}_1, \tilde{z}_0 \in Q^+_{S,1}.$$

By virtue of (28) and (29), we can replace $\tilde{z}_0$ and $\tilde{z}_1$ by $h^{-1}(z_0)$ and $h^{-1}(z_1)$, i.e., we obtain (23).

In case when $\kappa^2 \leq 4(T - t_0)$, from (35) and the conditions of the lemma it follows that $\tilde{u}$ is Hölder but on $\Sigma_{S,3}$, however, we also have $t_1 - t_0 \leq \rho_1^2 \leq (900)^{-1}\kappa^2$, $4^{-1}\kappa^2 - (t_1 - t_0) > 8^{-1}\kappa^2$, so that

$$h^{-1}(z_0), h^{-1}(z_1) \in \{(t, x) : t < S - \frac{1}{128}\rho_1^{-2}\kappa^2\} \tag{37}$$

and, applying Corollary 4.5.3, we again come to (23) for $\alpha_1 \in (0, \beta_3 \wedge \alpha]$. It remains to take $\beta_1 \wedge \beta_2 \wedge \beta_3$ as α_0. This proves the lemma.

We now prove the major result of this section.

4. THEOREM. Let $D = \{x : \psi(x) > 0\}$, where $\psi \in C^2(E_d)$, $||\psi||_{C^2(E_d)} \leq K_1$ and $||\psi_x|| \geq \varepsilon_1$ on ∂D (if $\partial D \neq \emptyset$). Let $\alpha \in (0, 1)$ and let $\varphi_1, \varphi_2 \in C(\bar{Q})$ be functions such that $\varphi_{ix} \in C(\bar{Q})$, $|\varphi_i|$, $|\varphi_{ix}| \leq K_1$ on $\bar{Q}$,

$$|\varphi_i(t_1, x_1) - \varphi_i(t_2, x_1)| \leq K_1|t_1 - t_2|^{(1+\alpha)/2},$$

$$|\varphi_{ix}(z_1) - \varphi_{ix}(z_2)| \leq K_1\rho^\alpha(z_1, z_2) \tag{38}$$

for any $i = 1, 2$, $z_i = (t_i, x_i) \in Q$. Suppose that $|u| \leq K_1$ on Q_1, $u = \varphi_1$ on $\Gamma \cap \partial' Q_1 \cap \partial_x Q$, $u = \varphi_2$ on $\Gamma \cap \partial' Q_1 \cap \partial_t Q$, $u_x \in C^2_{loc}(Q_1) \cap C(\bar{Q}_1)$.

(Any two functions are assumed to coincide on the empty set.) Finally, let

$$\left|F - F_{u_{ij}} u_{x^i x^j}\right| \le K_1(1 + |u_x|^2), \tag{39}$$

$$\left|F_{u_r}\right|(1 + |u_x|) + |F_u| + \left|F_{x^r}\right|(1 + |u_x|)^{-1}$$
$$\le K_1(1 + |u_x|^2 + ||u_{xx}||), \tag{40}$$

$$\varepsilon|\lambda|^2 \le F_{u_{ij}} \lambda^i \lambda^j \le K|\lambda|^2 \tag{41}$$

with the abbreviated notation introduced in Lemma 2 for all $\lambda \in E_d$ $r = 1, \ldots, d$ on Q_1. Then $u_x \in C^{\alpha 1}(\bar{Q}_0)$, where $\alpha_1 = \alpha \wedge \alpha_0$, $\alpha_0 = \alpha_0(d, K, \varepsilon) \in (0, 1)$, and the norm of u_x in this space is estimated by a constant which depends only on d, K, K_1, ε, ε_1, κ, α.

Proof. We first estimate $|u_x|$ in Q_0. We would like to apply Lemma 2, wherein K_2, Q_1, and κ are replaced by 1, Q', and $(1/2)\kappa$ respectively and the constant K in (8) is replaced by a constant $N(K_1, K, d)$. We now check the conditions of the lemma. The fact that Q_0 lies within Q' as desired follows from Lemma 1. From (39) and (40) it follows that

$$|F| \le \left|\frac{1}{2}\left(F_{u_{ij}} + F_{u_{ji}}\right)u_{x^i x^j}\right| + K_1(1 + |u_x|^2)$$
$$\le N(K_1, K, d)(1 + |u_x|^2 + ||u_{xx}||). \tag{42}$$

Therefore, conditions (8)—(10) are also satisfied for $|u_x| \ge 1$. We now estimate $|u_x|$ on $\partial' Q' \cap \Gamma$. Denote by N various constants depending only on d, K, K_1, ε_1, κ, α, ε.

Let $z_1 = (t_1, x_1) \in \partial' Q' \cap \Gamma$. We find $z_0 = (t_0, x_0) \in Q'$ such that $t_0 < t_1$, the points z_0 and z_1 could be connected by a continuous curve of the form $(t, x(t))$ lying in Q' for $t \in [t_0, t_1)$, and $\rho(z_0, z_1) \le \rho_1 d^{-1}$, where $\rho_1 = ((1/4)\rho_0) \wedge ((d/60)\kappa)$ is taken from the previous proof with κ replaced by $\frac{1}{2}\kappa$. If $d(x_0, \partial D) \ge 3\rho_1$, then we set $G = G(z_0, 3\rho_1)$. Since $d(x_1, \partial D) \ge 2\rho_1 > 0$, we have $z_1 \notin \partial_x Q$, $t_1 = T$, $T - t_0 \le \rho_1^2$, and $G = z_0 + C_{T-t_0, 3\rho_1}$. By Lemma 1, $\partial_t G \subset \Gamma \cap \partial' Q_1$, by hypothesis, $u = \varphi_2$, $u_x = \varphi_{2x}$ on $\partial_t G$ and $|u_x(z_1)| \le K_1$.

If $d(x_0, \partial D) < 3\rho_1$, then, as in the previous proof, we find the point x_2, straighten the boundary, and, using $(1/2)\kappa$ in place of κ, construct G, $\tilde{u}$, and $\tilde{L}$ from formulas (31)—(33). Then for $t_1 = T$ we obtain $|u_x(z_1)| \leq K_1$ from the conditions of the theorem, while for $t_1 < T$ we obtain $|u_x(z_1)| \leq N$ from Corollary 4.5.6 applied to (36). It is also worth mentioning that if $\kappa^2 > 16(T - t_0)$, one should use relationships (34) and (35) and keep in mind assertion (b) of Theorem 4.5.4 while applying Corollary 4.5.6, but if $\kappa^2 \leq 16(T - t_0)$, one should use relationships (35) and (37) and keep in mind assertion (a) of Theorem 4.5.4.

In order to apply Lemma 2, we now need to estimate the Hölder constant of order $\alpha_1 = \alpha_0 \wedge \alpha$ for the function u in Q', where α_0 is taken from Lemma 3. To establish this estimate, we replace Q_0 and κ in Lemma 3 by Q' and $(1/2)\kappa$ and note that for $z_1 = (t_1, x_1)$, $z_2 = (t_2, z_2) \in Q$ we obtain

$$|\varphi_i(z_1) - \varphi_i(z_2)| \leq |\varphi_i(t_1, x_1) - \varphi_i(t_2, x_1)| + $$
$$+ |\varphi_i(t_2, x_1) - \varphi_i(t_2, x_2)|.$$

By hypothesis, the first summand does not exceed $K_1|t_1 - t_2|^{\frac{1+\alpha}{2}} \leq K_1|t_1 - t_2|^{\frac{1}{2}\alpha_1}$ if $|t_1 - t_2| \leq 1$ and, obviously, it does not exceed $2K_1|t_1 - t_2|^{\frac{1}{2}\alpha_1}$, if $|t_1 - t_2| > 1$. The second summand can be estimated using Lemma 1.1.3. Lemma 3, therefore, completes the check of the conditions of Lemma 2, whereas Lemma 2 now enables us to assert that $|u_x|$ does not exceed a constant N on Q_0 and, by virtue of continuity, on $\bar{Q}_0$.

Since the domain Q' is as good as Q_0, we have $|u_x| \leq N$ on $\bar{Q}'$. Now recall that, by Lemma 1, the conditions of the theorem under consideration are satisfied if Q_1 is replaced by Q', therefore, passing, if necessary, from Q_1 to Q', replacing κ by $(1/2)\kappa$, and increasing the constant K_1, we can adopt the justified assumption

$$|u_x| \leq K_1 \text{ in } \bar{Q}_1 \tag{43}$$

for the rest of the proof.

It remains for us to prove the inequality

$$|u_x(z_0) - u_x(z_1)| \leq N\rho^{\alpha_1}(z_0, z_1) \tag{44}$$

for $z_0, z_1 \in \bar{Q}_0$ where $\alpha_1 = \alpha_0 \wedge \alpha$, $\alpha_0 = \alpha_0(d, K, \varepsilon) \in (0, 1)$. Following the proof of Lemma 3, we can assume without loss of generality that $z_0 = (t_0, x_0)$, $z_1 = (t_1, x_1) \in Q_0$, $t_0 \leq t_1$ and (see (24))

$$d\cdot\rho(z_0, z_1) \leq \left(\frac{1}{4}\rho_0\right) \wedge \left(\frac{1}{30d}\kappa\right) =: \rho_1.$$

Denote

$$u^j = u_{x^j}, \quad Lv = v_t + a^{ij}v_{x^ix^j}, \quad g^k = -F_{u_k},$$

$$p^j = -F_u u^j - F_{x^j}. \tag{45}$$

By differentiation of (4) with respect to x^j in Q_1, we obtain

$$Lu^i - u^j_{x^k} g^k = p^j, \quad j = 1, \ldots, d. \tag{46}$$

Here

$$|p^j| \leq N(1 + (u^r_x, u^r_x))^{1/2}, \quad g^k g^k \leq N(1 + (u^r_x, u^r_x)) \tag{47}$$

in Q_1 by virtue of (40) and (43).

We now assume that $d(x_0, \partial D) \geq 3\rho_1$ and $t_0 + 9\rho_1^2 < T$. Then $G(z_0, 3\rho_1) = z_0 + C_{9\rho_1^2, 3\rho_1}$, by Lemma 1 relationships (46) and (47) hold in $G(z_0, 3\rho_1)$ and by Theorem 4.3.5, inequality (44) is true for any $\alpha_1 \in (0, \beta_1]$, where $\beta_1 = \beta_1(d, K, \varepsilon) \in (0, 1)$.

If $d(x_0, \partial D) \geq 3\rho_1$, but $t_0 + 9\rho_1^2 \geq T$, then from (27) and Theorem 4.5.2 we obtain (44) with $\alpha_1 \in (0, \beta_2 \wedge \alpha]$, where $\beta_2 = \beta_2(d, K, \varepsilon) \in (0, 1)$.

When $d(x_0, \partial D) < 3\rho_1$, we should straighten the boundary according to the adopted approach. We would like readers to focus their attention on the fact that if equation (4) is expressed in terms of $\tilde{u}$, then, generally speaking, one cannot differentiate it with respect to x since ψ, f, and $\tilde{u}$ can only be twice differentiable with respect to x. Therefore, on introducing x_2, G, S, h, and $\tilde{L}$ according to (31)—(33) as in the proof of Lemma 3, we introduce the functions $\tilde{u}^j = u^j(h)$. Then in $Q^+_{S,3}$ (46) and (47) can be converted to analogous relationships for $\tilde{u}^j$ and $\tilde{L}$ (see (36)). Moreover, inequality (36) is also true and using (43) it can be rewritten as $|\tilde{L}\tilde{u}| \leq N$.

Furthermore, let $\kappa^2 > 4(T - t_0)$. In this case, by virtue of (34),

(35) and conditions (38) of the theorem, $\tilde{u}$ conicides with $\varphi_1(h)$ on $\Sigma_{S,3}$, with $\varphi_2(h)$ on $\partial_t Q^+_{S,3}$, for $\tilde{z}_1 = (\tilde{t}_1, \tilde{x}_1)$, $\tilde{z}_2 = (\tilde{t}_2, \tilde{x}_2) \in \Sigma_{S,3}$ $i = 2, \ldots, d$, we have

$$|\tilde{u}(\tilde{t}_1, \tilde{x}_1) - \tilde{u}(\tilde{t}_2, \tilde{x}_1)| \leq N|\tilde{t}_1 - \tilde{t}_2|^{\frac{1+\alpha}{2}},$$

$$|\tilde{u}_{x^i}(\tilde{z}_1) - \tilde{u}_{x^i}(\tilde{z}_2)| \leq N\rho^{\alpha}(\tilde{z}_1, \tilde{z}_2) \tag{48}$$

and finally, the function $\tilde{u}_{x^1} = (\varphi_2(h))_{x^1}$ satisfies the Hölder condition of order α with a constant N on $\partial_t Q^+_{S,3}$. By Theorem 1.8 applied to the 'equation' $\tilde{L}\tilde{u} + p = 0$ with $p := -\tilde{L}\tilde{u}$, $|p| = |\tilde{L}\tilde{u}| \leq N$, hence for $\beta_4 = \alpha \wedge \beta_3$ with $\beta_3 = \beta_3(d, K, \varepsilon) \in (0, 1)$ we conclude that the Hölder constant of order β_4 for the function $\tilde{u}_{x^1}$ on $\Sigma_{S,2}$ does not exceed a constant N. Such estimates for $\tilde{u}_{x^i}$ with $i \geq 2$ are presented in (48).

Consequently, on $\Sigma_{S,2} \cup \partial_t Q^+_{S,2}$, the Hölder constants of order β_4 for the functions $\tilde{u}_x$ and $\tilde{u}^j$ associated with $\tilde{u}_x$ in an obvious way are estimated. By Theorem 4.5.2 and the above argument concerning $\tilde{u}^j$, $\tilde{L}$ and (46), (47), we obtain the estimate of the Hölder constant of order $\alpha_1 \in (0, \beta_4 \wedge \beta_5]$ for the functions $\tilde{u}^j$ in $Q^+_{S,1}$, where $\beta_5 = \beta_5(d, K, \varepsilon) \in (0, 1)$. Finally, from (28) and (29) we derive (44) with $\alpha_1 \in (0, \beta_4 \wedge \beta_5]$.

In the last possible case $\kappa^2 \leq 4(T - t_0)$ and (34) is no longer valid. However, (35) still holds. From this relationship we obtain (48) and Theorems 1.8 and 4.5.2 again yield the estimate for the Hölder constant of order $\alpha_1 \in (0, \beta_4 \wedge \beta_5]$ for the functions $\tilde{u}^j$ in $Q^+_{S-\nu,1}$, where $\nu = 128^{-1}\rho_1^{-2}\kappa^2$. It remains to apply (37) and set $\alpha_0 = \beta_1 \wedge \beta_2 \wedge \beta_3 \wedge \beta_5$. This proves the theorem.

5. Remark. We did not use conditions (38) for $i = 2$ prior to formula (43); they can therefore be omitted (for $i = 2$) if we are interested only in the estimate of $|u_x|$ in Q_0.

6. Remark. There certainly is a possibility of using the results of this section in the study of elliptic equations (cf. Remark 1.13).

5.3. The Estimate of the Derivative with Respect to t

Let us again consider the situation described in Section 2 preceding Lemma 2.1. In addition to the assumptions there, suppose that F is continuously differentiable with respect to (u_{ij}, u_i, u, t, x) for $(t, x) \in Q_1$, $(u_{ij}, u_i, u) \in E_{d^2+d+1}$. According to Bernstein's method, prior to estimating $|u_t|$ one should estimate $|u_x|$ and then examine the points where the auxiliary function of the form $(u_t)^2 + \tau|u_x|^2$ attains its maximum. Since the estimate of $|u_x|$ should be available in advance, the rate of growth of F and its derivatives with respect to u_x is immaterial and it is even possible to permit the quadratic growth of F_t with respect to $||u_{xx}||$. Also, note that τ is introduced in order to 'cope' with F_t, but if $F_t = 0$, we can put $\tau = 0$. Finally, in the lemma below, the function $f(t)$ is introduced to make this lemma applicable to the case when the first-order matching conditions are abandoned.

1. LEMMA. Let $u_t, u_x \in C^2_{loc}(Q_1)$, $u_x \in C(\bar{Q}_1)$, $u_t \in C(\bar{Q}_1 \setminus \partial_{tx}Q)$, $|u_x| \le K$ in Q_1, $|u_t| \le K$ on $(\Gamma \cap \partial' Q_1) \setminus \partial_{tx}Q$ and let

$$|F| + |F_{u_r}| + |F_{x^r}| \le K(1 + ||u_{xx}||),$$

$$F_u \le K(1 + ||u_{xx}||), \quad |F_t| \le K(1 + ||u_{xx}||^2) + f(t), \tag{1}$$

$$\int_0^T f(t)\,dt \le K, \quad \varepsilon|\lambda|^2 \le F_{u_{ij}}\lambda^i\lambda^j \le K|\lambda|^2$$

on Q_1, for all $\lambda \in E_d$, $r = 1, \ldots, d$, where we use the abbreviated notation introduced in Lemma 2.2, and let $f(t)$ be a nonnegative function which is bounded on $[0, T]$. Finally, let (a) $D \in C^2$ and u_{xx} be bounded in Q_1, or (b) $u_t \in C(\bar{Q}_1)$. Then $|u_t| \le N(d, K, \varepsilon, \kappa)$ in Q_0.

Proof. We fix $z_0 = (t_0, x_0) \in Q_0$, $\rho = (1/4)\kappa$ and take the function ζ from the proof of Lemma 2.2 and a number $\sigma \ge 1$ which will be defined later. Also, we set $G = G(z_0, (1/4)\kappa)$ (see Lemma 2.1),

$$\tau = \sigma(1 + \sup(\zeta^{1/2}|u_t|, G)), \quad g(t) = \int_t^T f(s)\,ds.$$

We also need the function

$$\zeta((u_t - g)_+)^2 + \tau|u_x|^2. \tag{2}$$

We first consider the case when function (2) takes on a value equal to its supremum on G at a point $z_1 = (t_1, x_1)$ with $z_1 \in \bar{G} \setminus \overline{\partial' G}$ and $u_t(z_1) > g(t_1)$.

Then the derivative with respect to t and the matrix of second derivatives with respect to x of function (2) are negative at z_1, while the first derivatives of (2) with respect to x are equal to zero. Moreover, we can differentiate equation (2.4) with respect to t and x. Taking this into account, we obtain at the point z_1

$$\frac{1}{2}\zeta_t(u_t - g)^2 + \zeta(u_t - g)(u_{tt} + f) + \tau u_{x^i} u_{x^i} \leqq 0,$$

$$\frac{1}{2}\zeta_{x^k}(u_t - g)^2 + \zeta(u_t - g)u_{tx^k} + \tau u_{x^i} u_{x^i x^k} = 0.$$

$$\Big(\zeta(u_t - g)u_{tx^k x^r} + \tau u_{x^i} u_{x^k x^k x^r} + \zeta u_{tx^k} u_{tx^r} + \tau u_{x^i x^k} u_{x^i x^r} +$$

$$+ (u_t - g)\zeta_{x^r} u_{tx^k} + (u_t - g)\zeta_{x^k} u_{tx^r} +$$

$$+ \frac{1}{2}\zeta_{x^k x^r}(u_t - g)^2\Big) \leqq 0,$$

$$0 = u_{tx^i} + F_{u_{kr}} u_{x^k x^r x^i} + F_{u_k} u_{x^k x^i} + F_u u_{x^i} + F_{x^i},$$

$$0 = u_{tt} + F_{u_{kr}} u_{tx^k x^r} + F_{u_k} u_{tx^k} + F_u u_t + F_t. \tag{3}$$

Multiplying the two latter equations by τu_{x^i}, $\zeta(u_t - g)$, adding together the results, and using the previous relationships, we obtain, similarly to inequality (2.20),

$$0 \leqq -\varepsilon\zeta|u_{tx}|^2 - \varepsilon\tau||u_{xx}||^2 + h, \tag{4}$$

where

$$h = h_1 + \ldots + h_6,$$

$$h_1 = -\frac{1}{2}(u_t - g)^2(\zeta_t + a^{kr}\zeta_{x^k x^r}), \quad h_2 = -\frac{1}{2} F_{u_k}\zeta_{x^k}(u_t - g)^2,$$

$$h_3 = -2a^{kr}\zeta_{x^r}u_{tx^k}(u_t - g), \qquad h_4 = F_{x^i}\tau u_{x^i},$$

$$h_5 = F_u(\zeta u_t(u_t - g) + \tau|u_x|^2), \quad h_6 = \zeta(u_t - g)(F_t - f),$$

$$a^{kr} = \frac{1}{2}\left(F_{u_{kr}} + F_{u_{rk}}\right).$$

In the proof of the estimates of h_i and elsewhere, we shall denote by N (with and without subscripts) the constants depending only on d, K, ε, κ (but independent of τ and σ). From (2.4) and (1) it follows that

$$u_t = -F \leqq K(1 + ||u_{xx}||), \quad (u_t - g)^2 \leqq N(1 + ||u_{xx}||^2). \tag{5}$$

Moreover, note that $|\zeta_x|u_t \leqq N\zeta^{1/2}u_t \leqq N\sigma^{-1}\tau$. Using these inequalities, assumptions (1) and the fact that $u_t(z_1) - g(t_1) > 0$, we can easily obtain at the point z_1

$$|h_1| \leqq N(1 + ||u_{xx}||^2),$$

$$|h_2| \leqq |F_{u_k}\zeta_{x^k}|(u_t - g)N(1 + ||u_{xx}||) \leqq \sigma^{-1}\tau N(1 + ||u_{xx}||^2),$$

$$|h_3| \leqq N\zeta^{1/2}|u_{tx}|(u_t - g) \leqq \varepsilon\zeta|u_{tx}|^2 + N(1 + ||u_{xx}||^2),$$

$$|h_4| \leqq N\tau(1 + ||u_{xx}||),$$

$$h_5 \leqq K(1 + ||u_{xx}||)(\sigma^{-1}\tau(u_t - g) + \tau K^2)$$

$$\leqq \sigma^{-1}\tau(1 + ||u_{xx}||^2) + N\tau(1 + ||u_{xx}||),$$

$$h_6 \leqq \zeta(u_t - g)K(1 + ||u_{xx}||^2) \leqq \sigma^{-1}\tau K(1 + ||u_{xx}||^2).$$

Combining these inequalities with (4), dividing by τ, and taking note that $1 \leqq \sigma^{-1}\tau$, we conclude that

$$0 \leqq -(\varepsilon - N_1\sigma^{-1})||u_{xx}||^2 + N_2||u_{xx}|| + N_3.$$

For $\sigma = 2N_1\varepsilon^{-1}$, it follows that $||u_{xx}(z_1)|| \leqq N$, and (5) implies that

$$\sup_G[\zeta((u_t - g)_+)^2 + \tau|u_x|^2] = \zeta(z_1)[(u_t(z_1) - g(t_1))_+]^2 +$$

$$+ \tau|u_x(z_1)|^2 \leqq N + N\tau. \tag{6}$$

We have derived (6) in the case when $z_1 \in \bar{G} \setminus \overline{\partial' G}$, $u_t(z_1) > g(t_1)$. The relationships (6) are clear if $z_1 \in \bar{G} \setminus \overline{\partial' G}$, $u_t(z_1) \leq g(t_1)$. Let us show that the latter inequality in (6) is valid for all $z_1 \in \partial' G \setminus \partial_{tx} Q$ rather than for the z_1 associated with the supremum of function (2) over G. If $z_1 \in (\partial' G \setminus \partial_{tx} Q) \setminus (z_0 + C_{\rho^2,\rho})$, then $\zeta(z_1) = 0$ and this inequality is clear. But if $z_1 \in (\partial' G \setminus \partial_{tx} Q) \setminus Q$, then following the proof of Lemma 2.2 we obtain $z_1 \in (\Gamma \cap \partial' Q_1) \setminus \partial_{tx} Q$ and the inequality in (6) follows from the conditions of the lemma.

Recalling the definition of z_1, we draw the conclusion from the first equation in (6) that the relationships (6) are true for $z_1 \in \overline{\partial' G} \setminus \partial_{tx} Q$ and if u_t is continuous in $\bar{Q}_1$, then they are also true for $z_1 \in \overline{\partial' G}$. Note that if u_t is continuous in $\bar{Q}_1$, there always exists $z_1 \in \bar{G}$ such that the first equality in (6) holds and in this case, by examining all possible positions of z_1 on $\bar{G}$, we have proved

$$\sup_G [\zeta((u_t - g)_+)^2 + \tau |u_x|^2] \leq N + N\tau. \tag{7}$$

If u_t is continuous only in $\bar{Q}_1 \setminus \partial_{tx} Q$ and there still exists a point $z_1 \in \bar{G} \setminus \partial_{tx} Q$ such that the first equality in (6) holds true, the derivation of inequality (7) is also valid.

It remains for us to examine the latter case when u_t is continuous only in $\bar{Q}_1 \setminus \partial_{tx} Q$ and there exists no point $z_1 \in \bar{G} \setminus \partial_{tx} Q$ such that the first equality in (6) is valid. In this case, by hypothesis, u_{xx} is bounded in Q_1 and every convergent sequence $z_n = (t_n, x_n) \in G$ which provides a maximum of function (2) in G converges to a point $\tilde{z} = (T, \tilde{x}) \in \partial_{tx} Q$. It is clear that $\rho(z_0, z_n), \rho(z_0, \tilde{z}) \leq (1/2)\kappa$.

Since the boundary of D is smooth, it follows that for sufficiently small $\eta, \gamma > 0$ and large n the inequalities $t_n > T - \eta > t_0$, $|\tilde{x} - x_n| < \gamma$ are true and every point of the domain $G_{\eta,\gamma} := (\tilde{z} - C_{\eta,\gamma}) \cap Q$ can be connected with a z_n by a curve of the form $(t, x(t))$ which lies entirely in $G_{\eta,\gamma}$. If $\gamma + \eta^{1/2} \leq (1/4)\kappa$, then all points of such curves lie in Q_1 since $z_n \in G \subset Q_1$ and otherwise there would be points of $\partial' Q_1 \setminus \Gamma$ on them, whose parabolic distance to z_0 is no greater than $(3/4)\kappa$. Thus, $G_{\eta,\gamma} \subset Q_1$. In addition, the parabolic distance from z_0 to the points of $\partial' G_{\eta,\gamma} \cap \partial' Q$ is less than $(3/4)\kappa$. Therefore, $\partial' G_{\eta,\gamma} \cap \partial' Q$, which lies on $\partial' Q_1$ in view of $G_{\eta,\gamma} \subset Q_1$, also lies on $\Gamma \cap \partial' Q_1$

$$\partial' G_{\eta,\gamma} \cap \partial' Q \subset \Gamma \cap \partial' Q_1. \tag{8}$$

We now apply Lemma 1.1.6 to the function u_t in the domain $G_{\eta,\gamma}$ regarding this function as a solution to the equation (cf. (3))

$$v_t + F_{u_{kr}} v_{x^k x^r} + F_{u_k} v_{x^k} + F_u v + F_t = 0,$$

whose coefficients except F_u are bounded (in Q_1), while F_u is bounded from above (u_{xx} is bounded in Q_1). By this lemma in view of (8) and the assumption that $|u_t| \leq K$ on $(\Gamma \cap \partial' Q_1) \setminus \partial_{tx} Q$, we ultimately conclude that

$$\overline{\lim_{n\to\infty}}\, u_t(z_n) \leq K,$$

$$\sup_G [\zeta((u_t - g)_+)^2 + \tau|u_x|^2]$$

$$= \lim_{n\to\infty} [\zeta((u_t - g)_+)^2 + \tau|u_x|^2](z_n) \leq K^2 + K^2\tau.$$

We have thus proved inequality (7) in all cases. We now omit $\tau|u_x|^2$ in it and take advantage of the fact that $(a-b)_+ \geq a_+ - |b|$. Then from (7) it follows that $\zeta((u_t)_+)^2 \leq N + N\tau$ on G. The latter inequality (with the same constants N, σ, τ) remains valid if u is replaced by (-u) in it. In order to show this, it suffices to note that the function (-u) satisfies the obvious equation which is similar to (2.4). Hence, $\zeta((u_t)_-)^2 \leq N + N\tau$, $\zeta(u_t)^2 \leq N + N\tau$ which, by the definition of τ, yields $(t-\sigma)^2 \leq N + N\tau$, $\tau \leq N$, and $\zeta(u_t)^2 \leq N + N\tau \leq N$ in G and the latter implies that $|u_t(z_0)| \leq N$ at the point z_0 in view of continuity. Since z_0 is an arbitrary point of Q_0, the lemma is proved.

In the theorem to come this lemma is not exploited in all its strength. As we have mentioned above, the lemma as it is formulated will be needed later.

2. THEOREM. Let the conditions of Theorem 2.4 hold, let $u_t \in C^2_{loc}(Q_1) \cap C(\bar{Q}_1)$, let $|F_t|$ after substituting the arguments $(u_{x^i x^j}(z), u_{x^i}(z), u(z), z)$ do not exceed $M(u(z), u_x(z))(1 + \|u_{xx}(z)\|)$ on Q_1 with $M(u, u_i)$ being a continuous function of real variables $u, u_1, \ldots, u_d$. Also, let

$$\|u_t\|_{C^{\alpha}(\Gamma \cap \partial' Q_1 \cap \partial_t Q)} + \|u_t\|_{C^{\alpha}(\Gamma \cap \partial' Q_1 \cap \partial_x Q)} \leqq K_1. \tag{9}$$

Then the norm of u_t in $C^{\alpha_1}(\bar{Q}_0)$ with $\alpha_1 = \alpha_0 \wedge \alpha$, $\alpha_0 = \alpha_0(d, K, \varepsilon) \in (0, 1)$ is estimated by a constant depending only on d, K, K_1, ε, ε_1, κ, α and the function M.

Proof. Consider the auxiliary domain Q' from Lemma 2.1. This domain can be used as Q_0 in Theorem 2.4 (if κ is replaced by $(1/2)\kappa$) and using it one can conclude that $|u_x| \leqq N$ in Q', where (and in what follows) we denote by N various constants depending only on d, K, K_1, ε, ε_1, κ, α, M. We now take Q' as Q_1 in Lemma 1 and note that the conditions of this lemma are satisfied for $f = 0$ and a constant N in place of K on Q' in view of (2.39) – (2.42) and the above estimate of $|u_x|$ on Q'. By Lemma 1, we obtain $|u_t| \leqq N$ in Q_0.

Let us go a little further using these arguments. We have proved that $|u_t| \leqq N$ in Q_0, but we can always take Q' in place of Q_0. Therefore $|u_t| \leqq N$ in Q'. Moreover, by Theorem 2.4, in $\bar{Q}'$ there is an estimate not only for $|u_x|$, but also for the norm of u_x in C^{β_1}, where $\beta_1 = \alpha \wedge \beta_0$, $\beta_0 = \beta_0(d, K, \varepsilon) \in (0, 1)$. Taking, if necessary, Q' in place of Q_1 and increasing the constant K_1, we can assume without loss of generality that

$$\|u_t\|_{C(Q_1)}, \quad \|u_x\|_{C^{\beta_1}(\bar{Q}_1)} \leqq K_1. \tag{10}$$

We now estimate the Hölder constant for u_t in Q_0. By differentiation of (2.4) with respect to x^j and t, introducing notation (2.45) for $j \geqq 1$, and setting $u^0 = u_t$, $p^0 = -F_u u_t - F_t$, we obtain (2.46) and (2.47) for all $j = 0, 1, \ldots, d$. Then, by almost repeating the arguments following (2.47), we obtain the estimate of the Hölder constant simultaneously for u_t and u_x. Of course, there is certain simplification in repeating the arguments since we need not estimate in advance the Hölder constants of the functions $\tilde{u}^j := u^j(h)$ on the appropriate faces of $Q^+_{S,3}$ because these estimates are given in (9) and (10). The theorem is proved.

5.4. Estimation of the Solution in the Norms of C^2 and $W_2^{1,2}$

Let us continue the study of the subjects introduced in Section 2 prior to Lemma 2.1. Throughout this section, we shall assume that for any

$t \in (0, T)$ the function F is twice continuously differentiable with respect to (u_{ij}, u_i, u, x) on $E_{d^2+d+1} \times \{x : (t, x) \in Q_1\}$. The purpose of this section is estimating u_{xx} in the norm of C. Other derivatives of u involved in the norm of C^2 have already been estimated. Here we also estimate the norm of u in $W_2^{1,2}$. It will be used in the proof of the solvability of nonlinear equations when F cannot be differentiated with respect to t.

The estimate of u_{xx} in C is developed in two stages. First, by differentiation of (2.4), we estimate $u_{(\xi)(\xi)}$ from above and then using equation (2.4) itself and the estimates of $u_{(\xi)(\xi)}$ from above we deduce their estimates from below. We have already encountered this plan in Section 2.2. The technique of developing an upper estimate of $u_{(\xi)(\xi)}$ can be demonstrated using equation (2.1) as an example. Suppose that F is concave in (u_{ij}, u_i, u), $F_u \leqq 0$. Then by differentiating (2.1) twice, we obtain

$$
\begin{aligned}
0 = {} & F_{u_{ij}}(u_{(\xi)(\xi)})_{x^ix^j} + F_{u_i}(u_{(\xi)(\xi)})_{x^i} + F_u u_{(\xi)(\xi)} + \\
& + F_{u_{ij}u_{rs}}(u_{(\xi)})_{x^ix^j}(u_{(\xi)})_{x^rx^s} + 2F_{u_{ij}u_r}(u_{(\xi)})_{x^ix^j}(u_{(\xi)})_{x^r} + \\
& + 2F_{u_{ij}u}(u_{(\xi)})_{x^ix^j}u_{(\xi)} + F_{u_iu_j}(u_{(\xi)})_{x^i}(u_{(\xi)})_{x^j} + \\
& + 2F_{u_iu}(u_{(\xi)})_{x^i}u_{(\xi)} + F_{uu}(u_{(\xi)})^2 \\
\leqq {} & F_{u_{ij}}(u_{(\xi)(\xi)})_{x^ix^j} + F_{u_i}(u_{(\xi)(\xi)})_{x^i} + F_u u_{(\xi)(\xi)} .
\end{aligned}
$$

The latter expression is, thus, nonnegative and, by the maximum principle, to estimate $u_{(\xi)(\xi)}$ from above on the interior of the domain it suffices to estimate $u_{(\xi)(\xi)}$ from above on the boundary. In the general case, we do not assume that F is concave in u_i and u and permit its dependence upon t and x. The extra difficulties which arise in this case can be surmounted by using the monitored maximum method introduced in Section 2.

Recall that $v_{(\xi)} := \xi^i v_{x^i}$ and $v_{(\xi)(\xi)} := \xi^i\xi^j v_{x^ix^j}$ are called the first and the second derivative of v along the direction ξ. Similar notation is used not only for functions v defined on domains of E_d, but also for functions of the form $v(t, x)$ and $F(u_{ij}, u_i, u, t, x)$ which will be differentiated not only with respect to x, but also with respect to (u_{ij}, u_i, u, x) along some vectors $(\tilde{u}_{ij}, \tilde{u}_i, \tilde{u}, \tilde{x}) \in E_{d^2+2d+1}$. Note that if $\eta = (\tilde{u}_{ij}, \tilde{u}_i, \tilde{u}, \tilde{x})$, then

$$F_{(\eta)(\eta)} = F_{u_{ij}u_{rs}}\tilde{u}_{ij}\tilde{u}_{rs} + 2F_{u_{ij}u_r}\tilde{u}_{ij}\tilde{u}_r + 2F_{u_{ij}u}\tilde{u}_{ij}\tilde{u} +$$
$$+ 2F_{u_{ij}x^r}\tilde{u}_{ij}\tilde{x}^r + F_{u_iu_j}\tilde{u}_i\tilde{u}_j + 2F_{u_iu}\tilde{u}_i\tilde{u} + \tag{1}$$
$$+ 2F_{u_ix^j}\tilde{u}_i\tilde{x}^j + F_{uu}(\tilde{u})^2 + 2F_{ux^i}\tilde{u}\tilde{x}^i + F_{x^ix^j}\tilde{x}^i\tilde{x}^j.$$

1. THEOREM. Let $\alpha \in (0, 1)$, let $|u_x|$ and Hölder constants of order α for the functions u_{x^i} on Q_1 do not exceed K, let $u_x, u_{xx} \in C^2_{loc}(Q_1) \cap C(\bar{Q}_1 \setminus \partial_{tx}Q)$, and let $u_{(\xi)(\xi)} \leq K$ on $(\Gamma \cap \partial' Q_1) \setminus \partial_{tx}Q$ for all unit $\xi \in E_d$. Also, let for any $z \in Q_1$, $\lambda \in E_d$, $k = 1, \ldots, d$, and $\eta = (\tilde{u}_{ij}, \tilde{u}_i, \tilde{u}, \tilde{x})$ with $\tilde{u}_{ij} = \tilde{u}_{ji}$ the inequalities

$$\varepsilon|\lambda|^2 \leq F_{u_{ij}}\lambda^i\lambda^j \leq K|\lambda|^2, \tag{2}$$

$$(1 + ||u_{xx}||)\left|F_{u_k}\right| + |F_u| + \left|F_{x^k}\right| \leq K(1 + ||u_{xx}||^2), \tag{3}$$

$$K^{-1}F_{(\eta)(\eta)} \leq \sum_{i,j}|\tilde{u}_{ij}|\left[\sum_i|\tilde{u}_i| + (1 + ||u_{xx}||)^{1/2}(|\tilde{u}| + |x|)\right] +$$
$$+ \sum_i|\tilde{u}_i|^2(1 + ||u_{xx}||) +$$
$$+ (1 + ||u_{xx}||^2)(|\tilde{u}|^2 + |\tilde{x}|^2), \tag{4}$$

hold, where for the sake of simplicity the arguments $(u_{x^ix^j}, u_{x^i}, u, z)$ of the derivatives of F and the arguments z of u and its derivatives are omitted. Finally, let $u_{xx} \in C(\bar{Q}_1)$ or $D \in C^2$. Then $u_{(\xi)(\xi)} \leq N(\varepsilon, K, d, \kappa, \alpha)$ in Q_0 for all unit $\xi \in E_d$.

Preparatory to proving this theorem, we will first discuss its assumptions and do some preliminary work. Note that condition (4) calls for only an upper estimate of $F_{(\eta)(\eta)}$. The form of this estimate and formula (1) immediately imply that $F_{u_{ij}u_{rs}}\tilde{u}_{ij}\tilde{u}_{rs} \leq 0$. If is inequality were satisfied for all arguments of F rather than $(u_{x^ix^j}, u_{x^i}, u, z)$, it would follow that F is concave in (u_{ij}).

Furthermore, if $F = a^{ij}(u_k, u, z)u_{ij} + a(u_k, u, z)$, then it can be easily seen that conditions (3) and (4) are satisfied with a K whenever a^{ij}, a, and their derivatives with respect to (u_k, u, x) up to the second order are bounded. In the latter case $F_{(\eta)(\eta)}$ in (4), can be replaced by $|F_{(\eta)(\eta)}|$, and if u is replaced by (-u) and F by $(-F(-u_{ij}, -u_i, -u, z))$, then the theorem yields a lower estimate for $u_{(\xi)(\xi)}$. In particular, in this case we obtain an estimate of $|u_{xx}|$ in Q_0 from the estimate of $|u_{xx}|$ on $\Gamma \cap \partial' Q_1$.

It is convenient to introduce some notation here. In what follows we will use u_{ijrs}, u_{ijr}, u_{ij}, u_i and u and formulate some assertions concerning these objects. Only those u_{ijrs}, u_{ijr}, and u_{ij} which are invariant under any permutation of the subscripts will be considered. A statement involving these functions is said to be true for the solution u if this statement is true after the substitution

$$(u_{ijrs}, u_{ijr}, u_{ij}, u_i, u)$$

$$= (u_{x^ix^jx^rx^s}(z), u_{x^ix^jx^r}(z), u_{x^ix^j}(z), u_{x^i}(z), u(z)).$$

Furthermore, we fix a $z_0 = (t_0, x_0) \in Q_0$ and take a constant $\rho \in (0,(1/2)\kappa]$ and a nonnegative function $\zeta \in C^2(z_0 + C_{\rho^2,\rho})$ equal to zero on $z_0 + \partial' C_{\rho^2,\rho}$ and such that $\zeta \leqq 1$, $\zeta(z_0) = 1$, and

$$|\zeta_x|^2 \leqq N\rho^{-2}\zeta, \quad ||\zeta_{xx}|| \leqq N\rho^{-2}, \quad |\zeta_t| \leqq N\rho^{-2}$$

in $z_0 + C_{\rho^2,\rho}$, where $N = N(d)$ (cf. (2.11)). Also, denote $p_0 = u_x(z_0)$, take a number $\tau > 0$, consider the domain $G = G(z_0, \rho)$ from Lemma 2.1, and denote by $z_1 = (t_1, x_1)$ the point of $\bar{G} \setminus \partial_{tx}Q$ where the function

$$\zeta^2\Big[\max_{|\xi|=1}(u_{(\xi)(\xi)})_+\Big]^2 + \tau|u_x - p_0|^2 \tag{5}$$

attains its supremum on $\bar{G} \setminus \partial_{tx}Q$ if such a point exists. The form of function (5) is discussed in Remark 7.2.5.

We also denote

$$\lambda_1 = \max_{|\xi|=1}(u_{(\xi)(\xi)}(z_1))_+,$$

and let ξ_1 be a vector such that $|\xi_1|^2 = \lambda_1\zeta(z_1)$, $\lambda_1\xi_1^i = u_{x^ix^j}(z_1)\xi_1^j$. An appropriate vector ξ_1 always exists since if $\lambda_1\zeta(z_1) = 0$, then we can take $\xi_1 = 0$, and if $\lambda_1\zeta(z_1) \neq 0$, then $\lambda_1 > 0$, λ_1 is the greatest eigenvalue of the matrix $u_{xx}(z_1)$ and the associated eigenvector of appropriate length can be taken as ξ_1. Finally, let $A^2u_{ij} = u_{ijrs}\xi_1^r\xi_1^s$, $A^2u_i = u_{irs}\xi_1^r\xi_1^s$, $A^2u = u_{rs}\xi_1^r\xi_1^s$, $Au_{ij} = u_{ijr}\xi_1^r$, $Au_i = u_{ir}\xi_1^r$, and $Au = u_r\xi_1^r$.

2. LEMMA. (a) If $z_1 \subset \bar{G} \setminus \overline{\partial' G}$, then

$$\Big[\zeta A^2u_{ij} + \tau(u_r - p_0^r)u_{rij} + \tau u_{ri}u_{rj} + \zeta_{x^ix^j}A^2u + \zeta_{x^j}A^2u_i + \\ + \zeta_{x^i}A^2u_j + 2v_\gamma^{rs}\Big(\zeta Au_{ri} + \zeta_{x^i}Au_r\Big)\Big(\zeta Au_{sj} + \zeta_{x^j}Au_s\Big)\Big] \leqq 0, \tag{6}$$

for the solution u and any $\gamma > 0$ at the point z_1, where (v_γ^{rs}) is the inverse matrix of $(v_{rs} + \gamma\delta_{rs})$ and $v_{rs} := |\xi_1|^2\delta_{rs} + 2\xi_1^r\xi_1^s - \zeta u_{rs}$

(b) If $z_1 \in \bar{G} \setminus \overline{\partial' G}$, then

$$0 \leqq F_{u_{ij}}(\zeta A^2u_{ij} + \tau(u_r - p_0^r)u_{rij}) - A^2uF_{u_i}\zeta_{x^i} + \\ + F_u(\zeta A^2u + \tau|u_x - p_0|^2) + \tau(u_r - p_0^r)\Big(F_{x^r} + p_0^rF_u\Big) + \\ + \zeta F_{(\eta_1)(\eta_1)} - \zeta_t A^2u, \tag{7}$$

$$\zeta A^2u_i + \zeta_{x^i}A^2u + \tau u_{ir}(u_r - p_0^r) = 0, \tag{8}$$

for the solution u at the point z_1, where $F_{(\eta_1)(\eta_1)}$ is defined in (1) for $\eta_1 = (Au_{ij}, Au_i, Au, \xi_1)$.

Proof. (a) For any number a, we obtain

$$\zeta^2(a_+)^2 = \max(2\zeta aq^2 - q^4 : q \geqq 0). \tag{9}$$

Therefore, taking $q\xi$ as ξ, we can rewrite function (5) as

$$\max(2\zeta u_{(\xi)(\xi)} - |\xi|^4 + \tau|u_x - p_0|^2 : \ \xi \in E_d). \tag{10}$$

In (9) the maximum is attained at $q^2 = \zeta a_+$, in (10) for a fixed z the maximum is attained at every vector $\xi \in E_d$ satisfying the conditions

$$|\xi|^2 = \zeta(z)\lambda(z), \quad \lambda(z)\xi^i = u_{x^ix^j}(z)\xi^j,$$

where $\lambda(z) = \max[(u_{(\mu)(\mu)}(z))_+ : \mu \in E_d, |\mu| = 1]$. Comparing these formulas with the formulas defining λ_1 and ξ_1, we conclude that at the point (z_1, ξ_1) the auxiliary function $2\zeta u_{(\xi)(\xi)} - |\xi|^4 + \tau|u_x - p_0|^2$ attains its maximum on $(\bar{G} \setminus \partial_{tx}Q) \times E_d$.

Since $z_1 \in \bar{G} \setminus \overline{\partial' G}$, we obtain $z_1 + \bar{C}_{\sigma^2,\sigma} \subset Q$ for a sufficiently small $\sigma > 0$ and at the point z_1 the function u, by hypothesis, has four derivatives with respect to x. Therefore, the first derivatives with respect to x of the auxiliary function equal zero at the point (z_1, ξ_1), while the matrix of the second derivatives with respect to (x, ξ) is negative at this point. In other words, at the point z condition (8) holds true for the solution u, and for any $\tilde{x}, \tilde{\xi} \in E_d$ we obtain

$$\Big(\zeta A^2 u_{ij} + 2\zeta_{x^i} A^2 u_j + \zeta_{x^ix^j} A^2 u + \tau u_{ri} u_{rj} +$$

$$+ \tau u_{rij}(u_r - p_0^r)\Big)\tilde{x}^i\tilde{x}^j + 4\Big(\zeta_{x^i} A u_j + \zeta A u_{ij}\Big)\tilde{x}^i\tilde{\xi}^j -$$

$$- 2v_{ij}\tilde{\xi}^i\tilde{\xi}^j \leq 0. \tag{11}$$

Note that

$$v_{ij}\tilde{\xi}^i\tilde{\xi}^j \geq |\tilde{\xi}_1|^2|\xi|^2 - \zeta u_{ij}\tilde{\xi}^i\tilde{\xi}^j = \zeta(\lambda_1|\tilde{\xi}|^2 - u_{ij}\tilde{\xi}^i\tilde{\xi}^j) \geq 0.$$

Hence, $(v_{ij}) \geq 0$, while $(v_{ij} + \gamma\delta_{ij}) > 0$ and the inverse of the latter matrix (i.e., (v_γ^{ij})) is defined. It is obvious that if v_{ij} in (11) is replaced by $v_{ij} + \gamma\delta_{ij}$, then the inequality is only strengthened. Let us make this substitution and take the supremum of the left-hand side of (11) with respect to $\tilde{\xi}$ (or simply put $\tilde{\xi}^i = v_\gamma^{ij}(\zeta_{x^r} A u_j + \zeta A u_{rj})\tilde{x}^r$). We then obtain that the quadratic form associated with the left matrix in (6) is negative, which proves (6).

(b) At the point z_1 we can differentiate (2.4) twice with respect to x. The derivative of the function $2\zeta u_{(\xi)(\xi)} - |\xi|^4 + \tau|u_x - p_0|^2$ with respect to t is negative at the point (z_1, ξ_1) $(z_1 + \bar{C}_{\sigma^2,\sigma} \subset Q_1$ for small $\sigma > 0)$. Taking this into account, we obtain, for the solution u at the point z_1,

$$\zeta_t u_{(\xi_1)(\xi_1)} + \zeta u_{(\xi_1)(\xi_1)} + \tau u_{tx^r} u_{x^r} - p_0^r) \leq 0, \tag{12}$$

$$0 = u_{tx^r} + F_{u_{ij}} u_{ijr} + F_{u_i} u_{ir} + F_u u_r + F_{x^r}, \tag{13}$$

$$0 = u_{(\xi_1)(\xi_1)t} + F_{u_{ij}} A^2 u_{ij} + F_{u_i} A^2 u_i + F_u A^2 u + F_{(\eta_1)(\eta_1)}. \tag{14}$$

Let us multiply (13) by $\tau(u_r - p_0^r)$, sum over r, and add the resulting equality to (14) multiplied by ζ. Moreover, we use (12) and the above equality (8). We thus obtain (7). The lemma is proved.

3. LEMMA. There exists a uniformly parabolic operator

$$L = \frac{\partial}{\partial t} + a^{ij} \frac{\partial^2}{\partial x^i \partial x^j} + \sigma^{ij} \frac{\partial^2}{\partial x^i \partial \xi^j} + $$

$$+ \sigma_0 \delta^{ij} \frac{\partial^2}{\partial \xi^i \partial \xi^j} + b^i \frac{\partial}{\partial x^i} + c$$

on $Q_1 \times S_2$ such that the coefficients a^{ij}, σ^{ij}, σ_0, b^i and c are bounded and Lv is bounded from below, where $v(t, x, \zeta) = u_{(\xi)(\xi)}(t, x)$.

<u>Proof</u>. We denote $a^{ij} = (1/2)\left(F_{u_{ij}} + F_{u_{ji}}\right)$, $b^i = F_{u_i}$, $c = F_u$. Like (14),

$$0 = v_t + a^{ij} v_{x^i x^j} + b^i v_{x^i} + cv + F_{(\eta)(\eta)},$$

where $\eta = \left(u_{x^i x^j (\xi)}, u_{x^i (\xi)}, u_{(\xi)}, \xi\right)$ and in view of (4) and the boundedness condition for u_x, u_{xx} in Q_1 with $|\xi| \leq 2$, we obtain

$$F_{(\eta)(\eta)} \leq N_1 + N_1 \sum_{i,j} |v_{x^i \xi_j}|,$$

where N_1 is a constant. We put

$$\sigma^{ij} = N_1 \operatorname{sgn} v_{x^i \xi^j}.$$

Then $\sigma^{ij} v_{x^i\xi^i} = N_1 \Sigma |v_{x^i\xi^j}|$ and, to prove the lemma, it remains for us to choose the constants σ_0 so that the operator L is uniformly parabolic. The possibility of such a choice obviously follows from inequality (2). The lemma is proved.

4. Proof of Theorem 1. Above we took an arbitrary $\rho \in (0,(1/2)\kappa]$, now $\rho \in (0,(1/4)\kappa]$. Let $\delta \in (0, 1)$ and the exact values of ρ and δ will be obtained after developing some estimates wherein we denote by N various constants depending only on d, K, ε until these constants are allowed to explicitly depend on α and κ.

To begin with we assume that $(2\rho)^{2\alpha}K^2 \leqq (2d)^{-1}\delta$. According to our suppositions, $|u_x(z) - p_0|^2 = |u_x(z) - u_x(z_0)|^2 \leqq (1/2)\delta$ on G and, following the proof of Lemma 2.2, we set

$$M(\tau) = \sup\Big[\zeta^2(z)\Big(\max_{|\xi|=1}(u_{(\xi)(\xi)}(z))_+\Big)^2 + \\ + \tau|u_x(z) - p_0|^2 : z \in G\Big], \tag{15}$$

which yields $\tau \geqq 0$ as a solution of the equation $M(\tau) = \delta\tau$. Assume that the discussion preceding Lemma 2 involves exactly this τ.

Also, suppose that the supremum in (15) is attained at a point $z_1 \in \bar{G} \setminus \partial_{tx}Q$ (it is not necessary that $z_1 \in G$). Then

$$\zeta^2(z_1)\lambda_1^2 + \tau|u_x(z_1) - p_0|^2 = M(\tau) = \delta\tau. \tag{16}$$

Furthermore, consider the case when $z_1 \in \bar{G} \setminus \overline{\partial' G}$, $\zeta(z_1) > 0$. Estimating from above the first summand on the right of (7) by use of (6) and setting $a^{ij} = (1/2)(F_{u_{ij}} + F_{u_{ji}})$, we obtain

$$0 \leqq -a^{ij}\Big[\tau u_{ri}u_{rj} + 2v_\zeta^{rs}\Big(\zeta A u_{ri} + \zeta_{x^i}Au_r\Big)\Big(\zeta A u_{sj} + \zeta_{x^j}Au_s\Big) + \\ + 2\zeta_{x^j}A^2u_i + \zeta_{x^ix^j}A^2u\Big] - A^2uF_{u_i}\zeta_{x^i} + \\ + F_u(\zeta A^2u + \tau|u_x - p_0|^2) + \zeta F_{(\eta_1)(\eta_1)} + \\ + \tau(u_r - p_0^r)\Big(F_{x^r} + p_0^r F_u\Big) - \zeta_t A^2 u \tag{17}$$

at the point z_1 for the solution u.

We need some relationships between the objects involved. It is obvious that $A^2u = \lambda_1|\xi_1|^2 = \lambda_1^2\zeta$, $\zeta A^2u = \lambda_1^2\zeta^2$, therefore

$$\zeta A^2u + \tau|u_x - p_0|^2 = \delta\tau, \quad |A^2u| \leqq \zeta^{-1}\delta\tau, \quad \lambda_1 \leqq \zeta^{-1}(\delta\tau)^{1/2},$$

$$|\xi_1| = \lambda_1^{1/2}\zeta^{1/2} \leqq (\delta\tau)^{1/4}, \quad |Au_i| = \lambda_1|\xi_1^i| \leqq \zeta^{-1}(\delta\tau)^{3/4},$$

$$|Au| \leqq K|\xi_1| \leqq K(\delta\tau)^{1/4}. \tag{18}$$

To estimate A^2u_i, we take advantage of (8). We obtain

$$|A^2u_i| \leq N\rho^{-1}\zeta^{-3/2}\delta\tau + N\zeta^{-1}\tau||u_{xx}||.$$

On the other hand, it follows from (16) and the inequality $|u_x(z_1) - p_0|^2 \leqq (1/2)\delta$, that $2\zeta^2\lambda_1^2 \geqq \delta\tau$. Moreover, λ_1 is zero or the greatest eigenvalue of the matrix $(u_{xx}(z_1))$. Its square is no greater than the sum of squares of the eigenvalues, i.e.,

$$||u_{xx}||^2 \geqq \lambda_1 \geqq \frac{1}{2}\zeta^{-2}\delta\tau.$$

Thus, from (2) it follows for (17) that

$$\tau a^{ij}u_{ri}u_{rj} \geqq \varepsilon\tau||u_{xx}||^2 \geqq \frac{1}{4}\varepsilon\zeta^{-2}\delta\tau^2 + \frac{1}{2}\varepsilon\tau||u_{xx}||^2,$$

$$2|a^{ij}\zeta_{x^j}A^2u_i| \leqq N\rho^{-2}\zeta^{-1}\tau + \frac{1}{16}\varepsilon\tau||u_{xx}||^2 =: B,$$

$$|a^{ij}\zeta_{x^ix^j}A^2u| \leqq N\rho^{-2}\zeta^{-1}\tau, \quad |\zeta_tA^2u| \leqq N\rho^{-2}\zeta^{-1}\tau.$$

To estimate other summands of (17), we make use of supposition (3) and the fact that $|u_r - p_0^r| \leqq K\rho^{\alpha}2^{\alpha}$. Then

$$|A^2uF_{u_i}\zeta_{x^i}| \leqq N\zeta^{-1/2}\rho^{-1}\tau(1 + ||u_{xx}||) \leqq B,$$

$$F_u(\zeta A^2u + \tau|u_x - p_0|^2) = \delta\tau F_u \leqq K\delta\tau + K\delta\tau||u_{xx}||^2,$$

$$\tau(u_r - p_0^r)\left(F_{x^r} + p_0^rF_u\right) \leqq N\tau + N_1\rho^{\alpha}\tau||u_{xx}||^2.$$

If choosing δ and ρ we provide the condition

$$K\delta + N_1\rho^{\alpha} \leqq \frac{1}{8}\varepsilon, \tag{19}$$

then (17) would imply

$$0 \leqq -\frac{1}{4}\varepsilon\zeta^{-2}\delta\tau^2 - \frac{1}{4}\tau\varepsilon||u_{xx}||^2 - 2\varepsilon v_\zeta^{rs}\left(\zeta Au_{ri} + \zeta_{x^i}Au_r\right) \times$$
$$\times\left(\zeta Au_{si} + \zeta_{x^i}Au_s\right) + \zeta F_{(\eta_1)(\eta_1)} + N\rho^{-2}\zeta^{-1}\tau. \tag{20}$$

Of course, we assume that inequality (19) holds true. The appropriate choice of δ and ρ could be done at the outset of the proof if N_1 were found explicitly, but this presents no difficulties. We can now discuss some consequences of inequality (20). Let us estimate $\zeta F_{(\eta_1)(\eta_1)}$ from above. Recall that $\eta_1 = (Au_{ij}, Au_i, Au, \xi_1)$, some components of the vector η_1 were estimated in (18), and we have inequality (4). This yields the following inequalities:

$$\zeta F_{(\eta_1)(\eta_1)} \leqq N\sum_{i,r}|\zeta Au_{ir} + \zeta_{x^i}Au_r - \zeta_{x^i}Au_r| \times$$
$$\times [\zeta^{-1}(\delta\tau)^{3/4} + (\delta\tau)^{1/4}(1 + ||u_{xx}||)^{1/2}] +$$
$$+ N\zeta^{-1}(\delta\tau)^{3/2}(1 + ||u_{xx}||) +$$
$$+ \zeta(\delta\tau)^{1/2}N(1 + ||u_{xx}||^2)$$
$$\leqq N[\zeta^{-1}(\delta\tau)^{3/4} + (\delta\tau)^{1/4}(1 + ||u_{xx}||^{1/2})]\sum_{i,r}|\zeta Au_{ir} +$$
$$+ \zeta_{x^i}Au_r| + N\rho^{-1}\zeta^{-3/2}(\delta\tau)^{3/2} +$$
$$+ N\zeta^{-1/2}\rho^{-1}\delta\tau + N\zeta^{-1/2}\rho^{-1}\delta\tau||u_{xx}||^{1/2} +$$
$$+ N\zeta^{-1}(\delta\tau)^{3/2}||u_{xx}|| + N\zeta(\delta\tau)^{1/2} +$$
$$+ N\zeta(\delta\tau)^{1/2}||u_{xx}||^2. \tag{21}$$

All summands of the latter sum except the first can be estimated by the expression

$$N\zeta^{-2}(\rho^{-4} + (\delta\tau)^2) + N(1 + \delta\tau)||u_{xx}||^2.$$

This fact easily follows from Young's inequality. To estimate the first summand on the right of (21), note that for any $\mu \in E_d$

$$|\mu| \leq (v_\zeta^{rs}\mu^r\mu^s)^{1/2}(\mathrm{tr}(v_{rs} + \delta_{rs}\zeta))^{1/2},$$

and $\mathrm{tr}(v_{rs} + \delta_{rs}\zeta) = \zeta d + (d+2)|\zeta_1|^2 - \zeta \mathrm{tr}(u_{rs}) \leq N(\zeta + (\delta\tau)^{1/2} + \zeta||u_{xx}||)$. Specifically, the first summand on the right of (21) does not exceed

$$N\left[v_\zeta^{rs}\left(\zeta Au_{ir} + \zeta_{x^i}Au_r\right)\left(\zeta Au_{is} + \zeta_{x^i}Au_s\right)\right]^{1/2}(\zeta + (\delta\tau)^{1/2} +$$

$$+ \zeta||u_{xx}||)^{1/2} \quad (\zeta^{-1}(\delta\tau)^{3/4} + (\delta\tau)^{1/4}(1 + ||u_{xx}||^{1/2}))$$

$$\leq 2\varepsilon v_\zeta^{rs}\left(\zeta Au_{ir} + \zeta_{x^i}Au_r\right)\left(\zeta Au_{is} + \zeta_{x^i}Au_s\right) +$$

$$+ N\zeta^{-2}(1 + (\delta\tau)^2) + N(1 + \delta\tau)||u_{xx}||^2.$$

To derive the latter inequality, we again apply Young's inequality. Summarizing our discussion of (21) and returning to (20), we conclude

$$0 \leq \left(N_2\delta^2 - \frac{1}{4}\varepsilon\delta\right)\tau^2 +$$

$$+ \left[N_2 + \tau\left(N_2\delta - \frac{1}{4}\varepsilon\right)\right]\zeta^2||u_{xx}||^2 + N_2\rho^{-2}\tau + N_2\rho^{-4}. \qquad (22)$$

We now finally set δ and ρ, which could be chosen from the outset if we found the constants N_1 and N_2 (depending only on d, K, ε) in advance. Let δ and ρ satisfy inequality (19), let $(2\rho)^{2\alpha} \cdot K^2 \leq (2d)^{-1}\delta$ let $\delta \in (0, 1)$ and $\rho \in (0, (1/4)\kappa]$, and moreover let $N_2\delta \leq 8^{-1}\varepsilon$. Let us choose the appropriate δ and ρ and fix them till the end of the proof. We can obviously assume that δ and ρ depend only on d, K, ε, α, κ. In what follows, the symbol N will now denote a constant which depends only on d, K, ε, α, κ. It follows from (22) that

$$0 \leq -\tau^2 + N\tau + N + N\left[N_2 - \frac{1}{8}\varepsilon\tau\right]\zeta^2||u_{xx}||^2 \qquad (23)$$

at the point z_1.

Here two cases are possible. If $8^{-1}\varepsilon\tau \geq N_2$, then the latter summand in (23) can be omitted and we obtain $\tau \leq N$. But if $8^{-1}\varepsilon\tau \leq N_2$, then $\tau \leq 8\varepsilon^{-1}N_2$.

Thus, in the case when $z_1 \in \bar{G} \setminus \overline{\partial' G}$, $\zeta(z_1) > 0$, we have proved that $\tau \leq N$, hence $M(\tau) \leq N$ by virtue of (16), and $u_{(\xi)(\xi)}(z_0) \leq N$ for any unit $\xi \in E_d$ by definition (15) and the continuity of u_{xx} at z_0. If $\zeta(z_1) = 0$, then $\tau = M(\tau) = 0$, since from (16) and the inequality $(2\rho)^{2\alpha} \cdot K^2 \leq (2d)^{-1}\delta$ it follows that $(1/2)\delta\tau \geq \delta\tau$. In this case $u_{(\xi)(\xi)}(z_0) \leq 0$ for all $\xi \in E_d$.

We now consider the case when $z_1 \in \overline{\partial' G} \setminus \partial_{tx}Q$. Note that if $z \in \partial' G \setminus \partial_{tx}Q$, then either $z \notin z_0 + C_{\rho^2,\rho}$, or $z \notin Q_1$. If $z \notin Q_1$, then $z \in \partial' Q_1$ and $z \in \Gamma$, since, if otherwise, the parabolic distance from z_0 to $\partial' Q_1 \setminus \Gamma$ would be less than $\rho(z_0, z) \leq 2\rho < \kappa$. Therefore, for $z \notin Q_1$, we obtain $z \in \Gamma \cap \partial' Q_1 \setminus \partial_{tx}Q$ and, by hypothesis,

$$\zeta(z)u_{(\xi)(\xi)}(z) \leq K \tag{24}$$

for $|\xi| = 1$. Inequality (24) is also true for $z \notin z_0 + C_{\rho^2,\rho}$, since $\zeta(z) = 0$ for such a z. Inequality (24) established for $z \in \partial' G \setminus \partial_{tx}Q$ is also true for $z \in \overline{\partial' G} \setminus \partial_{tx}Q$ in view of continuity. This together with (16) and the condition $(2\rho)^{2\alpha}K^2 \leq (2d)^{-1}\delta$ for $z_1 \in \overline{\partial' G} \setminus \partial_{tx}Q$ yield

$$K^2 + \frac{1}{2}\delta\tau \geq \delta\tau, \quad \delta\tau \leq 2K^2, \quad M(\tau) \leq 2K^2, \quad u_{(\xi)(\xi)}(z_0) \leq 2K$$

for all unit vectors $\xi \in E_d$.

It remains for us to discuss the case when there is no appropriate point $z_1 \in \bar{G} \setminus \partial_{tx}Q$. Then any convergent sequence $z_n \in G$ which provides the maximum to (15) converges to a point $\tilde{z} \in \partial_{tx}Q$. Following the proof of Lemma 3.1, note that for small $\eta, \gamma > 0$, $D \in C^2$, we have $G_{\eta,\gamma} := (\tilde{z} - C_{\eta,\gamma}) \cap Q \subset Q_1$, and $\partial' G_{\eta,\gamma} \cap \partial' Q \subset \Gamma \cap \partial' Q_1$. Furthermore, we apply Lemma 3 and Lemma 1.1.6 to $G_{\eta,\gamma} \times S_2$. Then it can be seen that

$$\overline{\lim_{n\to\infty}}\, u_{(\xi)(\xi)}(z_n) \leq \sup[u_{(\lambda)(\lambda)}(z) : |\lambda| \leq 2,$$

$$z \in (\Gamma \cap \partial' Q_1) \setminus \partial_{tx}Q] \leq 4K$$

for all unit vectors $\xi \in E_d$, if $D \in C^2$ or, obviously, (after replacing z_n by their projections on $\partial_t Q$) if $u_{xx} \in C(\bar{Q}_1)$,

$$\delta\tau = M(\tau) \leqq \frac{1}{2}\,\delta\tau + \overline{\lim_{n\to\infty}}\, \zeta^2(z_n) \left(\max_{|\xi|=1} (u_{(\xi)(\xi)}(z_n))_+\right)^2$$

$$\leqq \frac{1}{2}\,\delta\tau + 16K^2, \quad \delta\tau \leqq 32K^2, \quad M(\tau) \leqq 32K^2, \quad u_{(\xi)(\xi)}(z_0) \leqq 6K.$$

Thus, the estimate $u_{(\xi)(\xi)}(z_0) \leqq N$ for any unit $\xi \in E_d$ is valid in every case, and since z_0 is an arbitrary point of Q_0, the theorem is proved.

5. Remark. We maintain all the conditions of Theorem 1 except for (4), where $(1 + ||u_{xx}||)^{1/2}$ and $||u_{xx}||^2$ are repalced by $1 + ||u_{xx}||$ and $||u_{xx}||^3$, respectively. We also require that the inequalities

$$u_t,\ F - F_{u_{ij}} u_{x^i x^j} \leqq K.$$

hold on Q_1. Then the assertion of Theorem 1 remains true.

Let us show this by repeating the previous proof. The new conditions turn out to enable us to rigidly relate τ and $||u_{xx}||$. On the one hand, we obtain from (16) $2\delta\tau \geqq 2\zeta^2\lambda_1^2 \geqq \delta\tau$ as above, moreover, $||u_{xx}|| \geqq \lambda_1$. On the other hand, $u_{xx} - \lambda_1(\delta_{ij}) \leqq 0$ and

$$u_t + a^{ij}u_{x^i x^j} \geqq -K, \quad a^{ij}\left(u_{x^i x^j} - \lambda_1\delta_{ij}\right) \geqq -2K - \lambda_1 Kd,$$

$$\varepsilon u_{(\xi)(\xi)} \geqq -2K - \lambda_1 Kd, \quad u_{(\xi)(\xi)} \leqq \lambda_1$$

for any unit vector $\xi \in E_d$. Hence, $||u_{xx}||^2 \leqq N(d,\ K,\ \varepsilon) \quad (1 + \lambda_1^2)$, which ultimately yields $2\zeta^2||u_{xx}||^2 \geqq \delta\tau \geqq \gamma\zeta^2||u_{xx}||^2 - \zeta^2$, where $\gamma = \gamma(d,\ K,\ \varepsilon) > 0$.

Furthermore, when proving the theorem we did not use (4) before formula (21) appeared. We now apply the estiamte $\delta\tau \leqq 2\zeta^2||u_{xx}||^2$ in order to estimate $\zeta F_{(\eta_1)(\eta_1)}$ from above under the new assumptions. Then (cf. (21)) we obtain

$$\zeta F_{(\eta_1)(\eta_1)} \leqq \zeta^{1/2}(1 + ||u_{xx}||^{3/2}) \sum_{i,r} |\zeta A u_{ir} +$$

$$+ \zeta_{x^i} A u_r| + N\rho^{-4} + N\zeta^2||u_{xx}||^4.$$

The first summand is estimated as before, while in (20) we also use the fact that $\tau \geqq \delta^{-1}\gamma\zeta^2||u_{xx}||^2 - \delta^{-1}\zeta$. From (20) it now follows that

$$0 \leqq -\varepsilon\delta^{-1}\zeta^2||u_{xx}||^4 + N\delta^{-1}\zeta||u_{xx}||^2 + N\rho^{-4} + N_2\zeta^2||u_{xx}||^4.$$

For $\varepsilon\delta^{-1} \geqq N_2 + 1$ we obtain $N \geqq 2\zeta^2||u_{xx}||^2 \geqq \delta\tau$ and the proof of Theorem 1 can be completed by the above arguments.

We now formulate the conditions which will be needed in the rest of this section. Of course, we assume that the conditions preceding Lemma 2.1 hold and suppose that F is twice continuously differentiable with respect to (u_{ij}, u_i, u, x). Moreover, let the following inequalities hold on Q_1:

$$\varepsilon|\lambda|^2 \leqq F_{u_{ij}}\lambda^i\lambda^j \leqq K|\lambda|^2, \tag{25}$$

$$|F - F_{u_{ij}}u_{x^ix^j}| \leqq K_1(1 + |u_x|^2), \tag{26}$$

$$|F_{u_r}|(1 + |u_x|) + |F_u| + |F_{x^r}|(1 + |u_x|)^{-1}$$
$$\leqq K_1(1 + |u_x|^2 + ||u_{xx}||), \tag{27}$$

$$\begin{aligned} M^{-1}(u, u_x)F_{(\eta)(\eta)} \leqq & \sum_{i,j}|\tilde{u}_{ij}|\Big[\sum_i|\tilde{u}_i| + \\ & + (1 + ||u_{xx}||)^{\frac{1+\theta}{2}}(|\tilde{u}| + |\tilde{x}|)\Big] + \\ & + \sum_i|\tilde{u}_i|^2(1 + ||u_{xx}||) + \\ & + (1 + ||u_{xx}||^{2+\theta})(|\tilde{u}|^2 + |\tilde{x}|^2), \end{aligned} \tag{28}$$

with the conventional abbreviated notation and for all $\lambda \in E_d$, $r = 1, \ldots, d$, $\eta = (\tilde{u}_{ij}, \tilde{u}_i, \tilde{u}, \tilde{x})$ such that $\tilde{u}_{ij} = \tilde{u}_{ji}$, where $M(u, u_i)$ is a continuous function defined for all real-valued $u, u_1, \ldots, u_d$; $M > 0$; $\theta = 0$ or 1.

6. THEOREM. Let Q_1 be a bounded domain, let $\Gamma = \emptyset$, let $|u| \leqq K_1$ in Q_1, let $u_x, u_{xx} \in C^2_{loc}(Q_1)$, and let $\theta = 0$. Then the norms of the functions u_t and u_{xx} in $L_2(Q_0)$ are bounded by a constant depending only on d, K,

ε, K_1, κ, the domains Q_0, Q_1, and the function M. Moreover, $u_{(\xi)(\xi)} \leq N(d, K, \varepsilon, K_1, \kappa, M)$ on Q_0 for all unit vectors $\xi \in E_d$. Finally, there exists an $\alpha = \alpha(d, K, \varepsilon) \in (0, 1)$ such that the norms of u and u_x in $C^{\alpha}(Q_0)$ are bounded by a constant $N(d, K, \varepsilon, K_1, \kappa)$.

Proof. The assertion concerning the estimate of the norm of u in $C^{\alpha}(Q_0)$ immediately follows from Corollary 4.3.6 and conditions (25) and (26). Furthermore, note that $z_0 + \bar{C}_{\rho^2,\rho} \subset Q_1$, if $z_0 \in Q_0$ and $\rho < (1/2)\kappa$. Therefore, the estimate of the norm of u_x in $C^{\alpha}(Q_0)$ results from Theorem 2.4 which is applied with $z_0 + C_{4\rho^2,2\rho}$ and $z_0 + C_{\rho^2,\rho}$ instead of $Q = Q_1$ and Q_0, respectively, where $\rho = (1/8)\kappa$ and z_0 is an arbitrary point of Q_0. On obtaining the estimates of the norms of u and u_x in $C^{\alpha}(Q_0)$, one can obviously prove the estimates of the norms of u and u_x in $C^{\alpha}(Q')$ where Q' is the domain in Lemma 2.1. Passing, if necessary, from κ and Q_1 to $(1/2)\kappa$ and Q', we assume that the norms of u and u_x in $C^{\alpha}(Q_1)$ are bounded by an available constant.

Then conditions (27) and (28) ensure conditions (3) and (4) with a constant N in place of K (we do not specially emphasize the dependence of N on d, K, ε, K_1, κ, M). Taking the domains $z_0 + C_{4\rho^2,2\rho}$ and $z_0 + C_{\rho^2,\rho}$ once again, we apply Theorem 1 and obtain $u_{(\xi)(\xi)}(z) \leq N$ for $z \in Q_0$ and $|\xi| = 1$. Of course, such an estimate exists in Q' also and substituting Q' for Q_1 we assume that it also exists in Q_1.

It remains for us to prove the first assertion of the theorem. It is obvious that it suffices to estimate the norms of u_t and u_{xx} in $\mathcal{L}_2(z_0 + C_{\rho^2,\rho})$ for any $z_0 \in Q_0$ and $\rho = (1/8)\kappa$. We fix a $z_0 \in Q_0$ and put $Q_2 = z_0 + C_{\rho^2,\rho}$, $Q_3 = z_0 + C_{4\rho^2,2\rho}$, $a^{ij} = (1/2)(F_{u_{ij}} + F_{u_{ji}})$. Since $\bar{Q}_3 \subset Q_1$, we have $u_{(\xi)(\xi)} \leq N|\xi|^2$ on Q_3 and $u - N|x|^2$ is concave in x on Q_3. The properties of symmetric nonnegative matrices imply that

$$K\Delta(u - N|x|^2) \leq a^{ij}(u - N|x|^2)_{x^ix^j} \leq \varepsilon\Delta(u - N|x|^2) \leq 0, \tag{29}$$

$$K\Delta u \leq a^{ij}u_{x^ix^j} + N \leq N \tag{30}$$

on Q_3.

Moreover, from (2.4) and (26) it follows that

$$-N \leqq u_t + a^{ij}u_{x^ix^j} \leqq N \tag{31}$$

on Q_3. This along with (30) yields

$$N - K\Delta u \geqq u_t \geqq -N_1, \quad N_2 - K\Delta v \geqq v_t \geqq 0 \tag{32}$$

on Q_3, where $v = u + N_1 t$. Furthermore, let $\psi \in C^2(Q_3)$, $1 \geqq \psi \geqq 0$, $\psi_t = 0$ in Q_3, $\psi = 1$ on Q_2, and $\psi = 0$ near $\partial_x Q_3$. Denoting $z_0 = (t_0, x_0)$ and $t_1 = t_0 + 4\rho^2$ we obtain from (32)

$$\int_{Q_2} v_t^2 \, dt\, dx \leqq \int_{Q_3} \psi v_t^2 \, dt\, dx \leqq N_2 \int_{Q_3} \psi v_t \, dt\, dx -$$

$$- K \int_{Q_3} \psi v_t \Delta v \, dt\, dx =$$

$$= N_2 \int_{x_0+S_{2\rho}} \psi v \, dx \Big|_{t=t_0}^{t_1} + K \int_{Q_3} \psi_{x^i} v_{x^i} v_t \, dt\, dx +$$

$$+ \frac{1}{2} K \int_{Q_3} \psi \frac{\partial}{\partial t} |v_x|^2 \, dt\, dx.$$

By virtue of (32), $v_t \geqq 0$ and

$$\int_{Q_3} \psi_{x^i} v_{x^i} v_t \, dt\, dx \leqq N \int_{Q_3} |v_t| \, dt\, dx = N \int_{x_0+S_{2\rho}} v \, dx \Big|_{t=t_0}^{t_1},$$

$$\int_{Q_3} \psi \frac{\partial}{\partial t} |v_x|^2 \, dt\, dx = \int_{x_0+S_{2\rho}} \psi |v_x|^2 \, dx \Big|_{t=t_0}^{t_1}.$$

These calculations show that the first assertion of the theorem concerning u_t is valid. At the same time we have obtained an estimate of the integrals of ψv_t^2 and ψu_t^2 over Q_3. Moreover, in view of (31), we obtain $|a^{ij}u_{x^ix^j}| \leqq |u_t| + N$, while (29) implies that $|\Delta u| \leqq N(1 + |u_t|)$ on Q_3. Thus,

$$\|\Delta(\psi u)\|_{L_2(Q_3)} \leqq \|\psi \Delta u\|_{L_2(Q_3)} + N$$

$$\leqq N\|\psi u_t\|_{L_2(Q_3)} + N \leqq N.$$

It remains for us to note that

$$\sum_{i,j}\int_{Q_2}\left(u_{x^ix^j}\right)^2 dt\,dx \le \sum_{i,j}\int_{Q_3}\left[(\psi u)_{x^ix^j}\right]^2 dt\,dx$$
$$= ||\Delta(\psi u)||^2_{L_2(Q_3)}.$$

The theorem is proved.

This theorem essentially completes our preparation for deriving the solvability in $W_2^{1,2}$ of nonlinear equations when F is only measurable in t by using the solvability in $C^{2+\alpha}$ for functions F which are smooth in t. In order to study the solvability in $C^{2+\alpha}$ we also need some other estimates.

7. THEOREM. Let $\alpha \in (0, 1)$, $D = \{x : \psi(x) > 0\}$, where $\psi \in C^{2+\alpha}(E_d)$, $||\psi||_{C^{2+\alpha}(E_d)} \le K_1$, and $|\psi_x| \ge \varepsilon_1$ on ∂D (if $\partial D \ne \emptyset$). Also, let the conditions preceding Theorem 6 hold for $\theta = 1$, let F be continuously differentiable with respect to (u_{ij}, u_i, u, t, x) for $(t, x) \in Q_1$, and let $|F_t|$ after substituting the arguments $(u_{x^ix^j}(z), u_{x^i}(z), u(z), z)$ do not exceed

$$f(t) + M(u, u_x)(1 + ||u_{xx}||^2),$$

on Q_1, where f(t) is a nonnegative bounded function on [0, T] such that $\int_0^T f(t)\,dt \le K$. Also, let $\varphi_1 \in C^{2+\alpha}(\bar{Q})$ and $\varphi_2 \in C^2(\partial_t Q)$ be functions such that

$$||\varphi_1||_{C^{2+\alpha}(Q)}, \quad ||\varphi_2||_{C^2(\partial_t Q)} \le K_1. \tag{33}$$

Finally, suppose that $u_x, u_{xx}, u_t \in C^2_{loc}(Q_1) \cap C(\bar{Q}_1 \setminus \partial_{tx}Q)$, $u_x \in C(\bar{Q}_1)$, $|u| \le K_1$ on Q_1, $u = \varphi_1$ on $\Gamma \cap \partial' Q_1 \cap \partial_x Q$, and let $u = \varphi_2$ on $\Gamma \cap \partial' Q_1 \cap \partial_t Q$. (Any two functions are assumed to coincide on the empty set.)

Then

$$||u||_{C^2(Q_0)} \le N(d, K, \varepsilon, K_1, \varepsilon_1, \alpha, \kappa, M). \tag{34}$$

Proof. We denote by N a constant like that involved in (34). We take the domain Q' from Lemma 2.1 and apply Theorem 2.4 to Q' rather than Q_0. One can assume that φ_2 does not depend on t and take advantage of Lemmas 1.1.3 and 1.1.4 to check conditions (2.38). By Theorem 2.4, the norm of u_x in $C^{\beta}(\overline{Q'})$ is bounded by a constant N for a $\beta = \beta(d, K, \varepsilon) \in (0, 1)$.

Furthermore, we estimate $|u_t|$ in Q_0 using Lemma 3.1, with Q' is taken as Q_1. Conditions (3.1) hold (in Q') with a constant N instead of K by virtue of (25)—(27), the above estimate of $|u_x|$ in Q', and the assumptions of this theorem concerning F_t. To prove the inequality $|u_t| \leq N$ on $(\Gamma \cap \partial' Q') \setminus \partial_{tx} Q$, we fix a point $z_1 = (t_1, x_1) \in (\Gamma \cap \partial' Q') \setminus \partial_{tx} Q$ and find a point $z_0 = (t_0, x_0) \in Q'$ such that $t_0 < t_1$, the points z_0 and z_1 can be connected by a continuous curve of the form $(t, x(t))$ lying in Q' with $t \in [t_0, t_1)$, and $\rho(z_0, z_1) \leq d^{-1}\rho_1$, where $\rho_1 = ((1/4)\rho_0) \wedge ((1/60)d\kappa)$ (cf. the beginning of the proof of Theorem 2.4). If $d(x_0, \partial D) \geq 3\rho_1$, then we take $G = G(z_0, 3\rho_1)$ of Lemma 2.1. Since $d(x_1, \partial D) \geq 2\rho_1 > 0$, then $z_1 \notin \partial_x Q$, $t_1 = T$, $G = z_0 + C_{T-t_0, 3\rho_1}$, and by Lemma 2.1, we obtain $\partial_t G \subset \partial' Q_1 \cap \Gamma \cap \partial_t Q$, $u = \varphi_2$ on $\partial_t G$. Moreover, from (2.4) and (26) it follows that $|u_t| \leq N(1 + |u_x|^2 + \|u_{xx}\|)$ in Q. Since $G \subset Q_1$ and $z_1 \in \partial_t G \setminus \partial_{tx} Q$, this inequality is true for the point z_1 (and its neighbourhood), where it means that

$$|u_t(z_1)| \leq N(1 + |\varphi_{2x}|^2 + \|\varphi_{2xx}\|) \leq N.$$

If $d(x_0, \partial D) < 3\rho_1$, then following the proof of Lemma 2.3, we find the point x_2 and straighten the boundary to determine G, S, and h from formulas (2.31) and (2.32), wherein we replace κ by $(1/2)\kappa$ in order to use these formulas for the point $z_0 \in Q'$. Then the point $h^{-1}(z_1)$ lies on $\Sigma_{S,3}$ or on $(\partial_t Q^+_{S,3}) \setminus \{x^1 = 0\}$, and in both cases the images of these points produced by the mapping h lie in $\Gamma \cap \partial' Q'$ (see (2.34) and (2.35)). It remains to take note that the equation $u(h) = \varphi_1(h)$ on $\Sigma_{S,3}$ can be differentiated with respect to t, and if $h^{-1}(z_1)$ lies on $(\partial_t Q^+_{S,3}) \setminus \{x^1 = 0\}$, then the estimate of $|u_t(z_1)|$ follows from (2.4), (26), and (33), just as above. Thus, $|u_t| \leq N$ on $(\Gamma \cap \partial' Q') \setminus \partial_{tx} Q$ and by Lemma 3.1, we obtain $|u_t| \leq N$ in Q_0.

The above estimates of u_x and u_t in the domain Q_0 are valid in Q' if different constants are chosen. As it has already been explained, passing from Q_1 and κ to Q' and $(1/2)\kappa$ we can and will assume that the norms of u_x in $C^{\beta}(Q_1)$, u_t in $C(Q_1)$ are bounded by constants N.

We further assume that we have managed to prove the inequality

$$||u_{xx}|| \leqq N \tag{34}$$

on $(\Gamma \cap \partial'Q') \setminus \partial_{tx}Q$.

In this case, we apply Remark 5 and Theorem 1, where Q' is taken as Q_1. Their conditions obviously hold by assumptions (25)—(28) and the estimate of u_x in $C^{\beta}(Q')$. Of course, the constant K involved in the formulation of Theorem 1 and Remark 5 should be replaced by a constant N. By Theorem 1, we obtain $u_{(\xi)(\xi)} \leqq N|\xi|^2$ in Q_0.

We derive a lower estimate for $u_{(\xi)(\xi)}$. As usual, let $a^{ij} = (1/2)(F_{u_{ij}} + F_{u_{ji}})$. From (2.4) and (26) we obtain $a^{ij}u_{x^ix^j} \geqq -N$ on Q_0. Since $u - N|x|^2$ is concave, it follows from (29) that $\Delta u \geqq -N$, i.e., $u_{x^1x^1} + \ldots + u_{x^dx^d} \geqq -N$ on Q_0. By what we have proved, $u_{x^2x^2}$, $\ldots$, $u_{x^dx^d} \leqq N$, therefore, $u_{x^1x^1} \geqq -N$ on Q_0. If the axis directed along a unit vector $\xi \in E_d$ is taken as the x^1-axis, we also obtain $u_{(\xi)(\xi)} \geqq -N$ on Q_0.

Ultimately, we have $|u_{(\xi)(\xi)}| \leqq N$ for $|\xi| = 1$ and $||u_{xx}|| \leqq N$ on Q_0. Thus, it remains for us to prove (35) on $(\Gamma \cap \partial'Q') \setminus \partial_{tx}Q$. An argument very similar to the one used in developing the estimate of $|u_t|$ on $(\Gamma \cap \partial'Q') \setminus \partial_{tx}Q$ readily enables us to derive (35) on $(\Gamma \cap \partial'Q' \cap \partial_t Q) \setminus \partial_{tx}Q$ from (33) and the equation $u = \varphi_2$ on $(\Gamma \cap \partial'Q_1) \cap \partial_t Q$. It therefore suffices to prove (35) on $\Gamma \cap \partial'Q' \cap \partial_x Q$.

Let $z_1 \in \Gamma \cap \partial'Q' \cap \partial_x Q$ and $z_1 = (t_1, x_1)$. We straighten the part of $\partial_x Q$ which is adjacent to z_1 using Lemma 1.1.2. In order to make use of available results, we define, as above, $z_0 = (t_0, x_0) \in Q'$ and ρ_1 and assume that $d(x_0, \partial D) \leqq 3\rho_1$. Furthermore, we find x_2 and assume that $(1/2)\kappa$ is taken in (2.28)—(2.33) in place of κ. Let g be the inverse of h defined on $h(\bar{Q}^+_{S,3})$. Note that here h and g are functions of the class $C^{2+\alpha}$ with respect to x in their domains of definition and their third derivatives with respect to x admit the established estimate (see Lemma 1.1.2).

Furthermore, for $z \in Q^+_{S,3}$ and real-valued u_{rs}, u_r, u we put

$$\tilde{F}(u_{rs}, u_r, u, z) = 16\rho_1^2 F\Big(u_{rs} g^r_{x^i}(h(z)) g^s_{x^j}(h(z)) + $$
$$+ u_r g^r_{x^i x^j}(h(z)), u_r g^r_{x^i}(h(z)), u, h(z)\Big),$$

$$\tilde{u} = u(h). \tag{36}$$

It is not difficult to check that (by (2.4) and the fact that $G \subset Q_1$)

$$\tilde{u}_t(z) + \tilde{F}\Big(\tilde{u}_{x^r x^s}(z), \tilde{u}_{x^r}(z), \tilde{u}(z), z\Big) = 0 \tag{37}$$

on $Q^+_{S,3}$.

We want to apply Theorem 1.9 to (37), checking its conditions and omitting clear values of the arguments of various functions. Condition 1.2 for $\tilde{F}$ in (36) holds since g is thrice continuously differentiable for $x^1 > 0$, while F and h are continuously differentiable with respect to the corresponding arguments. It is clear that

$$\tilde{F}_{u_{rs}} = 16\rho_1^2 F_{u_{ij}} g^r_{x^i} g^s_{x^j} \tag{38}$$

and from (25) and (2.30) it follows that condition 1.3 holds with $(1/4)\varepsilon$ and 4K instead of ε and K. Furthermore,

$$\tilde{F}_{x^k} = 16\rho_1^2 a^{ij}\Big(2\tilde{u}_{x^r x^s} g^r_{x^i x^p} h^p_{x^k} g^s_{x^j} + \tilde{u}_{x^r} g^r_{x^i x^j x^p} h^p_{x^k}\Big) +$$
$$+ 16\rho_1^2 F_{u_i} \tilde{u}_{x^r} g^r_{x^i x^p} h^p_{x^k} + 16\rho_1^2 F_{x^p} h^p_{x^k}. \tag{39}$$

$$\tilde{F}_{u_r} = 16\rho_1^2 a^{ij} g^r_{x^i x^j} + 16\rho_1^2 F_{u_i} g^r_{x^i}.$$

From (25) it follows that $|a^{ij}| \leqq K$. Moreover, $G \subset Q_1$ and, as we assumed above, the estimates $|u_x|$, $|u_t| \leqq N$ are valid on Q_1. Therefore, $|\tilde{u}_x|$, $|\tilde{u}_t| \leqq N$ in $Q^+_{S,3}$. Hence, we obtain from (26) and (27) that condition 1.4 with $\alpha = 1$ and condition 1.5 hold for $\tilde{u}$, $\tilde{F}$, if K_1 is replaced with another constant N. Finally, the inequality in (1.10) follows from (2.35) and (33).

When $16(T - t_0) \geqq \kappa^2$, we have $S = (256)^{-1}\rho_1^{-2}\kappa^2 \geqq 14$, while $t_1 - t_0 \leqq \rho_1^2$, so that $\tilde{z}_1 := h^{-1}(z_1) \in \Sigma_{S-1,1}$ and $||\tilde{u}_{xx}(\tilde{z}_1)|| \leqq N$ by Theorem 1.9. But if $\kappa^2 > 16(T - t_0)$, then from (2.34) and (33) we obtain $\tilde{u} = \varphi_2(h)$ on

$\partial_t Q^+_{S,3}$, condition (1.11) holds, and $||\tilde{u}_{xx}(\tilde{z}_1)|| \leqq N$ again by virtue of Theorem 1.9. In both cases this inequality and the estimates of $|u_x|$ lead to the inequality $||u_{xx}(z_1)|| \leqq N$. The point z_1 is arbitrary on $\Gamma \cap \partial' Q' \cap \partial_x Q$, which completes the proof.

8. Remark. The results of this section can easily be extended to elliptic equations. It suffices formally to add $u_t = 0$. Of course, it would be pointless to extend Theorem 6 to the elliptic case since its elliptic counterpart is much weaker than the respective counterpart of Theorem 7.

5.5. The Estimates of u in the Norm of $C^{2+\alpha}$

In this section, we basically complete the preparation of the proof of the solvability in $C^{2+\alpha}$ of nonlear equations defined by smooth functions F.

Given constants $K \geqq \varepsilon > 0$ and a domain $Q \subset E_{d+1}$, we adopt the following definition.

1. Definition. For all $(t, x) \in Q$ and real-valued u_{ij}, u_i $(i, j = 1, \ldots, d)$, and u, let a real-valued Borel function $F(u_{ij}, u_i, u, t, x)$ be defined. We shall denote $F \in \mathcal{F}_1(\varepsilon, K, Q)$ if for every t the function F is twice continuously differentiable with respect to $(u_{ij}, u_i, u, x) \in E_{d^2+d+1} \times \{x \in E_d : (t, x) \in Q\}$ and, for all $(t, x) \in Q$, λ, $\tilde{x} \in E_d$, $r = 1, \ldots, d$ and real-valued $u_{ij} = u_{ji}$, $\tilde{u}_{ij} = \tilde{u}_{ji}$, u_i, $\tilde{u}_i$, u, and $\tilde{u}$, the inequalities

$$\varepsilon|\lambda|^2 \leqq F_{u_{ij}} \lambda^i\lambda^j \leqq K|\lambda|^2, \tag{1}$$

$$|F - F_{u_{ij}} u_{ij}| \leqq M_1^F(u)\Big(1 + \sum_i (u_i)^2\Big), \tag{2}$$

$$\begin{aligned} &|F_{u_r}|\Big(1 + \sum_i |u_i|\Big) + |F_u| + |F_{x^r}|\Big(1 + \sum_i |u_i|\Big)^{-1} \\ &\qquad \leqq M_1^F(u)\Big(1 + \sum_i |u_i|^2 + \sum_{i,j} |u_{ij}|\Big), \end{aligned} \tag{3}$$

$$\begin{aligned} [M_2^F(u, u_k)]^{-1} F_{(\eta)(\eta)} \leqq {} & \sum_{i,j} |\tilde{u}_{ij}| \Big[\sum_i |\tilde{u}_i| + \\ & + \Big(1 + \sum_{i,j} |u_{ij}|\Big)(|\tilde{u}| + |\tilde{x}|)\Big] + \end{aligned}$$

$$+ \sum_i |\tilde{u}_i|^2\Big(1 + \sum_{i,j} |u_{ij}|\Big) +$$

$$+ \Big(1 + \sum_{i,j} |u_{ij}|^3\Big)(|\tilde{u}|^2 + |\tilde{x}|^2), \qquad (4)$$

hold, where the arguments (u_{ij}, u_i, u, t, x) of F and its derivatives are omitted, $\eta = (\tilde{u}_{ij}, \tilde{u}_i, \tilde{u}, \tilde{x})$, $F_{(\eta)(\eta)}$ is defined by (4.1), $M_1^F(u)$ and $M_2^F(u, u_k)$ are some continuous functions which grow with $|u|$ and $u_k u_k$ and $M_2^F \geqq 1$. We denote $F \in \mathcal{F}(\varepsilon, K, Q)$, if $F \in \mathcal{F}_1(\varepsilon, K, Q)$, the function F is continuously differentiable with respect to all arguments, and for $(t, x) \in Q$, $u_{ij} = u_{ji}$ and all u_i, u, we have

$$|F_t| \leqq M_3^F(u, u_k)\Big(1 + \sum_{i,j} |u_{ij}|^2\Big), \qquad (5)$$

where M_3^F is a continuous function growing with $|u|$ and $u_k u_k$.

This definition introduces the classes of operators $\mathcal{F}_1(\varepsilon, K, Q)$ and $\mathcal{F}(\varepsilon, K, Q)$ which are important for the following discussion. Of course, different elements of these classes can be associated with different functions M_1^F, M_2^F, and M_3^F. The common property of functions of $\mathcal{F}_1(\varepsilon, K, Q)$ is the concavity in u_{ij} on the set of all symmetric matrices (u_{ij}). This property follows from condition (4), where on the right there are no second-order terms with respect to $\tilde{u}_{ij}$. A broad class of functions belonging to $\mathcal{F}_1(\varepsilon, K, Q)$ are the functions $a^{ij}(u_k, u, t, x)u_{ij} + a(u_k, u, t, x)$ under natural conditions on a^{ij} and a. Examples of functions F which are nonlienar in u_{ij} will be presented later.

Throughout this section, we use the design preceding Lemma 2.1 in Section 2. According to this design, we have constants T, K, $\varepsilon > 0$, $\kappa \in (0, 1]$ a domain $D \subset E_d$, a domain $Q_1 \subset Q := (0, T) \times D$, a closed set $\Gamma \subset \overline{\partial' Q}$, and a nonempty domain $Q_0 \subset Q_1$, wherein every point has a neighbourhood in E_{d+1} such that the parabolic distance from any point of this neighbourhood to $\partial' Q_1 \setminus \Gamma$ is greater than κ. Also, let u be a function such that $u, u_x, u_t, u_{xx} \in C^2_{loc}(Q_1) \cap C(\bar{Q}_1)$. Assume that the equality

$$u_t(t, x) + F\Big(u_{x^i x^j}(t, x), u_{x^i}(t, x), u(t, x), t, x\Big) = 0 \qquad (6)$$

holds for a function F on Q_1 under conditions which will be specified late

2. THEOREM. Let $F \in \mathcal{F}(\varepsilon, K, Q_1)$, $\alpha \in (0, 1)$, and let $D = \{x : \psi(x) > 0\}$ so that $\psi \in C^{2+\alpha}(E_d)$, $||\psi||_{C^{2+\alpha}(E_d)} \le K_1$, $|\psi_x| \ge \varepsilon_1$ on ∂D (if $\partial D \ne \emptyset$). Let $\varphi \in C^{2+\alpha}(\bar{Q})$,

$$||u||_{C(Q_1)}, \quad ||\varphi||_{C^{2+\alpha}(Q)} \le K_1, \tag{7}$$

$u = \varphi$ on $\Gamma \cap \partial' Q_1$ (Any two functions are assumed to coincide on the empty set.) Then $u \in C^{2+\alpha_1}(\bar{Q}_0)$, where $\alpha_1 = \alpha_0 \wedge \alpha$, $\alpha_0 = \alpha_0(d, K, \varepsilon) \in (0, 1)$, and the norm of u in this space is bounded by a constant depending only on d, K, K_1, ε, ε_1, α, κ and the functions M_i^F, $i = 1, 2, 3$.

The proof of this theorem will be given later after some preliminaries. To illustrate the idea underlying this proof, we first consider the case when $F = \Phi(u_{11}, u_{22}, \dots, u_{dd}) + f(x)$ and $u(x)$ is a solution of the equation

$$\Phi\left(u_{x^1x^1}(x), \dots, u_{x^dx^d}(x)\right) + f(x) = 0, \quad x \in D. \tag{8}$$

As we have already mentioned, the function Φ is concave in $(u_{11}, \dots, u_{dd})$. Therefore, by differentiating (8) twice with respect to x^r and setting $u^r = u_{x^rx^r}$ for every r, we find $Lu^r + f_{x^rx^r} \ge 0$, where

$$Lv := \Phi_{u_{ij}} v_{x^ix^j}.$$

Moreover, from (8) and the Hadamard formula, it follows that

$$0 = a^r(u^r(x_1) - u^r(x_2)) + f(x_1) - f(x_2),$$

where

$$\varepsilon \le a^r \le K, \quad |f(x_1) - f(x_2)| \le M_1^F(0)|x_1 - x_2|d =: N|x_1 - x_2|$$

by conditions (1) and (3). Consequently,

$$\begin{aligned} 0 &= a^r(u^r(x_1) - u^r(x_2)) + f(x_1) - f(x_2) \\ &\ge -N|x_1 - x_2| + \varepsilon \sum_r [u^r(x_1) - u^r(x_2)]_+ - \\ &\quad - K \sum_r [u^r(x_1) - u^r(x_2)]_- . \end{aligned} \tag{9}$$

Application of Theorem 4.3.1 yields the estimate of the Hölder constant for $u_{x^r x^r}$ in the interior of D. The estimate of the Hölder constant for mixed derivatives can be obtained (however, in the general case we do not follow this route from the theory of linear equations since Δu has an already estimated Hölder constant). The case when Φ depends on a finite number of pure derivatives with respect to different directions is no more complex (see Evans [2]). In the general case, we introduce additional coordinates in order to obtain the inequality of the form $Lu^r + f_{x^r x^r} \geqq 0$ by differentiation of equation (6).

First of all, we estimate u, u_x, and u_t in C^α to be able to replace (2)—(4) by stronger conditions. The conditions of Theorem 2 are, of course, assumed to hold until we prove this theorem. We denote by Q' the domain introduced in Lemma 2.1.

3. LEMMA. There exists a constant $\alpha_0 = \alpha_0(d, K, \varepsilon) \in (0, 1)$ such that, for $\alpha_1 \in (0, \alpha_0 \wedge \alpha]$, the norms of u, u_x, and u_t in $C^{\alpha_1}(\bar{Q}_0)$ and the norm of u_{xx} in $C(\bar{Q}_0)$ are bounded by a constant which depends only on d, K, K_1, ε, ε_1, κ, α_1, α, $M_1^F(K_1)$, M_2^F, M_3^F.

Proof. It suffices to prove the assertion for $\alpha_1 = \alpha_0 \wedge \alpha$, where α_0 is the least constant denoted by α_0 in Theorems 2.4 and 3.2. The conditions of Theorem 2.4 hold by (2), (3), (7), and Lemma 1.1.4. By this theorem, the norm of u_x in $C^{\alpha_1}(\bar{Q}_0)$ is bounded by a constant depending only on the parameters permitted in the formulation of the lemma. All these constants are denoted by N. We can take Q' to be Q_0, so the norm of u_x in $C^{\alpha_1}(\bar{Q}')$ does not exceed N. By Lemma 2.1, we can now take Q' and $(1/2)\kappa$ in place of Q_1 and κ for the remaining proof. For the sake of simplicity, we assume that the norm of u_x in $C^{\alpha_1}(\bar{Q}_1)$ is bounded by a known constant N.

By Theorem 4.7, we obtain $|u_t|$, $||u_{xx}|| \leqq N$ in Q_0. Using an argument similar to the previous one, we can assume, for the rest of the proof, that $|u_t|$, $||u_{xx}|| \leqq N$ in Q_1 as well.

From the estimates of $|u_t|$ and $|u_x|$ we obtain $|u(z_0) - u(z_1)| \leqq N\rho(z_0, z_1)$ in Q_0. To prove this estimate, it is natural to use the Langrange theorem. The only difficulty arising in this case is that the line segment $[z_0, z_1]$ connecting the points z_0 and z_1 can move out of Q_1 in which case we cannot use the bounds of $|u_t|$ and $|u_x|$ in Q_1. This difficulty can be surmounted by considering the points z_0 and z_1 that are sufficiently close to one another, for example those satisfying condition (2.24), and further straightening the boundary of D if $\rho(x_0, \partial D) < 3\rho_1$ in order that either the points themselves lie in the convex domain (2.25), or their images under the mapping h^{-1} (see (2.28), (2.29), and (2.32)) lie in the convex domain $D^+_{S,1}$ (see (2.31)). This is sufficient to estimate $|u(z_0) - u(z_1)|$.

It remains for us to estimate the norm of u_t in $C^{\alpha_1}(Q_0)$. This will be done using Theorem 3.2, where Q' is taken as Q_1. Note that this theorem calls for a linear estimate of $|F_t|$ in terms of $||u_{xx}||$ as distinguished from (5). However, $||u_{xx}|| \leq N$ in Q_1 so that in the case under consideration we have $|F_t| \leq N$ in Q_1 and Q'. By Theorem 3.2, in order to prove the lemma, we only need to estimate the norm of u_t in $C^{\alpha}(\Gamma \cap \partial' Q' \cap \partial_x Q)$ and in $C^{\alpha}(\Gamma \cap \partial' Q' \cap \partial_t Q)$.

Of course, when considering $\Gamma \cap \partial' Q' \cap \partial_x Q$, we make use of (7) together with the fact that $u = \varphi$ on $\Gamma \cap \partial' Q_1 \cap \partial_x Q$. At interior points (in relative topology on $\partial_x Q$) of this set, $u_t = \varphi_t$. When deriving the inequality $|u_t| \leq N$ on $(\Gamma \cap \partial' Q') \setminus \partial_{tx} Q$ in the proof of Theorem 4.7, we saw that any point of $\Gamma \cap \partial' Q' \cap \partial_x Q$ is relatively interior for $\Gamma \cap \partial' Q_1 \cap \partial_x Q$. Therefore, the norm of u_t in $C^{\alpha}(\Gamma \cap \partial' Q' \cap \partial_x Q)$ does not exceed K_1 by virtue of (7).

To estimate u_t in the norm of $C^{\alpha}(\Gamma \cap \partial' Q' \cap \partial_t Q)$, we make use of equation (6). We take $z_0, z_1 \in \Gamma \cap \partial' Q' \cap \partial_t Q$, $z_i = (T, x_i)$, $i = 0, 1$ such that $2\rho(z_0, z_1)d \leq (1/4)\rho_0) \wedge ((1/60)d\kappa) =: \rho_1$ (cf. 2.24)) and choose points $\tilde{z}_1 = (\tilde{t}_1, \tilde{x}_1) \in Q'$ sufficiently close to z_i, e.g., such that $\rho(z_i, \tilde{z}_i) < \frac{1}{2}\rho_1$ and moreover $\tilde{t}_1 = \tilde{t}_2$. We first consider the case when $d(x_0, \partial D) \geq 3\rho_1$. In this case, the domain $G = G(\tilde{z}_0, 2\rho_1)$ in Lemma 2.1 coincides with $\tilde{z}_0 + C_{T-\tilde{t}_0, 2\rho_1}$, the latter lying in Q_1 by Lemma 2.1, so that on $\tilde{z}_0 + \partial_t C_{T-\tilde{t}_0, 2\rho_1}$ we have $u = \varphi$, $u_x = \varphi_x$, $u_{xx} = \varphi_{xx}$. Furthermore, from (6) it follows that

$$
\begin{aligned}
|u_t(\tilde{z}_0) - u_t(\tilde{z}_1)| \leq {} & \left|F\left(u_{x^ix^j}(\tilde{z}_0), u_{x^i}(\tilde{z}_0), u(\tilde{z}_0), \tilde{z}_0\right) - \right. \\
& \left. - F\left(u_{x^ix^j}(\tilde{z}_1), u_{x^i}(\tilde{z}_1), u(\tilde{z}_1), \tilde{z}_0\right)\right| + \\
& + \left|F\left(u_{x^ix^j}(\tilde{z}_1), u_{x^i}(\tilde{z}_1), u(\tilde{z}_1), \tilde{z}_0\right) - \right.
\end{aligned}
$$

$$- F\left(u_{x^i x^j}(\tilde{z}_1),\ u_{x^i}(\tilde{z}_1),\ u(\tilde{z}_1),\ \tilde{z}_1\right)\Big|. \tag{10}$$

To estimate the first summand, we apply the Hadamard formula and assumptions (1) and (3). This summand is estimated from above by the expression

$$N(\|u_{xx}(\tilde{z}_0) - u_{xx}(\tilde{z}_1)\| + |u_x(\tilde{z}_0) - u_x(\tilde{z}_1)| + |u(\tilde{z}_0) -$$
$$- u(\tilde{z}_1)|)(1 + |u_x(\tilde{z}_0)|^2 + |u_x(\tilde{z}_1)|^2 +$$
$$+ \|u_{xx}(\tilde{z}_0)\| + \|u_{xx}(\tilde{z}_1)\|).$$

The latter summand in (10) can be estimated using the Lagrange formula in terms of

$$N(1 + |u_x(\tilde{z}_1)|^3)(1 + \|u_{xx}(\tilde{z}_1)\|)|\tilde{x}_0 - \tilde{x}_1|.$$

On substituting these estimates into (10), letting $\tilde{z}_i \to z_i$, and recalling (7) and the equalities $u = \varphi$, $u_x = \varphi_x$, $u_{xx} = \varphi_{xx}$, which hold true at the points z_i, we obtain $|u_t(z_0) - u_t(z_1)| \leq N|x_0 - x_1|^{\alpha}$. The latter inequality is also true if $d(x_0, \partial D) < 3\rho_1$, which can be proved by the same argument based on straightening the boundary (see the proof of Lemma 2.3). It should only be noted that there is no need to express $u_{xx}(h)$ in terms of $[u(h)]_{xx}$. The lemma is proved.

The following lemma will enable us to derive relationships of the form (9) in the general case.

4. LEMMA. There exist ρ_2, $\delta \in (0, 1)$, an integer $n \geq 1$, and unit vectors $l_1, \ldots, l_n \in E_d$ depending only on d, K, ε such that the inequality

$$a^{ij}u_{ij} \geq \delta \sum_{r=1}^{n} (l_r^i l_r^j u_{ij})_+ - 3K \sum_{r=1}^{n} (\tilde{l}_r^i \tilde{l}_r^j u_{ij})_-$$

holds for all d × d matrices a, symmetric d × d matrices u, and vectors $\tilde{l}_i \in E_d$ satisfying the conditions

$$\varepsilon|\lambda|^2 \leq a^{ij}\lambda_i\lambda_j \leq K|\lambda|^2,\quad |\tilde{l}_r^k - l_r^k| \leq \rho_2,$$

for all $r = 1, \ldots, n$, $k = 1, \ldots, d$, $\lambda \in E_d$.

Proof. Replacing a by the matrix $\frac{1}{2}(a + a^*)$, we reduce our consideration to the case when a is symmetric. Furthermore, the set

$\Gamma(\varepsilon, K)$ of all symmetric $d \times d$-matrices a satisfying the condition

$$\varepsilon|\lambda|^2 \leqq (a\lambda, \lambda) \leqq K|\lambda|^2$$

for all $\lambda \in E_d$ is a closed convex set of the linear space of all symmetric $d \times d$-matrices. We enclose $\Gamma(\varepsilon, K)$ in an open polyhedron Π so that $\Pi \subset \Gamma(\frac{1}{2}\varepsilon, 2K)$. This can be done since $\Gamma(\varepsilon, K)$ lies in the interior of $\Gamma(\frac{1}{2}\varepsilon, 2K)$. Let $a_1, \ldots, a_{n_1}$ be vertices of Π, $\lambda_i(r)$ eigenvalues and $l_i(r)$ eigenvectors of a_r, $i = 1, \ldots, d$, $r = 1, \ldots, n_1$. Also, let us take sufficiently small ρ_2, $\delta > 0$ such that, for any vectors $\tilde{l}_i(r)$ satisfying the condition $|\tilde{l}_i^k(r) - l_i^k(r)| \leqq \rho_2$, the minimum closed convex polyhedron that contains all matrices

$$\sum_{k=1}^{d} \lambda_k(r)\tilde{l}_k^i(r)\tilde{l}_k^j(r), \quad r = 1, \ldots, n_1, \tag{11}$$

also contains the set

$$\Gamma(\varepsilon, K) - \delta \sum_{r=1}^{n_1} \sum_{k=1}^{d} \tilde{l}_k(r)\tilde{l}_k^*(r). \tag{12}$$

This choice of ρ_2 and δ is possible since, given $\tilde{l}_i(r) = l_i(r)$, the matrices in (11) coincide with a_r, and, therefore, whenever $\tilde{l}_i(r)$ is close to $l_i(r)$, they are close to a_r, the polyhedron spanned by these matrices is close to Π, set (12) is close to $\Gamma(\varepsilon, K)$ for small δ, and the closed set $\Gamma(\varepsilon, K)$ together with a neighbourhood is contained in the open polyhedron Π. For any matrix $a \in \Gamma(\varepsilon, K)$, we can now choose numbers $p_r \geqq 0$, $r = 1, \ldots, n_1$ depending only on a and $\tilde{l}_i(r)$ such that $\sum_{r=1}^{n_1} p_r = 1$ and

$$a - \delta \sum_{r=1}^{n_1} \sum_{k=1}^{d} \tilde{l}_k(r)\tilde{l}_k^*(r) = \sum_{r=1}^{n_1}, \sum_{k=1}^{d} p_r\lambda_k(r)\tilde{l}_k(r)\tilde{l}_k^*(r).$$

Here we make use of the well-known representation of polyhedron points in terms of its vertices. Hence

$$a^{ij}u_{ij} \geqq \delta \sum_{r=1}^{n_1} \sum_{k=1}^{d} \left(\sum_{i,j=1}^{d} \tilde{l}_k^i(r) l_k^j(r) u_{ij} \right)_+ -$$

$$- \sum_{r=1}^{n_1} \sum_{k=1}^{d} (\delta + p_r\lambda_k(r))\left(\sum_{i,j=1}^{d} \tilde{l}_k^i(r)\tilde{l}_k^j(r)u_{ij}\right)_-,$$

so that it remains for us to denote $n = n_1 d$, $\{l_1, \ldots, l_n\} = \{l_i(r) : i = 1, \ldots, d,\ r = 1, \ldots, n_1\}$ and note that $0 \leqq \lambda_k(r) \leqq 2K$ (since $a_r \in \Gamma(\frac{1}{2}\varepsilon, 2K)$). The lemma is proved.

Preparatory to the following lemma, we construct some auxiliary functions and operators satisfying an inequality similar to the inequality $Lu^r + f_{x^r x^r} \geqq 0$ which plays a certain role in the discussion following the statement of Theorem 2.

Using Lemma 3 (applied to Q' rather than Q_θ), we choose a number $\rho_3 > 0$ depending only on d, K, K_1, ε, ε_1, κ, α and the functions M_i^F such that the inequality

$$\rho_3^{-1} \geqq M_2^F(u(t, x), u_x(t, x))(1 + |u_x(t, x)|^2 + \|u_{xx}(t, x)\|^2)$$

holds on Q'.

Also, we denote

$$\rho_4 = \rho_2\rho_3,$$

$$u^r(t, x, \xi) = u_{x^i x^j}(t, x)(\xi^i + \rho_3 l_r^i)(\xi^j + \rho_3 l_r^j),$$

$$\xi \in E_d,\ r = 1, \ldots, n,$$

$$\sigma^{ij}(t, x, \xi) = \frac{1}{4} M_2^F(u(t, x), u_x(t, x))\Bigg[\sum_{r=1}^{d} |u_{x^r(\xi)}(t, x)| + \left(1 + \sum_{r,k=1}^{d} |u_{x^r x^k}(t, x)|\right)(|u_{(\xi)}(t, x)| + |\xi|)\Bigg] \operatorname{sgn} u_{x^i x^j(\xi)}(t,x)$$

$$a^{ij}(t, x) = \frac{1}{2}\left(F_{u_{ij}} + F_{u_{ji}}\right), \quad b^i(t, x) = F_{u_i},$$

where the arguments $\left(u_{x^i x^j}(t, x), u_{x^i}(t, x), u(t, x), t, x\right)$ of the derivatives of F are omitted,

$$L(t, x, \xi) = \frac{\partial}{\partial t} + a^{ij}(t, x)\frac{\partial^2}{\partial x^i \partial x^j} + \sigma^{ij}(t, x, \xi)\frac{\partial^2}{\partial x^i \partial \xi^j} +$$

$$+ \sigma^{ij}(t, x, \xi) \frac{\partial^2}{\partial \xi^i \partial x^j} + K_0 \delta^{ij} \frac{\partial^2}{\partial \xi^i \partial \xi^j} +$$

$$+ b^i(t, x) \frac{\partial}{\partial x^i},$$

where K_0 is a constant such that, for any $\tilde{x}$, $\tilde{\xi} \in E_d$, the inequality

$$a^{ij}\tilde{x}^i\tilde{x}^j + \sigma^{ij}\tilde{x}^i\tilde{\xi}^j + \sigma^{ij}\tilde{\xi}^i\tilde{x}^j + K_0|\tilde{\xi}|^2 \geq \frac{1}{2}\varepsilon(|\tilde{x}|^2 + |\tilde{\xi}|^2) \quad (13)$$

holds on $Q' \times D_{2\rho_3}$.

It is significant to note that $|\sigma^{ij}|$ are bounded in $Q' \times D_{2\rho_3}$ by a constant depending only on d. Therefore, the constant K_0 can be chosen to depend only on d and ε. It is also clear that the left-hand side of (13) does not exceed $N(d, K, \varepsilon) \cdot (|\tilde{x}|^2 + |\tilde{\xi}|^2)$ in $Q' \times D_{2\rho_3}$. It follows that, for the operators

$$L_i(t, x, \xi) := L(t, x, \xi + \rho_3 1_i), \quad i = 1, \ldots, n,$$

considered in the domain $Q' \times D_{\rho_4}$, the eigenvalues of the second-derivative coefficient matrix are bounded below by $\frac{1}{2}\varepsilon$ and above constant depending only on d, K, ε.

5. LEMMA. Let α_0 be the constant in Lemma 3 and let $\alpha_1 = \alpha_0 \wedge \alpha$. Then there exists a constant N_1 depending only on d, K, K_1, ε, ε_1, κ, α and the functions M_j^F such that $L_i u^i \geq -N_1$ (with no summation over i) in $Q' \times D_{\rho_4}$ with $i = 1, \ldots, n$. Moreover, there exists a constant $\delta_1 = \delta_1(d, K, \varepsilon) > 0$ such that, for any (z_1, ξ_1), $(z_2, \xi_2) \in Q' \times D_{\rho_4}$, we have

$$N_1(|\xi_1 - \xi_2| + \rho(z_1, z_2))^{\alpha_1}$$
$$\geq \delta_1 \sum_{i=1}^{n} [u^i(z_1, \xi_1) - u^i(z_2, \xi_2)]_+ -$$
$$- \sum_{i=1}^{n} [u^i(z_1, \xi_1) - u^i(z_2, \xi_2)]_-. \quad (14)$$

<u>Proof</u>. We denote by N various constants depending on initial parameters in the same manner as N_1. To prove the first assertion, it suffices, as can easily be seen, to prove the inequality

$$Lv \geqq -N \tag{15}$$

for the function $v(z, \xi) := u_{x^ix^j}(z)\xi^i\xi^j$ in $Q' \times D_{2\rho_3}$.

By differentiating (6) twice with respect to x along the direction ξ, we find

$$0 = v_t + a^{ij}v_{x^ix^j} + b^iv_{x^i} + F_uv + F_{(\eta)(\eta)}, \tag{16}$$

where, for the sake of simplicity, obvious values of the arguments are omitted and

$$\eta = \left(u_{x^ix^j(\xi)}, u_{x^i(\xi)}, u_{(\xi)}, \xi\right).$$

By virtue of assumption (4) and the definition of σ^{ij},

$$\begin{aligned} F_{(\eta)(\eta)} \leqq \sigma^{ij}v_{x^i\xi^j} + \sigma^{ij}v_{\xi^ix^j} + \\ + M_2^F\Big\{|u_{(\xi)x}|^2\Big(1 + \sum_{i,j}|u_{x^ix^j}|\Big) + \\ + (1 + ||u_{xx}||^3)(|u_{(\xi)}|^2 + |\xi|^2)\Big\}. \end{aligned}$$

The latter summand in this expression is bounded on $Q' \times D_{2\rho_3}$ by a constant N by Lemma 3 and the fact that $\rho_3 \leqq 1$ $(M_2^F \geqq 1)$. This together with (16) imply

$$0 \leqq Lv - 2K_0\Delta u + F_uv + N$$

on $Q' \times D_{2\rho_3}$. This together with (3) and Lemma 3 immediately yield (15).

We now prove the second assertion. Similarly to (10),

$$\begin{aligned} F\Big(u_{x^ix^j}(z_1), u_{x^i}(z_1), u(z_1), z_1\Big) - \\ - F\Big(u_{x^ix^j}(z_2), u_{x^i}(z_2), u(z_2), z_1\Big) \leqq |u_t(z_1) - u_t(z_2)| + \\ + \Big|F\Big(u_{x^ix^j}(z_2), u_{x^i}(z_2), u(z_2), z_1\Big) - \\ - F\Big(u_{x^ix^j}(z_2), u_{x^i}(z_2), u(z_2), z_2\Big)\Big|. \end{aligned} \tag{17}$$

The latter summand can be estimated in the same manner as $|u(z_1) - u(z_2)|$ was estimated in the proof of Lemma 3. Then, by Lemma 3, the right-hand side of (17) is smaller than $N\rho^{\alpha_1}(z_1, z_2)$.

Furthermore, the left-hand side of (17) is transformed by use of Hadamard's formula, and Lemma 3 is again made use of. As a result we obtain

$$\tilde{a}^{ij}\left|u_{x^ix^j}(z_1) - u_{x^ix^j}(z_2)\right| \leq N\rho^{\alpha_1}(z_1, z_2), \tag{18}$$

where the matrix $(\tilde{a}^{ij})$ is such that $\varepsilon|\lambda|^2 \leq \tilde{a}^{ij}\lambda^i\lambda^j \leq K|\lambda|^2$ for all $\lambda \in E_d$. By Lemma 4 with $\xi_1, \xi_2 \in D_{\rho_4}$, we obtain from (18)

$$\begin{aligned}
\rho_3^2 N\rho^{\alpha_1}(z_1, z_2) &\geq \delta \sum_i [u^i(z_1, \xi_1) - u^i(z_2, \xi_1)]_+ - \\
&\quad -3K \sum_i [u^i(z_1, \xi_1) - u^i(z_2, \xi_1)]_- \\
&\geq \delta \sum_i [u^i(z_1, \xi_1) - u^i(z_2, \xi_2)]_+ - \\
&\quad -3K \sum_i [u^i(z_1, \xi_1) - u^i(z_2, \xi_2)]_- - \\
&\quad - N \sum_i |u^i(z_2, \xi_1) - u^i(z_2, \xi_2)|.
\end{aligned}$$

This together with Lemma 3 and the inequalities $|\xi_1^i - \xi_2^i| \leq 2\rho_4 \leq 2$, $|\xi_1 - \xi_2| \leq 2|\xi_1 - \xi_2|^{\alpha_1}$, and $\rho_3 \leq 1$ imply (14) with $\delta_1 = \delta(3K)^{-1}$. The lemma is proved.

6. <u>Proof of Theorem 2</u>. Owing to Lemma 3, we need only estimate $||u_{xx}(z_0) - u_{xx}(z_1)||$ in terms of $\rho^{\alpha_1}(z_0, z_1)$. We shall consider different positions of z_0 and z_1 in Q_0 and find the appropriate $\alpha_1 = \alpha_0 \wedge \alpha$ with $\alpha_0 = \alpha_0(d, K, \varepsilon) \in (0, 1)$ in every case. Then it will remain for us to take the minimum α_0 among those considered, as follows from Lemma 3.

Note again that the assertions of Lemma 3 are obviously true if Q_0 is replaced by Q'. We shall follow the proof of Lemma 2.3 and, without loss of generality, consider only $z_0 = (t_0, x_0)$, $z_1 = (t_1, x_1) \in Q_0$ such that $t_0 \leq t_1$

$$\rho(z_0, z_1)d \leq \left(\frac{1}{4}\rho_0\right) \wedge \left(\frac{1}{60d}\kappa\right) =: \rho_1.$$

First, let $d(x_0, \partial D) \geq 3\rho_1$. In this case it suffices to repeat the corresponding fragment of the proof of Lemma 2.3, where Theorems 4.3.1 and 4.5.1 (with $\nu = \alpha$, $f_r(t) = t$) and Lemma 5 are used instead of Corollaries 4.3.6 and 4.5.3 in order to prove the inequality

$$\sum_{i=1}^{n} |u^i(z_0, \xi) - u^i(z_1, \xi)| \leq N\rho^{\alpha_1}(z_0, z_1) \tag{19}$$

for any $\xi \in D_{(1/3)\rho_4}$. Here $\alpha_1 = \alpha_0 \wedge \alpha$, $\alpha_0 \in (0, 1)$, α_0 depends only on n, d, δ_1, ε, and the constant that bounds the eigenvalues of the second-derivative coefficient matrix of the operators L_i on $Q' \times D_{\rho_4}$ (as was said before in Lemma 5, this constant depends only on d, K, and ε), the constant N depends only on the same parameters as well as α, κ, ρ_4, a constant bounding $|b^i|$ on Q', N_1 (see Lemma 5), and a constant bounding $|u^i|$ on $Q' \times D_{\rho_4}$. By Lemmas 3 to 5 and condition (3), we can assume that the four latter constants depend only on d, K, ε, ε_1, K_1, κ, α and the functions M_j^F. All these constants, which are generally large, are denoted by N. From (19) it follows that

$$|u^1(z_0, \xi) - u^1(z_1, \xi)| \leq N\rho^{\alpha_1}(z_0, z_1) \tag{20}$$

for $|\xi^i| \leq (1/3)\rho_4$, where $\alpha_1 = \alpha_0 \wedge \alpha$, $\alpha_0 = \alpha_0(d, K, \varepsilon) \in (0, 1)$. The expression $u^1(z_0, \xi) - u^1(z_1, \xi)$ is a quadratic function of ξ and

$$[u^1(z_1, \xi) - u^1(z_0, \xi)] - 2[u^1(z_1, 0) - u^1(z_0, 0)] + \\ + [u^1(z_1, -\xi) - u^1(z_0, -\xi)] = 2\left(u_{x^ix^j}(z_1) - \right. \\ \left. - u_{x^ix^j}(z_0)\right)\xi^i\xi^j.$$

By (20), the latter expression admits an estimate similar to (20). Using the 'polarization formula'

$$\left(u_{x^ix^j}(z_1) - u_{x^ix^j}(z_0)\right)\xi^i\eta^j = \left(u_{x^ix^j}(z_1) - \right. \\ \left. - u_{x^ix^j}(z_0)\right) \frac{\xi^i + \eta^i}{2} \cdot \frac{\xi^j + \eta^j}{2} -$$

$$- \left(u_{x^ix^j}(z_1) - u_{x^ix^j}(z_0)\right)\frac{\xi^i - \eta^i}{2} \cdot \frac{\xi^j - \eta^j}{2},$$

we can see that for $|\xi^i|$, $|\eta^i| \leqq (1/3\rho)_4$ the value $|u_{(\xi)(\eta)}(z_0) - u_{(\xi)(\eta)}(z_1)|$ does not exceed the right-hand side of (20) multiplied by 4. If ξ and η are chosen to have the length of $(1/3)\rho_4$ and to lie on coordinate axes, then we can see that

$$|u_{x^ix^j}(z_0) - u_{x^ix^j}(z_1)| \leqq N\rho^{\alpha_1}(z_0, z_1). \tag{21}$$

We now consider the second case when $d(x_0, \partial D) < 3\rho_1$. In this case, using formulas (2.28) to (2.37) wherein κ and Q_1 are replaced by $\frac{1}{2}\kappa$ and Q', respectively, and putting

$$\tilde{u}^i(t, x, \xi) = u^i(h(t, x), 4\rho_1\xi),$$

$$\tilde{L}_i v = 16\rho_1^2\left(L_i\left(v\left(h^{-1}, \tfrac{1}{4}\rho_1^{-1}\xi\right)\right)\right)(h, 4\rho_1\xi),$$

we bring the inequalities $L_i u^i \geqq -N_1$ and (14) involved in Lemma 5 to the domain $Q^+_{S,3} \times D_{\rho_5}$, where $\rho_5 = 4^{-1}\rho_1^{-1}\rho_4$. Thereafter, by Theorem 4.5.1 (with $\nu = \alpha = \alpha_1$, $f_r(t) = t$), we obtain (19) and (20) for $|\xi^i| \leqq (1/3)\rho_4$ if we make a preliminary estimate of Hölder constants of order $\alpha_1 = \alpha_0 \wedge \alpha$ with $\alpha_0 = \alpha_0(d, K, \varepsilon) \in (0, 1)$ for the functions $\tilde{u}^i$ on $(\Sigma_{S,2} \cup \partial_t Q^+_{S,2}) \times D_{\frac{1}{2}\rho_5}$ in the case when it is uncertain whether the points $h^{-1}(z_j)$ are separated from $\partial_t Q^+_{S,2}$ and if we estimate the Hölder constants on $\Sigma_{S^1,2} \times D_{\frac{1}{2}\rho_5}$ when, for example, $h^{-1}(z_j) \in Q^+_{S^1,2}$, where $S^1 = S - 10^{-3}\rho_1^{-2}\kappa^2$.

The latter inclusions are valid (cf. (2.37)) for $16(T - t_0) \geqq \kappa^2$. Thus to derive (20), we only need to show that for $16(T - t_0) \geqq \kappa^2$, $\tilde{z}_j \in \Sigma_{S^1,2}$, $D_{\frac{1}{2}\rho_5}$, the inequality

$$|\tilde{u}^i(\tilde{z}_0, \xi_0) - \tilde{u}^i(\tilde{z}_1, \xi_1)| \leqq N(\rho(\tilde{z}_0, \tilde{z}_1) + |\xi_0 - \xi_1|)^{\alpha_1},$$

$$i = 1, \ldots, n \tag{22}$$

holds and this inequality is also true for $\tilde{z}_j \in \Sigma_{S,2} \cup \partial_t Q^+_{S,2}$,

$\xi_j \in D_{\frac{1}{2}\rho_5}$, whenever $16(T - t_0) < \kappa^2$. If this is proved, we shall obtain not only (20) but, as we could see above, (21) as well, so that the proof of the theorem will be complete.

Since $G \subset Q'$, $h(Q^+_{S,3}) \subset G$, therefore, by virtue of Lemma 3, the values $u_{x^ix^j}(h)$ are bounded on $Q^+_{S,3}$. The functions $\tilde{u}^i$ can readily be expressed in terms of these values. It follows that, estimating the left-hand side of (22), we need only estimate $|u_{x^ix^j}(h(\tilde{z}_0)) - u_{x^ix^j}(h(\tilde{z}_1))|$ in terms of $\rho^{\alpha_1}(\tilde{z}_0, \tilde{z}_1)$. In turn, $u_{x^ix^j}(h)$ are readily related to $(u(h))_{x^ix^j}$ and $(u(h))_{x^i}$ and to the first and second derivatives of the mapping h and its inverse g. Note that the latter derivatives satisfy the Hölder condition of order α with an available constant. Therefore, to prove the theorem, it remains for us to show that

$$||\tilde{u}_{xx}(\tilde{z}_0) - \tilde{u}_{xx}(\tilde{z}_1)|| \leq N\rho^{\alpha_1}(\tilde{z}_0, \tilde{z}_1) \tag{23}$$

for $\tilde{z}_j \in \Sigma_{S^1,2}$ when $16(T - t_0) \geq \kappa^2$ and for $\tilde{z}_j' \in \Sigma_{S,2} \cup \partial_t Q^+_{S,2}$ when $16(T - t) < \kappa^2$.

The proof of Theorem 4.7 involves equation (4.37) for $\tilde{u}$ where $\tilde{F}$ is intorduced by formula (4.36). This equation acts in $Q^+_{S,3}$ and, to derive (23), it is natural to apply Theorem 1.11. The conditions of this theorem on $\tilde{u}$ and $\tilde{F}$ can easily be checked using Lemma 3, (2.35), (4.38), (4.39), and Lemma 1.1.2. Moreover, if $16(T - t_0) < \kappa^2$, then (2.34) is also true which together with (7) ensure assumption (b) of Theorem 1.11. By Theorem 1.11, (23) holds true and our theorem is proved.

We now extend the assertion of Theorem 2 to the case of elliptic equations.

7. Definition. We shall write $F \in \mathcal{F}(\varepsilon, K, D)$ if $F \in \mathcal{F}(\varepsilon, K, (0, 1) \times D)$ and F does not depend on t.

For elliptic equations, Theorem 2 takes the following form.

8. THEOREM. Let $D = \{x : \psi(x) > 0\}$, let $\alpha \in (0, 1)$, let $\psi \in C^{2+\alpha}(E_d)$ be a function such that $||\psi||_{C^{2+\alpha}(E_d)} \leq K_1$ and $|\psi_x| \geq \varepsilon_1$ on ∂D (if $\partial D \neq \emptyset$), let $D^0 \subset D^1 \subset D$ be some domains, let $G \subset \partial D$ be a closed set, and let the distance from any point of D^0 to $\partial D^1 \setminus G$ be greater than κ. Let $F \in \mathcal{F}(\varepsilon, K, D^1)$, let $u, u_x, u_{xx} \in C^2_{loc}(D^1) \cap C(\overline{D^1})$, $|u| \leq K_1$, and let

$$F\left(u_{x^ix^j}(x),\ u_{x^i}(x),\ u(x),\ x\right) = 0 \tag{24}$$

on D^1. Finally, suppose that $u = \varphi$ on $G \cap \partial D^1$, where $||\varphi||_{C^{2+\alpha}(\overline{D^1})} \le K_1$. Then $u \in C^{2+\alpha_1}(\overline{D^0})$, where $\alpha_1 = \alpha_0 \wedge \alpha$, $\alpha_0 = \alpha_0(d, K, \varepsilon) \in (0, 1)$, and the norm of u in this space is bounded by a constant depending only on d, K, K_1, ε, ε_1, κ, α and the functions M_1^F and M_2^F.

Theorem 8 formally follows from Theorem 2 if the equality $u_t = 0$ is added to (24) and $Q = (0, 2) \times D$, $Q_1 = (0, 2) \times D^1$, $\Gamma = [0, 2] \times G$, and $Q_0 = (0, 1) \times D^0$.

5.6. Discussion of the First-Order Matching Conditions

If $Q_1 = Q$, then the continuity of u_t and u_{xx} in $\bar{Q}_1$ required in Section 5 cannot be provided by any smoothness conditions on boundary data since such a continuity property would imply that boundary data on $\partial_{tx}Q$ satisfy the first matching condition. In what follows, we shall prove the solvability of nonlinear parabolic equations without this condition. Therefore, the question naturally arises whether it is possible to derive the continuity of u_t and u_{xx} in $\bar{Q}_1$ from the first-order matching condition. Readers who have no interest in the influence of matching conditions on the smoothness of a solution in $\bar{Q}_1$ can skip over this section, which has no effect on subsequent material.

Sometimes, when $Q_1 = Q$, the affirmative answer to the above question follows from the theorem asserting the existence of a $C^{2+\alpha}(\bar{Q})$-solution using the parametric extension method. For example, this is the case (see Theorem 1.3.1) when boundary data satisfy the matching condition not only for the nonlinear equation under consideration, but also for the heat conduction equation. In the general case, an affirmative answer will be given in this section using a different method.

Throughout this section we assume that all conditions preceding Theorem 5.2 are satisfied except the condition $u, u_t, u_x, u_{xx} \in C^2_{loc}(Q_1) \cap C(\bar{Q}_1)$. In place of this condition we assume that

$$u \in C^2(\bar{Q}_1 \setminus \partial_{tx}Q) \cap C(\bar{Q}_1),\ u_t,\ u_x,\ u_{xx} \in C^2_{loc}(Q_1).$$

Moreover, we assume that all conditions of Theorem 5.2 hold. In order for the first-order matching condition to make sense, we also assume that F is defined not only for $(t, x) \in Q_1$, but also for $(t, x) \in \bar{Q}_1$ and it is continuous in (t, x).

THEOREM 1. Let the above conditions be satisfied and let

$$\varphi_t(z) + F\left(\varphi_{x^i x^j}(z),\ \varphi_{x^i}(z),\ \varphi(z),\ z\right) = 0 \tag{1}$$

on $\Gamma \cap \partial'Q_1 \cap \partial_{tx}Q$. Then $u \in C^2(\bar{Q}_0)$.

This theorem as applied to Q' in Lemma 2.1 together with Theorem 5.2 show that u really belongs to $C^{2+\alpha_1}(\bar{Q}_0)$. The assertion of the theorem means that u_x, u_t, and u_{xx} as regarded only on Q_0 admit continuous extension to $\bar{Q}_0$. Of course, we only need to prove the case when $\bar{Q}_0 \cap \partial_{tx}Q \neq \emptyset$. In this case, it is not difficult to see that it suffices to prove the following. If $z_1 \in \partial Q_0 \cap \partial_{tx}Q$, $z_n \in Q_0$, $n \geq 2$, and $z_n \to z_1$, then $u_x(z_n)$, $u_t(z_n)$, $u_{xx}(z_n)$ have limits independent of the choice of z_n. We fix $z_1 = (T, x_1) \in \partial Q_0 \cap \partial_{tx}Q$ and choose $z_0 = (t_0, x_0) \in Q_0$ such that $4(T - t_0) < \kappa^2$,

$$\rho(z_1, z_0)d \leq \left(\frac{1}{4}\rho_0\right) \wedge \left(\frac{1}{30d}\kappa\right) =: \rho_1,\quad d(x_0, \partial D) < 3\rho_1.$$

We again want to apply the results established in the proof of Lemma 2.3 following (2.28). According to these results, we define $x_2 \in \partial D$ and other objects. For convenience, let $x_2 = x_1$. Note that

$$S = 16^{-1}\rho_1^{-2}(T - t_0),\quad z_1 := h^{-1}(z_1) \in \bar{\Sigma}_{S,1} \cap \partial_t Q^+_{S,1},$$

$$x_1 \in (x_1 + S_{4\rho_1}) \cap \bar{D},\ [(t_0, T) \times (x_1 + S_{4\rho_1})] \cap Q \subset h(Q^+_{S,3})$$

$$\subset G \subset Q_1. \tag{2}$$

The following discussion is partly summarized in the lemma.

2. LEMMA. Let $z_n \in Q_1$, $n \geq 2$, $z_n \to z_1$, then $u_t(z_n) \to \varphi_t(z_1)$ and $u_x(z_n) \to \varphi_x(z_1)$. If the functions $u_{ij}(z)$ which are equal to $\varphi_{x^i x^j}(z)$ on $\Gamma \cap \partial'Q_1 \cap \partial_t Q$ and equal to $u_{x^i x^j}(z)$ on $\Gamma \cap \partial'Q_1 \cap \partial_x Q$, as considered only on $\Gamma \cap \partial'Q_1$ are also continuous in the relative topology on this set at the point z_1, then $u_{xx}(z_n) \to \varphi_{xx}(z_1)$.

<u>Proof</u>. First, note that $h^{-1}(z_n) \in Q^+_{S,3}$ for all sufficiently large n (by virtue of (2)). Furthermore, by Lemma 4.3, there exists a parabolic

operator L_1 on $Q_1 \times S_2$ such that $L_1 v$ is bounded below where $v(z, \xi) = u_{(\xi)(\xi)}(z)$. Using the mapping h, we can bring the function v and the operator L_1 to $Q^+_{S,3} \times S_2$ and then, using (2.34) and (2.35), one can see that $v(h, \xi) = u_{ij}(h)\xi^i\xi^j$ on $\Sigma_{S,3} \cup (\partial_t Q^+_{S,3} \setminus \{x^1 = 0\})$. By Lemma 1.1.6, it follows that if u_{ij} is continuous at z_1, then

$$\overline{\lim_{n\to\infty}} u_{(\xi)(\xi)}(z_n) \leq u_{ij}(z_1)\xi^i\xi^j = \varphi_{x^ix^j}(z_1)\xi^i\xi^j. \tag{3}$$

By differentiation of equation (5.6) with respect to t, we construct an operator L_2 such that $L_2(u_t)$ is bounded in Q_1. We bring u_t and L_2 to $Q^+_{S,3}$. Since $F(u_{ij}, u_i, u, h(z))$ is continuous in z in $\overline{Q^+_{S,3}}$ and satisfies the Lipschitz condition with respect to (u_{ij}, u_i, u) for symmetric matrices (u_{ij}) on any bounded set uniformly in z, therefore, it is continuous in (u_{ij}, u_i, u, z) and from (5.6) we obtain

$$\begin{aligned} u_t(h) &= -F\left(u_{x^ix^j}(h), u_{x^i}(h), u(h), h\right) \\ &= -F\left(\varphi_{x^ix^j}(h), \varphi_{x^i}(h), \varphi(h), h\right) \end{aligned}$$

on $\partial_t Q^+_{S,3} \setminus \{x^1 = 0\}$.

This together with (1) and Lemma 1.1.6 applied to $\pm u_t(h)$ and L_2 imply

$$\lim_{n\to\infty} u_t(z_n) = \varphi_t(z_1).$$

The fact that $u_x(z_n) \to \varphi_x(z_1)$ straightforwardly follows from the boundedness of $(u(h))_{xx}$, the continuity of φ_x and u, and the equality $u(h) = \varphi(h)$ on $\partial_t Q^+_{S,3}$.

We now return to $u_{xx}(z_n)$. Let a matrix $\bar{u} = (\bar{u}_{ij})$ be a limit point of the sequence $u_{xx}(z_n)$. Then, by what we have proved, from (5.6) it follows that

$$\varphi_t(z_1) = F\left(\bar{u}_{ij}, \varphi_{x^i}(z_1), \varphi(z_1), z_1\right). \tag{4}$$

Here $\bar{u} \leq \varphi_{xx}(z_1)$, by (3). Moreover, from (5.1) and the Hadamard formula it readily follows that

$$F\left(\bar{u}_{ij}, \varphi_{x^i}(z_1), \varphi(z_1), z\right) - F\left(\varphi_{x^ix^j}(z_1), \varphi_{x^i}(z_1), \varphi(z_1), z\right)$$

$$\leqq \varepsilon\, \mathrm{tr}(\bar{u} - \varphi_{xx}(z_1))$$

for $z \in Q_1$ and, by continuity, this is also true for $z = z_1$. Comparison between (1) and (4), by virtue of the latter inequality, yields $\bar{u} = \varphi_{xx}(z_1)$. The lemma is thus proved.

3. Proof of Theorem 1. Lemma 2 shows that it suffices to prove the continuity of u_{ij} in the relative topology on $\Gamma \cap \partial' Q_1$ at the point z_1. Using the mapping h, we extend equation (5.6) to $Q^+_{S,3}$. Then, for $\tilde{u} = u(h)$, we obtain equation (4.37), where $\tilde{F}$ is introduced by formula (4.36) and $g = h^{-1}$. Computations (4.38), (4.39) together with the properties of the third derivatives of g established in Lemma 1.1.2 and the assumed boundedness of u_x, u_t, u_{xx}, $\tilde{u}_x$, $\hat{u}_t$, and $\tilde{u}_{xx}$ show that the conditions of Theorem 1.10 hold. By this theorem, $\tilde{u}_{x^i x^1}$ are continuous in the relative topology on $\Sigma_{S,1} \cup \partial_t Q_{S,1}$, $i \geqq 2$ (whether or not condition (1) holds). A similar assertion obviously holds for $\tilde{u}_{x^i x^j} = \tilde{\varphi}_{x^i x^j}$ with $i, j \geqq 2$, where $\tilde{\varphi} = \varphi(h)$. Now let $\bar{u}_{ij} = \tilde{\varphi}_{x^i x^j}(\tilde{z}_1)$ with $i \cdot j \geqq 2$, and let $\bar{u}_{11}$ be any limit point of the sequence $\tilde{u}_{x^1 x^1}(\tilde{z}_n)$, where $\tilde{z}_n = g(z_n) \in \Sigma_{S,1} \cup \partial_t Q_{S,1}$, and $\tilde{z}_n \to \tilde{z}_1$. By continuity, equation (4.37) is valid on $\Sigma_{S,1} \cup \partial_t Q_{S,1} \setminus \{x^1 = 0, t = T\}$. Therefore, if $\tilde{z}_n \notin \{x^1 = 0, t = T\}$ for infinitely many values of n, then

$$\tilde{\varphi}_t(\tilde{z}_1) + \tilde{F}(\bar{u}_{ij}, \tilde{\varphi}_{x^i}(\tilde{z}_1), \bar{\varphi}(\tilde{z}_1), \tilde{z}_1) = 0. \tag{5}$$

This inequality is also true in the case when $\tilde{z}_n \in \{x^1 = 0, t = T\}$, for all sufficiently large n as implied by (1). Relationship (1) can be rewritten at the point $\tilde{z}_1$ as

$$\tilde{\varphi}_t(\tilde{z}_1) + \tilde{F}\left(\tilde{\varphi}_{x^i x^j}(\tilde{z}_1), \tilde{\varphi}_{x^i}(\tilde{z}_1), \tilde{\varphi}(\tilde{z}_1), \tilde{z}_1\right) = 0. \tag{6}$$

We now compare (5) and (6), taking into account the definition of $\bar{u}_{ij}$ and the fact that, by (4.38) and condition (5.1), the function $\tilde{F}(u_{ij}, u_i, u, z)$ strictly grows in u_{11} for symmetric (u_{ij}) and $z \in Q^+_{T,3}$ which, by continuity, is also true for $z = z_1$. Then we obtain $\bar{u}_{11} = \tilde{\varphi}_{x^1 x^1}(\tilde{z}_1)$.

Thus, $\tilde{u}_{x^i x^j}(\tilde{z}_n) \to \tilde{\varphi}_{x^i x^j}(\tilde{z}_1)$ for all i, j. Moreover, Lemma 2 surely implies $\tilde{u}_x(\tilde{z}_n) \to \tilde{\varphi}_x(\tilde{z}_1)$. Using interconnections between $\tilde{u}_{xx} = (u(h))_{xx}$, $u_{xx}(h)$, and $u_x(h)$, we conclude that $u_{xx}(z_n) \to \varphi_{xx}(z_1)$. It remains to note that, by virtue of (2.34) and (2.35), we obtain $u_{x^ix^j}(z_n) = u_{ij}(z_n)$ and $u_{ij}(z_n) \to \varphi_{x^ix^j}(z_1) = u_{ij}(z_1)$. The theorem is proved.

5.7. Comments

The main results of this chapter are presented in Krylov [24].

Section 1. The approach to studying solution derivatives based on the equation for $(x^1)^{-1} u$ was suggested by the author in Krylov [17]. This approach was used to estimate the Hölder constant for the derivatives of a solution only on the boundary (Krylov [20, 21]). Safonov [6] suggested a different approach to studying $(x^1)^{-1}$ u for elliptic equations, which yields the same results. Theorem 9 develops estimating the maxima of the second derivatives only on the boundary (cf. Ivanov [1], Krylov [21]). There are a number of papers (see Il'in [1, 2], Kohn and Nirenberg [1], Oleinik and Radkevich [1], Safanov [3], Lions [4]) where such an estimate is proved only together with an interior estimate in C^2. In Krylov [21], this estimate is based on consideration of equation (17) which can equally be used for degenerate equations when $a^{11} \geq \varepsilon$. Note that equation (17) has a simple probabilistic meaning which entails the desired estimate without any computation. Theorem 10 involves conditions guaranteeing mixing derivatives to be continuous on $\bar{\Sigma}_{T,1}$. Among those conditions there is no first-order matching condition and the result established seems to be new even for linear equations.

Section 2. The monitored maximum method is first introduced in this book. In the proof of Lemma 2, its action begins with formula (21). Relationship (14) lacking, the term $\varepsilon\delta\tau^2$ would be replaced by $\tau\varepsilon\zeta|u_x|^2$ that would not 'oppress' $\zeta^2|u_x|^4$ if we did not know from (14) that τ 'has the order' $\zeta|u_x|^2$. The interior estimates of u_x in C^α for the case under consideration have also been established by Trudinger [2]. Theorem 4 for quasi-linear equations is attributed to Ladyzhenskaya and Ural'tseva.

Section 3. The method based on auxiliary functions like (2) where τ is associated with the quantity to be estimated as in the proof of Lemma 1, was used by the author in Krylov [19].

Section 4. Here the method of the monitored maximum is combined with some constructions in Krylov [19], whence, for example, Lemma 2 is taken. One can also find there a specific case of Theorem 5. The papers by Evans and Friedman [1], Evans and Lions [1], Lions [4], and Trudinger [2] are related to the material of this section.

Section 5. First results on the possibility of interior estimates in $C^{2+\alpha}$ for the solution of a general nonlinear equation were independently established by Evans [2] (elliptic equations) and Krylov [19] (elliptic and parabolic equations). They were extended by Trudinger [2] (elliptic equations) and Krylov [24]. Estimates in $C^{2+\alpha}$ up to the boundary were proved by the author in Krylov [20, 21]. Here we follow the Krylov's papers [19] and [21].

Chapter 6

EXISTENCE THEOREMS FOR SOLUTIONS OF NONDEGENERATE EQUATIONS

In this chapter, existence theorems are, in a more or less standard way, derived from the results established in the previous chapter. In Section 1, we introduce a class $\bar{F}$ of operators which covers Bellmann equations in the form $\inf_{\omega}(L(\omega)u + f(\omega)) = 0$, where $L(\omega)$ is a linear operator, which are important in the theory of controlled diffusion processes. Furthermore, we prove existence theorems for equations specified by smooth nonlinear functions F, and then using various limiting processes, we consider operators $F \in \bar{F}$ and $F \in \bar{F}_0$.

6.1. The Class $\bar{F}$, the Uniqueness of a Solution, and the Estimate of $|u|$.

We fix constants $K \geq \varepsilon > 0$, $T > 0$, a domain $D \subset E_d$ and set $Q = (0, T) \times D$, $Q(K) = (0, T - K) \times D(K)$. Furthermore, we take the class defined in Definition 5.5.1 and recall that

$$P(M, Q) = \left\{(u_{ij}, u_i, u, z): \sum_{i,j=1}^{d} |u_{ij}| + \sum_{i=1}^{d} |u_i| + \right.$$

$$\left. + |u| \leq M, u_{ij} = u_{ji}, z \in Q\right\}.$$

1. Definition. Let a function $F(u_{ij}, u_i, u, z)$ be defined for $z \in Q$ and real-valued $u_{ij} = u_{ji}$, $u_i (i, j = 1, \dots, d)$, u. We write $F \in \bar{F}(\varepsilon, K, Q)$ if there exists a sequence $F_n \in F(\varepsilon, K, Q)$ converging to F as $n \to \infty$ at every point of the set $P(\infty, Q)$ such that:

(a) $M_i^{F_1} = M_i^{F_2} = \dots =: \tilde{M}_i^F$, $i = 1, 2, 3$;

(b) for any $n = 1, 2, \dots$, the function F_n is infinitely differentiable with respect to (u_{ij}, u_i, u, z) and the derivative of any order of the function F_n with respect to (u_{ij}, u_i, u, z) is bounded on $P(M, Q)$ for any $M < \infty$;

(c) there exist constants $\delta_0 =: \delta_0^F > 0$ and $M_0 =: M_0^F > 0$ such that

$$F_n(u_{ij}, 0, -M_0, z) \geq \delta_0, \quad F_n(-u_{ij}, 0, M_0, z) \leq -\delta_0 \tag{1}$$

for all $n \geq 1$, $z \in Q$, and symmetric nonnegative matrices (u_{ij}).

Moreover, if F and F_n are independent of t, then we write $F \in \tilde{F}(\varepsilon, K, D)$.

2. Definition. We write $F \in \bar{F}(\varepsilon, K, Q)$ if $F \in \tilde{F}(\varepsilon, K, Q(\kappa))$ for any $\kappa > 0$*), and the funcitons $\tilde{M}_1^F(u)$, $\tilde{M}_2^F(u, u_k)$, and $\tilde{M}_3^F(u, u_k)$ as well as δ_0^F and M_0^F in Definition 1 are independent of κ. The class $\bar{F}(\varepsilon, K, D)$ is defined in a similar manner.

Some functions of $\bar{F}$ will be encountered later in this section (see Example 8). Here we only note that, obviously, $\tilde{F}(\varepsilon, K, Q) \subset \bar{F}(\varepsilon, K, Q)$. It is also clear that the set $F(\varepsilon, K, Q)$ does not belong to $\bar{F}(\varepsilon, K, Q)$ in view of condition (c) in Definition 1. We now show that in fact this is the only difficulty.

3. LEMMA. Let $F \in F(\varepsilon, K, Q)$, $\delta_0 > 0$, $M_0 > 0$, and for any $z \in Q$ and symmetric nonnegative matrices (u_{ij}), let

$$F(u_{ij}, 0, -M_0, z) \geqq \delta_0, \quad F(-u_{ij}, 0, M_0, z) \leqq -\delta_0. \tag{2}$$

Then $F \in \bar{F}(\varepsilon, K, Q)$ and we can take $\delta_0^F = (1/3)\delta_0$, $M_0^F = M_0$, while $\tilde{M}_1^F$, $\tilde{M}_2^F$, and $\tilde{M}_3^F$ can be taken depending only on their arguments, on M_1^F, M_2^F, and M_3^F respectively, and on d.

Proof. We first prove that there exists a constant $\gamma > 0$ such that for any $z \in Q$, symmetric nonnegative (u_{ij}), symmetric (v_{ij}), and real-valued v_i and v satisfying the inequality $|v_{ij}|$, $|v_i|$, $|v| \leqq \gamma$, the inequalities

$$\pm F(\pm u_{ij} + v_{ij}, v_{ij}, v_i, \mp M_0 + v, z) \geqq \frac{1}{2}\delta_0 \tag{3}$$

hold.

Let us consider only upper signs and suppose that our assertion is not true. Then, for $\gamma = 1/n$ and any n, there exist $z^n \in Q$, symmetric $(u_{ij}^n) \geqq 0$, (v_{ij}^n), and scalars v_i^n and v^n such that

$$|v_{ij}^n|, \ |v_i^n|, \ |v^n| \leqq 1/n,$$

$$F\left(u_{ij}^n + v_{ij}^n, v_i^n, -M_0 + v^n, z^n\right) < \frac{1}{2}\delta_0. \tag{4}$$

*) Of course, if $Q(\kappa) \neq \emptyset$. However, if we assume that all conditions hold on the empty set, this observation is excessive.

By condition (5.5.1), we obtain

$$F_{u_{ij}} u^n_{ij} = \frac{1}{2}\left(F_{u_{ij}} + F_{u_{ji}}\right)u^n_{ij} \geq \varepsilon u^n_{ij}\delta^{ij} \geq 0,$$

$$\left|F_{u_{ij}} v^n_{ij}\right| \leq K \sum_{i,j} |v^n_{ij}| \leq \frac{1}{n} Kd^2.$$

This together with (5.5.2) and (4) imply that the sequence $\delta^{ij}u^n_{ij}$ is bounded and since $(u^n_{ij}) \geq 0$ we can see that all the numbers $|u^n_{ij}|$ are bounded. We now transform the expression

$$F(u^n_{ij}, 0, -M_0, z^n) - F(u^n_{ij} + v^n_{ij}, v^n_i, -M_0 + v^n, z^n), \qquad (5)$$

which, by (2) and (4), is greater than $\delta_0/2$, to $F_{u_{ij}} v^n_{ij} + F_{u_i} v^n_i + F_u v^n$, using, for example, Lagrange's theorem. We next make use of condition (5.5.3). It can be seen that (5) tends to zero. We have thus derived the desired contradiction.

We now directly prove that $F \in \bar{F}(\varepsilon, K, Q)$. We should check that $F \in \tilde{F}(\varepsilon, K, Q(\kappa))$ for any κ and $\tilde{M}^F_i$, δ^F_0, M^F_0 in Definition 1 are independent of κ and can be chosen as given in the statement. We do this using an average operation. Take a nonnegative function $\zeta \in C^\infty_0(E_1)$ which is equal to zero for $|t| \geq 1$, equal to 1 near zero, and such that $\int\zeta(t)dt = 1$.

We put

$$F^* = F\left(\frac{1}{2}(u_{ij} + u_{ji}), u_i, u, z\right)$$

with $z \in Q$, note that, for any compactum $\Gamma \subset D$, the function F^*, by Definition 5.5.1, is uniformly continuous on any set of the form

$$\left\{(u_{ij}, u_i, u, z): \sum|u_{ij}| + \sum|u_i| + |u| \leq M, z \in (0, T) \times \Gamma\right\},$$

continuously extend F^* to the closure of such a set, and put

$$F^*(u_{ij}, u_i, u, t, x) = F^*(u_{ij}, u_i, u, 0, x) \text{ for } t \leq 0,$$

$$F^*(u_{ij}, u_i, u, t, x) = F^*(u_{ij}, u_i, u, T, x) \text{ for } t \geq T$$

for $x \in D$.

Furthermore, we fix a $\kappa > 0$ and define, for $n > 2\kappa^{-1}$, F_n, by convolving F^* with the function $\xi_n = \chi_n\eta_n$ with respect to u_{ij}, u_i, u, z, where

$$\chi_n = \left(\prod_{i,j} n\zeta(nu_{ij})\right)\left(\prod_i n\zeta(nu_i)\right)n\zeta(nu),$$

$$\eta_n = cn^{d+1}\zeta(nt)\zeta(n|x|), \quad c^{-1} = \int_{E_d} \zeta(|x|)\, dx.$$

It is obvious that F_n are defined and infinitely differentiable with respect to all arguments for $z \in E_1 \times D(\kappa)$. Moreover, by the continuity of F^*, we obtain $F_n \to F^* = F$ for $z \in Q(\kappa)$, $u_{ij} = u_{ji}$, and all u_i, u.

We now take advantage of the fact that the derivatives of the convolution of F^* with ξ_n are equal to the convolutions of ξ_n with the respective derivatives of F^*. Then, for example, it follows from (5.5.4) with $z \in Q(\kappa)$, $u_{ij} = u_{ji}$, $\tilde{u}_{ij} = \tilde{u}_{ji}$ that

$$\begin{aligned}
F_{n(\eta)(\eta)} &\leq \int M_2^F(u - v, u_k - v_k) \\
&\times \Big\{\sum_{i,j}|\tilde{u}_{ij}|\Big[\sum_i|\tilde{u}_i| + \Big(1 + \sum_{i,j}|u_{ij} - \\
&- \tfrac{1}{2}(v_{ij} + v_{ji})|\Big)(|\tilde{u}| + |\tilde{x}|)\Big] + \\
&+ \sum_i|\tilde{u}_i|^2\Big(1 + \sum_{i,j}|u_{ij} - \tfrac{1}{2}(v_{ij} + v_{ji})|\Big) + \\
&+ \Big(1 + \sum_{i,j}|u_{ij} - \tfrac{1}{2}(v_{ij} + v_{ji})|^3\Big)(|\tilde{u}|^2 + \\
&+ |\tilde{x}|^2)\Big\}\chi_n(v_{ij}, v_i, v)\, dv \prod_{i,j,k} dv_{ij}\, dv_k \\
&\leq N(d) \sup_{|v|,\,|v_k|\leq 1} M_2^F(u - v, u_k - v_k)\Big\{\sum_{i,j}|\tilde{u}_{ij}|\Big[\sum_i|\tilde{u}_i| + \\
&+ \Big(1 + \sum_{i,j}|u_{ij}|\Big)(|\tilde{u}| + |\tilde{x}|)\Big] + \sum_i|\tilde{u}_i|^2\Big(1 + \\
&+ \sum_{i,j}|u_{ij}|\Big) + \Big(1 + \sum_{i,j}|u_{ij}|^3\Big)(|\tilde{u}|^2 + |\tilde{x}|^2)\Big\}.
\end{aligned}$$

Similarly, it can be checked that the functions F_n, for $n > 2\kappa^{-1}$ satisfy conditions (5.5.1), (5.5.3), (5.5.5) on $Q(\kappa)$ with functions $\tilde{M}_1^F$ and $\tilde{M}_3^F$ computed from d, M_1^F and M_3^F and independent of κ. To check conditions (5.5.2), we can use the same arguments combined with the fact that $F_{u_{ij}}(v_{ij} + v_{ji}) = \frac{1}{2}(F_{u_{ij}} + F_{u_{ji}})(v_{ij} + v_{ji})$, while $\frac{1}{2}(F_{u_{ij}} + F_{u_{ji}})$ is bounded by virtue of (5.5.1).

It remains to check condition (c) in Definition 1 for $\delta_0^F = (1/3)\delta_0$, $M_0^F = M_0$. By the continuity of F^* and the mean-value theorem for $z \in Q(\kappa)$, symmetric $(u_{ij}) \geq 0$, and $n > 2\kappa^{-1}$, the expression $F_n(u_{ij}, 0, -M_0, z)$ is equal to $F(u_{ij} + v_{ij}, v_i, -M_0 + v, z')$, where $v_{ij} = v_{ji}$, $|v_{ij}|$, $|v_i|$, $|v| \leq 1/n$, $z' \in \bar{Q}(\kappa - 1/n)$. The latter expression does not change too much if z' is replaced by a near point of $Q(\kappa - 1/n) \subset Q$, and then, for $n > \gamma^{-1}$ from (3), we can see that $F_n(u_{ij}, 0, -M, z) \geq (1/3)\delta_0$ Similarly, $F_n(-u_{ij}, 0, M_0, z) \leq -(1/3)\delta_0$.

Thus, for $n > \gamma^{-1} \vee (2\kappa^{-1})$, the function F_n fits Definition 1 to be used in consideration of $Q(\kappa)$. It remains, if necessary, to renumber F_n.

4. LEMMA. Let $F_1, \ldots, F_k \in \mathcal{F}(\varepsilon, K, Q)$, $M_i^{F_n}$ be independent of n with $i = 1, 2, 3$, let the functions F_n be infinitely differentiable with respect to (u_{ij}, u_i, u, z), and let the derivatives of any order of F_n with respect to (u_{ij}, u_i, u, z) be bounded on $P(M, Q)$ for every $M < \infty$. Also, let there exist constants $\delta_0 > 0$, $M_0 > 0$ such that inequalities (1) hold for all $n = 1, \ldots, k$, $z \in Q$, $(u_{ij}) = (u_{ij})^* \geq 0$.

Then $F := F_1 \wedge \ldots \wedge F_k \in \tilde{\mathcal{F}}(\varepsilon, K, Q) (\subset \bar{\mathcal{F}}(\varepsilon, K, Q))$ and we can take $1 + M_i^{F_1}$, $\frac{1}{2}\delta_0$, and M_0 instead of $\tilde{M}_i^F$, δ_0^F, M_0^F, respectively.

Proof. We take ζ from the previous proof and let $m \geq 1$ and Ψ_m be the convolution of $\Psi = f^1 \wedge \ldots \wedge f^k$ with $\prod_{i \leq k} m\zeta(mf^i)$ with respect to the variables $f^1, \ldots, f^k$. We should note at once that Ψ has the first Sobolev derivatives

$$\Psi_{f^i} \geq 0, \quad \sum_{i=1}^{k} \Psi_{f^i} = 1, \quad \Psi = f^i \Psi_{f^i} \quad \text{(a.e.)}$$

Therefore,

$$\Psi_{mf^i} \geqq 0, \quad \sum_{i=1}^{k} \Psi_{mf^i} = 1,$$

$$\left|\Psi_m - f^i \Psi_{mf^i}\right| = \left|\frac{1}{m}\int g^i \Psi_{f^i}\left(f^j - \frac{1}{m} g^j\right) \prod_{j=1}^{k} \zeta(g^j)\, dg^j\right| \tag{6}$$

$$\leqq \frac{1}{m}\int \sum_i \Psi_{f^i}\left(f^j - \frac{1}{m} g^j\right) \prod_{i=1}^{k} \zeta(g^j)\, dg^j = \frac{1}{m}.$$

We now put $\Phi_m = \Psi_m(F_1, \ldots, F_k)$. It follows from (6) and the concavity of functions $\Psi(f^1, \ldots, f^k)$ and $\Psi_m(f^1, \ldots, f^k)$ in $(f^1, \ldots, f^k)$ that the infinitely differentiable function Φ_m satisfies conditions (5.5.1) to (5.5.5) with $M_i^{\Phi_m} = 1 + M_i^{1}$, $i = 1, 2, 3$. Moreover, $\Psi_m \to \Psi$ and $\Phi_m \to F$. Finally, for $m \geqq 2\delta_0^{-1}$, $z \in Q$, and symmetric $(u_{ij}) \geqq 0$, we obtain $F_n(u_{ij}, 0, -M_0, z) \geqq \delta_0$, which, by (6), yields

$$\Phi_m(u_{ij}, 0, -M_0, z) \geqq \delta_0 - \frac{1}{m} \geqq \frac{1}{2}\delta_0.$$

Similarly, $\Phi_m(-u_{ij}, 0, M_0, z) \leqq -\frac{1}{2}\delta_0$. The lemma is proved.

5. THEOREM. (a) Let $F \in \bar{\mathfrak{F}}(\varepsilon, K, Q)$, $M < \infty$, $(u_{ij}^1, u_i^1, u^1, z)$, $(u_{ij}^2, u_i^2, u^2, z) \in P(M, Q)$. Then, for some numbers $a^{ij} = a^{ji}$, b^i, and c, we obtain

$$F(u_{ij}^1, u_i^1, u^1, z) - F(u_{ij}^2, u_i^2, u^2, z)$$

$$= a^{ij}(u_{ij}^1 - u_{ij}^2) + b^i(u_i^1 - u_i^2) + c(u^1 - u^2),$$

and $\varepsilon|\lambda|^2 \leqq a^{ij}\lambda^i\lambda^j \leqq K|\lambda|^2$ for all $\lambda \in E_d$; $|b^i|$, $|c| \leqq N(M, M_1^F, d)$. Moreover, $|F(u_{ij}^1, u_i^1, u^1, z) - \tilde{a}^{ij}u_{ij}^1| \leqq \tilde{M}_1^F(u^1)(1 + u_i^1u_i^1)$, where $\tilde{a}^{ij} = \tilde{a}^{ji}$, $\varepsilon|\lambda|^2 \leqq \tilde{a}^{ij}\lambda^i\lambda^j \leqq K|\lambda|^2$ for all $\lambda \in E_d$.

(b) If Ω is a set of indices, a function $F_\omega \in \bar{\mathfrak{F}}(\varepsilon, K, Q)$ is defined for every $\omega \in \Omega$, and $\tilde{M}_1^{F_\omega}$ is independent of ω, then the family of functions $\{F_\omega : \omega \in \Omega\}$ is uniformly bounded and equicontinuous on any compactum lying in $P(\infty, Q)$.

(c) If $F_n \in \bar{\mathfrak{F}}(\varepsilon, K, Q)$, $n = 1, 2, \ldots$, $\tilde{M}_i^{F_n}$, $\delta_0^{F_n}$, $M_0^{F_n}$ are independent of n and $F_n \to F$ on $P(\infty, Q)$ as $n \to \infty$, then $F \in \bar{\mathfrak{F}}(\varepsilon, K, Q)$. Moreover, we can replace $\tilde{M}_i^F$, δ_0^F, M_0^F by $\tilde{M}_i^{F_1}$, $\delta_0^{F_1}$, $M_0^{F_1}$.

(d) If Ω is a set of indices, a function $F_\omega \in \bar{F}(\varepsilon, K, Q)$ is defined for every $\omega \in \Omega$, and $\tilde{M}_i^{F_\omega}$, $\delta_0^{F_\omega}$, $M_0^{F_\omega}$ are independent of ω, then $F := \inf\{F_\omega : \omega \in \Omega\} \in \bar{F}(\varepsilon, K, Q)$. Moreover, $\tilde{M}_i^F$, δ_0^F, M_0^F can be replaced by $1 + \tilde{M}_i^{F_\omega}$, $\frac{1}{2}\delta_0^{F_\omega}$, $M_0^{F_\omega}$ (for any $\omega \in \Omega$).

Proof. (a) Let $\kappa > 0$ be such that $z \in Q(\kappa)$. Since $F \in \tilde{F}(\varepsilon, K, Q(\kappa))$ by Definition 1, we can choose an appropriate sequence F_n. By the Hadamard formula, we have

$$F_n(u^1_{ij}, u^1_i, u^1, z) - F_n(u^2_{ij}, u^2_i, u^2, z)$$
$$= \frac{1}{2}(a_n^{ij} + a_n^{ji})(u^1_{ij} - u^2_{ij}) + b_n^i(u^1_i - u^2_i) + c_n(u^1 - u^2), \tag{7}$$

where

$$a_n^{ij} = \int_0^1 F_{nu_{ij}}(\theta u^1_{pq} + (1 - \theta)u^2_{pq}, \theta u^1_p +$$
$$+ (1 - \theta)u^2_p, \theta u^1 + (1 - \theta)u^2, z)\, d\theta,$$

and b_n^i and c_n are defined similarly. It follows from inequalities (5.5.1) and (5.5.3) applied to F_n that $(a_n^{ij} + a_n^{ji})$, b_n^i, and c_n are bounded with respect to n, while any limit point of the sequence $(1/2))(a_n^{ij} + a_n^{ji})$, b_n^i, c_n satisfies the conditions required in the first assertion. The second assertion can be proved in a similar way.

(b) The equicontinuity of F_ω on a compactum lying in $P(\infty, Q)$ can easily be established using approximation of F_ω by functions $F_{\omega n}$ and a formula similar to (7) when the difference between values of $F_{\omega n}$ is calculated at points which also differ in z-components but such that the line segment connecting z_1 and z_2 lies entirely in $Q(\kappa)$ for a $\kappa > 0$. To prove the uniform boundedness of F_ω (even in $P(M, Q)$), it suffices to make use of the second assertion in (a).

Assertion (c) follows from the fact that if $\kappa > 0$, F_{nm} is a functional sequence associated with F_n according to Definition 1 in the domain $Q(\kappa)$, then, by assertion (b), the family of functions $\{F_{nm}\}$ is compact in the sense of uniform convergence on compacta of $P(\infty, Q(\kappa))$. We can, therefore, select a subsequence $F_{nm(n)}$ such that $F_{nm(n)} \to F$ at every point of $P(\infty, Q(\kappa))$. It remains to note that M_1, M_2, M_3, δ_0, M_0

corresponding to F_{nm}, by Definition 1, coincide with $\tilde{M}_1^{F_1}$, $\tilde{M}_2^{F_1}$, $\tilde{M}_3^{F_1}$, $\delta_0^{F_1}$, $M_0^{F_1}$.

To prove (d), we first note that, by the separability of the Banach space of continuous functions on a compactum and assertion (b), the infimum over Ω can be replaced by the infimum over a countable subset of Ω.

The infimum over a countable set is the limit of infima over finite subsets. It follows from this fact and assertion (c) that the set Ω can be regarded as a finite one.

Let $\Omega = \{1, \ldots, k\}$. If $F_1, \ldots, F_k$ satisfy the conditions of Lemma 4, then F belongs to $\tilde{F}(\varepsilon, K, Q)$. In the general case, we take F_{mn} as functions associated, according to Definition 1, with F_m in the domain $Q(\kappa)$ for a $\kappa > 0$. By Lemma 4 and assertion (c) of this theorem, $F_{1n} \wedge \ldots \wedge F_{kn} \in \bar{F}(\varepsilon, K, Q(\kappa))$, $F \in \bar{F}(\varepsilon, K, Q(\kappa))$. The latter is true for all $\kappa > 0$. However, since $\bar{F}(\varepsilon, K, Q(\kappa)) \subset \tilde{F}(\varepsilon, K, Q(2\kappa))$ we have $F \in \tilde{F}(\varepsilon, K, Q(\kappa))$ for all $\kappa > 0$. Lemma 4 shows, finally, which $\tilde{M}_i^F$, δ_0^F, M_0^F can be taken. The theorem is proved.

6. Remark. If ∂D is sufficiently regular, then assertion (b) of Theorem 5 remains true if the compacta in it are replaced by bounded sets lying in $P(\infty, Q)$. This is the case, for example, if, for a neighbourhood of every point $x_0 \in \partial D$, there exists a one-to-one C^1-mapping of this neighbourhood into S_1 such that the overlapping of this neighbourhood and D is mapped into $S_1 \cap \{x^1 > 0\}$. This remark can easily be verified in the following manner. Conducting the proof of assertion (b) in the case when the points x_1 and x_2 for $z_1 = (t_1, x_1)$ and $z_2 = (t_2, x_2)$, and sufficiently close to each other and the line segment $[x_1, x_2]$ intersects ∂D, one can note that x_1 and x_2 fall in an appropriate neighbourhood of a point $x_0 \in \partial D$, and hence one can change variables x, map x_i into $S_1 \cap \{x^1 > 0\}$ and connect the images of x_i by a line segment.

7. Remark. Assertions 3 — 6 can easily be extended to functions F independent of t and they are still valid if Q is everywhere replaced by D.

Theorem 5 enables us to construct a number of examples of functions belonging to the class $\bar{F}(\varepsilon, K, Q)$.

8. Example. Let Ω be a set of indices and for every $\omega \in \Omega$ and $z \in Q$ and real-valued $u_1, \dots, u_d$, u, let functions $a^{ij}(\omega, u_k, u, z)$, $i, j = 1, \dots, d$, $b(\omega, u_k, u, z)$ be defined. Suppose that, for every fixed $\omega \in \Omega$, the functions a^{ij} and b are continuously differentiable with respect to (u_k, u, z), and for every t, they are twice continuously differentiable with respect to (u_k, u, x). In addition, let the second derivatives of a^{ij} and b with respect to (u_k, u, x) and the first derivatives with respect to t be bounded on every set $\tilde{P}(M) = \{(\omega, u, u, z) : \omega \in \Omega$, $\Sigma|u_k| + |u| \leq M$, $z \in Q\}$ and for all $\omega \in \Omega$, $z \in Q$, $s, r, p = 1, \dots, d$, $\lambda \in E_d$, $u, u_1, \dots, u_d$, the inequalities

$$\varepsilon|\lambda|^2 \leq a^{ij}\lambda^i\lambda^j \leq K|\lambda|^2, \quad |b| \leq M_1(u)\Big(1 + \sum_i |u_i|^2\Big),$$

$$|a^{sp}_{u_r}|\Big(1 + \sum_i |u_i|\Big) + |a^{sp}_u| + |a^{sp}_{x^r}|\Big(1 + \sum_i |u_i|\Big)^{-1} \leq M_1(u),$$

$$|b_{u_r}|\Big(1 + \sum_i |u_i|\Big) + |b_u| + |b_{x^r}|\Big(1 + \sum_i |u_i|\Big)^{-1} \leq M_1(u)\Big(1 + \sum_i |u_i|^2\Big)$$

hold, where $M_1(u)$ is a continuous function.

Finally, assume that the inequalities

$$b(\omega, 0, -M_0, z) \geq \delta_0, \quad b(\omega, 0, M_0, z) \leq -\delta_0 \tag{8}$$

hold for some constants $\delta_0 > 0$, $M_0 > 0$ with $\omega \in \Omega$, $z \in Q$.

Then it can easily be seen that, by Lemma 3 and Theorem 5(d), we have

$$F := \inf_{\omega \in Q} [a^{ij}(\omega, u_k, u, z)u_{ij} + b(\omega, u_k, u, z)] \in \bar{F}(\varepsilon, K, Q),$$

and we can put $M_0^F = M_0$ and $\delta_0^F = \frac{1}{2}\delta_0$, while $\tilde{M}_i^F$ can be chosen depending only on d, K, M_1 and the functions of M which dominate the second derivatives of a^{ij}, b with respect to (u_k, u, x) and their first derivatives with respect to t on $\tilde{P}(M)$.

9. Remark. Condition (8) (and the similar condition in Definition 1) does not impose any requirements on the behaviour of $b(\omega, 0, u, z)$ as $u \to \pm\infty$. It may occur that $b(\omega, 0, u, z) \to \infty$ as $u \to \infty$ and nevertheless

the second inequality in (8) holds for some M_0. As an example, we can take $b = u^2 - u$, $M_0 \in (0, 1)$. In this context, note also that we use condition (c) in Definition 1 to estimate the maximum modulus of the solution.

We now come to discuss this estimate and the uniqueness of the solution.

10. THEOREM. (a) Let D be a bounded domain, let $F \in \bar{F}(\varepsilon, K, Q)$, let $u \in C^2_{loc}(Q) \cap C(\bar{Q})$, $|u| < M_0^F$ on $\partial'Q$, and let

$$u_t + F\left(u_{x^i x^j}, u_{x^i}, u, z\right) = 0 \tag{9}$$

in Q. Then $|u| \leq M_0^F$ in $\bar{Q}$.

(b) Let D be a bounded domain, let $u, v \in C^2(Q) \cap C(\bar{Q})$, $u = v$ on $\partial'Q$, let $F \in \bar{F}(\varepsilon, K, Q)$, and let

$$u_t + F\left(u_{x^i x^j}, u_{x^i}, u, z\right) = v_t + F\left(v_{x^i x^j}, v_{x^i}, v, z\right) \tag{10}$$

in Q. Then $u = v$ in $\bar{Q}$.

Proof. (a) Denote $\Gamma = \{z \in Q : u(z) > M_0^F\}$. We now show that $\Gamma = \emptyset$. If this is not true, then there exists a point $z_0 = (t_0, x_0)$ in $\bar{\Gamma}$, which has the greatest possible coordinate t. Since $t_0 > 0$ and $u < M_0^F$ on $\partial'Q$, one can see that $z_0 \notin \partial'Q$, equality (9) holds at the point z_0, $u(z_0) = M_0^F$, and the function u as considered only on $Q \cap \{t \geq t_0\}$ attains its maximum at the point z_0. Therefore, $u_x(z_0) = 0$, $u_t(z_0) \leq 0$, the matrix $u_{xx}(z_0)$ is nonpositive, and by (9),

$$F\left(u_{x^i x^j}(z_0), 0, M_0^F, z_0\right) \geq 0,$$

which is a contradition by virtue of Definitions 1 and 2. Thus, $\Gamma = \emptyset$, $u \leq M_0^F$ on Q and $\bar{Q}$. Similarly, $u \geq -M_0^F$.

(b) Assume the contrary. Then either $u > v$ somewhere in Q, or $v > u$. We consider the first situation. Let $\gamma > 0$ be sufficiently large and let $\delta > 0$ be sufficiently small. Also, let z_0 be a point of $\bar{Q} \cap \{t \geq \delta\}$ at which $(u - v)\exp(\gamma t)$ attains its greatest value. If δ is sufficiently small, then this value is strictly positive and, thus, $z_0 \notin \partial'Q$. At the point z_0, we obtain $(u - v)_x = 0$, $(u - v)_{xx} \leq 0$, $(u - v)_t + (u - v) \leq 0$.

Hence, by Theorem 5(a) and (10),

$$0 = (u - v)_t + a^{ij}(u - v)_{x^i x^j} + b^i(u - v)_{x^i} + c(u - v)$$

$$\leqq (c - \gamma)(u - v) \tag{11}$$

at the point z_0. Here $(u - v)(z_0) > 0$, while γ can be chosen greater than c since the first and the second derivatives of u and v are bounded in Q so that c is bounded by a constant which is independent of z_0 and γ. For such a γ, relationship (11) yields a contradiction, therefore, $u \leqq v$. The theorem is proved.

In the next section, we will need a comparison theorem and a theorem on the solution stability relative to data variation on $\partial'Q$. We should attract the reader's attention to the fact that the comparison theorem does not require the solution derivatives to be bounded in the whole domain Q.

11. THEOREM. (a) Let D be a bounded domain, let $u, v \in C^2_{loc}(Q) \cap C(\bar{Q})$, $u < v$ on $\partial'Q$, let $F \in \bar{F}(\varepsilon, K, Q)$, and let the left-hand side of (10) be no smaller than the right-hand side in $\bar{Q}$. Then $u \leqq v$ in $\bar{Q}$.

(b) Let D be a bounded domain, $u, v \in C^2(Q) \cap C(\bar{Q})$, let ν be a positive number, let $u < v + \nu$ on $\partial'Q$, let $F \in \bar{F}(\varepsilon, K, Q)$, and let the left-hand side of (10) be no smaller than the right-hand side in Q. Then $u \leqq v + N\nu$ in $\bar{Q}$, where N depends only on T, d, $\tilde{M}^F_1$, and the norms of u and v in $C^2(Q)$.

Proof. Assertion (a) can be proved by repeating the proof of assertion (b) of the previous theorem; we should only add that that those points of $\bar{Q} \cap \{t \geqq \delta\}$, where $u - v > 0$, lie at a positive distance from $\partial'Q$, therefore, c in (11) is bounded by an available constant.

To prove (b), we consider the function $(u - v - \nu)\exp(\gamma t)$. If this function is negative on Q, the proof is obvious. But if it is strictly positive at a point of Q, then, just as in the previous proof, we obtain

$$(u - v)_x = 0, \quad (u - v)_{xx} \leqq 0, \quad (u - v)_t + \gamma(u - v - \nu) \leqq 0,$$

$$0 \leqq (u - v)_t + a^{ij}(u - v)_{x^i x^j} + b^i(u - v)_{x^i} + c(u - v)$$

$$\leqq (c - \gamma)(u - v - \nu) + \nu c$$

at the maximum point $z_0 \in Q$ of this function. When $\gamma > 2|c|$, we obtain $u - v - \nu \leqq \nu$ at the point $z_0 = (t_0, x_0)$, thus $(u - v - \nu)\exp(\gamma t) \leqq \nu \exp(\gamma t_0) \leqq \nu \exp(\gamma T)$, $u \leqq v + \nu(1 + \exp(\gamma T))$ in Q. It remains to note that from the beginning we can choose γ dependin only on d, $\tilde{M}_1^F$, and the norms of u and v in $C^2(Q)$ (see Theorem 5(a)). The theorem is proved.

6.2. The Existence of a Solution for $F \in \mathfrak{F}$. The First Boundary-Value Problem.

We fix constants $K \geqq \varepsilon > 0$, $T > 0$, and $\alpha \in (0, 1)$ and a function $\psi \in C^{2+\alpha}(E_d)$ such that $D := \{x \in E_d: \psi(x) > 0\}$ is a non-void *bounded* domain and $|\psi_x| \geqq 1$ on ∂D. Denote $Q = (0, T) \times D$. Throughout this section, we consider a function $F \in \mathfrak{F}(\varepsilon, K, Q)$ (see Definition 5.5.1) such that its first derivatives with respect to all arguments and second derivatives with respect to (u_{ij}, u_i, u, x) satisfy a Hölder condition of order α on every set $P(M, Q)$ relative to distance (1.3.1). Moreover, suppose that there exist constants M_0 and $\delta_0 > 0$ such that

$$F(u_{ij}, 0, -M_0, z) \geqq \delta_0, \quad F(-u_{ij}, 0, M_0, z) \leqq -\delta_0 \tag{1}$$

for any $z \in Q$ and symmetric nonnegative (u_{ij}).

As in Section 1.3, we define, by continuity, F for $z \in \bar{Q}$. Finally, suppose that a function $\varphi \in C^{2+\alpha}(\bar{Q})$ is given.

In this section, the most severe problems are involved in estimating the norm of a solution in C^2 when there are no matching conditions. To surmount these difficulties, we need four lemmas.

1. LEMMA. Let $|\varphi| < M_0$ on $\partial' Q$ and

$$\varphi_t + \Delta\varphi = 0, \quad \varphi_t + F\left(\varphi_{x^i x^j}, \varphi_{x^i}, \varphi, z\right) = 0 \text{ on } \partial_{tx} Q. \tag{2}$$

Then the problem

$$u_t + F\left(u_{x^i x^j}, u_{x^i}, u, z\right) = 0 \text{ in } Q, \tag{3}$$

$$u = \varphi \text{ on } \partial' Q \tag{4}$$

has a unique solution $u \in C^{2+\beta}(\bar{Q})$, where $\beta = \alpha_0 \wedge \alpha$, $\alpha_0 = \alpha_0(d, K, \varepsilon) \in (0, 1)$. Moreover, $u_x, u_t, u_{xx} \in C^{2+\alpha}_{loc}(Q)$.

Proof. By Theorem 1.3.1 (and Theorem 1.10), it suffices to find constants $\alpha_0 = \alpha_0(d, K, \varepsilon) \in (0, 1)$ and $N_1 < \infty$ such that, for $\beta = \alpha_0 \wedge \alpha$ the estimate

$$\|u\|_{C^{2+\beta}(Q)} \leq N_1 \tag{5}$$

holds whenever $u \in C^{2+\beta}(\bar{Q})$, u_x, u_t, $u_{xx} \in C^2_{loc}(Q)$, and, for an $\eta \in [0, 1]$,

$$u_t + \eta F\left(u_{x^i x^j}, u_{x^i}, u, z\right) + (1 - \eta)\Delta u = 0 \text{ in } Q,$$

$$u = \varphi \text{ on } \partial' Q.$$

Let us put $Q_0 = Q_1 = Q$. $\Gamma = \overline{\partial' Q}$, and $\kappa = 1$ in Theorem 5.5.2. Then we can see that, to prove (5), it suffices to show that $|u| \leq N_2$ in Q for the same functions u, where N_2 is independent of u and η. When $\eta > 0$, by Theorem 1.10(a) and Lemma 1.3, we can take M_0 as N_2. Obviously, it can also be taken as N_2 if $\eta = 0$. The lemma is proved.

We now abandon the first matching condition in (2).

2. LEMMA. Let $|\varphi| < M_0$ on $\bar{Q}$, let φ_x, φ_t, $\varphi_{xx} \in C^{2+\alpha}(\bar{Q})$, and let the second equality in (2) hold on $\partial_{tx} Q$. Then the assertion of Lemma 1 remains in force.

Proof. We fix a sufficiently large $\lambda > 0$, whose exact value will be discussed later, denote $f = \varphi_t + \Delta\varphi - \lambda\varphi$, and first prove estimate (5) under the conditions $u \in C^{2+\beta}(\bar{Q})$ $(\beta = \alpha_0 \wedge \alpha)$, u_x, u_t, $u_{xx} \in C^2_{loc}(Q)$, and for an $\eta \in [0, 1]$,

$$u_t + \eta F\left(u_{x^i x^j}, u_{x^i}, u, z\right) + (1 - \eta)(\Delta u - \lambda u - f) = 0 \text{ in } Q,$$

$$u = \varphi \text{ on } \partial' Q. \tag{6}$$

By Theorem 5.5.2, it again suffices to prove that $|u| \leq M_0$ in Q. By hypothesis, $M_0 > |\varphi|$ and, for nonnegative symmetric (u_{ij}),

$$\eta F(u_{ij}, 0, -M_0, z) + (1 - \eta)(\delta^{ij} u_{ij} + \lambda M_0 - f(z))$$

$$\geq \eta\delta_0 + (1 - \eta)(\lambda(M_0 + \varphi(z)) - \varphi_t(z) - \Delta\varphi(z)) \geq \delta_0,$$

if

$$\lambda = (M_0 - \sup|\varphi|)^{-1}(\delta_0 + \sup|\varphi_t + \Delta\varphi|).$$

For such a λ, we also have

$$\eta F(-u_{ij}, 0, M_0, z) + (1 - \eta)(-\delta^{ij}u_{ij} - \lambda M_0 - f(z)) \leqq -\delta_0.$$

Thus, by Theorem 1.10(a), $|u| \leqq M_0$.

Once the a priori estimate (5) is established for solutions to problems (6), we can repeat Proof 1.3.4 with the only difference that Δv in (1.3.9) is replaced by $(\Delta v - \lambda v - f)$. A reader who prefers a more formal approach to the use of Theorem 1.3.1 can employ the fact that the formula $v = (u - \varphi)\exp(-\lambda t)$ sets a one-to-one correspondence between the solutions of problems (6) and the solutions of problems

$$v_t + \eta\Big[e^{-\lambda t}F\Big(e^{\lambda t}v_{x^ix^j} + \varphi_{x^ix^j}, e^{\lambda t}v_{x^i} + \varphi_{x^i}, e^{\lambda t}v + \varphi, z\Big) + + e^{-\lambda t}\varphi_t + \lambda v\Big] + (1 - \eta)\Delta v = 0 \text{ in } Q, \tag{7}$$

$$v = 0 \text{ on } \partial' Q.$$

As in the case (6), there is an *a priori* estimate of the solution of these problems. By Theorem 1.3.1, (7) and hence (6) are solvable. The uniqueness of a solution to (6) and (7) follows from Theorem 1.10. The lemma is proved.

The next step consists in abandoning both matching conditions (2).

3. LEMMA. Let $|\varphi| < M_0$ on $\bar{Q}$, $\varphi_x, \varphi_t, \varphi_{xx} \in C^{2+\alpha}(\bar{Q})$. Then problem (3) and (4) has a unique solution in $C(\bar{Q}) \cap C^2(Q)$.

Proof. The uniqueness of a solution again follows from Theorem 1.10. To prove the solvability for every $n = 1, 2, \ldots$, we define infinitely differentiable functions $\chi_n(t)$ on $[0, T]$ such that the conditions $\chi_n(t) = 0$ for $t \leqq T - 1/n$, $\chi_n(t) = 1$ for $t \in [T - 1/2n, T]$, $\chi_n' \geqq 0$ hold. We put

$$\lambda = \sup_Q \Big|\varphi_t + F\Big(\varphi_{x^ix^j}, \varphi_{x^i}, \varphi, z\Big)\Big|\Big(M_0 - \sup_Q|\varphi|\Big)^{-1}, \tag{8}$$

$$F_n(u_{ij}, u_i, u, z) = F(u_{ij}, u_i, u, z) - - \chi_n(t)\Big[\lambda u + \varphi_t(z) - \lambda\varphi(z) + + F\Big(\varphi_{x^ix^j}(z), \varphi_{x^i}(z), \varphi(z), z\Big)\Big], \tag{9}$$

where, naturally, $z = (t, x)$. It is not difficult to check that, for every n, the functions F_n satisfy the conditions in the beginning of the section and the constants δ_0 and M_0 fit F_n too. Moreover,

$$\varphi_t + F_n\left(\varphi_{x^ix^j}, \varphi_{x^i}, \varphi, z\right) = 0 \text{ in } \partial_{tx}Q.$$

By Lemma 2, the problem

$$u_t + F_n\left(u_{x^ix^j}, u_{x^i}, u, z\right) = 0 \text{ in } Q, \tag{10}$$

$$u = \varphi \text{ on } \partial'Q \tag{11}$$

for every n, has a solution $u^n \in C^{2+\beta}(\bar{Q})$ such that $u^n_x, u^n_t, u^n_{xx} \in C^{2+\alpha}_{loc}(Q)$. By Theorem 1.10, the functions u^n are uniformly bounded in $\bar{Q} : |u^n| \leq M_0$.

Furthermore, we apply Theorem 5.4.7 to u^n setting $Q_0 = Q_1 = Q$, $\Gamma = \overline{\partial'Q}$, and $\kappa = 1$. Note that conditions (5.4.25) to (5.4.28) are satisfied at the solution u^n with the same ε, K, K_1, and M for all functions F_n. The condition of Theorem 5.4.7 concerning F_t is satisfied with a function M independent of n and with

$$f_n(t) = \chi_n'(t)\left[\lambda M_0 + \sup\left|\varphi_t - \lambda\varphi + F\left(\varphi_{x^ix^j}, \varphi_{x^i}, \varphi, z\right)\right|\right] +$$

$$+ \chi_n(t)\sup\left|\frac{\partial}{\partial t}\left[\varphi_t - \lambda\varphi + F\left(\varphi_{x^ix^j}, \varphi_{x^i}, \varphi, z\right)\right]\right|.$$

It is obvious that $\int_0^T \chi_n'(t)\, dt = 1$. It follows that, by Theorem 5.4.7, the norms of u^n in $C^2(Q)$ are uniformly bounded.

Finally, it is clear that equation (10) coincides with (3) on $Q \cap \{t < T - 1/m\}$ with $n > m$. By Theorem 5.5.2 (where we take

$$Q \cap \left\{t < T - \frac{1}{2m}\right\},\quad Q \cap \left\{t < T - \frac{1}{m}\right\},\quad \overline{\partial_x Q} \cap \left\{t \leq T - \frac{1}{2m}\right\}$$

as $Q_1 = Q$, Q_0, and Γ) with $n \geq 2m$, the norms of u^n in $C^{2+\beta}(Q \cap \{t < T - 1/m\})$ are bounded by a constant independent of n and $\beta \in (0, 1)$, β is independent of n. Thus there exists a subsequence $n' \to \infty$ such that $u^{n'}$, $u^{n'}_x$, $u^{n'}_t$, $u^{n'}_{xx} \to u, u_x, u_t, u_{xx}$ in Q, u, u_x, u_t, u_{xx} are continuous in Q and u satisfies equation (3). Uniform estimates of the norms of u^n in $C^2(Q)$ show that $u \in C^2(Q)$, u^n are equicontinuous in $\bar{Q}$, and $u \in C(\bar{Q})$, $u = \varphi$ on $\partial'Q$. The lemma is proved.

We now consider the derivatives of the solution to problem (3) and (4) constructed in Lemma 3 and refine their properties.

4. LEMMA. Let $|\varphi| < M_0$ on $\bar{Q}$, let $\varphi_x, \varphi_t, \varphi_{xx} \in C^{2+\alpha}(\bar{Q})$, and let u be the solution to problem (3) and (4) in the class $C(\bar{Q}) \cap C^2(Q)$. Then, for every $n \geq 1$, the function u belongs to the space $C^{2+\beta}([0, T] \times \bar{D}(\frac{1}{n}) \cup [0, T - 1/n] \times \bar{D})$, where $\beta = \alpha_0 \wedge \alpha$, $\alpha_0 = \alpha_0(d, K, \varepsilon) \in (0, 1)$, and the norm of u in this space is bounded by a constant depending only on n, d, K, ε, α, M_0, the functions M_i^F, i = 1, 2, 3, and the norms of the functions ψ and φ in $C^{2+\alpha}(E_d)$ and $C^{2+\alpha}(Q)$ respectively. Moreover, the norm of u in $C^2(Q)$ is bounded by a constant depending only on the same quantities except n.

Proof. We first sketch the proof. The construction used in the proof of Lemma 3 provided an estimate of the norm of u in $C^2(Q)$. As we could see at the end of the proof of Lemma 3, this construction provides an estimate of the norm of u in $C^{2+\beta}([0, T - 1/n] \times \bar{D})$. To estimate the norm of u in $C^{2+\beta}([0, T] \times \bar{D}(1/n))$, we construct another sequence of functions u^n converging to a solution to problem (3) and (4). For these new functions u^n, we can prove a uniform in n estimate of the norm in $C^{2+\beta}([0, T] \times \bar{D}(1/n))$. This yields an estimate of this norm for the limit function. Furthermore, using Theorem 1.11 (rather than Theorem 1.10) we can prove that the new limit function coincides with the solution to (3) and (4) constructed in Lemma 3. Then we prove the last assertion of the lemma.

Let us construct a sequence of infinitely differentiable functions $\chi_n(z)$ such that $0 \leq \chi_n \leq 1$ in Q, $\chi_n = 1$ in $[T - 1/2n, T] \times (D \setminus D(1/2n))$, $\chi_n = 0$ in $Q_n := Q \setminus ([T - 1/n, T] \times (D \setminus D(1/n)))$. Using formulas (8) and (9), we find λ and F_n and denote by u^n the solutions of problems (10) and (11). By Lemma 2, we obtain $u^n \in C^{2+\beta}(\bar{Q})$, where $\beta = \alpha_0 \wedge \alpha$, $\alpha_0 = \alpha_0(d, K, \varepsilon) \in (0, 1)$, $u^n_x, u^n_t, u^n_{xx} \in C^{2+\alpha}_{loc}(Q)$. Moreover, $|u^n| \leq M_0$, by Theorem 1.10.

Since equation (10) coincides with (3) on Q_n, it, therefore, follows that, by Theorem 5.5 (with obvious Q_1, Q_0, Γ) for $n \geq 2m$, the norms of u^n in $C^{2+\beta}(\bar{Q}_m)$ are bounded by a constant depending only on m, d, K, ε, α, M_0, the functions M_i^F and the norms of ψ and φ in $C^{2+\alpha}(E_d)$ and $C^{2+\alpha}(Q)$ respectively. Specifically, the sequence u^n is compact in the sense of uniform convergence on every set $\bar{Q}_n$. It easily follows from conditions (5.5.1), (5.5.2) and Corollary 4.5.3 by use of straightening the boundary that u^n have a uniformly bounded Hölder constant with an exponent in $[0, T] \times (D \setminus D(1/m))$. Therefore, there exists a subsequence $n' \to \infty$ and a function $v \in C(\bar{Q})$ such that $u^{n'}$ uniformly converges to v in $\bar{Q}$. It is clear that $v = \varphi$ on $\partial' Q$. Moreover, the estimates of the norms of u^n in $C^{2+\beta}(\bar{Q}_m)$ imply that $v \in C^{2+\beta}(\bar{Q}_m)$ and the norm of v in this space can be estimated as desired.

We now show that thus constructed function v coincides with the (unique) solution of problem (3) and (4) in the class $C(\bar{Q}) \cap C^2(Q)$. We cannot yet apply Theorem 1.10 since we do not know whether v_x, v_t, and v_{xx} are bounded in Q. Therefore, we make use of Theorem 1.11. Let us denote by $v[\varphi]$ any function v which is obtained by the above method from boundary condition (4). We cannot assert in advance that the sequence converges to a limit, therefore, there can be different v for the φ in question and we take one of them as $v[\varphi]$. Also, let $u[\varphi]$ be a solution of (3) and (4) belonging to $C(\bar{Q}) \cap C^2(Q)$.

By Theorem 1.11(a), for small $\nu > 0$ we obtain $u[\varphi - \nu] \leqq v[\varphi] \leqq u[\varphi + \nu]$, $u[\varphi - \nu] \leqq u[\varphi] \leqq u[\varphi + \nu]$. By Theorem 1.11(b), we obtain $u[\varphi + \nu] - u[\varphi - \nu] \leqq N\nu$, where N depends only on T, d, M_1^F and the norms of $u[\varphi + \nu]$, $u[\varphi - \nu]$ in $C^2(Q)$. The latter, as it easily follows from the proof of Lemma 3, can be estimated irrespective of ν (although their estimates would involve, for example, the maxima of the second derivatives of F with respect to u_{ij}, u_i, u, the maxima of the fourth derivatives of φ with respect to x, etc).

Thus, $|v[\varphi] - u[\varphi]| \leqq N\nu$ in Q and we obtain $v[\varphi] = u[\varphi]$ as $\nu \to 0$.

It remains for us to prove the last assertion of the lemma. It seems to be difficult to derive this proof from that of Lemma 3 since we do not permit the dependence of the constant on the properties of fourth derivatives of φ with respect to x. However, by what we have proved above, $u \in C^2(\bar{Q} \setminus \partial_{tx} Q) \cap C^{2+\beta}_{loc}(Q)$, by Lemma 1.3.2, u_x, u_t, $u_{xx} \in C^{2+\alpha}_{loc}(Q)$. An elementary theorem of mathematical analysis shows that $u_x \in C(\bar{Q})$ in view of the continuity of u in $\bar{Q}$ and the boundedness of u_{xx}. Now Theorem 5.4.7 (with $f = 0$, $Q_0 = Q_1 = Q$, $\Gamma = \overline{\partial' Q}$, $\kappa = 1$) immediately yields the desired estimate. The lemma is proved.

One of the major results of this section is as follows.

5. THEOREM. Let $|\varphi| \leqq M_0$ on $\bar{Q}$ (and the conditions introduced in the beginning of this section hold). Then problem (3) and (4) has a unique solution $u \in C^2(Q) \cap C(Q)$. This solution satisfy the following conditions:

(a) $|u| \leqq M_0$ in Q, u_x, u_t, $u_{xx} \in C^{3+\alpha}_{loc}(Q)$;

(b) for any $n \geqq 1$, the function u belongs to the space

$C^{2+\beta}([0, T] \times \bar{D}(1/n) \cup [0, T - 1/n] \times \bar{D})$, where $\beta = \alpha_0 \wedge \alpha$, $\alpha_0 = \alpha_0(d, K, \varepsilon) \in (0, 1)$, and the norm of u in this space is bounded by a constant depending only on n, d, K, ε, α, M_0, the functions M_i^F and the norms of functions ψ and φ in $C^{2+\alpha}(E_d)$ and $C^{2+\alpha}(Q)$ respectively;

(c) if $\varphi_1 \in C^{2+\alpha}(\bar{Q})$ and $\varphi_1 = \varphi$ on $\partial_x Q$, then the norm of u in $C^2(Q)$ is bounded by a constant depending only on d, K, ε, α, M_0, the functions M_i^F and the norms of ψ, φ_1, φ in $C^{2+\alpha}(E_d)$, $C^{2+\alpha}(Q)$, and $C^2(\partial_t Q)$ respectively;

(d) if $\varphi_t + F(\varphi_{x^ix^j}, \varphi_{x^i}, \varphi, z) = 0$ on $\partial_{tx} Q$, then $u \in C^{2+\beta}(\bar{Q})$ with the same β as above and the norm of u in this space is bounded by a constant depending only on d, K, ε, α, M_0, the functions M_i^F and the norms of functions ψ and φ in $C^{2+\alpha}(E_d)$ and $C^{2+\alpha}(Q)$ respectively.

Proof. The uniqueness of a solution in the class $C^2(Q) \cap C(\bar{Q})$ follows from Theorem 1.10. Assertion (d) follows from (a), (b), Theorem 5.6.1 (with $Q_0 = Q_1 = Q$, $\Gamma = \overline{\partial' Q}$, $\kappa = 1$) and Theorem 5.5.2. Assertion (c) can be obtained from (a), (b), and Theorem 5.4.7. Except for the estimate $|u| \leq M_0$, the assertions of (a) follow from Lemma 1.3.2 and (b). Thus, it remains for us to construct a solution to problem (3) and (4) in the class $C^2(Q) \cap C(\bar{Q})$ such that assertion (b) and the inequality $|u| \leq M_0$ hold. In view of the fact that we can easily pass from $(1 - 1/n)\varphi$ to φ by using a limiting process, we assume that $|\varphi| < M_0$ in $\bar{Q}$.

For integer $n \geq 1$, we denote $D^n = \{x : 2\psi(x) > 1/n\}$, $Q^n = (0, T - 1/n) \times D^n$. We only consider sufficiently large n such that $Q^n \neq \emptyset$, $|2\psi_x| \geq 1$ on ∂D^n. Furthermore, we fix an appropriate n and let φ^m, $m \geq 1$ be a sequence of functions infinitely differentiable on $\bar{Q}$ such that φ^m uniformly converges to φ on $\bar{Q}^n$, $|\varphi^m| < M_0$ on $\bar{Q}^n$, and the norms of φ^m in $C^{2+\alpha}(Q^n)$ do not exceed the doubled norm of φ in $C^{2+\alpha}(Q)$. Such functions φ^m surely exist, as can be shown by averaging.

By Lemma 3, the problem

$$u_t + F\left(u_{x^jx^j}, u_{x^i}, u, z\right) = 0 \text{ in } Q^n,$$

$$u = \varphi^m \text{ on } \partial' Q^n$$

has a unique solution $u^{nm} \in C^2(Q^n) \cap C(\bar{Q}^n)$. By Lemma 4, the norms of u^{nm} in $C^2(Q^n)$ are bounded by a constant independent of n and m.

Specifically, these functions are equicontinuous and their every limit point in $\bigcap_i C(Q^i)$ associated with any subsequence $m' \to \infty$, $n' \to \infty$ admits continuous continuation to $\bar{Q}$ which is obviously equal to φ on $\partial' Q$. Lemma 4 immediately implies that, if u is a limit point, then the norm of u in $C^{2+\beta}([0, T) \times D(1/n) \cup [0, T - 1/n] \times D)$ is bounded as in assertion (b). It remains for us to note that, by the uniform continuity of the derivatives of u in $[0, T) \times D(1/n) \cup [0, T - 1/n] \times D$, they can be continuously extended to the closure of this set so that they are equal to the respective derivatives on the boundary and the norm of u in $C^{2+\beta}$ on the closure remains equal to the norm of u in $C^{2+\beta}$ in the initial set. Finally, note that, by virtue of the uniform estimate of the norms of u^{nm} in $C^2(Q^n)$, $u \in C^2(Q)$. The theorem is proved.

The following theorem immediately follows from Theorem 5 and Theorem 5.4.6.

6. THEOREM. Let $|\varphi| \leq M_0$ in $\bar{Q}$, let u be the solution to problem (3) and (4) involved in Theorem 5. Let $\kappa \in (0, 1]$. Let $Q_0 = (0, T - \kappa^2) \times D(\kappa)$ and suppose that $Q_0 \neq \emptyset$. Then the norms of u_t, u_{xx} in $L_2(Q_0)$ are bounded by a constant depending only on d, K, ε, M_0, T, κ, the domain D, and the functions M_1^F and M_2^F. Moreover, $u_{(\xi)(\xi)} \leq N(d, K, \varepsilon, M_0, \kappa, M_1^F, M_2^F)$ on Q_0 for all unit $\xi \in E_d$. Finally, there exists an $\alpha_0 = \alpha_0(d, K, \varepsilon) \in (0, 1)$ such that the norms of u, u_x in $C^{\alpha_0}(Q_0)$ are bounded by a constant $N(d, K, \varepsilon, M_0, \kappa, M_1^F)$.

We can extend the results that have been proved to elliptic equations.

7. THEOREM. Let φ and F be independent of t, let $|\varphi| \leq M_0$ on D, and let the assumptions in the beginning of the section hold true. Then the problem

$$F\left(u_{x^ix^j}, u_{x^i}, u, x\right) = 0 \text{ in } D, \tag{12}$$

$$u = \varphi \text{ on } \partial D \tag{13}$$

has a solution $u \in C^{2+\beta}(\bar{D})$, where $\beta = \alpha_0 \wedge \alpha$, $\alpha_0 = \alpha_0(d, K, \varepsilon) \in (0, 1)$, such that:

(a) the norm of u in $C^{2+\beta}(\bar{D})$ is bounded by a constant depending only on d, K, ε, α, M_0, the functions M_1^F and M_1^F, and the norms of ψ and φ in $C^{2+\alpha}(E_d)$ and $C^{2+\alpha}(D)$ respectively.

(b) $|u| \le M_0$ in D, u_x, $u_{xx} \in C^{2+\alpha}_{loc}(D)$.

Proof. Let $\chi(t)$ be an infinitely differentiable function on $(-\infty, 0]$ such that $0 \le \chi \le 1$, $\chi' \ge 0$, $\chi = 0$ for $t \le -1$, $\chi(0) = 1$. For $T > 0$, we set $Q^T = (0, T) \times D$, $\varphi^T(t, x) = (1 - \chi(t - T))\varphi(x) + \chi(t - T)M_0$. Denote by u^T the solution of the class $C^2(Q^T) \cap C(\bar{Q}^T)$ to the problem

$$u^T_t(t, x) + F\left(u^T_{x^i x^j}(t, x), u^T_{x^i}(t, x), u^T(t, x), x\right) = 0 \text{ in } Q^T, \tag{14}$$

$$u^T = \varphi^T \text{ on } \partial' Q^T. \tag{15}$$

By Theorem 5, this problem has a unique solution since $|\varphi^T| \le M_0$ in Q^T. Note that the uniqueness of the solution immediately implies that $u^T(t, x) = u^{T-t}(0, x)$ for $t \in [0, T]$. Furthermore, by Theorem 5 with $T \ge 2$, the norms of the functions u^T in $C^{2+\beta}(Q^1)$ are bounded by the constant involved in assertion (a) of the current theorem. The norms of $u^T(0, x)$ in $C^{2+\beta}(\bar{D})$ can obviously be estimated in the same manner. Moreover, $u^T(0, x) = \varphi(x)$ for $T \ge 2$ on ∂D in view of (15).

By differentiation of equation (14) with respect to t, we obtain a linear equation for u^T_t with zero absolute term and coefficients bounded according to Theorem 5(c). The differentiation is legitimate by virtue of Theorem 5(a). The function u^T_t is bounded and continuous in $\bar{Q}^T \setminus \partial_{tx} Q^T$; we have

$$u^T_t(t, x) = (M_0 - \varphi(x))\chi'(t - T) \ge 0,$$

on $\partial_x Q^T$, and, by (14) and (15),

$$u^T_t(T, x) = -F(0, 0, M_0, x) \ge 0$$

on $\{t = T\} \times D$. By the maximum principle, it follows that $u^T_t \ge 0$ in Q^T. It now follows from equality $u^T(t, x) = u^{T-t}(0, x)$ that $u^T(0, x)$ decreases in T. Since $|u^T| \le M_0$ in Q^T, $u^T(0, x)$ has a finite limit as $T \to \infty$. Denote this limit by $u(x)$. It is clear that $u = \varphi$ on ∂D, $|u| \le M_0$ in D, and, by virtue of the estimates of u in $C^{2+\beta}(\bar{Q}^1)$, assertion (a) of the theorem in question is true. From these estimates and the equality $u^T(t, x) = u^{T-t}(0, x)$ it evidently follows that $u^T(t, x)$ converges to $u(x)$ as $T \to \infty$ (uniformly on Q^1) and the derivatives u^T_x, u^T_t, and u^T_{xx} converge to the respective derivatives of u. Since $u_t = 0$, $u^T_t \to 0$ and letting $T \to \infty$ in (14), we conclude that u is a solution of equation (12).

The inequality $|u| \leq M_0$ in assertion (b) was proved above and the inclusions u_x, $u_{xx} \in C^{2+\alpha}_{loc}(D)$ follow from Lemma 1.3.2 (if we put $\eta = 1$ in equation (1.3.3.)). The theorem is proved.

8. Remark. In contrast to Theorem 5, Theorem 7 cannot assert the uniqueness of a solution in the class $C^2(D)$. For example, the functions 0 and sin x satisfy the equation $u'' + u = 0$ on $[-\pi, \pi]$ and are equal to zero at the extremities of this segment. The function $u_{11} + u$ does not satisfy condition (1). However, if u is replaced by f(u), where $f(u) = u$ for $|u| \leq 1$ and f behaves, for example, as (-u) for $|u| \geq 2$, then condition (1) is satisfied with $M_0 = 3$, $\delta_0 = 1$, while the non-uniqueness remains.

In order to provide the uniqueness of a solution in the class of functions $u \in C^2(D)$ satisfying the condition $|u| \leq M_0$ on D, it suffices, by Hadamard's formula and the maximum principle, to impose the requirement that, for all $x \in D$, symmetric (u_{ij}), $u \in [-M_0, M_0]$, and real-valued u_i, the inequality $F_u(u_{ij}, u_i, u, x) \leq 0$ hold.

6.3. The Existence of a Solution for $F \in \mathcal{F}$ in a Non-Smooth Domain and Cauchy's problem

In this section, we consider problem (2.3) and (2.4) when the domain D is nonsmooth. The basic method consists of approximating the domain D from inside by smooth domains followed by passage to the limit. In this case, the interior estimates in the norm of $C^{2+\alpha}$ guarantee that the functions involved in the limiting process are compact in C^2 and the continuity of the limit funciton in $\bar{Q}$ can be proved by use of barriers. In order to construct simple barriers, we assume that D is a domain of E_d and for any point $x_0 \in \partial D$ there exist a $\rho_0 > 0$ and a closed ball of radius ρ_0 lying in $E_d \setminus D$ and containing x_0 (on its boundary). It is not assumed that D is bounded, therefore, for $D = E_d$, we consider Cauchy's problem.

Let $T > 0$, $K \geq \varepsilon > 0$, $\alpha \in (0, 1)$, $Q = (0, T) \times D$, and let $F \in \mathcal{F}(\varepsilon, K, Q)$. Let the first derivatives of F with respect to all arguments and the second derivatives with respect to (u_{ij}, u_i, u, x) satisfy a Hölder condition of order α on every compactum lying in the set $P(\infty, Q)$ relative to distance (1.3.1). Moreover, suppose that there exist constants M_0, $\delta_0 > 0$ such that condition (2.1) holds for any $z \in Q$ and symmetric nonnegative matrices (u_{ij}).

Throughout this section these conditions are assumed to hold.

1. THEOREM. Let $\varphi \in C(\bar{Q})$, $|\varphi| \le M_0$ in Q. Then problem (2.3) and (2.4) has a solution possessing the following properties:

(a) $u \in C(\bar{Q})$, $|u| \le M_0$ on Q,

(b) for every $\kappa \in (0, 1)$, the function u belongs to the space $C^{2+\alpha_0}((0, T-\kappa^2) \times D(\kappa))$, where $\alpha_0 = \alpha_0(d, K, \varepsilon) \in (0, 1)$, and the norm of u in this space is bounded by a constant depending only on d, K, ε, M_0, κ and the functions M_i^F, $i = 1, 2, 3$.

Proof. Setting $\varphi(t, x) = \varphi(0, x)$ with $t \le 0$ and next, more or less arbitrarily smoothing φ in $(-\infty, T) \times D$, we reduce our consideration to the case when $\varphi \in C^{2+\alpha}([0, T - \kappa^2] \times \Gamma)$ with any $\kappa \in (0, 1)$ and for any compactum $\Gamma \subset D$. Furthermore, we define a sequence of expanding bounded domains $D^n \uparrow D$ such that ∂D^n be infinitely differentiable and denote $Q^n = (1/n, T - 1/n) \times D^n$. By Theorem 2.5, the problem

$$u^n_t + F\left(u^n_{x^ix^j}, u^n_{x^i}, u^n, z\right) = 0 \text{ in } Q^n,$$

$$u^n = \varphi \text{ on } {}'Q^n$$

has a solution $u^n \in C^2(Q^n) \cap C(\bar{Q}^n)$ such that $u^n_x, u^n_t, u^n_{xx} \in C^{2+\alpha}_{loc}(Q^n)$, $|u^n| \le M_0$ on Q^n.

If $z_0 + C_{4\rho^2,2\rho}$ together with its closure lie in Q, then this domain also lies in all Q^n beginning with an n. In Theorem 5.5.2, we take the domains $z_0 + C_{4\rho^2,2\rho}$ and $z_0 + C_{\rho^2,\rho}$ as $Q_1 = Q$ and Q_0 respectively (with obvious ψ, Γ, κ). Then, by this theorem, we can estimate the norms of u^n in $C^{2+\alpha_0}(z_0 + C_{\rho^2,\rho})$ uniformly in n beginning with an n. It follows that, for a sequence $n' \to \infty$, the functions $u^{n'}$ converge, at every point $z \in Q$, to a function u that possesses property (b) of the theorem statement and satisfies equation (2.3). Obviously, $|u| \le M_0$ on Q and it remains for us to prove that u can be extended to a function which is continuous in $\bar{Q}$ and equal to φ on $\partial' Q$. It is clear that, for every point $z_0 = (t_0, x_0) \in \overline{\partial' Q}$, it suffices to show that $u(z_m) \to \varphi(z_0)$ whenever $z_m = (t_m, x_m) \in Q$, $z_m \to z_0$.

First, let $z_0 = (T, x_0)$, $x_0 \in D$, $\rho > 0$, $x_0 + S_\rho \subset D$. Based on condition (5.5.2), we obtain

$$|u^n_t + F_{u_{ij}} u^n_{x^ix^j}| \le N(1 + |u^n_x|^2),$$

in Q^n, where (and hereafter) N is independent of n. Hence, as in the proof of Theorem 4.3.5 (there we took n = 1 and g^k from Corollary 4.3.6) for a constant $\gamma > 0$ and all n, we obtain that in Q

$$v^n_t + F_{u_{ij}} v^n_{x^ix^j} \ge -N, \tag{1}$$

where $v^n(z) = \exp\{\gamma(u^n(z) - \varphi(z_0))^2\} - \exp\gamma\delta$ and δ is any constant. Put

$$\delta = \delta(\rho) \doteq \sup\{(\varphi(z) - \varphi(z_0))^2 : z_0 - z \in C_{\rho^2,\rho}\}.$$

It is clear that $v^n(z) \le 0$ on $(z_0 - C_{\rho^2,\rho}) \cap \partial_t Q^n$ for all sufficiently large n. Also, note that if we omit u(T, 0) everywhere in the formulation of Lemma 4.4.2, then it remains valid (in fact, it was proved in this form). It easily follows from this lemma that $v^n(t, x) \le N(T - t)$ for $|x - x_0| \le \frac{1}{2}\rho$ whenever T - t is sufficiently small (independent of n) while n is sufficiently large. Thus, for the same z = (t, x),

$$\exp\{\gamma(u(z) - \varphi(z_0))^2\} - \exp \gamma\delta \le N(T - t).$$

Substituting here $z = z_m$ and setting $\xi = \overline{\lim}(u(z_m) - \varphi(z_0))^2$, we conclude that $\xi \le \delta(\rho)$. Letting $\rho \downarrow 0$, we obtain $\xi = 0$.

We now consider the second case $z_0 \in \partial_x Q$. Let us take the same functions v^n and define $\delta = \delta(\rho)$ in the following way. Let the ball $x_1 + \bar{S}_{\rho_0}$ lie in $E_d \setminus D$ and $x_0 \in x_1 + \bar{S}_{\rho_0}$. When $\rho > \rho_0$, we put

$$G_\rho = \{z = (t, x) \in Q : |t - t_0| < \rho - \rho_0, |x - x_1| < \rho\},$$

$$\delta = \delta(\rho) = \sup\{(\varphi(z) - \varphi(z_0))^2 : z \in G_\rho\}.$$

With no loss of generality, we can assume that x_0 is a unique point of $\bar{D}$ lying in $x_1 + \bar{S}_{\rho_0}$, and then $\delta(\rho) \to 0$ as $\rho \downarrow \rho_0$, by the continuity of φ. Simple calculations carried out in polar coordinates show that there exists a ρ_1 which is strictly greater than ρ_0, sufficiently close to ρ_0 (independent of n), and such that, for $\rho \in (\rho_0, \rho_1]$, the function

$$w^\rho := \frac{1}{4}(\rho - \rho_0)^2 - \left(|x - x_1| - \frac{1}{2}(\rho_0 + \rho)\right)^2$$

in $G_\rho \cap Q^n$ satisfies the inequality

$$w_t^\rho + F_{u_{ij}} w^\rho_{x^i x^j} + \varepsilon \leqq 0,$$

where, as in (1) and hereafter, the arguments of $F_{u_{ij}}$ are $u^n_{x^i x^j}$, $u^n_{x^i}$, u^n, and z. In $G_\rho \cap Q^n \cap \{w^\rho(t, x) < \varepsilon(t_0 + \rho - t)\}$, we obtain

$$\left(\frac{w^\rho}{t_0 + \rho - t}\right)_t + F_{u_{ij}}\left(\frac{w^\rho}{t_0 + \rho - t}\right)_{x^i x^j}$$

$$\leqq \frac{w^\rho}{(t_0 + \rho - t)^2} - \varepsilon\frac{1}{t_0 + \rho - t} \leqq 0.$$

Fixing ρ_1, it is easy to choose a sufficiently large $r > 0$ (cf. the proof of Theorem 4.5.4) such that the function $\pi^\rho := (\rho_0^{-r} - \rho^{-r})^{-1}$ $(\rho_0^{-r} - |x - x_1|^{-r})$ with $\rho \in (\rho_0, \rho_1]$ satisfy the inequality

$$\pi_t^\rho + F_{u_{ij}} \pi^\rho_{x^i x^j} \leqq 0$$

in $G_\rho \cap Q^n$.

We now compare the function v^n in $G_\rho \cap Q^n$ with the function

$$h^\rho = w^\rho + \pi^\rho + w^\rho(t_0 + \rho - t)^{-1}.$$

It is clear that $v^n \leqq Nh^\rho$ on $G_\rho \cap Q^n \cap \{w^\rho \geqq \varepsilon(t_0 + \rho - t)$. To estimate v in terms of h^ρ on the remaining part of $G_\rho \cap Q^n$, we apply the maximum principle. For a fixed ρ such that $\rho \in (\rho_0, \rho_1]$, $\rho - \rho_0 < T - t_0$ and sufficiently large n, it is obvious that

$$G_\rho \cap Q^n \cap \{w^\rho < \varepsilon(t_0 + \rho - t)\}$$

$$= G_\rho \cap [(0, T) \times D^n] \cap \{w^\rho < \varepsilon(t_0 + \rho - t)\}$$

and, for every point $\tilde{z} = (\tilde{t}, \tilde{x})$ lying on the parabolic boundary of this domain, one of the following conditions holds: (i) $\tilde{z}$ does not lie in G_ρ and hence $\tilde{z} \in \partial' G_\rho$; (ii) $\tilde{z}$ lies in G_ρ but it does not belong to $(0, T) \times$

D^n and, thus, $\tilde{x} \in \partial D^n$; (iii) $\tilde{z}$ lies in $G_\rho \cap [(0, T) \times D^n]$, but it does not belong to $\{w^\rho < \varepsilon(t_0 + \rho - t)\}$ and since it can be approximated by points of the latter set, we obtain $w^\rho(\tilde{z}) = \varepsilon(t_0 + \rho - \tilde{t})$. In the first case ($w^\rho(\tilde{z}) \leqq \varepsilon(t_0 + \rho - t)$), we obtain $\pi^\rho(\tilde{z}) = 1$, in the second case $v^n(\tilde{z}) \leqq 0$, and in the third case, $(t_0 + \rho - \tilde{t})^{-1} w^\rho(\tilde{z}) = \varepsilon$. It follows that, for $\rho \in (\rho_0, \rho_1]$, $\rho - \rho_0 < T - t_0$, and all sufficiently large n, the inequality $v^n \leqq Nh^\rho$ holds on $G_\rho \cap Q^n$. Therefore,

$$\exp\{\gamma(u(z) - \varphi(z_0))^2\} - \exp \gamma\delta(\rho) \leqq Nh^\rho(z)$$

in G_ρ and for the same ξ we obtain $\xi \leqq \delta(\rho)$ and $\xi = 0$.

It remains for us to study the third case, i.e., $t_0 = T$ and $x_0 \in \partial D$. In this case, it suffices to repeat the above proof where the third summand in (2) is omitted. The theorem is proved.

Unfortunately, we do not know whether or not the solution to problem (2.3) and (2.4) with the properties described in Theorem 1 is unique. Therefore, the next theorem does not assert that the solution in Theorem 1 possesses some additional properties under additional conditions, but only that there exists a more regular solution under additional conditions.

2. THEOREM. Let $\varphi \in C(\bar{Q})$, let $|\varphi| \leqq M_0$ on Q, let D^0, D^1, and D be domains such that $D^0 \subset D^1 \subset D$, and let $\rho := \text{dist}(\partial D^0, \partial D^1) > 0$. Then:

(a) if $\varphi(T, \cdot) \in C^2(D^1)$, then there exists a solution to problem (2.3) and (2.4) which possesses properties (a) and (b) of Theorem 1 and such that $u \in C^2((0, T) \times D^0)$ and the norm of u in this space is bounded by a constant depending only on d, K, ε, M_0, ρ, the functions M_i^F, i = 1, 2, 3, and the norm of $\varphi(T, \cdot)$ in $C^2(D^1)$;

(b) if $\varphi(T, \cdot) \in C^{2+\alpha}(D^1)$, then there exists a solution to problem (2.3) and (2.4) possessing properties (a) and (b) of Theorem 1 and such that $u \in C^{2+\beta}([0, T] \times D^0)$, where $\beta = \alpha \wedge \alpha_0$, $\alpha_0 = \alpha_0(d, K, \varepsilon) \in (0, 1)$, and the norm of u in this space is bounded by a constant which depends only on d, K, ε, α, M_0, ρ, the functions M_i^F, i = 1, 2, 3, and the norm of $\varphi(T, \cdot)$ in $C^{2+\alpha}(D^1)$.

Proof. The arguments of the previous proof make it clear that to prove asserion (a), it suffices to additionally apply Theorem 5.4.7 and construct a function $\tilde{\varphi}$ such that $\tilde{\varphi} \in C^{2+\alpha}_{loc}(Q)$,

$$|\tilde{\varphi}| \leq M_0 \text{ in } Q, \quad \tilde{\varphi} \in C(\bar{Q}), \quad \tilde{\varphi} = \varphi \text{ on } \partial' Q,$$

$$\tilde{\varphi} \in C^2\left((0, T) \times D^1\left(\frac{1}{2}\rho\right)\right).$$

and the norm of $\tilde{\varphi}$ in the latter space is bounded by a constant depending only on d, M_0, ρ, and the norm of φ in $C^2(D^1)$. The domain $D^1(\frac{1}{2}\rho)$ can be taken as a new domain D^0 in which case we see that assertion (a) need be proved only for the solution to the problem $\tilde{\varphi}_t + \Delta\tilde{\varphi} = 0$ in Q, $\tilde{\varphi} = \varphi$ on $\partial'Q$. Note that, for a continuous in $\bar{Q}$ solution to this problem, the inequality $|\tilde{\varphi}| \leq M_0$ in Q follows from the maximum principle.

Existence of the desired function $\tilde{\varphi}$ is well known in the theory of linear equations (e.g., see Ladyzhenskaya, Solonnikov, Ural'tseva [1], Friedman [1]).

Similar arguments can be applied to assertion (b). The theorem is proved.

We have made use of Theorems 5.4.7 and 5.5.2 only partly. We may for example, study the case when a part of ∂D is sufficiently smooth and the Cartesian product of this part by (0, T) contains a domain where φ is sufficiently smooth. In such a case, we may assert that a solution is smooth up to the considered domain on $\partial_x Q$. This case can be combined with the case when φ is sufficiently smooth in a subdomain of $\partial_t Q$. Moreover, 'sufficiently smooth' may be interpreted as something weaker than belonging to C^2 or $C^{2+\alpha}$, in which case it is possible to apply the results of Sections 5.2 and 5.3. In short, there are a large number of different situations, but we do not direct attention to them. Readers who have worked through the proof of Theorems 1 and 2 can do this for themselves. Moreover, Theorems 1, 2, 2.5, and 2.6 seem to present the most important information on the operators $F \in \mathcal{F}$.

We complete the section by extending Theorem 1 to the case of elliptic operators. Recall that the conditions introduced at the beginning of the section are assumed to hold, while the domain D can be unbounded and coincide with E_d.

3. THEOREM. Suppose that φ and F are independent of t and $|\varphi| \leq M_0$ on D. Then problem (2.12) and (2.13) has a solution u which possesses the following properties:

(a) $u \in C(\bar{D})$, $|u| \leq M_0$ on D;

(b) for any $\kappa \in (0, 1)$, the function u belongs to the space $C^{2+\alpha_0}(D(\kappa))$, where $\alpha_0 = \alpha_0(d, K, \varepsilon) \in (0, 1)$, and the norm of u in this space is bounded by a constant depending only on d, K, ε, M_0, κ, and the functions M_1^F and M_2^F.

This theorem can be proved in the same manner as Theorem 1, but using Theorem 2.7, and considering the behaviour of the solution on the boundary (in the unique case possible), one should use (2) as a barrier, where the last summand should be omitted.

6.4. The Existence of a Solution for $\bar{F}$. Examples.

In this section, we show how the solvability theorems for problem (2.3) and (2.4) with $F \in F$ yield the solvability theorems for $F \in \bar{F}$. Only Theorems 2.5, 2.7, 3.1 to 3.3 are extended to the case $F \in \bar{F}$. The extension method is based on simple passage to the limit, which is admissible in view of Theorem 1.5. It will be clear that this method is also applicable to other cases when there are existence theorems for functions of F. In this connection, we would like to attract the reader's attention to the remark following the proof of Theorem 3.2.

We fix constants $K \geqq \varepsilon > 0$, $T > 0$, $\alpha \in (0, 1)$, a domain $D \subset E_d$, $Q = (0, T) \times D$, and a function $F \in \bar{F}(\varepsilon, K, Q)$.

1. THEOREM. Let $D = \{x : \psi(x) > 0\}$ be a bounded domain such that $\psi \in C^{2+\alpha}(E_d)$ and $|\psi_x| \geqq 1$ on ∂D, let $\varphi \in C^{2+\alpha}(\bar{Q})$ and let $|\varphi| \leqq M_0^F$ on Q. Then problem (2.3) and (2.4) has a unique solution $u \in C^2(Q) \cap C(\bar{Q})$. This solution possesses the following properties:

(a) $|u| \leqq M_0^F$ in $\bar{Q}$, for every $n \geqq 1$ the function u belongs to the space $C^{2+\beta}([0, T] \times \bar{D}(1/n) \cup [0, T - 1/n] \times \bar{D})$, where $\beta = \alpha_0 \wedge \alpha$, $\alpha_0 = \alpha_0(d, K, \varepsilon) \in (0, 1)$, and the norm of u in this space is bounded by a constant depending only on n, d, K, ε, α, M_0^F, the functions $\tilde{M}_i^F$ and the norms of ψ and φ in $C^{2+\alpha}(E_d)$ and $C^{2+\alpha}(Q)$ respectively;

(b) if $\varphi_1 \in C^{2+\alpha}(\bar{Q})$ and $\varphi_1 = \varphi$ on $\partial_x D$, then the norm of u in $C^2(Q)$ is bounded by a constant depending only on d, K, ε, α, M_0^F, the functions $\tilde{M}_i^F$ and the norms of ψ, φ_1, φ in $C^{2+\alpha}(E_d)$, $C^{2+\alpha}(Q)$, $C^2(\partial_t Q)$ respectively.

Proof. We put $D^n = \{x : \psi(x) > 1/n\}$, $Q^n = (0, T - 1/n) \times D^n$, and consider only those n such that $2|\psi_x| \geqq 1$ on ∂D^n. By definition, for every n, there exists a sequence of infinitely differentiable $F_{nm} \in F(\varepsilon, K, Q^n)$ that converges to F at every point of $P(\infty, Q)$, the functions $M_i^{F_{nm}}$ being independent of n and m and equal to $\tilde{M}_i^F$, moreover, all F_{nm} satisfying inequalities (1.1) on Q^n with $M_0 = M_0^F$ and $\delta_0 = \delta_0^F$. By Theorem 2.5, the problem

$$u_t^{nm} + F_{nm}\left(u_{x^ix^j}^{nm},\ u_{x^i}^{nm},\ u^{nm},\ z\right) = 0 \text{ in } Q^n, \tag{1}$$

$$u^{nm} = \varphi \text{ on } \partial' Q^n \tag{2}$$

has a solution possessing well-known properties. In equation (1), by Theorem 1.5(b), we can pass to the limit over an appropriate sequence $m' \to \infty$. Furthermore, in the same equation, by Theorem 1.5(a), we can pass to the limit over a sequence $n' \to \infty$ and obtain a solution to problem (2.3) and (2.4) which belongs to $C^2(Q) \cap C(\bar{Q})$ and possesses property (a) of the current theorem. The uniqueness of a solution belonging to $C^2(Q) \cap C(\bar{Q})$ is asserted in Theorem 1.10 so that it remains for us to construct a solution of problem (2.3) and (2.4) which belongs to the class $C^2(Q) \cap C(\bar{Q})$ and possesses property (b).

It is not convenient to follow this route in considering problem (1) and (2). We therefore use another approach to approximating the solution of the initial problem. Let $Q^{rn} = (0, T - 1/r) \times D^n$, $n \geqq r \geqq 1$ and let u^{rnm} be a solution to problem (1) and (2) such that Q^n is replaced by a smaller domain Q^{rn}. If $\varphi_1^{rn} \in C^{2+\frac{1}{2}\alpha}(\overline{Q^{rn}})$ and $\varphi_1^{rn} = \varphi$ on $\partial_x Q^{rn}$, then, by Theorem 2.5, the norm of u^{rnm} in $C^2(Q^{rn})$ is bounded by a constant depending only on d, K, ε, α, M_0, the functions $\tilde{M}_i^F$ and the norms of ψ, φ_1^{rn}, φ in the spaces $C^{2+\alpha}(E_d)$, $C^{2+\frac{1}{2}\alpha}(Q^{rn})$ and $C^2(\partial_t Q^{rn})$ respectively. Moreover, as above it follows from Theorems 2.5 and 1.5 that u^{rnm} converges to a solution of problem (2.3) and (2.4) over the appropriate sequences. Consequently, assertion (b) will be proved if we manage to construct functions φ_1^{rn} such that $\varphi_1^{rn} = \varphi$ on $\partial_x Q^{rn}$, $\varphi_1^{rn} \in C^{2+\frac{1}{2}\alpha}(\overline{Q^{rn}})$, and, for every r and sufficiently large n, the norms of φ_1^{rn} in $C^{2+\frac{1}{2}\alpha}(Q^{rn})$ do not exceed a constant multiplied by the norm of φ_1 in $C^{2+\alpha}((0, T - 1/r) \times D)$, the constant being dependent only on d, α, and the norm of ψ in the space $C^{2+\alpha}(E_d)$.

The parameter r is no longer significant (nor are the properties of the function F) and, replacing T by $T - 1/r$, we come to the following problem of function theory. Given functions $\varphi_1, \varphi \in C^{2+\alpha}(\bar{Q})$ such that $\varphi_1 = \varphi$ on $\partial_x Q$, it is required to construct functions $\varphi^n \in C^{2+\frac{1}{2}\alpha}(\bar{Q}_n)$ in $Q_n := (0, T) \times D^n$ such that $\varphi^n = \varphi$ on $\partial_x Q_n$ and the norms of φ^n in $C^{2+\frac{1}{2}\alpha}(Q_n)$ with large n are bounded by a constant (depending only on d, α, and the norm of ψ in $C^{2+\alpha}(E_d)$), multiplied by the norm of φ_1 in $C^{2+\alpha}(Q)$.

Using the function φ, we find a function $\tilde{\varphi} \in C^{2+\alpha}(\tilde{Q})$ with $\tilde{Q} = (0, T + 1) \times D$ such that $\tilde{\varphi} = \varphi$ on Q, $\tilde{\varphi}(T + 1, \cdot) = 0$, and the norm of $\tilde{\varphi}$ in $C^{2+\alpha}(\tilde{Q})$ does not exceed the norm of φ in $C^{2+\alpha}(Q)$ multiplied by a constant depending only on d, α, and the norm of ψ in $C^{2+\alpha}(E_d)$. Such a function $\tilde{\varphi}$ can be constructed, for example, by first extending φ with respect to x to E_d to produce a function v, then extending $f := v_t - \Delta v$ into $(0, T + 1) \times E_d$ to produce a function g, then solving the problem $w_t - \Delta w = g$, $w(0, \cdot) = v(0, \cdot)$, and, finally, cutting w near $t = T + 1$. We can treat φ_1 similarly and, to avoid introducing additional notation, we assume that $\varphi, \varphi_1 \in C^{2+\alpha}(\tilde{Q})$, $\varphi(t, x) = \varphi_1(t, x) = 0$ for t close to $T + 1$.

We now define φ^n as solutions to the problems $v_t + \Delta v = 0$ in $(0, T + 1) \times D^n$, $v = \varphi$ on $\partial'(0, T + 1) \times D^n$.

It is clear that the norms of φ^n in $C^{2+\alpha}((0, T + 1) \times D^n)$ are uniformly bounded and, in $\tilde{Q}$, a subsequence $\{\varphi^{n'}\}$ converges to a solution of the problem

$$\tilde{\varphi}_t + \Delta\tilde{\varphi} = 0 \text{ in } \tilde{Q}, \quad \tilde{\varphi} = \varphi_1 \text{ on } \partial'\tilde{Q}$$

as $n' \to \infty$ and, by the uniqueness of the limit, this is also true for $\{\varphi^n\}$.

It remains for us to prove that

$$\varlimsup_{n\to\infty} ||\varphi^n||_{C^{2+\frac{1}{2}\alpha}(Q_n)} \leq N||\tilde{\varphi}||_{C^{2+\alpha}(Q)} \leq N||\varphi_1||_{C^{2+\alpha}(\tilde{Q})}, \tag{3}$$

where N depends only on d, α, and the norm of ψ in $C^{2+\alpha}(E_d)$. We already know that the second inequality is true, so let us prove the first one. Let $z_n \in Q_n$:

$$|\varphi^n_x(z_n)| - \sup\{|\varphi^n_x|, \ Q_n\} \to 0. \tag{4}$$

Since the norms of φ^n in $C^{2+\alpha}(Q_n)$ are uniformly bounded, we can find a sequence $\tilde{z}_n(\delta)$ for any $\delta > 0$ that does not approach ∂Q_n, converges to a point $z(\delta) \in Q$, and such that

$$\lim_{n\to\infty}|\varphi^n_x(\tilde{z}_n(\delta))| > \varlimsup_{n\to\infty}|\varphi^n_x(z_n)| - \delta. \tag{5}$$

The expression on the left is equal to $|\tilde{\varphi}_x(z(\delta))|$ and, by (4) and (5),

$$\lim_{n\to\infty} \sup\{|\varphi^n_x|, \ Q_n\} \leq \sup\{|\tilde{\varphi}_x|, \ Q\}.$$

It follows that

$$\overline{\lim_{n\to\infty}} \|\varphi^n\|_{C^2(Q_n)} \leq \|\tilde{\varphi}\|_{C^2(Q)} \tag{6}$$

Finally, let

$$|\varphi_t^n(z_n^1) - \varphi_t^n(z_n^2)|\rho^{-\alpha/2}(z_n^1, z_n^2) -$$

$$- \sup\{|\varphi_t^n(z^1) - \varphi_t^n(z^2)|\rho^{-\alpha/2}(z^1, z^2) : z^1, z^2 \in Q_n\} \to 0.$$

Since

$$|\varphi_t^n(z^1) - \varphi_t^n(z^2)| \leq N\rho^{\alpha}(z^1, z^2)$$

for all z^1, $z^2 \in Q_n$, where N is independent of n, we can assume that $\underline{\lim}_{n\to\infty} \rho(z_n^1, z_n^2) > 0$, but then an argument similar to the previous one shows that

$$\overline{\lim_{n\to\infty}} \sup\{|\varphi_t^n(z^1) - \varphi_t^n(z^2)|\rho^{-\alpha/2}(z^1, z^2) : z^1, z^2 \in Q_n\}$$

$$\leq \sup\{|\tilde{\varphi}_t(z^1) - \tilde{\varphi}_t(z^2)|\rho^{-\alpha/2}(z^1, z^2) : z^1, z^2 \in Q\}.$$

Similar inequalities are true for φ^n, φ_x^n, and φ_{xx}^n. These inequalities together with (6) yield (3). The theorem is proved.

In the case of elliptic equations, the second part of the proof of the theorem is not needed and, just as above, we can derive the following theorem from Theorem 2.7.

2. THEOREM. Let the conditions of Theorem 1 hold, let φ be independent of t, and let $F \in \bar{F}(\varepsilon, K, D)$. Then problem (2.12) and (2.13) has a solution u belonging to $C^{2+\beta}(\bar{D})$, where $\beta = \alpha_0 \wedge \alpha$, $\alpha_0 = \alpha_0(d, K, \varepsilon) \in (0, 1)$, and the norm of u in this space is bounded by a constant depending only on d, K, ε, α, M_0^F, the functions $\tilde{M}_1^F$, $\tilde{M}_2^F$, and the norms of ψ and φ in $C^{2+\alpha}(E_d)$ and $C^{2+\alpha}(D)$ respectively. Moreover, $|u| \leq M_0^F$ on D.

Recall that the second assertion of Theorem 1.5(a) holds true for functions $F \in \bar{F}$. Therefore, the method used in the first part of the proof of Theorem 1 (see (1) and (2)) and the methods used to prove Theorems 3.1 to 3.3 readily yield the following results, where the domain D can be unbounded and coincide with E_d.

3. THEOREM. For any point $x_0 \in \partial D$, let there exist a closed ball of radius $\rho_0 > 0$ lying in $E_d \setminus D$ and containing x_0. Suppose that $\varphi \in C(\bar{Q})$ and $|\varphi| \leq M_0^F$ on Q. Then problem (2.3) and (2.4) has a solution u possessing the following properties:

(a) $u \in C(\bar{Q})$, $|u| \leq M_0^F$ on Q;

(b) for every $\kappa \in (0, 1)$, the function u belongs to the space $C^{2+\alpha_0}((0, T - \kappa^2) \times D(\kappa))$ where $\alpha_0 = \alpha_0(d, K, \varepsilon) \in (0, 1)$ and the norm of u in this space is bounded by a constant depending only on d, K, ε, M_0^F, κ, and the functions $\tilde{M}_i^F$, $i = 1, 2, 3$.

4. THEOREM. Let the conditions of Theorem 3 hold, let D^0, D^1, and D be domains such that $D^0 \subset D^1 \subset D$, and let $\rho := \text{dist}(\partial D^0, \partial D^1) > 0$. Then:

(a) if $\varphi(T, \cdot) \in C^2(D^1)$, then there exists a solution of problem (2.3) and (2.4) possessing properties (a) and (b) of Theorem 3 and such that $u \in C^2((0, T) \times D^0)$ and the norm of u in this space is bounded by a constant depending only on d, K, ε, M_0^F, ρ, the functions $\tilde{M}_i^F$, $i = 1, 2, 3$, and the norm of $\varphi(T, \cdot)$ in $C^2(D^1)$;

(b) if $\varphi(T, \cdot) \in C^{2+\alpha}(D^1)$, then there exists a solution of problem (2.3) and (2.4) possessing properties (a) and (b) of Theorem 3 and such that $u \in C^{2+\beta}([0, T] \times \bar{D}^0)$ where $\beta = \alpha_0 \wedge \alpha$, $\alpha_0 = \alpha_0(d, K, \varepsilon) \in (0, 1)$, and the norm of u in this space is bounded by a constant depending only on d, K, ε, α, M_0^F, ρ, the functions $\tilde{M}_i^F$, $i = 1, 2, 3$ and the norm of $\varphi(T, \cdot)$ in $C^{2+\alpha}(D^1)$.

5. THEOREM. Let the conditions of Theorem 3 hold, let φ be independent of t, and let $F \in \bar{\mathfrak{F}}(\varepsilon, K, D)$. Then problem (2.12) and (2.13) has a solution enjoying the following properties:

(a) $u \in C(\bar{D})$, $|u| \leq M_0^F$ on D;

(b) for every $\kappa \in (0, 1)$, the function u belongs to the space $C^{2+\alpha_0}(D(\kappa))$, where $\alpha_0 = \alpha_0(d, K, \varepsilon) \in (0, 1)$, and the norm of u in this space is bounded by a constant depending only on d, K, ε, M_0^F, κ, $\tilde{M}_1^F$, $\tilde{M}_2^F$.

We illustrate these results by examples. Note that the following examples are based on Example 1.8 and equivalent transformations of the equations involving the function F in this example.

6. Example. The most important equation of the theory presented in this book is equation (3.2.19). To an extent, this equation plays the same role of a reference point as does the heat conduction equation in the

linear theory. It was the study of this equation that enabled us to establish Akeksandrov's estimates, a Hölder condition for solutions to linear equations with measurable coefficients, and *a priori* estimates in $C^{2+\alpha}$ of solutions to nonlinear equations. Using the notation and assumptions of Theorem 3.2.4, we now show that v_t, v_x, and v_{xx} are not only bounded, but also Hölder continuous on every compactum of E_{d+1} where $f > 0$.

For example, let $f \geqq \varepsilon$ in $Q := (0, 1) \times S_1$. We know that $\lambda v - v_t \geqq 0$, the matrix $(\lambda\delta_{ij}v - v_{x^ix^j})$ is nonnegative, and v_t, v_{xx} are bounded in E_{d+1}. Now the form of equation (3.2.19) implies that $\lambda v - v_t$ and the eigenvalues of the matrix $(\lambda\delta_{ij}v - v_{x^ix^j})$ are bounded away from zero by a strictly positive constant on Q (a.e.). Moreover, v satisfies equation (3.2.18) and the above properties of $\lambda v - v_t$ and $(\lambda\delta_{ij}v - v_{x^ix^j})$ imply that the supremum in (3.2.18) can be taken on Q under the additional condition $r(\omega) \geqq \mu$, $\omega \geqq \mu(\delta_{ij})$ with a constant $\mu > 0$ and v remains to be a solution to equation (3.2.18) thus modified on Q (a.e.). We now take advantage of this and resolve the modified equation with respect to v_t. We then obtain that v satisfies the equation

$$u_t + \max\left(\frac{1}{r(\omega)}\,\omega^{ij}u_{x^ix^j} - \frac{\lambda}{r(\omega)}\,u + \right.$$

$$\left. + \frac{1}{r(\omega)}\sqrt[d+1]{r(\omega)\det\omega\, f}\right) = 0 \tag{7}$$

on Q (a.e.), where the maximum is taken over all $\omega \in \Omega_0$ such that $r(\omega) \geqq \mu$, $\omega \geqq \mu(\delta_{ij})$.

Equation (7) involves maximum, whereas Example 1.8 considers infimum. This disagreement can easily be settled by replacing u in (7) with, say, $(-w)$. Then it is clear that all theorems of this section can be applied to the operator on the left of (7). Specifically, equation (7) with the boundary condition $u = v$ on $\partial'Q$ has a continuous solution whose derivatives u_t, u_x, u_{xx} satisfy a Hölder condition e.g. in $(0, 1/2) \times S_{1/2}$. The derivatives v_t, v_x, and v_{xx} possess the same property since, by Corollary 3.4.3, we have $u = v$ in Q.

7. Example. Let $D \subset E_d$ be a bounded domain and let $d \geqq 2$. Consider the problem

$$\det\left(u_{x^ix^j} + \delta_{ij}\Delta u\right) = f^d\left(u_{x^k}, u, x\right) \text{ in } D, \tag{8}$$

$$u = \varphi \text{ on } \partial D, \tag{9}$$

whose solution we are going to seek among functions belonging to $C^2_{loc}(D) \cap C(\bar{D})$ such that $(u_{x^i x^j} + \delta_{ij}\Delta u) \geq 0$. Assume that f is non-negative, twice continuously differentiable with respect to (u_k, u, x), its second derivatives are bounded on every set of the form

$$\{(u_k, u, x) : |u| + \Sigma|u_k| \leq M, \quad x \in D\},$$

and

$$f_u \geq 0, \quad f \leq M_1(u)(1 + u_k u_k),$$

$$|f_{u_r}|(1 + \Sigma|u_k|) + f_u + |f_{x^r}|(1 + \Sigma|u_k|)^{-1}$$

$$\leq M_2(u)(1 + u_k u_k),$$

where $M_1(u)$ and $M_2(u)$ are some continuous functions of u, $M_1(u)$ increases in u, and

$$\lim_{u \to -\infty} M_1(u) = 0. \tag{10}$$

By Lemma 3.2.2, equation (8) together with the condition $\left(u_{x^i x^j} + \delta_{ij}\Delta u\right) \geq 0$ are equivalent to the equation

$$\min\left[a^{ij}\left(u_{x^i x^j} + \delta_{ij}\Delta u\right) - d\sqrt[d]{\det a}\, f\left(u_{x^k}, u, x\right) d\right] = 0, \tag{11}$$

where the minimum is taken over all nonnegative $d \times d$ matrices a such that tr a = 1. In view of the obvious equality

$$a^{ij}\left(u_{x^i x^j} + \delta_{ij}\Delta u\right) = (a^{ij} + \delta^{ij})u_{x^i x^j},$$

the operator on the left of (11) satisfies all conditions of Example 1.8 except condition (1.8).

In order that the latter condition hold, we modify equation (11) using assumption (10) so that the solution of the problem with the new equation and condition (9) simultaneously is a solution of problem (11) and (9) and, consequently, problem (8) and (9). Let $D \subset S_R$. We choose a large constant $N_1 \geq R^2$ such that the inequalities

$$M_1(R^2 - 2N_1)(1 + 4R^2) \leq 2(d + 1), \quad N_1 \geq ||\varphi||_{C(D)} \tag{12}$$

hold. Then we somehow define a sufficiently smooth function $c(u)$ so that $c(u) = 0$ for $-2N_1 \leqq u \leqq 2N_1$, $c(-3N_1) - M_1(-3N_1) \geqq 1$, $c(3N_1) \leqq -1$, $c(u) \leqq 0$ for $u \geqq 0$, and $c(u) \geqq 0$ for $u \leqq 0$. It is clear that the left-hand side of the equation

$$\min\left[(a^{ij} + \delta^{ij})u_{x^ix^j} + c(u) - \sqrt[d]{\det a}\, f\left(u_{x^k}, u, x\right)^d\right] = 0 \quad (13)$$

satisfies all conditions of Example 1.8 including condition (1.8) with $M_0 = 3N_1$, $\delta_0 = 1$. Therefore, we can apply, for example, Theorems 2 and 5 to problem (13) and (9). Finally, we show that every solution $u \in C^2_{loc}(D) \cap C(\bar{D})$ of problem (13) and (9) also is a solution of problem (11) and (9). In order to do that, it obviously suffices to show that $|u| \leqq 2N_1$ in D.

If $u > 2N_1$ somewhere in D, then, for sufficiently small $\delta > 0$, the function $u + \delta|x|^2$ attains its maximum on the interior of D, say, at a point x_0 and $u(x_0) \geqq 0$. It is clear that $(u_{x^ix^j}(x_0)) \leqq -2\delta(\delta_{ij})$, $c(u(x_0)) \leqq 0$, $f(u_x(x_0), u(x_0), x_0) \geqq 0$, and equation (13) cannot be satisfied at this point. Consequently, $u \leqq 2N_1$ in D.

If $u < -2N_1$ somewhere in D, then the function $u + N_1 - |x|^2$, which is strictly smaller than $(-N_1)$ at some points of D and exceeds $(-N_1)$ on the boundary, attains its maximum at a point x_0. At this point, $(u_{x^ix^j}) \geqq 2(\delta_{ij})$, $u_x = -2x_0$, $u(x_0) \leqq |x_0|^2 - 2N_1 \leqq R^2 - 2N_1 \leqq 0$, and the expression in the brackets of (13) is greater than

$$2(d + 1) - f(u_x, u, x_0) \geqq 2(d + 1) -$$

$$- (1 + 4|x_0|^2)M_1(R^2 - 2N_1).$$

The latter is strictly positive, by virtue of (12), and equation (13) cannot be satisfied at x_0 either. Thus, $u \geqq -2N_1$ in D.

8. Example. Using the same Lemma 3.2.2, we can study the first boundary-value problem for the equation

$$(u_t + \Delta u) \det\left(u_{x^ix^j} + \delta_{ij}(u_t + \Delta u)\right)$$

$$= f^{d+1}(u_x, u, t, x),$$

where f satisfies the conditions of the previous example and is once continuously differentiable with respect to all arguments, while the solution is sought for in the class of functions such that $u_t + \Delta u \geqq 0$ and the matrix $(u_{x^i x^j} + \delta_{ij}(u_t + \Delta u))$ is nonnegative.

9. Example. The one-dimensional equation $u'' - |u'| - 1 = 0$ fits our theory despite the fact that $|u_1|$ is not a smooth function of u_1. This is due to the fact that the equation can be rewritten as

$$\min(u'' - u' - 1,\ u'' + u' - 1) = 0.$$

Note that this equation on $(-1, 1)$ with the boundary condition $u(\pm 1) = 0$ can easily be solved and we then see that its solution $\exp(|x|) - e + 1 - |x|$ has the third derivative which is discontinuous at zero. Therefore, our results concerning the fact that $u \in C^{2+\alpha}$ are quite reasonable.

A large number of examples can be constructed by considering minima of a finite or infinite number of operators. For example,

$$\Delta u + |u_x|, \quad \Delta u + \frac{1}{2}(u_{x^1 x^1})_-,$$

$$\Delta u - \frac{1}{2}(u_{x^1 x^1})_- - \frac{1}{2}(u_{x^2 x^2})_- . \tag{14}$$

The operator

$$\Delta u - \frac{1}{2}(u_{x^1 x^1})_- - \frac{1}{2}(u_{x^2 x^2})_- + \frac{1}{2}(u_{x^3 x^3})_+$$

resembles the latter in (14), but it does not fit our theory and we know nothing about the solvability of equations with such an operator.

6.5. The Existence of a Solution for $F \in \bar{\bar{F}}_0$

In this section we prove theorems on solvability of problem (2.3) and (2.4) with no assumptions concerning the smoothness of F with respect to t. The corresponding results and, in the first place, the methods used to establish them will be needed in the following chapter where we shall consider degenerate equations. In contrast with the usual situation, the solution of a degenerate equation is not smooth with respect to x and the smoothness cannot be ensured by the differentiability of initial data with respect to t. Therefore, the smoothness with respect to t is, to an extent, excessive, and we show what happens if the smoothness assumption is abandoned.

Given constants $K \geqq \varepsilon > 0$, $T > 0$, a domain $D \subset E_d$, $Q = (0, T) \times D$, suppose that, for every point $x_0 \in \partial D$, there exists a closed ball of radius $\rho_0 > 0$ lying in $E_d \setminus D$ and containing x_0. Obviously, if $\partial D = \emptyset$, i.e., if $D = E_d$, this condition holds.

1. Definition. Let a function $F(u_{ij}, u_i, u, z)$ be defined for $z \in Q$ and real-valued u_{ij}, u_i $(i, j = 1, \ldots, d)$, and u. We write $F \in \bar{F}_0(\varepsilon, K, Q)$ if there exist a countable set Ω and, for every $\omega \in \Omega$, Borel functions $a^{ij}(\omega, u_k, u, t, x)$, $i, j = 1, \ldots, d$, $b(\omega, u_k, u, t, x)$ defined for $(t, x) \in Q$ and all real-valued $u_1, \ldots, u_d$, u such that:

(a) $F(u_{ij}, u_i, u, z) = \inf_{\omega\in\Omega}[a^{ij}(\omega, u_k, u, z)u_{ij} + b(\omega, u_k, u, z)]$ on the set of definition of F;

(b) for all $\omega \in \Omega$, $t \in (0, T)$, the functions a^{ij} and b are twice continuously differentiable with respect to (u_k, u, x) and their second derivatives with respect to (u_k, u, x) are bounded on every set of the form

$$\{(\omega, u_k, u, z) : \omega \in \Omega, \Sigma|u_k| + |u| \leqq q, z \in Q\}$$

by a constant $\bar{M}_2^F(q)$;

(c) for all $\omega \in \Omega$, $z \in Q$, $s, r, p = 1, \ldots, d$, $\lambda \in E_d$, $u_1, \ldots, u_d$, u the conditions

$$a^{sr} = a^{rs}, \quad \varepsilon|\lambda|^2 \leqq a^{ij}\lambda^i\lambda^j \leqq K|\lambda|^2,$$

$$|b| \leqq \bar{M}_1^F(u)(1 + u_iu_i),$$

$$\left|a^{sp}_{u_r}\right|\left(1 + \sum_i|u_i|\right) + |a^{sp}_u| + |a^{sp}_{x^r}|\left(1 + \sum_i|u_i|\right)^{-1} \leqq \bar{M}_1^F(u),$$

$$\left|b_{u_r}\right|\left(1 + \sum_i|u_i|\right) + |b_u| + |b_{x^r}|\left(1 + \sum_i|u_i|\right)^{-1} \leqq \bar{M}_1^F(u)(1 + u_iu_i),$$

hold, where $\bar{M}_1^F(u)$ is a continuous function of u;

(d) there exist constants $\bar{\delta}_0^F$ and $\bar{M}_0^F > 0$ such that

$$b(\omega, 0, -\bar{M}_0^F, z) \geqq \bar{\delta}_0^F, \quad b(\omega, 0, \bar{M}_0^F, z) \leqq -\bar{\delta}_0^F$$

for all $\omega \in \Omega$, $z \in Q$.

It is clear that the constants $\bar{\delta}_0^F$ and $\bar{M}_0^F$ and the functions $\bar{M}_1^F$ and $\bar{M}_2^F$ are associated with the function F under consideration, as can be seen in the adopted denotation. Moreover, we can and will assume that the functions $\bar{M}_1^F$ and $\bar{M}_2^F$ increase in q and $|u|$.

The major result of this section is the following.

2. THEOREM. Let $F \in \bar{F}_0(\varepsilon, K, Q)$, $\varphi \in C(Q)$, and $|\varphi| \leq \bar{M}_0^F$ in Q. Then there exists a function $u \in C(\bar{Q})$ possessing the following properties:

(a) $u \in W^{1,2}_{2,loc}(Q)$, u satisfies equation (2.3) almost everywhere on Q, $u = \varphi$ on $\partial' Q$, and $|u| \leq \bar{M}_0^F$ on Q;

(b) for every bounded domain Q_0 of the form $(0, T^0) \times D^0$ with $T^0 < T$ and $\bar{D}^0 \subset D$, and unit vectors $\xi \in E_d$, the norms of u_t, u_{xx} in $L_2(Q_0)$, the function $u_{(\xi)(\xi)}$ (a.e. Q_0), and the norms of u, u_x in $C^\alpha(Q_0)$, where $\alpha = \alpha(d, K, \varepsilon) \in (0, 1)$, are bounded from above by a constant depending only on d, K, ε, the domains Q_0, Q, the number $\bar{M}_0^F$, and the functions $\bar{M}_1^F$, $\bar{M}_2^F$.

To prove the theorem, we need two lemmas. Before formulating these lemmas, we extend a^{ij} and b to $E_1 \times D$ using formulas like $v(t, x) = v(\frac{1}{2}T, x)$ for $t \leq 0$ and $t \geq T$; furthermore, we define a nonnegative function $\xi \in C_0(E_1)$ such that $\xi(t) = 0$ for $|t| \geq 1$ and $\int \zeta \, dt = 1$. Also, we take a sequence of numbers $\nu_n \to \infty$, put $\zeta^n(t) = \nu_n \zeta(\nu_n t)$, and denote

$$a_n^{ij}(\omega, u_k, u, t, x) = a^{ij}(\omega, u_k, u, t, x) * \zeta^n(t),$$

$$b_n(\omega, u_k, u, t, x) = b(\omega, u_k, u, t, x) * \zeta^n(t),$$

where the functions are convolved only with respect to the variable t.

3. LEMMA. Let $q \geq 0$ and let $\Gamma \subset D$ be a compactum. Then the family of functions a_n^{ij}, b_n, a^{ij}, and b regarded as functions only of (u_k, u, x) are uniformly (in $n \geq 1$, $\omega \in \Omega$, $t \in (0, T)$) bounded and equicontinuous on the set $\{(u_k, u, x) : x \in \Gamma, \Sigma|u_k| + |u| \leq q\}$. Moreover, for every $\omega \in \Omega$, on $(0, T)$ there exists a set I of full Lebesgue measure such that, for $t \in I$, we obtain $a_n^{ij} \to a^{ij}$, $b_n \to b$ uniformly on $\{(u_k, u, x) : x \in \Gamma, \Sigma|u_k| + |u| \leq q\}$.

Proof. The first assertion immediately follows from Definition 1(c). By virtue of the first assertion, we can, while calculating the supremum of $|a_n^{ij} - a^{ij}|$ and $|b_n - b|$ over the specified set of (u_k, u, x), restrict our consideration to the maximum of these values over a finite set of values (u_k, u, x) making as small error as desired. The intersection of a finite or a countable number of sets of full measure is a set of full measure itself. Therefore, we should study the convergence of a_n^{ij}, b_n to a^{ij}, b only for fixed (u_k, u, x). It now remains to recall that, by Lebesgue's theorem, for all ω, u_k, u, x, we obtain $a_n^{ij} \to a^{ij}$ $b_n \to b$ for almost all $t \in (0, T)$. The lemma is proved.

4. LEMMA. Let the set Ω consist of a finite number of points. Then the assertions of Theorem 2 hold true.

Proof. Let $\Omega = \{\omega_1, \ldots, \omega_r\}$. We denote

$$F_s(u_{ij}, u_k, u, z) = a^{ij}(\omega_s, u_k, u, z)u_{ij} + b(\omega_s, u_k, u, z),$$

$$F_s^n(u_{ij}, u_k, u, z) = a_n^{ij}(\omega_s, u_k, u, z)u_{ij} + b_n(\omega_s, u_k, u, z),$$

$$F^n = F_1^n \wedge \ldots \wedge F_r^n.$$

It is clear that $F_s^n \in F(\varepsilon, K, Q)$ and the functions $M_1^{F_s^n}$ and $M_2^{F_s^n}$ (see Definition 5.5.1) can be defined as depending only on d, $\bar{M}_1^F$, $\bar{M}_2^F$. Of course, the function $M_3^{F_s^n}$ depends on n, but this is of no importance. Also, by Lemma 1.3, $F_s^n \in \bar{F}(\varepsilon, K, Q)$ and $\tilde{M}_1^{F_s^n}$ and $\tilde{M}_2^{F_s^n}$ can again be defined using only d, $\bar{M}_1^F$, and $\bar{M}_2^F$. Moreover, we can put $\delta_0^{F_s^n} = (1/3)\bar{\delta}_0^F$, $M_0^{F_s^n} = \bar{M}_0^F$. Finally, by Theorem 1.5, we obtain $F^n \in \bar{F}(\varepsilon, K, Q)$ and Theorem 4.3 yields a function u^n possessing the properties specified in it and such that

$$u_t^n + F^n\left(u_{x^ix^j}^n, u_{x^i}^n, u^n, z\right) = 0 \text{ in } Q,$$

$$u^n = \varphi \text{ on } \partial'Q. \tag{1}$$

Furthermore, the method in the proof of Theorem 3.1 together with Theorem 5.4.6 enable us to assert that the solution constructed in Theorem 3.1 possesses property (b) of the current theorem. We could, therefore, assert this property in Theorem 4.3 too. In short, the norms of u_t^n, u_{xx}^n in $L_2(Q_0)$, the functions $u_{(\xi)(\xi)}^n$ $(|\xi| = 1)$, and the norms of u^n, u_x^n in $C^\alpha(Q_0)$ are bounded from above by a constant depending only on d, K, ε, Q_0, Q, $\bar{M}_0^F$, $\bar{M}_1^F$, $\bar{M}_2^F$. It obviously follows that, for a subsequence $n' \to \infty$, the functions $u^{n'}$ converge uniformly on every set of the form $[0, T^0] \times \Gamma$, where $T^0 < T$ and the compactum Γ belongs to D, to a function u such that assertion (b) holds true. The barriers constructed in the proof of Theorem 3.1 enable us to estimate the difference $|u^n(z) - \varphi(z_0)|$ for $z_0 \in \overline{\partial' Q}$ uniformly in n. Consequently, the function u can be extended to a function which is continuous in $\bar{Q}$ and equal to φ on $\partial' Q$.

It remains for us to prove that u satisfies equation (2.3) almost everywhere in Q. We prove this using the theorem on passage to the limit under the nonlinear operator in Seciton 3.6. From (1) with any $m_0 \geqq 1$ and $n \geqq m_0$ we obtain

$$\inf_{m>m_0,\ s\leqq r} \left[a_m^{ij}(\omega_s, u^m_{x^k}, u^m, z) u^n_{x^i x^j} + u^n + \right.$$
$$\left. + b_m(\omega_s, u^m_{x^k}, u^m, z) \right] \leqq 0$$

in Q. With no loss of generality, we can assume that the above subsequence $\{n'\}$ coincides with $\{n\}$. By Theorem 3.6.3, it follows that

$$\inf_{m\geqq m_0, s\leqq r} \left[a_m^{ij}(\omega_s, u^m_{x^k}, u^m, z) u_{x^i x^j} + u_t + \right.$$
$$\left. + b_m(\omega_s, u^m_{x^k}, u^m, z) \right] \leqq 0 \qquad (2)$$

almost everywhere in Q. On a set of full measure, these inequalities hold for all m_0 at once.

Moreover, by Lemma 3, on a set of full measure in Q, $a_m^{ij}(\omega_s, u_k, u, z)$ and $b_m(\omega_s, u_k, u, z)$ converge to $a^{ij}(\omega_s, u_k, u, z)$ and $b(\omega_s, u_k, u, z)$ uniformly with repsect to any compactum in the domain of variation of u_k and u. We now fix an appropriate point z, let $m_0 \to \infty$ in (2), and take advantage of the convergence of $u^m_x(z)$ and $u^m(z)$ to $u_x(z)$ and $u(z)$. We then obtain

$$u_t + F\left(u_{x^i x^j}, u_{x^i}, u, z\right) \leqq 0 \text{ (a.e.Q)}. \qquad (3)$$

Let us prove the opposite inequality. From (1) with $s = 1, \ldots, r$ we obtain

$$u^n_t + a_n^{ij}\left(\omega_s, u^n_{x^k}, u^n, z\right) u^n_{x^i x^j} + b_n\left(\omega_s, u^n_{x^k}, u^n, z\right) \geqq 0. \qquad (4)$$

In every domain of the form Q_0, the functions u^n_t and u^n_{xx} weakly converge to u_t and u_{xx} in L_2, while, by Lemma 3, the functions $a_n^{ij}(\omega_s, u^n_{x^k}, u^n,$

z) and $b_n(\omega_s, u^n_{x^k}, u^n, z)$ strongly converge to $a^{ij}(\omega_s, u_{x^k}, u, z)$ and $b(\omega_s, u_{x^k}, u, z)$ in L_2. Such a situation is known to admit passage to the limit in inequality (4). Thus,

$$u_t + a^{j}\left(\omega_s, u_{x^k}, u, z\right)u_{x^ix^j} + b\left(\omega_s, u_{x^k}, u, z\right) \geq 0 \text{ (a.e.}Q)$$

for any s. This yields the inequality opposite to (3) and the lemma is proved.

5. Proof of Theorem 1. Let $\Omega = \{\omega_1, \omega_2, \ldots\}$. Using Lemma 4, we construct function u_n such that $u_n \in C(\bar{Q})$, $u_n = \varphi$ on $\partial' Q$,

$$\min_{s \leq n}\left[u_{nt} + a^{ij}\left(\omega_s, u_{nx^k}, u_n, z\right)u_{nx^ix^j} + \right.$$
$$\left. + b\left(\omega_s, u_{nx^k}, u_n, z\right)\right] = 0 \text{ (a.e.}Q).$$

The argument of the previous proof shows that it suffices to pass to the limit for a subsequence $n' \to \infty$ in the latter relationship assuming that $u_{n'} \to u$ and $u_{n'x} \to u_x$ uniformly on compacta of Q and $u_{n't} \to u_t$ and $u_{n'xx} \to u_{xx}$ weakly in L_2 in every domain of the form Q_0.

It is clear that, for any $m_0 \geq 1$ with $r \in \{n'\}$, $r \geq m_0$,

$$\inf_s \inf_{m \geq m_0,\, m \in \{n'\}}\left[u_{rt} + a^{ij}\left(\omega_s, u_{mx^k}, u_m, z\right)u_{rx^ix^j} + \right.$$
$$\left. + b\left(\omega_s, u_{mx^k}, u_m, z\right)\right] \leq 0.$$

Hence, letting $r \to \infty$ and using Theorem 3.6.3, then letting $m_0 \to \infty$ and using Lemma 3, we obtain inequality (3) for u. The opposite inequality can be derived as in the previous proof. The theorem is proved.

6.6. Comments

Section 1. In general, the sequence of arguments of this section are taken from Krylov [21], where the author introduced classes similar to $\tilde{F}$ and $\bar{F}$ and proved assertions like 4 and 5.

Section 2. The assertions of this section of different strength and generality were earlier proved by Evans [2, 3], Krylov [19 – 21], and Trudinger [2]. In their most general form, these results seem to be presented here for the first time.

Section 3. Krylov [19] described a method for constructing a solution to a nonlinear parabolic equation in a nonsmooth noncylindric domain on the basis of solving Cauchy's problem. Theorem 3 is close to the corresponding result established in Trudinger [2].

Section 4. Theorems like Theorems 1 to 5 were elaborated in Krylov [21]. Saforov [8] developed a remarkable technique for nonlinear equations which, even in the case of linear equations, yields unimprovable results on the solvability in $C^{2+\alpha}$. The class of the equations he studied is narrower, but the results are much stronger than those in this book. He examined (see Krylov [27]) not only the Dirichlet problem, but also the Neumann problem. As for the Neumann problem, see also Lieberman and Trudinger [1].

Chapter 7

DEGENERATE NONLINEAR EQUATIONS IN THE WHOLE SPACE

This chapter deals with Bellman equations in the whole space. The equations in question are of the form $\inf_{\omega}(L(\omega)u + f(\omega)) = 0$, where $L(\omega)$ are linear, generally speaking degenerate, elliptic or parabolic operators. The solvability of such equations in the class of smooth functions cannot generally be proved and we therefore make considerable use of functions which can be represented as a sum of convex and smooth functions. Here the notions of the derivatives/measures and the measure lattice come to play an increasingly important role and all the power of the corresponding results of Sections 1.2, 3.6, and 3.7 is used. It should be said that Sections 3.6 and 3.7 dealt with operators of the form $\sup_{\omega}(L(\omega)u + f(\omega))$. Here it is more convenient to consider inf rather than sup. It is obvious that $\inf(L(-u)-f) = -\sup(Lu + f)$.

Also, note that this section partly aims at preparing some techniques for Chapter 8; we would like to show that these techniques are simple, natural, and enable us to study equations in the whole space easily.

7.1. Permutation of a Differential Operator with an Elliptic Operator.

The purpose of this section is to prove two formulas which imply that the first and second derivatives of the result produced by an elliptic operator acting on a function can be written as results of applying an elliptic operator to the first and second derivatives of the original function. The benefits of such formulas are clear. Actually, if the original function satisfies an elliptic equation, then its derivatives also satisfy elliptic equations and they can readily be estimated using the maximum principle.

It should be said that the formulas in question have, in fact, long been widely used in the probabilistic version of the proof of the smoothness of solutions to degenerate equations (see, e.g. Gikhman [1], Blagoveshchenskii and Freidlin [1], Freidlin [1], Krylov [10], and others). Therefore, we need only translate the corresponding formulas into the language of differential equations. These formulas involve a parameter π, and if $\pi = 0$, then we translate conventional formulas of the probability theory into another language, but if $\pi \neq 0$ then comparatively recent formulas are translated which can be found in Hausman [1] and Bismut [1]. Such parameters have arisen in connection with the well-known Girsanov theorem in the theory of random processes and the fact that the change of measure does not commute with differentiation of a solution of a stochastic equation with respect to initial data. From the point of view of the theory of differential equations the introduction of the parameters π and Girsanov's theorem resembles (and Girsanov's theorem itself contains) the method of reducing differential equations with lower terms to differential equations having only higher derivatives, i.e., the method suggested in Kruzhkov and Oleinik [1].

Let $\sigma^{ik}(x)$, $i = 1, \ldots, d$, $k = 1, \ldots, d_1$, $b^i(x)$, $i = 1, \ldots, d$, and $c(x)$ be functions in E_d which are continuously differentiable with respect to x. The number d_1 may differ from d and this fact is very important for certain applications. We put

$$a^{ij}(x) = \frac{1}{2}\sigma^{ik}(x)\sigma^{jk}(x),$$

$$Lu(x) = a^{ij}(x)u_{x^ix^j}(x) + b^i(x)u_{x^i}(x) + c(x)u(x). \tag{1}$$

On twice differentiating the latter expression with respect to x and completing the result to obtain an elliptic operator applied to the second derivative of u, we obtain a monstrously long formula which is very inconvenient to deal with. We therefore adopt the following notation. We introduce d-dimensional vectors $b = (b^i)$ and $\sigma^k = (\sigma^{ik}$, $i = 1, \ldots, d)$ and recall that for $\xi, \eta \in E_d$

$$u_{(\xi)(\eta)} := \xi^i\eta^j u_{x^ix^j}.$$

Then equality (1) is rewritten as

$$Lu(x) = \frac{1}{2}\delta^{kr}u_{(\sigma^k(x))(\sigma^r(x))}(x) + u_{(b(x))}(x) + c(x)u(x). \tag{2}$$

Note that any operator written in this form is automatically an elliptic one since for $\xi \in E_d$

$$a^{ij}\xi^i\xi^j = \frac{1}{2}\sum_{k=1}^{d_1}\left(\sum_{i=1}^{d}\xi^i\sigma^{ik}\right)^2 \geq 0.$$

Furthermore, we consider the space E_{d+1} whose points $\tilde{\xi}$ are viewed as pairs (ξ, ξ^{d+1}), where $\xi \in E_d$ and $\xi^{d+1} \in E_1$. Fix a vector $\pi_1 \in E_{d_1}$ and introduce functions $\tilde{\sigma}^k(x, \tilde{\xi})$, $\tilde{b}(x, \tilde{\xi}) \in E_{2d+1}$, $k = 1, \ldots, d_1$ defined on $E_d \times E_{d+1}$ using the formulas

$$\tilde{\sigma}^{ik}(x, \tilde{\xi}) = \sigma^{ik}(x), \quad \tilde{b}^{i}(x, \tilde{\xi}) = b^{i}(x), \quad i \leq d,$$

$$\tilde{\sigma}^{d+i,k}(x, \tilde{\xi}) = \sigma^{ik}_{(\xi)}(x), \quad \tilde{b}^{d+i}(x, \tilde{\xi}) = b^{i}_{(\xi)}(x) - \sigma^{ik}(x)\pi^{k}_{1}, \quad i \leq d, \tag{3}$$

$$\tilde{\sigma}^{2d+1,k}(x, \tilde{\xi}) = \pi^{k}_{1}, \quad \tilde{b}^{2d+1}(x, \tilde{\xi}) = c_{(\xi)}(x).$$

For smooth $v(x, \tilde{\xi})$, we conventionally define $v_{(\zeta)}(x, \tilde{\xi})$ with $\zeta \in E_{2d+1}$ as

$$v_{x^i}(x, \tilde{\xi})\zeta^{i} + v_{\xi^i}(x, \tilde{\xi})\zeta^{d+i} = \sum_{i=1}^{d} v_{x^i}(x, \tilde{\xi})\zeta^{i} + \sum_{i=1}^{d+1} v_{\xi^i}(x, \tilde{\xi})\zeta^{d+i},$$

then introduce $v_{(\zeta)(\eta)}(x, \tilde{\xi})$ by a similar formula, and denote

$$\tilde{L}v(x, \tilde{\xi}) = \frac{1}{2}\delta^{kr} v_{(\tilde{\sigma}^{k}(x, \tilde{\xi}))(\tilde{\sigma}^{r}(x, \tilde{\xi}))}(x, \tilde{\xi}) + v_{(\tilde{b}(x, \tilde{\xi}))}(x, \tilde{\xi}) + c(x)v(x, \tilde{\xi}) \tag{4}$$

for $(x, \tilde{\xi}) \in E_d \times E_{d+1}$.

We have described the rule of constructing the operator L acting on functions of $(x, \tilde{\xi})$ using an operator L defined on functions of x. To calculate the new operator $\tilde{L}$, one also needs to know π and σ^k in addition to formula (1), and therefore the denotation of this operator should involve π and σ^k. However, for the sake of brevity, we omit the indices and hope that the reader will understand which π and σ^k are meant from the context. Prior to formulating basic properties of the operator $\tilde{L}$, we set

$$\partial(\tilde{\xi})u = u_{(\xi)} + \xi^{d+1}u = \xi^{1}u_{x^1} + \dots + \xi^{d}u_{x^d} + \xi^{d+1}u \tag{5}$$

for

$$\tilde{\xi} = (\xi, \xi^{d+1}) \in E_{d+1}.$$

1. LEMMA. (a) If $u \in C^{2}_{loc}(E_d)$, $u = u(x)$, then $\tilde{L}u(x, \tilde{\xi}) = Lu(x)$.

(b) If $u^{i} \in C^{2}_{loc}(E_d)$, $u^{i} = u^{i}(x)$, $i = 1, \dots, d + 1$, then

$$\tilde{L}[\xi^i u^i] = \xi^i L u^i + u^i(b^i_{(\xi)} - \sigma^{ik}\pi^k_i) +$$

$$+ u^{d+1} c_{(\xi)} + u^i_{(\sigma^k)} \sigma^{ik}_{(\xi)} + u^{d+1}_{(\sigma^k)} \pi^k_1, \tag{6}$$

where

$$\tilde{\xi} = (\xi, \xi^{d+1}) = (\xi^1, \ldots, \xi^{d+1}),$$

the sum on the left is taken over all possible values of i, i.e., $i = 1, \ldots, d + 1$, while on the right, the sum is also taken over all $k = 1, \ldots, d_1$, the sum in the second and the fourth member on the right of (6) being taken only for $1 \leqq i \leqq d$ since b^{d+1} and σ^{d+1} are not defined.

(c) If $u \in C^3_{loc}(E_d)$ and $u = u(x)$, then

$$\tilde{L}[\partial(\tilde{\xi})u] = \partial(\tilde{\xi})Lu.$$

(d) If a function $v(x, \tilde{\xi})$ of the class $C^2_{loc}(E_{2d+1})$ attains its local maximum at a point $(x_0, \tilde{\xi}_0)$, then $\tilde{L}v - cv \leqq 0$ at this point.

Proof. Assertion (a) immediately follows from (4). To prove (b), let us note that for any operator of the form (2)

$$L(uw) = uLw + wLu + u_{(\sigma^k)} w_{(\sigma^k)} - cuw. \tag{7}$$

It remains for us to use (a) and the fact that

$$u^i_{(\tilde{\sigma}^k)} = u^i_{(\sigma^k)}, \quad (\xi^i)_{(\tilde{\sigma}^k)} = \sigma^{ik}_{(\xi)} \text{ for } i \leqq d, \quad (\xi^{d+1})_{(\tilde{\sigma}^k)} = \pi^k_1,$$

$$(\xi^i)_{(\tilde{\sigma}^k)(\tilde{\sigma}^r)} = 0, \quad (\xi^i)_{(\tilde{b})} = b^i_{(\xi)} - \sigma^{ik}\pi^k_1 \text{ for}$$

$$i \leqq d, \quad (\xi^{d+1})_{(\tilde{b})} = c_{(\xi)}.$$

We now prove (c). By virtue of (b),

$$\tilde{L}[\partial(\tilde{\xi})u] = \xi^i L(u_{x^i}) + \xi^{d+1}Lu + u_{x^i}(b^i_{(\xi)} - \sigma^{ik}\pi^k_1) +$$

$$+ uc_{(\xi)} + u_{x^i x^j}\sigma^{jk}\sigma^{ik}_{(\xi)} + u_{(\sigma^k)}\pi^k_1.$$

This expression coincides with $\partial(\tilde{\xi})Lu$ since

$$u_{x^i}\sigma^{ik} = u_{(\sigma^k)}, \quad \sigma^{jk}\sigma^{ik}_{(\xi)}u_{x^ix^j} = \frac{1}{2}(\sigma^{ik}\sigma^{jk})_{(\xi)}u_{x^ix^j} = a^{ij}_{(\xi)}u_{x^ix^j}.$$

Assertion (d) follows from the conventional maximum principle applied to the elliptic operator $\tilde{L}$. The lemma is proved.

In what follows, we shall also need the following formulas resulting from (3) and rule (7):

$$\tilde{L}^0(\xi^i\xi^j) = \xi^i(b^j_{(\xi)} - \sigma^{jk}\pi_1^k) + \xi^j(b^i_{(\xi)} - \sigma^{ik}\pi_1^k) + \\ + \sigma^{ik}_{(\xi)}\sigma^{jk}_{(\xi)}, \quad i, j \leqq d, \tag{8}$$

$$\tilde{L}^0[(\xi^{d+1})^2] = 2\xi^{d+1}c_{(\xi)} + \pi_1^k\pi_1^k, \tag{9}$$

$$\tilde{L}^0[(\xi^{d+1})^4] = 2(\xi^{d+1})^2[2\xi^{d+1}c_{(\xi)} + |\pi_1|^2] + \\ + 4(\xi^{d+1})^2|\pi_1|^2, \tag{10}$$

where $\tilde{L}^0 = \tilde{L} - c$, and if $B^{ij} = B^{ji} \in C^2_{loc}(E_d)$, $B^{ij} = B^{ij}(x)$, $i, j = 1, \dots, d$, $(B\xi, \xi) = \sum_{i,j=1}^{d} B^{ij}\xi^i\xi^j$, and $L^0 = L - c$, then

$$\tilde{L}^0(B\xi, \xi) = L^0(B\xi, \xi) + 2(B\xi, b_{(\xi)} - \sigma^k\pi_1^k) + \\ + (B\sigma^k_{(\xi)}, \sigma^k_{(\xi)}) + 2(B_{(\sigma^k)}\xi, \sigma^k_{(\xi)}), \tag{11}$$

$$\tilde{L}^0[(B\xi, \xi)^2] = 2(B\xi, \xi)[L^0(B\xi, \xi) + 2(B\xi, b_{(\xi)} - \sigma^k\pi_1^k) + \\ + (B\sigma^k_{(\xi)}, \sigma^k_{(\xi)}) + 2(B_{(\sigma^k)}\xi, \sigma^k_{(\xi)})] + \\ + \sum_{k=1}^{d_1}(B_{(\sigma^k)}\xi + 2B\sigma^k_{(\xi)}, \xi)^2. \tag{12}$$

We now consider the question of the commutability of the double differentiation operator and L. It is hereafter assumed that σ, b, and c are twice continuously differentiable. Along with the vector π_1, we also fix a vector $\pi_2 \in E_{d+1}$ and, for $x \in E_d$, $\tilde{\xi} = (\xi, \xi^{d+1})$, and $\tilde{\eta} = (\eta, \eta^{d+1}) \in E_{d+1}$ we define vectors $\tilde{\sigma}^k(x, \tilde{\xi}, \tilde{\eta})$, $\tilde{b}(x, \tilde{\xi}, \tilde{\eta}) \in E_{3d+2}$, $k = 1, \ldots, d_1$, supplementing the above vectors $\tilde{\sigma}^k(x, \tilde{\xi})$ and $\tilde{b}(x, \tilde{\xi})$ (see (3)) by coordinates $i = 2d + 2, \ldots, 3d + 2$ defined as

$$\tilde{\sigma}^{2d+1+i,k} = \sigma^{ik}_{(\xi)(\xi)} + \sigma^{ik}_{(\eta)},$$

$$\tilde{b}^{2d+1+i} = b^i_{(\xi)(\xi)} + b^i_{(\eta)} - 2\pi^k_1\sigma^{ik}_{(\xi)} - \pi^k_2\sigma^{ik}, \quad i \leq d,$$

$$\tilde{\sigma}^{3d+2,k} = \pi^k_2, \quad \tilde{b}^{3d+2} = -|\pi_1|^2 + c_{(\xi)(\xi)} + c_{(\eta)}. \tag{13}$$

For the functions $v(x, \tilde{\xi}, \tilde{\eta})$ with $x \in E_d$ and $\tilde{\xi}, \tilde{\eta} \in E_{d+1}$, we define the expression $\tilde{L}v(x, \tilde{\xi}, \tilde{\eta})$ using the same formula (4) where we replace $\tilde{\sigma}(x, \tilde{\xi})$ and $\tilde{b}(x, \tilde{\xi})$ by $\tilde{\sigma}(x, \tilde{\xi}, \tilde{\eta})$ and $\tilde{b}(x, \tilde{\xi}, \tilde{\eta})$. It is clear that the two definitions of $\tilde{L}$ match for functions of $(x, \tilde{\xi})$, so we do not introduce a new notation for the new operator $\tilde{L}$. Also, we define the operator $\partial(\tilde{\xi}, \tilde{\eta})$ acting on functions $u = u(x)$ using the formula

$$\partial(\tilde{\xi}, \tilde{\eta})u = \partial^2(\tilde{\xi})u + \partial(\tilde{\eta})u, \tag{14}$$

where $\partial^2(\tilde{\xi})$ is the square of the operator $\partial(\tilde{\xi})$ in (5) which, of course, relates only to variables x.

2. LEMMA. (a) If $u \in C^4_{loc}(E_d)$ and $u = u(x)$, then

$$\tilde{L}[\partial(\tilde{\xi}, \tilde{\eta})u] = \partial(\tilde{\xi}, \tilde{\eta})Lu.$$

(b) If a function $v(x, \tilde{\xi}, \tilde{\eta})$ of the class $C^2_{loc}(E_{3d+2})$ attains its local minimum at a point $(x_0, \tilde{\xi}_0, \tilde{\eta}_0)$, then $\tilde{L}v - cv \leq 0$ at this point.

Proof. Assertion (b) again follows from the fact that $\tilde{L}$ is an elliptic operator. To prove (a), we note that, by (7) to (9), (11), and Lemma 1,

$$\tilde{L}[\partial(\tilde{\xi}, \tilde{\eta})u] = \tilde{L}[\partial^2(\tilde{\xi})u] + \tilde{L}[\partial(\tilde{\eta})u],$$

$$\tilde{L}[\partial^2(\tilde{\xi})u] = \tilde{L}\left[u_{x^ix^j}\xi^i\xi^j + 2\xi^{d+1}\partial(\tilde{\xi})u - (\xi^{d+1})^2u\right]$$

$$= L^0u_{(\xi)(\xi)} + 2u_{x^i(\xi)}\left(b^i_{(\xi)} - \sigma^{ik}\pi^k_i\right) +$$

$$+ u_{x^ix^j}\sigma^{ik}_{(\xi)}\sigma^{jk}_{(\xi)} + 2u_{x^ix^j(\xi)}\sigma^{ik}\sigma^{jk}_{(\xi)} + cu_{(\xi)(\xi)} +$$

$$+ 2\xi^{d+1}\partial(\tilde{\xi})Lu + 2(u_{(\xi)} + \xi^{d+1}u)c_{(\xi)} +$$

$$+ 2\pi_1^k\left(u_{x^i}\sigma^{ik}_{(\xi)} + u\pi_1^k + u_{x^i(\xi)}\sigma^{ik} + \xi^{d+1}u_{x^i}\sigma^{ik}\right) -$$

$$- u(2\xi^{d+1}c_{(\xi)} + |\pi_1|^2) - (\xi^{d+1})^2 Lu -$$

$$- u_{x^i}\sigma^{ik}2\xi^{d+1}\pi_1^k$$

$$= L^0 u_{(\xi)(\xi)} + [2\xi^{d+1}\partial(\tilde{\xi})Lu - (\xi^{d+1})^2 Lu] +$$

$$+ 2u_{x^i(\xi)}b^i_{(\xi)} + \left[\left(a^{ij}u_{x^ix^j}\right)_{(\xi)(\xi)} -\right.$$

$$\left.- a^{ij}u_{(\xi)(\xi)x^ix^j} - \sigma^{ik}_{(\xi)(\xi)}\sigma^{jk}u_{x^ix^j}\right] +$$

$$+ [(cu)_{(\xi)(\xi)} - uc_{(\xi)(\xi)}] + 2u_{x^i}\sigma^{ik}_{(\xi)}\pi_1 + u|\pi_1|^2.$$

Furthermore, using (7) once more, we obtain

$$\tilde{L}[\partial(\tilde{\eta})u] = \eta^i Lu_{x^i} + \eta^{d+1}Lu + u_{x^i}\left(b^i_{(\xi)(\xi)} + b^i_{(\eta)} -\right.$$

$$\left.- 2\pi_1^k\sigma^{ik}_{(\xi)} - \pi_2^k\sigma^{ik}\right) + u(c_{(\xi)(\xi)} + c_{(\eta)} -$$

$$- |\pi_1|^2) + u_{x^ix^j}\sigma^{jk}\left(\sigma^{ik}_{(\xi)(\xi)} + \sigma^{ik}_{(\eta)}\right) + u_{x^j}\sigma^{jk}\pi_2^k.$$

Adding the latter relationships and noting that

$$\partial^2(\tilde{\xi})Lu = (Lu)_{(\xi)(\xi)} + 2\xi^{d+1}\partial(\tilde{\xi})Lu - (\xi^{d+1})^2 Lu,$$

$$(Lu)_{(\xi)(\xi)} = \left(a^{ij}u_{x^ix^j}\right)_{(\xi)(\xi)} + \left(b^iu_{x^i}\right)_{(\xi)(\xi)} + (cu)_{(\xi)(\xi)},$$

$$\partial(\tilde{\eta})Lu = (Lu)_{(\eta)} + \eta^{d+1}Lu = \eta^i Lu_{x^i} + \eta^{d+1}Lu +$$

$$+ \sigma^{ik}_{(\eta)}\sigma^{jk}u_{x^ix^j} + b^i_{(\eta)}u_{x^i} + c_{(\eta)}u,$$

we obtain our assertion (a). The lemma is proved.

We introduce some more useful formulas. They follow from (3), (13), and (7). For $\xi, \eta \in E_d$, we denote

$$\partial(\xi, \eta)u = u_{(\xi)(\xi)} + u_{(\eta)}.$$

Then

$$\tilde{L}^0(\eta^i\eta^j) = \eta^j\left(\partial(\xi, \eta)b^i - 2\pi_1^k\sigma^{ik}_{(\xi)} - \pi_2^k\sigma^{ik}\right) +$$

$$+ \eta^i\left(\partial(\xi, \eta)b^j - 2\pi_1^k\sigma^{jk}_{(\xi)} - \pi_2^k\sigma^{jk}\right) +$$

$$+ [\partial(\xi, \eta)\sigma^{ik}][\partial(\xi, \eta)\sigma^{jk}], \quad i, j \leq d,$$

$$\tilde{L}^0[(\eta^{d+1})^2] = 2\eta^{d+1}(\partial(\xi, \eta)c - |\pi_1|^2) + |\pi_2|^2, \tag{15}$$

and if $B^{ij} = B^{jk} \in C^2_{loc}(E_d)$, $B^{ij} = B^{ij}(x)$, $i, j = 1, \ldots, d$, $(B\eta, \eta) = \sum_{i,j=1}^{d} B^{ij}\eta^i\eta^j$,

then

$$\tilde{L}^0[(B\eta, \eta)] = L^0(B\eta, \eta) + 2(B\eta, \partial(\xi, \eta)b - 2\pi_1^k\sigma^k_{(\xi)} - \pi_2^k\sigma^k) +$$

$$+ (B\partial(\xi, \eta)\sigma^k, \partial(\xi, \eta)\sigma^k) + 2\left(B_{(\sigma^k)}\eta, \partial(\xi, \eta)\sigma^k\right). \tag{16}$$

3. Remark. Knowing $\tilde{L}^0(\xi^i\xi^j)$ and $\tilde{L}^0(\eta^i\eta^j)$, we can readily obtain the coefficients of the corresponding derivatives in $\tilde{L}$, and the established formulas show how long the formula expressing $\tilde{L}$ directly through the partial derivatives should be. However, it turns out that this formula will never be needed.

4. Remark. We have seen that the operators $\partial(\tilde{\xi})$ and $\partial(\tilde{\xi}, \tilde{\eta})$ commute well with L in the sense that the result of such a commutation can again be written as an elliptic operator applied to $\partial(\tilde{\xi})$ and $\partial(\tilde{\xi}, \tilde{\eta})$. It should be said that if σ and b are not linear functions of x, then for example $\partial^2(\xi)$ does not commute well with L. In this connection, we present (without proof) a rule of constructing operators of an order

higher than 2 which commute well with elliptic operators of the second order. To this end one subsequently has to take $u(x(t))$, where $u(x)$ and $x(t)$ are sufficiently smooth functions of their arguments $x \in E_d$ and $t \in E_1$, to differentiate u n times with respect to t by the chain rule of differentiation if nth-order operators are in question, to write the result as a sum of members of the form

$$x^{(k_1)i_1}(t) \ldots x^{(k_r)i_r}(t) \frac{\partial^r}{\partial x^{i_1} \cdot \ldots \cdot \partial x^{i_r}} u(x(t)), \tag{17}$$

where $x^{(j)}(t) = \frac{d^j}{dt^j} x(t)$, and then to compose the operator as the same sum where (17) is replaced by

$$\partial\left(\tilde{\xi}^{(k_1)}\right) \ldots \partial\left(\tilde{\xi}^{(k_r)}\right).$$

The resulting operator acts in the space of functions whose argu- arguments are $(x, \tilde{\xi}^{(1)}, \ldots, \tilde{\xi}^{(n)})$, where $x \in E_d$ and $\tilde{\xi}^{(j)} \in E_{d+1}$. For example, for n = 3, we obtain the operator

$$\partial^3(\tilde{\xi}) + 3\partial(\tilde{\xi})\partial(\tilde{\eta}) + \partial(\tilde{\zeta}).$$

7.2. *A priori* Estimates of the First and the Second Derivatives

Let $\delta \in (0, 1)$, $K \geq 0$, and $T > 0$. Let $d, d_1 \geq 1$ be integers, Ω be a set, $Q = (0, T) \times E_d$, and, for $\omega \in \Omega$, $i = 1, \ldots, d$, $k = 1, \ldots, d_1$, $(t, x) \in Q$, real-valued functions $\sigma^{ik}(\omega, t, x)$ and $b^i(\omega, t, x)$, $c(\omega, t, x)$, and $f(\omega, t, x)$ be defined so that they are bounded in ω for every (t, x) and $c \leq 0$. Denote

$$a^{ij} = \frac{1}{2}\sigma^{ik}\sigma^{jk}, \quad a = (a^{ij}), \quad \sigma^k = (\sigma^{ik}), \quad b = (b^i), \tag{1}$$

$$L = L(\omega) = L(\omega, z) = a^{ij}(\omega, z)\frac{\partial^2}{\partial x^i \partial x^j} + b^i(\omega, z)\frac{\partial}{\partial x^i} + c(\omega, z), \tag{2}$$

$$F(u_{ij}, u_i, u, z) = \inf_{\omega \in \Omega}[a^{ij}(\omega, z)u_{ij} + b^i(\omega, z)u_i + c(\omega, z)u + f(\omega, z)] \tag{3}$$

and assume that whatever u_{ij}, u_i, u, z the latter infimum is attained at an $\omega_0 = \omega_0(u_{ij}, u_i, u, z) \in \Omega$. Throughout this section we also assume that a function $u \in C^2(\bar{Q})$ is fixed such that

$$u_t + F\left(u_{x^i x^j}, u_{x^i}, u, z\right) = 0 \tag{4}$$

in Q.

In this section, we make use of the results of the previous section concering elliptic operators whose coefficients depend only on x. Here the coefficients of L in (2) also depend on ω and t. Therefore, it should be noted that, deriving $\tilde{L}$ from L (under the corresponding smoothness condition on σ, b, and c with resepct to x), we consider ω and t as parameters, always use the above formulas for every ω and t, keep in mind that σ^k are the same as specified above, and take vectors of E_{d_1} arbitrarily depending on ω, t, x, $\tilde{\xi}$, $\tilde{\eta}$ as π_1 and π_2. Recall that, to define $\tilde{L}$ on functions $v(x, \tilde{\xi})$, it suffices that σ, b, and c be once continuously differentiable with respect to x for every ω and t, whereas, to define $\tilde{L}$ on functions $v(x, \tilde{\xi}, \tilde{\eta})$, it suffices that σ, b, and c be twice continuously differentiable with respect to x.

The following lemma demonstrates the major approach to estimating derivatives of u.

1. LEMMA. Consider a compactum $\Gamma \subset Q$.

(a) Let two functions $\varphi_1, \varphi_2 \in C^2_{loc}(Q)$ be defined in Q such that

$\varphi_2 \geqq u \geqq \varphi_1$ in $Q \setminus \Gamma$,

$$\varphi_{1t} + L(\omega)\varphi_1 + f(\omega) \geqq 0 \geqq \varphi_{2t} + L(\omega_1)\varphi_2 + f(\omega_1) \tag{5}$$

on Γ for all $\omega \in \Omega$ and some $\omega_1 = \omega_1(z) \in \Omega$. Then $\varphi_2 \geqq u \geqq \varphi_1$ in Q.

(b) Let $u_x \in C^2_{loc}(Q)$, for all ω and t, let functions σ^k, b, c and f be once continuously differentiable with respect to x, and let a function $\varphi(z, \tilde{\xi})$ be defined in $Q \times E_{d+1}$ such that $z \in Q$, $\tilde{\xi} = (\xi, \xi^{d+1})$, $\varphi \in C^2_{loc}(Q \times E_{d+1})$, $\partial(\tilde{\xi})u(z) \leqq \varphi(z, \tilde{\xi})$ in $(Q \setminus \Gamma) \times E_{d+1}$, and

$$\lim_{|\tilde{\xi}|\to\infty} |\tilde{\xi}|^{-1} \inf_{z\in\Gamma} \varphi(z, \tilde{\xi}) = \infty, \tag{6}$$

$$\varphi_t + \tilde{L}(\omega)\varphi + \partial(\tilde{\xi})f(\omega) \leqq 0 \tag{7}$$

on $\Gamma \times E_{d+1}$ for all $\omega \in \Omega$ (and some $\pi_1(\omega, z, \tilde{\xi})$). Then $\partial(\tilde{\xi})u(z) \leqq \varphi(z, \tilde{\xi})$ in $Q \times E_{d+1}$. Specifically,

$$|u(z)| \leq \varphi(z, 0, \ldots, 0, 1) \vee \varphi(z, 0, \ldots, 0, -1),$$

$$|u_x(z)| \leq \sup\{\varphi(z, \tilde{\xi}) : \xi^{d+1} = 0, |\xi| = 1\}. \tag{8}$$

(c) Let u_x, $u_{xx} \in C^2_{loc}(Q)$, for all ω and t, let functions σ^k, b, c and f be twice continuously differentiable with respect to x, and let a function $\varphi(z, \tilde{\xi}, \tilde{\eta})$ be defined in $Q \times E_{2d+2}$ uch that $z \in Q$, $\tilde{\xi} = (\xi, \xi^{d+1})$, $\tilde{\eta} = (\eta, \eta^{d+1}) \in E_{d+1}$, and

$$\varphi \in C^2_{loc}(Q \times E_{2d+2}), \ \partial(\tilde{\xi}, \tilde{\eta})u(z) \leq \varphi(z, \tilde{\xi}, \tilde{\eta})$$

$$\text{in } (Q \setminus \Gamma) \times E_{2d+2}, \tag{9}$$

$$\lim_{|\tilde{\xi}|+|\tilde{\eta}|\to\infty} (|\tilde{\xi}|^2 + |\tilde{\eta}|)^{-1} \inf_{z\in\Gamma} \varphi(z, \tilde{\xi}, \tilde{\eta}) = \infty,$$

$$\varphi_t + \tilde{L}(\omega)\varphi + \partial(\tilde{\xi}, \tilde{\eta})f(\omega) \leq 0 \tag{10}$$

on $\Gamma \times E_{2d+2}$ for all $\omega \in \Omega$ (and some $\pi_i(\omega, z, \tilde{\xi}, \tilde{\eta})$). Then $\partial(\tilde{\xi}, \tilde{\eta}) \quad u(z) \leq \varphi(z, \tilde{\xi}, \tilde{\eta})$ in $Q \times E_{2d+2}$. Specifically

$$|u(z)| \leq \varphi(z, 0, \ldots, 0, 1) \vee \varphi(z, 0, \ldots, 0, -1),$$

$$|u_x(z)| \leq \sup\{\varphi(z, 0, \tilde{\eta}) : \eta^{d+1} = 0, \ |\eta| = 1\},$$

$$u_{(\xi)(\xi)}(z) \leq \varphi(z, \xi, 0, \ldots, 0). \tag{11}$$

Proof. Assertion (a) immediately follows from the maximum principle, condition (4), and the fact that $u_t + L(\omega_1)u + f(\omega_1) \geq 0 = u_t + L(\omega_0)u + f(\omega_0)$, where $\omega_0 = \omega_0(u_{x^ix^j}(z), u_{x^i}(z), u(z), z)$. To prove assertion (b), we suppose that inequality $\partial(\tilde{\xi})u \leq \varphi$ is violated somewhere in $Q \times E_{d+1}$. Then it is violated at some points of $\Gamma \times E_{d+1}$, and, by virtue of (6), the strictly positive maximum value of the function $(\partial(\tilde{\xi})u - \varphi)e^t$ on $\Gamma \times E_{d+1}$ (and therefore on $Q \times E_{d+1}$) is attaiend at a point $(z_0, \tilde{\xi}_0) \in \Gamma \times E_{d+1}$. By Lemma 1.1(d), at this point

$$0 \geq \left(\frac{\partial}{\partial t} + \tilde{L}(\omega)\right)[e^t(\partial(\tilde{\xi})u - \varphi)] = e^t(\partial(\tilde{\xi})u - \varphi) +$$

$$+ e^t\left(\frac{\partial}{\partial t} + \tilde{L}(\omega)\right)(\partial(\tilde{\xi})u - \varphi),$$

$$0 > \left(\frac{\partial}{\partial t} + \tilde{L}(\omega)\right)\partial(\tilde{\xi})u - \left(\frac{\partial}{\partial t} + \tilde{L}(\omega)\right)\varphi \tag{12}$$

for any $\omega \in \Omega$. We put $\omega = \omega_0 = \omega_0(u_{x^ix^j}(z_0), u_{x^i}(z_0), u(z_0), z_0)$ and note that, by hypothesis (see (4)), the function

$$u_t(z) + L(\omega_0, z)u(z) + f(\omega_0, z) \tag{13}$$

attains its infimum with respect to z at the point $z = z_0$ and this infimum equals zero. Hence, by Lemma 1.1, we obtain at the point $(z_0, \tilde{\xi}_0)$

$$\begin{aligned} 0 &= \partial(\tilde{\xi})(u_t(z) + L(\omega_0, z)u(z) + f(\omega_0, z)) \\ &= \left(\frac{\partial}{\partial t} + \tilde{L}(\omega_0)\right)\partial(\tilde{\xi})u + \partial(\tilde{\xi})f(\omega_0). \end{aligned} \tag{14}$$

This together with (12) immediately yields a contradiction to condition (7). Thus, $\partial(\tilde{\xi})u \leqq \varphi$ in $Q \times E_{d+1}$. Inequalities (8) are obvious consequences of the latter inequality.

Assertion (c) can be proved by essentially repeating the previous proof. It only has to be noted that not only are function (13) and its first derivatives with resepct to x equal to zero at z_0, but also its second derivatives along any direction in the space of x are non-negative. Therefore, if we replace $\partial(\tilde{\xi})$ in (14) by $\partial(\tilde{\xi}, \tilde{\eta})$ and take advantage of Lemma 1.2, then at the corresponding point $(z_0, \tilde{\xi}_0, \tilde{\eta}_0)$ we obtain

$$0 \leqq \left(\frac{\partial}{\partial t} + \tilde{L}(\omega_0)\right)\partial(\tilde{\xi}, \tilde{\eta})u + \partial(\tilde{\xi}, \tilde{\eta})f(\omega_0).$$

The lemma is proved.

This lemma reduces the proof of estimates (8) and (11) to solution of inequalities (7) and (10). Let us show that they can be solved if $|c|$ is 'sufficiently large' as compared with the first and second derivatives of σ and b.

2. THEOREM. Let $u_x, u_{xx} \in C^2_{loc}(Q)$, let $u(t, x)$, $|u_x(t, x)|$, $(u_{(\xi)(\xi)}(t, x))_+ \to 0$ as $|x| \to \infty$, $(t, x) \in Q$ and $\xi \in E_d$, $|\xi| = 1$. Also, let, for every ω and t, the functions σ^k, b, c, and f be twice continuously differentiable with respect to x, and for all unit $\xi \in E_d$ the inequalities

$$\begin{aligned} &\sum_k |\sigma^k_{(\xi)(\xi)}|^2 + |b_{(\xi)(\xi)}| + |c_{(\xi)}| + |c_{(\xi)(\xi)}| + |f| + \\ &\quad + |f_{(\xi)}| + (f_{(\xi)(\xi)})_+ \leqq -Kc, \end{aligned} \tag{15}$$

$$2(\xi, b_{(\xi)}) + (1 + \delta)\sum_k |\sigma^k_{(\xi)}|^2 + 2\sum_k (\xi, \sigma^k_{(\xi)})^2 \leqq -(1 - \delta)c \tag{16}$$

hold on $\Omega \times Q$.

Finally, let $|u(T, x)| + |u_{(\xi)}(T, x)| + (u_{(\xi)(\xi)}(T, x))_+ \le K$ on E_d for all unit $\xi \in E_d$. Then $|u|, |u_x|, u_{(\xi)(\xi)} \le N(\delta, K)$ in $\bar{Q}$ for all unit $\xi \in E_d$.

Proof. We take a (large) constant $\lambda \ge 1$ and denote $\gamma = (2 - \delta)^{-1}$.

$$\varphi(\tilde{\xi}, \tilde{\eta}) = [\lambda^4|\xi|^4 + \lambda^3|\eta|^2 + \lambda^3|\xi^{d+1}|^4 + \lambda^2|\eta^{d+1}|^2 + \lambda]^\gamma.$$

It is clear that $\gamma < 1$. For any elliptic operator L and a smooth function $\psi > 0$, we have

$$L\psi^\gamma \le \gamma\psi^{\gamma-1}L\psi, \tag{17}$$

whenever $L1 \le 0$, therefore, by formulas (1.10), (1.12), (1.15), and (1.16) with $\pi_1 = \pi_2 = 0$, we obtain

$$\begin{aligned}
\gamma^{-1}\varphi^{\frac{1}{\gamma}-1}\tilde{L}\varphi &= \gamma^{-1}\varphi^{\frac{1}{\gamma}-1}(\tilde{L}^0\varphi + c\varphi) \\
&\le \gamma^{-1}\varphi^{\frac{1}{\gamma}-1}\left(c\varphi + \gamma\varphi^{1-\frac{1}{\gamma}}\tilde{L}^0\varphi^{\frac{1}{\gamma}}\right) \\
&= \tilde{L}^0\varphi^{\frac{1}{\gamma}} + c\gamma^{-1}\varphi^{\frac{1}{\gamma}} \\
&= \left\{2\lambda^4|\xi|^2\left[2(\xi, b_{(\xi)}) + \sum_k |\sigma^k_{(\xi)}|^2\right] + \right. \\
&\quad \left. + 4\lambda^4\sum_k(\xi, \sigma^k_{(\xi)})^2\right\} + 4\lambda^3(\xi^{d+1})^3c_{(\xi)} + \\
&\quad + 2\lambda^3(\eta, \partial(\xi, \eta)b) + \lambda^3\sum_k|\partial(\xi, \eta)\sigma^k|^2 + \\
&\quad + 2\lambda^2\eta^{d+1}\partial(\xi, \eta)c + c\gamma^{-1}\varphi^{1/\gamma}.
\end{aligned} \tag{18}$$

By hypothesis (16), the expression in the braces does not exceed $2\lambda^4 \times (1 - \delta)|c|\cdot|\xi|^4$. Furthermore, for $\varepsilon \in (0, 1)$,

$$\begin{aligned}
4(\xi^{d+1})^3 c_{(\xi)} &\le 4K|\xi^{d+1}|^3\cdot|c|\cdot|\xi| \le \varepsilon|c|\cdot|\xi^{d+1}|^4 + \\
&\quad + N\varepsilon^{-3}|c|\cdot|\xi|^4,
\end{aligned}$$

$$\sum_k|\partial(\xi, \eta)\sigma^k|^2 \le \left(1 + \frac{1}{\delta}\right)\sum_k|\sigma^k_{(\xi)(\xi)}|^2 + (1 + \delta)\sum_k|\sigma^k_{(\eta)}|^2,$$

$$2(\eta, \partial(\xi, \eta)b) + \sum_k |\partial(\xi, \eta)\sigma^k|^2$$

$$\le (1 - \delta)|c|\cdot|\eta|^2 + 2(\eta, b_{(\xi)(\xi)}) +$$

$$+ \left(1 + \frac{1}{\delta}\right) \sum_k |\sigma^k_{(\xi)(\xi)}|^2$$

$$\le (1 - \delta)|c|\cdot|\eta|^2 + \varepsilon|c|\cdot|\eta|^2 + N\varepsilon^{-1}|c|\cdot|\xi|^4,$$

$$2\eta^{d+1}\partial(\xi, \eta)c \le \varepsilon|c|\cdot|\eta^{d+1}|^2 + N\varepsilon^{-1}|c|(|\xi|^4 + |\eta|^2),$$

where the constant N depends only on δ and K. This together with (18) implies

$$\gamma^{-1}\varphi^{1/\gamma}\tilde{L}\varphi \le 2|c|\left\{|\xi|^4\left[\lambda^4(1 - \delta) + N\varepsilon^{-3}\lambda^3 - \left(1 - \frac{\delta}{2}\right)\lambda^4\right] +\right.$$

$$+ |\xi^{d+1}|^4\lambda^3\left(\varepsilon - 1 + \frac{\delta}{2}\right) +$$

$$+ |\eta|^2\lambda^3\left(N\varepsilon^{-1}\lambda^{-1} + \varepsilon - \frac{1}{2}\right) +$$

$$\left. + |\eta^{d+1}|^2\lambda^2\left(\varepsilon - 1 + \frac{\delta}{2}\right) - \lambda\left(1 - \frac{\delta}{2}\right)\right\}.$$

It is now clear that we can choose $\varepsilon = \varepsilon(\delta, K)$, $\lambda_0 = \lambda_0(\delta, K)$, and $\kappa = \kappa(\delta, K) > 0$ such that, for all $\lambda \ge \lambda_0$,

$$\gamma^{-1}\varphi^{1/\gamma}\tilde{L} \le -|c|\kappa^{1/\gamma}, \quad \tilde{L}\varphi \le -\kappa\gamma|c|\varphi.$$

By virtue of (15), $\partial(\tilde{\xi}, \tilde{\eta})f \le \kappa\gamma|c|\varphi$ for a sufficiently large $\lambda = \lambda(\delta, K) \ge \lambda_0$, therefore, for this λ, inequality (10) holds on $Q \times E_{2d+2}$ for all $\omega \in \Omega$. If necessary, we can increase the obtained $\lambda = \lambda(\delta, K)$ so that the inequality $\partial(\tilde{\xi}, \tilde{\eta})u(T, x) + 1 \le \varphi(\tilde{\xi}, \tilde{\eta})$ hold. We then apply Lemma 1(c) to the function $\varphi + \mu t^{-1}$. By choice of λ and our assumptions, the inequality $\partial(\tilde{\xi}, \tilde{\eta})u \le \varphi + \mu t^{-1}$ can be violated only on a set $\Gamma \times E_{2d+2}$, where Γ is a compactum such that $\Gamma \subset Q$. Moreover, the function $\varphi + \mu t^{-1}$ satisfies inequality (10) and condition (9). Therefore, inequalities (11) hold if φ is replaced by $\varphi + \mu t^{-1}$. Since μ is arbitrary, the theorem is proved.

3. Remark. If nonzero parameters π_1 and π_2 were used, condition (16) could be replaced by a less rigid one (cf. (8.5.40)), which would automatically be satisfied, for example, for uniformly nondegenerate (a^{ij}) and bounded σ_x and b_x. This condition, which has yet to be considered,

is deduced in a more complicated manner and the difficulties involved cannot be justified from the standpoint of the purpose of this chapter.

4. Remark. Although the methods that have led us to Theorem 2 are rather rough, condition (16) cannot be weakened in certain cases. Specifically, it is impossible to reduce the coefficient of $(\xi, b_{(\xi)})$ without violating the statement of the theorem. The corresponding example is presented in Section 5.

5. Remark. Putting $\gamma = 1$ in the proof of Theorem 2, we should replace $(1 - \delta)c$ in condition (16) by $\frac{1}{2}(1 - \delta)c$. On the one hand, this would make the condition rougher, but on the other hand, it would provide a possibility of doing without the results established in Section 1 altogether. Indeed, the proof of Lemma 1 with

$$\varphi = \lambda^4|\xi|^4 + \lambda^3|\eta|^2 + \lambda^3|\xi^{d+1}|^4 + \lambda^2|\eta^{d+1}|^2 + \lambda$$

is based on studying the points at which the function

$$\partial(\tilde{\xi}, \tilde{\eta})u - (\lambda^4|\xi|^4 + \lambda^3|\eta|^2 + \lambda^3|\xi^{d+1}|^4 + \lambda^2|\eta^{d+1}|^2 + \lambda)$$

attains its maxima with respect to $(t, x, \tilde{\xi}, \tilde{\eta})$.

For fixed (t, x), we can find the maximum of this function with respect to $\tilde{\xi}$ and $\tilde{\eta}$ and explicitly express it in terms of u, u_x, and u_{xx} (cf. (5.4.5) and (5.4.10)). Then we can estimate the maximum over (t, x) of the latter maximum over $(\tilde{\xi}, \tilde{\eta})$ without invoking L (cf. Lemma 5.4.2 and Proof 5.4.4). Exactly this consideration gave rise to the idea that function (5.4.5) can be 'well' estimated using the maximum principle.

We further prove a theorem on estimating the first derivatives with respect to (t, x). In this theorem, it is required that σ^k, b, c, and f be differentiable with respect to t and derivatives be appropriately bounded in t. These conditions imply that σ, b, c, and f can be continuously extended to $\bar{Q}$ in a unique fashion.

6. Theorem. Let $u_x, u_t \in C^2_{loc}(Q)$ and let $u(t, x)$, $|u_x(t, x)|$, $u_t(t, x) \to 0$ as $|x| \to \infty$ for $(t, x) \in Q$. For every ω, let the functions σ^k, b, c, and f be continuously differentiable with respect to (t, x) in Q and, for all unit vectors $\xi \in E_d$, the inequalities

$$\sum_k|\sigma^k_t|^2 + |b_t| + |c_t| + |c_{(\xi)}| + |f| + |f_t| + |f_{(\xi)}| \leqq -Kc, \tag{19}$$

$$2(\xi, b_{(\xi)}) + (1 + \delta)\sum_k|\sigma^k_{(\xi)}|^2 \leqq -2(1 - \delta)c \tag{20}$$

hold on $\Omega \times Q$. Also, let

$$|u(T, x)| + |u_x(T, x)| + |L(\omega, T, x)u(T, x) + f(\omega, T, x)| \leq K \tag{21}$$

for all $x \in E$, $\omega \in \Omega$.

Then

$$|u|, |u_x|, |u_t| \leq N(\delta, K) \text{ in } \bar{Q}.$$

Proof. Following the discussion of Section 1, we consider $L(\omega) + \frac{\partial}{\partial t}$ as an elliptic operator, from which we construct an operator $\hat{L}(\omega)$ in the same manner as the operator $\tilde{L}$ in Section 1. The operator $L(\omega) + \frac{\partial}{\partial t}$ acts on functions of (t, x). Here we have a variable whose denotation differs from the denotation x^i of other variables. Accordingly, operator $\hat{L}(\omega)$ is defined on functions of $(t, x, \tau, \tilde{\xi})$ with $(t, x) \in Q$, $\tau \in E_1$, $\tilde{\xi} = (\xi, \xi^{d+1}) \in E_{d+1}$ as

$$\hat{L}(\omega)v(t, x, \tau, \tilde{\xi}) = \frac{1}{2}\delta^{kr}v_{(\hat{\sigma}^k(t, x, \tau, \tilde{\xi}))(\hat{\sigma}^r(t, x, \tau, \tilde{\xi}))} \times$$

$$\times (t, x, \tau, \tilde{\xi}) +$$

$$+ v_{(\hat{b}(t, x, \tau, \tilde{\xi}))}(t, x, \tau, \tilde{\xi}) +$$

$$+ c(t, x)v(t, x, \tau, \tilde{\xi}), \tag{22}$$

where the result of differentiation of $v(t, x, \tau, \tilde{\xi})$ along the vector ζ with $2d + 3$ coordinates is defined as

$$v_{(\zeta)} = v_t\zeta^1 + \sum_{i=1}^{d} v_{x^i}\zeta^{i+1} + v_t\zeta^{d+2} + \sum_{i=1}^{d+1} v_{\xi^i}\zeta^{d+2+i},$$

$v_{(\zeta_1)(\zeta_2)}$ is defined in the same manner and donated by

$$\hat{\sigma}^{1k} = 0, \qquad \hat{b}^1 = 1,$$

$$\hat{\sigma}^{i+1,k} = \sigma^{ik}, \qquad \hat{b}^{i+1} = b^i, \ i = 1, \ldots, d,$$

$$\hat{\sigma}^{d+2,k} = 0, \qquad \hat{b}^{d+2} = 0,$$

$$\hat{\sigma}^{d+2+i,k} = \sigma_t^{ik}\tau + \sigma_{(\xi)}^{ik}, \qquad \hat{b}^{d+2+i} = b_t^i\tau + b_{(\xi)}^i, \ i = 1, \ldots, d,$$

$$\hat{\sigma}^{2d+3,k} = 0, \qquad \hat{b}^{2d+3} = c_t\tau + c_{(\xi)}.$$

By Lemma 1.1(c) we obtain

$$\hat{L}[u_t\tau + \partial(\tilde{\xi})u] = \tau\frac{\partial}{\partial t}\left(\frac{\partial}{\partial t} + L\right)u + \partial(\tilde{\xi})\left(\frac{\partial}{\partial t} + L\right)u.$$

As in the proof of Lemma 1(b), it follows from this equality that, whenever $\varphi \in C^2_{loc}(Q \times E_1 \times E_{d+1})$, a compactum Γ belongs to Q,

$$\tau u_t(z) + u_{(\xi)}(z) + \xi^{d+1}u(z) \leq \varphi(z, \tau, \tilde{\xi}) \tag{23}$$

for $z \in Q \setminus \Gamma$ and arbitrary τ and $\tilde{\xi}$, and also if

$$\lim_{|\tilde{\xi}|+|\tau|\to\infty} (|\tilde{\xi}| + |\tau|)^{=1} \inf_{z\in\Gamma} \varphi = \infty, \tag{24}$$

$$\hat{L}(\omega)\varphi(z, \tau, \tilde{\xi}) + \tau f_t(\omega, z) + \partial(\tilde{\xi})f(\omega, z) \leq 0 \tag{25}$$

for $z \in \Gamma$ and all ω, τ, $\tilde{\xi}$, then $\tau u_t(z) + \partial(\tilde{\xi})u(z) \leq \varphi(z, \tau, \tilde{\xi})$ for all values of arguments. In this connection we take a constant $\lambda \geq 1$ and denote $\gamma = (2 - \delta)^{-1}$, $\varphi = (\lambda^3\tau^2 + \lambda^2|\xi|^2 + \lambda|\xi^{d+1}|^2 + \lambda)^\gamma$. From (17), (22) (cf. (18)) we obtain

$$\begin{aligned}\gamma^{-1}\varphi^{\frac{1}{\gamma} - 1}\hat{L}\varphi &\leq (\hat{L} - c)\varphi^{\frac{1}{\gamma}} + c\gamma^{-1}\varphi^{\frac{1}{\gamma}} \\ &= 2\lambda^2(\xi, b_t\tau + b_{(\xi)}) + \lambda^2 \sum_k|\sigma^k_t\tau + \sigma^k_{(\xi)}|^2 + \\ &\quad + 2\lambda\xi^{d+1}(c_t\tau + c_{(\xi)}) + \\ &\quad + c(2 - \delta)(\lambda^3\tau^2 + \lambda^2|\xi|^2 + \lambda|\xi^{d+1}|^2 + \lambda).\end{aligned}$$

Elementary estimation based on Cauchy–Buniakovski's inequality and assumption (19) and (20) show that

$$\gamma^{-1}\varphi^{\frac{1}{\gamma} - 1}\hat{L}\varphi \leq -|c|\kappa\varphi^{\frac{1}{\gamma}}, \quad \hat{L}\varphi \leq -\kappa\gamma|c|\varphi$$

for $\kappa = \kappa(\delta, K) > 0$, $\lambda_0 = \lambda_0(\delta, K)$ and all $\lambda \geq \lambda_0$.

This together with (19) implies that, for a sufficiently large $\lambda = \lambda(\delta, K) \geq \lambda_0$, the function φ satisfies condition (25) for all z. Condition (24) obviously holds. It follows from (21) that increasing $\lambda = \lambda(\delta, K)$ we can assume that

$$1 + K|\tau| + |u_{(\xi)}(T, x)| + |\xi^{d+1}u(T, x)| \leq \varphi(\tau, \tilde{\xi}).$$

Moreover, for every $z = (t, x) \in Q$, there exists an $\omega_0 \in \Omega$ such that $|u_t(z)| = |L(\omega_0, z)u(z) + f(\omega_0, z)|$. If t is close to T, then the left-hand side is close to $|u_t(T, x)|$ whereas the right-hand side is uniformly close to $|L(\omega_0, T, x)u(T, x) + f(\omega_0, T, x)| \leq K$ with respect to ω_0. (The inequality $|c_t| \le K|c|$ and the boundedness of $c(\omega, z)$ in ω imply that $c(\omega, t, x)$ is bounded in (ω, t), hence $|\sigma_t^k|$, $|b_t|$, $|c_t|$, $|f_t|$ are also bounded in (ω, t) for a fixed x). Therefore, $|u_t(T, x)| \le K$ and

$$1 + \tau u_t(T, x) + \partial(\tilde{\xi})u(T, x) \le \varphi(\tau, \tilde{\xi}),$$

which together with the conditions on u_t and u_x enables us to find a compactum Γ for any $\mu > 0$ so that the left-hand side of (23) does not exceed $\varphi + \mu t^{-1}$ for $z \in \Gamma$. Thus, $\tau u_t + \partial(\tilde{\xi})u \le \varphi(\tau, \xi) + \mu t^{-1}$ for all values of the arguments and every $\mu > 0$. This obviously proves the theorem.

7. Remark. If we omit u_t, σ_t^k, b_t, c_t, and f_t everywhere in the formulation of Theorem 6, abandon the requirement that σ_t^k, b_t, c_t, f_t must exist, and omit $L(\omega)u + f(\omega)$ in (21), then the assertions of the theorem remain true. In this case we can replace $1 + \delta$ in (20) by 1. This can easily be shown by repeating the proof of Theorem 6, recalling Lemma 1(b), and using the function $(\lambda^2|\xi|^2 + \lambda|\xi^{d+1}|^2 + \lambda)^\gamma$.

7.3. The Existence of a Solution in the Class of Concave Functions

Let δ, $\delta_1 \in (0, 1)$, let $K, K_1, K_2 \ge 0$, $T > 0$, let $d, d_1 \ge 1$ be integers, let Ω be a finite or a countable set, let $Q = (0, T) \times E_d$, and let, for $\omega \in \Omega$, $i = 1, \dots, d$, $k = 1, \dots, d_1$ real-valued Borel functions $\sigma^{ik}(\omega, t, x)$, $b^i(\omega, t, x)$, $c(\omega, t, x)$, $f(\omega, t, x)$ be defined on Q such that they are twice continuously differentiable with respect to x for every ω and t and $c \le 0$.

We shall use the notation of (2.1) — (2.3). Recall some necessary notions. Given a function u in Q, we denote by $u_t(z)$, $u_{x^i}(z)$, and $u_{x^ix^j}(z)$ its ordinary or Sobolev derivatives of the corresponding order (if such derivatives exist) and by $u_t(dz)$ and $u_{x^ix^j}(dz)$ we denote the differentials of the measures that are generalized derivatives of the corresponding order of u in the sense of the theory of generalized functions (if these derivatives are measures). Given a measure μ in Q, we denote by $\mu^{(0)}$ the function which is the Radon—Nikodym derivative with respect to the Lebesgue measure of the absolutely continuous component of μ.

The basic result of this section is the following.

1. THEOREM. Let inequalities (2.15) and (2.16) hold on $\Omega \times Q$ for all

unit vectors $\xi \in E_d$, let a function $\varphi \in C^2(E_d)$ be defined, and let

$$|\varphi(x)|,\ |\varphi_x(x)|,\ ||\varphi_{xx}(x)|| \leq K, \tag{1}$$

$$|c(\omega, t, x)| \leq K_1(1 + |x|),\quad \sum_k |\sigma^k(\omega, t, 0)|^2 + |b(\omega, t, 0)| +$$

$$+ \sum_k |\sigma^k_{(\xi)}(\omega, t, x)|^2 + |b_{(\xi)}(\omega, t, x)| \leq K_1 \tag{2}$$

for all $\omega \in \Omega$, $(t, x) \in Q$, and unit $\xi \in E_d$. Then there exists a function u on $\bar{Q}$ such that:

(a) $u \in C(\bar{Q})$, $u(T, x) = \varphi(x)$, in Q there exist Sobolev derivatives u_x, u_t and measures u_{xx};

(b) for all $R > 0$, unit $\xi \in E_d$, $t, s \in [0, T]$, $|x| \leq R$, we obtain $|u(t, x)| + |u_x(t, x)| \leq N(\delta, K)$ (a.e.), $\xi^i\xi^j u_{x^ix^j}(dt\,dx) \leq N(\delta, K)dt\,dx$, and

$$|u(t, x) - u(s, x)| \leq N(d, \delta, K, K_1, R, T)|t - s|^{1/2}, \tag{3}$$

$$||u_t||_{\mathcal{L}_2(C_{T,R})} \leq N(d, \delta, K, K_1, R, T); \tag{4}$$

(c) $a^{ij}(\omega)u_{x^ix^j}(dt\,dx) << dt\,dx$ for all $\omega \in \Omega$,

$$u_t + F\left(u^{(0)}_{x^ix^j}, u_{x^i}, u, z\right) = 0 \text{ a.e. } Q. \tag{5}$$

The function u satisfying the above properties is unique.

This theorem will be proved after some preliminary analysis wherein the hypotheses of the theorem are assumed to be true. Note that, by virtue of (2),

$$\sum_k |\sigma^k(\omega, t, x)|^2 + |b(\omega, t, x)|^2 \leq N(K_1)(1 + |x|)^2, \tag{6}$$

therefore, the uniqueness of u follows from Remark 3.7.11 and Corollary 3.7.10. We now start proving the existence of u.

2. LEMMA. Let $\varepsilon \in (0, d^{-1})$, let Ω be a finite set, and let σ^k, b, c, and f be infinitely differentiable functions with respect to z on $\bar{Q}$ for all $\omega \in \Omega$, let $\varphi \in C^\infty(E_d)$, and let

$$\sum_k |\sigma^k_t|^2 + |b_t| + |c_t| + |f_t| \leq -K_2 c, \tag{7}$$

$$|L(\omega, T, x)\varphi(x) + f(\omega, T, x)| \leq K_2. \tag{8}$$

Finally, suppose that $c \leq -\delta_1$ and the functions σ^k, b, $c + \delta_1$, f, and φ have compact support, i.e., there exists an $R_0 > 0$ such that all of the functions vanish for $|x| \geq R_0$. Then the Cauchy problem

$$u_t + \varepsilon\Delta u + F\left(u_{x^ix^j}, u_{x^i}, u, z\right) = 0 \text{ in } Q, \tag{9}$$

$$u(T, x) = \varphi(x) \text{ in } E_d \tag{10}$$

has a unique solution $u \in C^2(\bar{Q})$. Moreover, for all unit vectors $\xi \in E_d$, $R > 0$, $t, s \in [0, T]$, and $|x| \leq R$, this solution satisfies the following conditions:

$$|u_t(t, x)| \leq N(\delta, K, K_2), \tag{11}$$

$$|u(t, x)|, |u_x(t, x)|, u_{(\xi)(\xi)}(t, x) \leq N(\delta, K), \tag{12}$$

$$|u(t, x) - u(s, x)| \leq N(d, \delta, K, K_1, R, T)|t - s|^{1/2}, \tag{13}$$

$$||u_t||_{L_2(C_{T,R})} \leq N(d, \delta, K, K_1, R, T), \tag{14}$$

$$||u||_{W_2^{1,2}(C_{T,R})} \leq N(\varepsilon, d, \delta, K, K_1, R, T). \tag{15}$$

Proof. To prove the uniqueness of the solution of problem (9) and (10), it suffices, by Theorem 3.4.6, to construct a function $\psi \in C^2_{loc}(\bar{Q})$ such that $\psi \geq 0$, $(\partial/\partial t + \varepsilon\Delta + L(\omega))\psi \leq 0$, and $\psi \to \infty$ as $|x| \to \infty$. It is almost obvious that we can take $(1 + |x|^2)\exp(-\gamma t)$ as ψ, if γ is sufficiently large.

As it was shown in the previous section, to prove estimate (11) and (12), it suffices to construct a solution of problem (9) and (10) such that u, $|u_x|$, u_t, $||u_{xx}|| \to 0$ as $|x| \to \infty$ and $u_x, u_t, u_{xx} \in C^2_{loc}(Q)$. The latter inclusions seem to be false. Therefore, we consider approximation of equation (9) by equations whose solutions are surely smooth enough.

Let $\Omega = \{\omega_1, \ldots, \omega_r\}$, and let Ψ_p be a convolution of the function $g^1 \wedge \ldots \wedge g^r$ with the function $\Pi p\zeta(pg^i)$ with respect to $g^1, \ldots, g^r$, where $\zeta \in C_0^\infty(E_1)$, $\zeta \geq 0$, $\int\zeta\, dt = 1$, $\zeta(t) = 0$ for $|t| \geq 1$. As mentioned above (cf. (6.1.6)),

$$\Psi_{pg^i} \geq 0, \quad \sum_i \Psi_{pg^i} = 1, \quad \left|\Psi_p - g^i\Psi_{pg^i}\right| \leq \frac{1}{p}. \tag{16}$$

Furthermore, for $g \in E_r$, we put

$$\xi_i(p, g) = \Psi_{pg^i}(g), \quad \eta(p, g) = \Psi_p(g) - g^i\Psi_{pg^i}(g).$$

Since Ψ_p with $p = 1, 2, \ldots$ is a smooth concave function, we obtain

$$\Psi_p(g) = \min_{h \in E_r} [g^i\xi_i(p, h) + \eta(p, h)], \tag{17}$$

where the minimum is attained at $h = g$. We now consider Cauchy's problem with the boundary condition (10) for the equation

$$u_t + \Psi_p(\varepsilon\,\Delta u + L(\omega_1)u + f(\omega_1), \ldots, \varepsilon\,\Delta u + L(\omega_r)u + $$
$$+ f(\omega_r)) - \Psi_p(0) = 0 \tag{18}$$

in Q.

Let us transform this equation using (17). For $h \in E_r$ we denote

$$\bar{L}_p(h, z)u(z) = \xi_i(p, h)L(\omega_i, z)u(z) + \varepsilon\,\Delta u(z),$$

$$\bar{f}_p(h, z) = \xi_i(p, h)f(\omega_i, z) + \eta(p, h) - \Psi_p(0).$$

Then equation (18) is converted to the form

$$u_t + \min_{h \in E_r} [\bar{L}_p(h)u + \bar{f}_p(h)] = 0, \tag{19}$$

where, as in (17), the minimum is attainable. Since $c \leqq -\delta_1$, it follows from Theorem 6.1.5(d) that equation (19) is of the form $0 = u_t + \bar{F}(u_{x^ix^j}, u_{x^i}, u, z)$ and $\bar{F} \in \bar{F}(\varepsilon, N, Q)$ for a constant N. We can set $\delta_0^{\bar{F}}$ equal to 1 and take any sufficiently large constant as $M_0^{\bar{F}}$. By Theorem 6.4.4, problem (19) and (10) has a solution $u^p \in C^{2+\alpha}(\bar{Q})$ such that $\alpha \in (0, 1)$, α is independent of p, and the norms of u^p in $C^{2+\alpha}(Q)$ are bounded by a constant independent of p.

It is clear that, by the uniqueness of the solution to problem (9) and (10) and the mentioned boundedness of the norms of u^p in $C^{2+\alpha}(Q)$, the functions u^p together with the derivatives with respect to t, x^i, x^ix^j uniformly converge to a solution of problem (9) and (10) on every compactum as $p \to \infty$. Therefore, to prove (11) and (12) it remains for us to prove these estimates for the functions u^p with sufficiently large p. We can do it using the results of the previous section, equation (19), and the fact that u_t^p, u_x^p, $u_{xx}^p \in C_{loc}^{2+\alpha}(Q)$ in view of the smoothness of Ψ and the coefficients of $L(\omega)$ and Lemma 1.3.2 applied to (18). We have thus reduced the study of problem (9) and (10) to the study of problem (19) and (10) whose solution has the advantageous property u_t^p, u_x^p, $u_{xx}^p \in C_{loc}^2(Q)$.

Furthermore, note that, by hypothesis, σ, b, c + δ_1, and f have compact support whence

$$\min_{h\in E_r} [\bar{L}_p(h, z)u(z) + \bar{f}_p(h, z)]$$

$$= \varepsilon\, \Delta u(z) - \delta_1 u(z) + \min_{h\in E_r} [\eta(p, h) - \Psi_p(0)]$$

$$= \varepsilon\, \Delta u(z) - \delta_1 u(z),$$

whenevery $z \notin C_{T,R_0}$. Specifically, u^p satisfies the equation $u_t + \varepsilon\Delta u - \delta_1 u = q$ and $q = 0$ outside C_{T,R_0}. The available explicit expression of u^p in terms of φ and q together with the fact that φ has compact support immediately imply that u^p, $|u_x^p|$, $|u_t^p|$, $||u_{xx}^p|| \to 0$ as $|x| \to \infty$.

We now have to convert the operators $\bar{L}_p$ to the form (2.2) and to find the corresponding $\bar{d}_1$, $\bar{\sigma}^k$, $\bar{b}$, $\bar{c}$. We put

$$\bar{\sigma}_p^{i+kr}(h, z) = (\xi_i(p, h))^{1/2}\sigma^{k+1}(\omega_i, z) \text{ with } k = 0, \ldots, d_1-1;\ i = 1, \ldots, r$$

$$\bar{\sigma}_p^{i,d_1 r+j}(h, z) = (2\varepsilon)^{1/2}\delta^{ij} \text{ with } i, j \leqq d,$$

$$\bar{b}_p(h, z) = \xi_i(p, h)b(\omega_i, z),\ \bar{c}_p(h, z) = \xi_i(p, h)c(\omega_i, z),$$

$$\bar{a}_p^{ij}(h, z) = \frac{1}{2}\sum_{k=1}^{d_1 r+d} \bar{\sigma}_p^{ik}\bar{\sigma}_p^{jk}(h, z) \text{ with } i, j \leqq d. \tag{20}$$

(There is no summation over i.) It is not difficult to see that

$$\bar{L}(h, z)u(z) = \bar{a}_p^{ij}(h, z)u_{x^i x^j}(z) + \bar{b}_p^i(h, z)u_{x^i}(z) + + \bar{c}_p(h, z)u(z).$$

It follows from (20), (16), (1), and formulas like

$$\sum_{k=1}^{d_1 r+d} |\bar{\sigma}_p^k(h, z)|^2 = \sum_{i=1}^{r} \sum_{k=1}^{d_1} |\sigma^k(\omega_i, z)|^2 \xi_i(p, h) + 2\varepsilon d$$

that condition (2.16) holds for $\bar{\sigma}_p$, $\bar{b}_p$, and $\bar{c}_p$, and also conditions (7) and (8) are satisfied with $N(K_2, K)$ in place of K_2 in the latter. Moreover, the assumption that $c \leq -\delta_1$ implies that $|\eta(p, h) - \Psi_p(0)| \leq 2p^{-1} \leq 2\delta_1 \leq -2\bar{c}_p$ with $p \geq \delta_1^{-1}$. Therefore, condition (2.15) also holds with $N(K)$ instead of K for $p \geq \delta_1^{-1}$. (It is essential that the constants involved in inequalities (2.15), (2.16), (7), and (8) associated with $\bar{\sigma}_p$, $\bar{b}_p$, and $\bar{c}_p$ are independent of the cardinality of Ω.) From Theorems 2.2 and 2.6, we obtain estimates (11) and (12) for u whenever p is sufficiently large.

It remains to prove estimates (13) to (15) for the solution of problem (9) and (10). In section 4.4, we constructed barriers for solutoins of linear equations and, using them in Lemma 4.4.2, estimated the difference between the solution and the boundary data on $\partial_t Q_{T,R}$. Since Ω is finite, the infimum in the definition of F is attainable and the function u satisfying equation (9) also satisfies a linear equation (whose coefficients are determined from u). This equation can be considered in $Q_{s,R}$, where $s \leq T$, and in Lemma 4.4.2, we can take s and 1 in place of T and α. Then we easily obtain (13) for $0 \leq t \leq s \leq T$, $|x| \leq R - 1$.

Almost literally repeating the corresponding fragment of the proof of Theorem 5.4.6 and using (6) and the established upper bounds on $|u|$, $|u_x|$, and $u_{(\xi)(\xi)}$, we can prove estimates (14) and (15). The lemma is proved.

We now abandon the assumption on infinite differentiability of σ, b, c, f, and φ.

3. LEMMA. Let $\varepsilon \in (0, d^{-1})$, let Ω be a finite set, let $c \leq -\delta_i$, and let the condition in Lemma 2 that σ, b, $c + \delta_1$, f, and φ have finite support hold. Then there exists a function $u \in C(\bar{Q}) \cap W_{2,loc}^{1,2}(Q)$, which satisfies (9) almost everywhere in Q, is equal to φ for $t = T$, and such that estimates (12) to (15) hold true for this function.

Proof. We extend σ, b, $c + \delta_1$, and f to the exterior of Q, where they are set to zero; we then take a nonnegative function $\zeta \in C_0^\infty(E_{d+1})$ such that $\int\zeta \, dt \, dx = 1$, set $\zeta_p(t, x) = p^{d+1}\zeta(pt, px)$ for $p \geq 1$, and define

σ^{kp}, b^p, c^p, f^p, and φ^p by convolving σ^k, ..., φ with ζ_p. It is clear that σ^{kp}, ..., φ^p are infinitely differentiable and satisfy the conditions of Theorem 1 if K and K_1 are replaced by N(K) and $N(K_1)$ respectively. Moreover, these functions satisfy condition (7) and (8) with a constant N_p for every p, since Ω is finite, σ, b, c + δ_1, f, and φ have compact support, the functions are appropriately bounded (cf. (6)), and the inequality $c \leq -\delta_1$ holds. By Lemma 2, if F is constructed from σ^{kp}, b^p, c^p, and f^p, then equation (9) with the boundary condition $u = \varphi^p$ has a solution $v^p \in C^2(\bar{Q})$ such that estimates (12) to (15) hold true. Note that the constant appearing in inequalities (7) and (8) is not involved in these estimates. This justifies the assertion that there exists a subsequence $p' \to \infty$ such that, on every cylinder $\bar{C}_{T,R}$, $v^{p'}$ uniformly converges to a function u obviously possessing all desired properties except, possibly, the one associated with equation (9).

We now show that u satisfies equation (9) almost everywhere in Q. We put

$$a^{ijp} = \frac{1}{2}\sum_k \sigma^{ikp}\sigma^{jkp}$$

and note that, for every p_0 and $p \geq p_0$, we have

$$\inf_{q\geq p_0,\ \omega\in\Omega} [v_t^p + (\varepsilon\delta^{ij} + a^{ijq})v^p_{x^ix^j} + b^{iq}v^p_{x^i} + c^q v^p + f^q] \leq 0.$$

By Theorem 3.6.3, it follows that

$$\inf_{q\geq p_0,\ \omega\in\Omega} [u_t + (\varepsilon\delta^{ij} + a^{ijq})u_{x^ix^j} + b^{iq}u_{x^i} + c^q u + f^q] \leq 0 \tag{21}$$

almost everywehre in Q. Here Ω is finite and a^{ijq}, b^{iq}, c^q, $f^q \to a^{ij}$, b^i, c, f almost everywhere on Q as $q \to \infty$. Therefore, letting $p_0 \to \infty$, we obtain from (21) that the left-hand side of (9) is no greater than zero (a.e.Q). The opposite inequality can be obtained as in the proof of Theorem 6.5.4. The lemma is proved.

At the next step, we let $\varepsilon \downarrow 0$ in equation (9).

4. LEMMA. Let Ω be a finite set, let $c \leq -\delta_1$, and let the assumption of

Lemma 2 that σ, b, c + δ_1, f, φ have compact support hold. Then all assertions of Theorem 1 hold true.

Proof. Given an $\varepsilon \in (0, d^{-1})$, we denote by w^{ε} the solution to problem (9) and (10) in Lemma 3. The function w^{ε} satisfies inequalities (12) and (13) which imply that, for a subsequence $\varepsilon' \downarrow 0$, the functions $w^{\varepsilon'}$ uniformly converge on every cylinder $\bar{C}_{T,R}$ to a function $u \in C(\bar{Q})$ such that $u(T, x) = \varphi(x)$ and $|u|, |u_x| \leq N(\delta, K)$ (a.e.Q). For the selected subsequence, by (14), $w_t^{\varepsilon'}$ also weakly converge in $\mathcal{L}_2(C_{T,R})$, their limit obviously being the Sobolev derivative of u with respect to t, which therefore exists and satisfies inequality (14).

We further denote by N constants depending only on δ and K and note that, by the inequality $w^{\varepsilon}_{(\xi)(\xi)} \leq N|\xi|^2$, the functions $w^{\varepsilon} - \frac{1}{2}N|x|^2$ are concave in x. The concavity is preserved as we pass to the limit, therefore, on the right of

$$-u = \frac{1}{2}\left(N|x|^2 - u\right) - \frac{1}{2}N|x|^2 \tag{22}$$

the first simmand is convex in x, while the second one is smooth. By the definition of $\tilde{W}^{bv,+}(C_{T,R})$ adopted in Section 3.6, we conclude that $(-u) \in \tilde{W}^{bv,+}(C_{T,R})$ for any $R > 0$. As we already know (see Section 3.6), the second generalized derivatives of convex functions are measures, the pure second derivatives being positive measures. This yields the existence of the measures $u_{x^i x^j}$ which together with the above properties of (22) leads to the inequality

$$\xi^i \xi^j u_{x^i x^j}(dt\, dx) \leq N|\xi|^2\, dt\, dx.$$

We now pass to the limit in equation (9). For any $\omega \in \Omega$ in Q, we obtain $w_t^{\varepsilon} + \varepsilon \Delta w^{\varepsilon} + L(\omega)w^{\varepsilon} + f(\omega) \geq 0$. Multiplying this inequality by a nonnegative $\eta \in C_0^{\infty}(Q)$, integrating by parts, and passing to the limit over the subsequence ε', we can easily obtain

$$\int (a^{ij}(\omega)\eta)_{x^i x^j} u\, dt\, dx + \int \eta(u_t + b^i(\omega)u_{x^i} + c(\omega)u + f(\omega))\, dt\, dx \geq 0. \tag{23}$$

Were $a^{ij}(\omega)$ infinitely differentiable with respect to (t, x) in this expression, we could use the definition of generalized derivatives and write

$$\int (a^{ij}(\omega)\eta)_{x^ix^j} u \, dt \, dx = \int \eta a^{ij}(\omega) u_{x^ix^j} (dt \, dx) \tag{24}$$

furthermore, by virtue of the arbitrariness of η, we could obtain from (23)

$$a^{ij}(\omega) \circ u_{x^ix^j} + (u_t + b^i(\omega) u_{x^i} + c(\omega)u + f(\omega)) \circ \Lambda \geqq 0 \tag{25}$$

in Q for any $\omega \in \Omega$. Equality (24) turns out also valid for a^{ij} which are twice continuously differentiable with respect to x and Borel with respect to t, although this fact does not straightforwardly follow from the definitions any longer. It follows from Lemma 4.2.5 of Krylov [10]. Having proved (25), we obtain

$$\bigwedge_{\omega\in\Omega} [a^{ij}(\omega) \circ u_{x^ix^j} + (u_t + b^i(\omega) u_{x^i} + c(\omega)u + f(\omega)) \circ \Lambda] \geqq 0. \tag{26}$$

We now prove the opposite inequality. Given any $\varepsilon \in (0, \varepsilon_0)$, we have $w_t^\varepsilon + \varepsilon_0 \Delta w^\varepsilon + F(w^\varepsilon_{x^ix^j}, w^\varepsilon_{x^i}, w^\varepsilon, z) = (\varepsilon_0 - \varepsilon)\Delta w^\varepsilon \leqq \varepsilon_0 N$, which in terms of the measures $w^\varepsilon_{x^ix^j}$ yields

$$\bigwedge_{\omega\in\Omega} [(a^{ij}(\omega) + \varepsilon_0\delta^{ij}) \circ w^\varepsilon_{x^ix^j} +$$

$$+ (w_t^\varepsilon + b^i(\omega) w^\varepsilon_{x^i} + c(\omega) w^\varepsilon + f^{\varepsilon_0}(\omega)) \circ \Lambda] \leqq 0,$$

where $f^{\varepsilon_0}(\omega) = f(\omega) - \varepsilon_0 N$. By Theorem 3.6.3, we can replace w^ε in the latter inequality by u. Consequently, the sum of the left-hand side of (26) and $\varepsilon_0 \Delta u - \varepsilon_0 N\Lambda$ is a nonnegative measure. As $\varepsilon_0 \downarrow 0$, we obtain an inequality which is opposite to (26). By Lemma 3.6.2, we have thus proved assertion (c) of Theorem 1. The lemma is proved.

5. LEMMA. The assertions of Theorem 1 are true if the set Ω is finite.

Proof. We shall need the sequence $R_n \to \infty$ and the sequence of functions $\chi_n(t)$ on $[0, \infty)$ which are infinitely differentiable, nonnegative, equal to 1 for $t \leqq n$, equal to zero for $t \geqq R_n$, decreasing, and such that

$$|\chi_n'(t)| \leqq \frac{2}{nt}, \quad |\chi_n''(t)| \leqq \frac{5}{nt^2}. \tag{27}$$

Such a sequence can be constructed in the following way. Given $n \geqq 1$ and $t \geqq 0$, for example, we set

$$\zeta_n(t) = 0 \qquad \text{for } t \leq n,\ \zeta_n(t) = \frac{1}{2}n^{-3} \text{ for } n \leq t < 2n,$$

$$\zeta_n(t) = -n^{-1}t^{-2} \qquad \text{for } 2n \leq t < 2ne^{n-1/2},$$

$$\zeta_n(t) = -\frac{1}{2}n^{-3}e^{-2n+1} \qquad \text{for } 2ne^{n-1/2} \leq t < 3ne^{n-1/2},$$

$$\zeta_n(t) = 0 \qquad \text{for } t \geq 3ne^{n-1/2}.$$

It is not difficult to see that

$$\int_0^\infty \zeta_n(t)\,dt = 0, \quad -(nt)^{-1} \leq \int_t^\infty \zeta_n(s)\,ds \leq 0 \text{ for } t \geq 0,$$

$$1 + \int_0^t ds \int_s^\infty \zeta_n(r)\,dr = 0 \text{ for } t \geq R_n := 3ne^{n-1/2},$$

$$1 + \int_0^t ds \int_s^\infty \zeta_n(r)\,dr = 1 \text{ for } 0 \leq t \leq n, \ |\zeta_n(t)| \leq 5n^{-1}t^{-2}.$$

It remains to smooth $\zeta_n(t)$ somewhat and determine $\chi_n(t)$ using the equation $\chi_n'' = \zeta_n$. Furthermore, we denote

$$\sigma^{kn}(\omega, t, x) = \chi_n(|x|)\sigma^k(\omega, t, x),$$

$$b^n(\omega, t, x) = \chi_n^2(|x|)b(\omega, t, x),$$

$$c^n(\omega, t, x) = \chi_n^2(|x|)c(\omega, t, x) - \frac{1}{\sqrt{n}},$$

$$f^n(\omega, t, x) = \chi_n^2(|x|)f(\omega, t, x),$$

$$\varphi^n(x) = \chi_n^2(|x|)\varphi(x). \tag{28}$$

The functions σ^{kn}, b^n, $c^n + 1/\sqrt{n}$, f^n, and φ^n thus defined have compact support and $c^n \leq -1/\sqrt{n}$. We now check that these functions also satisfy the conditions of Theorem 1 for all sufficiently large n. We denote by N various constants independent of ω and n. For $|\xi| = 1$, it follows from (27), (2), and (6) that

$$b^n_{(\xi)} = \chi_n^2 b_{(\xi)} + 2\chi_n\chi_n' \frac{(x,\xi)}{|x|} b, \quad (\xi, b^n_{(\xi)}) \leq \chi_n^2(\xi, b_{(\xi)}) + \frac{1}{n} N,$$

$$\sum_k (\xi, \sigma^{kn}_{(\xi)})^2 = \chi_n^2 \sum_k (\xi, \sigma^k_{(\xi)})^2 +$$

$$+ 2\chi_n\chi_n' \frac{(x, \xi)}{|x|} \sum_k (\xi, \sigma^k_{(\xi)})(\xi, \sigma^k) +$$

$$+ (\chi_n')^2 \frac{(x, \xi)^2}{|x|^2} \sum_k (\xi, \sigma^k)^2$$

$$\leq \chi_n^2 \sum_k (\xi, \sigma^k_{(\xi)})^2 + \frac{1}{n} N.$$

Estimating $\sum|\sigma^{kn}_{(\xi)}|^2$ in a similar manner, we can see that σ^{kn}, b^n, and c^n satisfy condition (2.16) for all sufficiently large n. Also, note that

$$|c^n_{(\xi)}| \leq \chi_n^2 |c_{(\xi)}| + 2|\chi_n' c| \leq K\chi_n^2 |c| + \frac{1}{n} N \leq K|c^n|,$$

$$|f^n_{(\xi)}| \leq K\chi_n^2 |c| + 2|\chi_n' f| \leq K|c^n|,$$

$$f^n_{(\xi)(\xi)} = \chi_n^2 f_{(\xi)(\xi)} + 4\chi_n\chi_n' \frac{(x, \xi)}{|x|} f_{(\xi)} +$$

$$+ 2f\left[(\chi_n')^2 \frac{(x, \xi)^2}{|x|^2} + \chi_n\chi_n'' \frac{(x, \xi)^2}{|x|^2} + \right.$$

$$\left. + \chi_n\chi_n' \frac{1}{|x|}\left(1 - \frac{(x, \xi)^2}{|x|^2}\right)\right]$$

$$(f^n_{(\xi)(\xi)})_+ \leq K\chi_n^2 |c| + \frac{1}{n} N \leq K|c^n|$$

for large n and $|\xi| = 1$, while $\sigma^{kn}_{(\xi)(\xi)}$, $b^n_{(\xi)(\xi)}$, and $c^n_{(\xi)(\xi)}$ can be estimated in the same fashion. From these estimates we obtain that condition (2.15) is satisfied for large values of n with N(K) in place of K. This is also true for conditions (1) and (2).

We now denote by v^n the function which, by Lemma 4, exists for all sufficiently large n if we take σ^{kn}, b^n, c^n, f^n, and φ^n as σ^k, b, c, f, and φ respectively. It is clear that, on every cylinder $\bar{C}_{T,R}$, a subsequence of v^n uniformly converges to a function u possessing properties (a) and (b) of Theorem 1. Moreover, for $n \geq R$ in $C_{T,R}$, we have

$$\bigwedge_{\omega\in\Omega}\Big[a^{ij}(\omega) \circ v^n_{x^i x^j} + (v^n_t + b^i(\omega) v^n_{x^i} + c(\omega) v^n -$$

$$- \frac{1}{\sqrt{n}} v^n + f(\omega))\circ\Lambda\Big] = 0. \tag{29}$$

The argument of the previous proof concerning formula (24) allows us to derive inequality (26). It also follows from (29) and Theorem 1 applied to v^n that, for $n \geq m \geq R$ and $\varepsilon > 0$ in $C_{T,R}$,

$$\bigwedge_{\omega\in\Omega}\Big[(a^{ij}(\omega) + \delta^{ij}\varepsilon)\circ v^n_{x^ix^j} +$$

$$+ (v^n_t + b^i(\omega)v^n_{x^i} + c(\omega)v^n + f^{\varepsilon,m}(\omega))\circ \Lambda\Big] \leq 0,$$

where $f^{\varepsilon,m} = -N(\delta, K)(m^{-1/2} + \varepsilon) + f$. By Theorem 3.6.3, we can replace v^n by u, in which case the left-hand side of (26) turns out to be no greater than the measure

$$N(\delta, K)(m^{-1/2} + \varepsilon)\Lambda - \varepsilon\Delta u$$

on $C_{T,R}$ for any $\varepsilon > 0$ and $m \geq R > 0$. Consequently, the inequality opposite to (26) holds. The lemma is proved.

6. Proof of Theorem 1. Let Ω be a countable set, $\Omega = \{\omega_1, \omega_2, \ldots\}$. Taking the set $\{\omega_1, \ldots, \omega_n\}$ instead of Ω in Lemma 5, we construct the corresponding function u^n. On every cylinder $\bar{C}_{T,R}$, a subsequence of u uniformly converges to a function u possessing properties (a) and (b). Moreover, for $n \geq m$,

$$\bigwedge_{r\leq m}\Big[a^{ij}(\omega_r)\circ u^n_{x^ix^j} + (u^n_t + b^i(\omega_r)u^n_{x^i} + c(\omega_r)u^n + f(\omega_r))\circ\Lambda\Big] \geq 0.$$

Hence the argument of the proof of Lemma 4 concerning formula (24) yields inequality (26). To prove the opposite inequality, we note that

$$\bigwedge_{r}\Big[a^{ij}(\omega_r)\circ u^n_{x^ix^j} + (u^n_t + b^i(\omega_r)u^n_{x^i} + c(\omega_r)u^n + f(\omega_r))\circ\Lambda\Big] \leq 0.$$

Adding $\varepsilon\Delta u^n - \varepsilon N(\delta, K)\Lambda$ to the left-hand side and letting $n \to \infty$ and $\varepsilon \downarrow 0$ as in the previous proof, we obtain the inequality opposite to (26). The theorem is proved.

Let us discuss Theorem 1 using the following example.

7. Example. Let $d = d_1 = 1$, $T = 2$, $\Omega = \{+1\} \cup \{-1\}$, let $\sigma = 1$ if $t \in [0, 1)$ and $\sigma = x$ if $t \in [1, 2]$, let $b_0 = 0$ if $t \in [0, 1)$ and $b_0 = \omega$ if $t \in [1, 2]$. Also, let $b = 2\sigma^2 x(x^2 + 1)^{-1} + b_0$, $c = (\sigma^2 + 2b_0 x)$

$(x^2 + 1)^{-1}$, $f = 0$, and $\varphi = -x^2(x^2 + 1)^{-1}$.

Formally, Theorem 1 is inapplicable in this case, since c may be positive and inequalities (2.15) and (2.16) are violated. However, if c and φ were replaced by c-N and φe^{2N} and we would take a sufficiently large constant N, then the conditions of Theorem 1 would be satisfied and hence there would exist the corresponding function u so that ue^{-Nt} would be the function corresponding to the initial c and φ by Theorem 1. In other words, the standard change of the unknown function shows that the assertions of Theorem 1 remain valid.

Let u be the function corresponding to the initial σ, b, c, f, and φ. Using the assertion of uniqueness, we can show that the function $v := (x^2 + 1)u$ is defined as

$$v(t, x) = -x^2e^{2-t} - 2|x|(e^{2-t} - 1) - 2(e^{2-t} - 1 - (2 - t))$$

for $t \in [1, 2]$,

$$v(t, x) = -(x^2 + 1 - t)e - \frac{2(e-1)}{\sqrt{2\pi(1-t)}} \int_{-\infty}^{\infty} |y| \exp\left(- \frac{(y-x)^2}{2(1-t)}\right) dy -$$

$$- 2(e - 2) \text{ for } t \in [0, 1).$$

If $t \geq 1$, it follows from assertion (c) that $x^2 \circ u_{xx} \ll \Lambda$, hence the measure u_{xx} must be absolutely continuous on $\{t > 1\} \cap Q \setminus \{x = 0\}$ with respect to Λ, i.e., the Sobolev derivative u_{xx} must exist on $\{t > 1\} \cap Q \setminus \{x = 0\}$. The above formulas show that this is exactly our case. Moreover, it can be seen that some mass of the measure u_{xx} is concentrated on the segment $[1, 2] \times \{0\}$ and there is no Sobolev derivative u_{xx} in $\{t > 1\} \cap Q$. By assertion (4), we obtain $u_t \in L_2(C_{T,R})$ for any R. The derivative u_t can easily be found and it is seen that in fact here $u_t \in L_{2+\gamma}(C_{2,R})$ for $\gamma \in [0, 1)$, although this is not true for $\gamma = 1$. We do not know whether it is possible to strengthen (4) in the general case. However, (3) cannot be strengthened since in our example $u(1, 0) - u(t, 0) \geq (1 - t)^{1/2}\varepsilon$ for $t \leq 1$ and an appropriately chosen $\varepsilon > 0$.

In the example under consideration, equation (5) can be rewritten in terms of the function v as

$$v_t + \frac{1}{2} x^2 v_{xx}^{(0)} - |v_x| = 0 \text{ for } t \in (1. 2),$$

$$v_t + \frac{1}{2} v_{xx}^{(0)} = 0 \text{ for } t \in (0, 1).$$

The above discussion shows that, whenever $u_{xx}^{(0)}$ enters an equation, i.e.

the coefficient of $u_{xx}^{(0)}$ is not zero, the Sobolev derivative u_{xx} exists and, therefore, $u_{xx}^{(0)} = u_{xx}$. This suggests that, in general, all second derivatives of u actually entering equation (5) should be Sobolev. We have already encountered such a situation in Section 2.2, so when studying this case we shall follow Theorem 2.2.6.

We put

$$F_1(u_{ij}, z) = \sup_{\omega} a^{ij}(\omega, z)u_{ij},$$

$$\mu(l) = \mu(l, z) = \inf_{\xi:(l,\xi)=1} F_1(\xi^i\xi^j, z), \quad l \in E_d, \quad l \neq 0, \tag{30}$$

$$\mu = \mu(z) = \inf_{|\xi|=1} \sup_{\omega} a^{ij}(\omega, z)\xi^i\xi^j = \inf_{|l|=1} \mu(l, z). \tag{31}$$

Recall that the relation of $\mu(l)$ to the second derivative actually entering equation (5) has been discussed in Remark 2.2.8.

8. THEOREM. Let the conditions of Theorem 1 hold and let u be the function in Theorem 1, $l \in E_d$, $|l| = 1$, let Q_0 be a domain contained in Q, let ε be a strictly positive constant, and let $\mu(l) \geqq \varepsilon$ on Q_0. Then the Sobolev derivative $u_{(l)(l)}$ exists and belongs to $\mathcal{L}_{2,loc}(Q_0)$. But if $\mu \geqq \varepsilon$ on Q_0, then $u \in W^{1,2}_{2,loc}(Q_0)$.

Proof. We choose z_0, τ, ρ such that $z_0 + C_{\tau,\rho} \subset Q_0$ and follow the proof of Theorem 2.2.6. For any $\omega_0 \in \Omega$, we obtain

$$a^{ij}(\omega_0)u_{x^ix^j}(dt\,dx) + (u_t + b^i(\omega_0)u_{x^i} +$$
$$+ c(\omega_0)u + f(\omega_0))\,dt\,dx \geqq 0,$$

$$a^{ij}(\omega_0)u_{x^ix^j}(dt\,dx) \geqq -(|u_t| + |u_x|\sup_{\omega}|b(\omega)| +$$
$$+ |u|\sup_{\omega}|c(\omega)| + \sup_{\omega}|f(\omega)|)\,dt\,dx$$

on Q.

This inequality is similar to (2.2.8). Hence using the inequality $u_{x^ix^j}(dt\,dx)\xi^i\xi^j \leqq N\,dt\,dx$, which is true for $\xi \in E_d$, $|\xi| = 1$, and applying Lemma 2.2.5, we obtain the inequality

$$-N\,dt\,dx \leqq u_{x^ix^j}(dt\,dx)l^il^j$$

$$\leqq N\mu^{-1}(l)(1 + |u_t| + |u_x| + |u|)\, dt\, dx,$$

on $z_0 + C_{\tau,\rho}$, which is similar to (2.2.9). In this inequality, N depends only on δ, K, z_0, ρ, K_1. The proof is completed by referring to (4) and using the same argument as in the proof of Theorem 2.2.6. The theorem is proved.

Sometimes it is possible to obtain the boundedness of u_t and u_{xx}.

9. THEOREM. Let the hypotheses of Theorem 1 hold and let the functions σ, b, c, f be differentiable with respect to (t, x) in Q. Also let inequalities (7) and (8) hold. Then the function u in Theorem 1 has the following additional properties:

(a) inequality (11) holds true;
(b) if $Q_0 \subset Q$ is a bounded domain, $l \in E_d$, $|l| = 1$, and $\mu(l) \geqq \varepsilon$ on Q_0 with $\varepsilon > 0$, then the Sobolev derivative $u_{(l)(l)}$ is bounded on Q_0, whereas if $\mu \geqq \varepsilon$ on Q_0, then all Sobolev's derivatives u_{xx} are bounded on Q_0.

Proof. Assertion (b) can be obtained from (a) in the same manner as in the proof of Theorem 8. To prove (a), we should first prove the corresponding counterparts of Lemmas 3 to 5 verifying estimate (11), which was established in Lemma 2, throughout the proofs. This can easily be done in the case of Lemma 3; one need only note that σ, b, c, and f can be continuously extended to $\bar{Q}$ and, rather than put them equal to zero outside Q as in the proof of Lemma 3, we use formulas like $\sigma(t, x) = \sigma(T, x)$ with $t \geqq T$ and $\sigma(t, x) = \sigma(0, x)$ with $t \leqq 0$. The corresponding counterpart of Lemma 4 can be proved using the above arguments. To prove Lemma 5, it suffices to make only one change, namely to replace the latter formula in (28) by $\varphi^n(x) = \chi_m^2(|x|)\varphi(x)$, where m is the smallest integer greater than R_n. Then if the operator L^n is constructed from σ^{kn}, b^n, and c^n, we obtain

$$L^n\varphi^n = \chi_n^2 L\varphi - \frac{1}{\sqrt{n}}\varphi^n, \quad |L^n\varphi^n + f^n| \leqq \chi_n^2|L\varphi + f| + \frac{1}{\sqrt{n}} K$$

and condition (8) for σ^{kn}, b^n, c^n, and f^n remains valid with $2K_2$ instead of K_2 for large n. The proof of the current theorem can be completed by repeating Proof 6.

7.4. The Existence of a Solution in the Class of Concave Functions for the Normalized Bellman Equation

In the previous section we studied the Bellman equation in the case when

σ, b, c, and f were uniformly bounded in ω (locally in (t, x)). It is known in the theory of optimal control of diffusion processes (see e.g. Krylov [5, 10]) that, if we abandon this condition, then we should consider the normalized Bellman equation. The properties of solutions to such equations are studied in this section.

We consider the same objects as in the beginning of Section 3, but now it is assumed that σ, b, c and f are not only twice continuously differentiable with respect to x for every t and ω, but also once continuously differentiable with respect to (t, x) in Q for every $\omega \in \Omega$. This assumption allows us to preserve the corresponding estimate of u_t if (3.2) is abandoned. As in Section 3 we shall use notation (2.1) to (2.3) and suppose that a function $\nu(\omega, z) \geqq 0$ is defined on $\Omega \times Q$. Denote

$$G(u_0, u_{ij}, u_i, u, z) =$$

$$= \inf_{\omega\in\Omega} \nu(\omega, z)[u_0 + a^{ij}(\omega, z)u_{ij} + b^i(\omega, z)u_i + $$

$$+ c(\omega, z)u + f(\omega, z)]. \tag{1}$$

We now formulate the major result of this section concerning parabolic equations.

1. THEOREM. Let $\varphi \in C^2(E_d)$, let inequalities (2.15) and (2.16) hold on $\Omega \times Q$ for all unit $\xi \in E_d$, and let

$$\sum_k |\sigma_t^k|^2 + |b_t| + |c_t| + |f_t| \leqq -Kc, \tag{2}$$

$$|L(\omega, T, x)\varphi(x) + f(\omega, T, x)| \leqq K,$$

$$|\varphi| + |\varphi_x| + ||\varphi_{xx}|| \leqq K. \tag{3}$$

Also, for every $\omega \in \Omega$, let there exist a constant $K_1(\omega) < \infty$ such that

$$|c(\omega, t, x)| \leqq K_1(\omega)(1 + |x|), \quad \sum_k |\sigma^k(\omega, t, 0)|^2 +$$

$$+ |b(\omega, t, 0)| + \sum_k |\sigma^k_{(\xi)}(\omega, t, x)|^2 +$$

$$+ |b_{(\xi)}(\omega, t, x)| \leqq K_1(\omega) \tag{4}$$

for all $(t, x) \in Q$, $\xi \in E_d$, $|\xi| = 1$. Then there exists a function $u \in C(\bar{Q})$ such that:

(a) there exist Sobolev derivatives u_x, u_t, and measures u_{xx} in Q;

(b) $|u| + |u_x| + |u_t| \leq N(\delta, K)$ (a.e.Q), $u_{(\xi)(\xi)}(dt\,dx) \leq N(\delta, K)\,dt\,dx$ in Q for $\xi \in E_d$, $|\xi| = 1$;

(c) $a^{ij}(\omega)u_{x^ix^j}(dt\,dx) << dt\,dx$ for all $\omega \in \Omega$, and if G is a finite-valued function for all values of its arguments, then

$$G\left(u_t,\ u^{(o)}_{x^ix^j},\ u_{x^i},\ u,\ z\right) = 0 \quad (\text{a.e.}Q). \tag{5}$$

Prior to proving the theorem, we discuss its assertion (c).

2. <u>Definition</u>. The function ν is called a *scaling factor* if G is a finite-valued function for all $z \in Q$ and real-valued u_0, u_{ij}, u_i, and u.

The simplest scaling factor is obviously

$$\nu = \nu_0 := (1 + \operatorname{tr} a + |b| + |c| + |f|)^{-1}. \tag{6}$$

If $\nu = \nu_0$, the right-hand side of (1) is denoted by G_0. Generally speaking, various scaling factors yield various equations (5) some of which can be of the simplest form. It is clear that sometimes there is a scaling factor which identically equals unity, in which case equation (5) coincides with the standard Bellman equation. The greatest amount of information on u is provided by equation (5) when $\nu = \nu_0$. This follows from a simple lemma which is formulated here without proof (see Lemma 6.3.8 in Krylov [10]).

3. LEMMA. Let ν be a scaling factor. Then every solution (u_0, u_{ij}, u_i, u, z) to the equation (inequality) $G_0(u_0, u_{ij}, u_i, u, z) = 0$ (≤ 0) also is a solution to the equation (inequality) $G(u_0, u_{ij}, u_i, u, z) = 0 (\leq 0)$.

4. <u>Proof of Theorem</u> 1. If the set Ω is finite, then the theorem follows from Theorem 3.9 because (5) obviously follows from (3.5). Let Ω be a countable set and let $\Omega = \{\omega_1, \omega_2 \ldots\}$. Replacing Ω in Theorem 3.9 by the set $\{\omega_1 \ldots \omega_n\}$, we construct the corresponding function u^n. A subsequence of u^n uniformly converges to a function u on every cylinder $\bar{C}_{T,R}$. It is clear that $u \in C(\bar{Q})$, $u(T, x) = \varphi(x)$, and u possesses properties (a) and (b) (cf. the proof of Lemma 3.4). Furthermore, for any $\varepsilon > 0$, by Theorem 3.9 (and Lemma 3.6.2), we obtain

$$\bigwedge_{r\leq n}\Big[\nu_0 a^{ij}(\omega_r)\cdot u^n_{x^ix^j} + \nu_0(\omega_r)\circ u^n_t + \nu_0(\omega_r)\Big(u^n_t + b^i(\omega_r)u^n_{x^i} + $$
$$+ c(\omega_r)u^n + f(\omega_r)\Big)\circ\Lambda\Big] = 0, \tag{7}$$

$$\bigwedge_{\omega}\Big[\nu_0 a^{ij}(\omega)\circ u^n_{x^i x^j} + \nu_0(\omega)\Big(u^n_t + b^i(\omega)u^n_{x^i} + c(\omega)u^n + $$

$$+ f(\omega)\Big)\circ\Lambda \leqq 0, \tag{8}$$

$$\bigwedge_{\omega}\Big[(\nu_0 a^{ij}(\omega) + \varepsilon\delta^{ij})\circ u^n_{x^i x^j} + (\nu_0(\omega) + \varepsilon)u^n_t\circ\Lambda +$$

$$+ (\nu_0 b^i(\omega)u^n_{x^i} + \nu_0 c(\omega)u^n + f^\varepsilon(\omega))\circ\Lambda\Big] \leqq 0. \tag{9}$$

where $f^\varepsilon(\omega) = \nu_0 f(\omega) - \varepsilon N_1$ and $N_1 = N_1(d, \delta, K)$ is such that

$$\varepsilon(\Delta u^n + u^n_t \circ \Lambda - N_1\Lambda) \leqq 0. \tag{10}$$

We need inequality (9) to apply Theorem 3.6.3. In this theorem, the coefficient of u_t must be bounded away from zero, therefore, unlike the proof of 3.6, we should add (10) to the left-hand side of (8) and use (2) and (3) to ensure the boundedness of u^n_t and the possibility of choosing a constant N_1 which is suited for (10). By Theorem 3.6.3, we can replace u^n in (9) by u, and letting $\varepsilon \downarrow 0$, we then obtain

$$\bigwedge_{\omega}\Big[\nu_0 a^{ij}(\omega)\circ u_{x^i x^j} + \nu_0(\omega)(u_t + b^i(\omega)u_{x^i} +$$

$$+ c(\omega)u + f(\omega))\circ\Lambda\Big] \leqq 0. \tag{11}$$

The opposite inequality can be derived from (7) using formula (3.24) as in the proof of Lemma 3.4. It remains to note that the equality in (11) is, by Lemma 3.6.2, equivalent to equality (5) with $G = G_0$ and to the fact that $\nu_0 a^{ij}\circ u_{x^i x^j} \ll \Lambda$ and then to make use of Lemma 3 and the the fact that $\nu_0 > 0$. Theorem is proved.

5. Remark. In Theorem 1 u may be nonunique, which immediately follows if we take $\nu \equiv 0$. The nonuniqueness may also occur in the case of a more reasonable choice of ν from formula (6). For example, if $d = d_1 = T = 1$, Ω is a countable, everywhere dense subset of $[0, \infty)$, $\sigma^k = f = \varphi = 0$, $b(\omega) = \omega$, and $c = -1$, then it is not difficult to see that any function $e^t\psi(t)$, where $\psi' \geqq 0$, $\psi(1) = 0$, possesses properties (a) to (c) of Theorem 1 for $G = G_0$. The uniqueness under consideration can sometimes be studied using Corollary 3.7.8.

6. Remark. It easily follows from Theorem 3.7.9 that the functions u^n which were constructed in Proof 4 as solutions to the

equations

$$u_t^n \circ \Lambda + \bigwedge_{r \leq n} \Big[a^{ij}(\omega_r) \circ u^n_{x^i x^j} + (b^i(\omega_r) u^n_{x^i} + c(\omega_r) u^n +$$

$$+ f(\omega_r)) \circ \Lambda \Big] = 0 \tag{12}$$

with the boundary condition $u^n(T, x) = \varphi(x)$, decrease in n. Therefore, not only a subsequence of u^n, but the whole sequence u^n, converges to u. In this connection, it is important to note that u may fail to be a solution to the 'limit' equation of (12) (of course, in such a case 1 is not a scaling factor). For example, let $d = d_1 = T = 1$, $\Omega = \{1, 2, \ldots\}$, $\sigma = b = \varphi = 0$, $c = -\omega$, and $f = 1$. Then, for $\nu = 1$, equation (5) is of the form

$$u_t - u + 1 + \inf_{n=0,1 \ldots} (-n)u = 0. \tag{13}$$

Here the finiteness of the infimum means that the infimum is equal to zero, $u \leq 0$, and $u_t - u + 1 = 0$. However, the two latter relationships are incompatible with the boundary condition $u(1, x) = 0$, and equation (13) with this boundary condition has no solutions at all. The reader can easily prove that in this example the function u taken from Theorem 1 is identically equal to zero.

Imposing some nondegeneracy conditions on the matrix $a(\omega)$, we can obtain more information on the derivatives of u in Theorem 1.

7. THEOREM. Let the hypotheses of Theorem 1 hold, let $Q_0 \subset Q$, and let u be the function constructed in Proof 4. For $l \in E_d$, $|l| = 1$, let

$$\mu_0(l) = \mu_0(l, z) = \inf_{\xi:(\xi,l)=1} \sup_{\omega\in\Omega} \nu_0 a^{ij}(\omega, z)\xi^i\xi^j, \tag{14}$$

$$\mu_0 = \mu_0(z) = \inf_{|\xi|=1} \sup_{\omega\in\Omega} \nu_0 a^{ij}(\omega, z)\xi^i\xi^j = \inf_{|l|=1} \mu_0(l), \tag{15}$$

where ν_0 is defined by formula (6). Then if $\mu_0(l) \geq \varepsilon$ in Q_0, where ε is a strictly positive constant, for an $l \in E_d$ with $|l| = 1$, then the generalized derivative $u_{(l)(l)}$ is bounded in Q_0. But if $\mu_0 \geq \varepsilon$ in Q_0, then all generalized derivatives u_{xx} are bounded in Q_0.

The proof of this theorem follows from the equality in (11) by repeating the corresponding arguments of the proof of Theorem 2.2.6.

We now extend Theorems 1 and 7 to elliptic equations.

8. THEOREM. Let σ, b, c, and f be independent of t, let inequalities (2.15) and (2.16) hold on $\Omega \times E_d$ for all unit vectors $\xi \in E_d$, and let the condition of Theorem 1 involving (4) hold. Then there exists a function $u \in C(E_d)$ such that:

(a) there exist Sobolev derivatives u_x and u_{xx} in E_d;

(b) $|u| + |u_x| \leq N(\delta, K)$ (a.e.E_d), $\xi^i\xi^j u_{x^ix^j}(dx) \leq N(\delta, K)$ in E_d for $\xi \in E_d$, $|\xi| = 1$;

(c) $a^{ij}(\omega)u_{x^ix^j}(dx) << dx$ for all $\omega \in \Omega$ and if $\nu = \nu(\omega, x)$ and $G(0, u_{ij}, u_i, u, x) > -\infty$ for all u_{ij}, u_i, u, x, then

$$G\left(0, u^{(0)}_{x^ix^j}, u_{x^i}, u, x\right) = 0 \text{ (a.e.}E_d). \tag{16}$$

Proof. It is clear that if u is a solution to equation (16), then $u - K$ is a solution of the same equation where f is replaced by $f + cK$. The converse is also true and, moreover, $f + cK \leq 0$ by virtue of (2.15). Therefore, without loss of generality we can assume that $f \leq 0$.

Let $\Omega = \{\omega_1, \omega_2, \ldots\}$. For $T > 0$, $\varphi \equiv 0$, and integer numbers $n \geq 1$, we define functions $u^{T,n}$ in $[0, T] \times E_d$ by Theorem 3.1 so that

$$\bigwedge_{r\leq n}\Big[a^{ij}(\omega_r)\circ u^{T,n}_{x^ix^j} + (u^{T,n}_t + b^i(\omega_r)u^{T,n}_{x^i} + c(\omega_r)u^{T,n} + f(\omega_r))\cdot\Lambda\Big] = 0 \tag{17}$$

in $(0, T) \times E_d$. Since the funcitons $u^{T,n}$ are unique, it easily follows that, for fixed n and x, they depend only on $T - t$. Furthermore, we can apply Theorem 3.7.9 to equation (17) (see Remark 3.7.11). By this theorem, comparing $u^{T,n}$ and 0 and using the inequality $f \leq 0$, we obtain $u^{T,n} \leq 0$. Moreover, the solution to equation (17) with negative boundary data at $t = T$ is smaller than $u^{T,n}$ in $[0, T] \times E_d$. Since $u^{T+\gamma,n}(T, x) \leq 0$ where $\gamma \geq 0$, we obtain $u^{T+\gamma,n}(t, x) \leq u^{T,n}(t, x)$ for $t \leq T$. Thus, $u^{T,n}(t, x)$ decrease in T and their limit, as $T \to \infty$, which is equal to the limit of $u^{T-t,n}(0, x)$, is independent of t. Denote

$$u^n(x) = \lim_{T\to\infty} u^{T,n}(t, x).$$

Theorem 3.1 implies that the functions $u^{T,n}$ enjoy well-known properties. It follows that the function $u(x)$ on $(0, 1) \times E_d$, as a function of (t, x), has Sobolev derivatives u^n_x and u^n_t and measures u^n_{xx}. Since u is independent of t, we obtain $u^n_t = 0$ and u^n has Sobolev derivatives u^n_x and measures u^n_{xx} on E_d. Thus we have measures u^n_{xx} on $(0, 1) \times E_d$ and E_d. To distinguish between them, we will specify them by their differentials $u^n_{xx}(dt\, dx)$ and $u^n_{xx}(dx)$. It immediately follows from the definition of generalized derivatives that $u^n_{x^ix^j}(dt\, dx) = u^n_{x^ix^j}(dx)dt$. Clearly, the functions u^n satisfy not only assertion (a) of the current theorem, but also assertion (b).

Now, considering (17) only on $(0, 1) \times E_d$, we let $T \to \infty$. As has already been shown, theorem 3.6.3 leads us to the fact that

$$\bigwedge_{r\leq n}\Big[a^{ij}(\omega_r)u^n_{x^ix^j}(dt\, dx) + (u^n_t + b^i(\omega_r)u^n_{x^i} + c(\omega_r)u^n + f(\omega_r))\, dt\, dx\Big] = 0.$$

on $(0, 1) \times E_d$.

The above discussion enables us to rewrite this relationship as

$$\bigwedge_{r\leq n}\Big[a^{ij}(\omega_r)u^n_{x^ix^j}(dx) + (b^i(\omega_r)u^n_{x^i} + c(\omega_r)u^n + f(\omega_r))\, dx\Big] = 0 \tag{18}$$

in E_d. Furthermore, we set

$$\bar{\nu}_0 = (\operatorname{tr} a + |b| + |c| + |f|)^{-1},$$
$$\text{if } \operatorname{tr} a + |b| + |c| + |f| > 0,$$
$$\bar{\nu}_0 = 1, \quad \text{if } \operatorname{tr} a + |b| + |c| + |f| = 0. \tag{19}$$

It is clear that $\bar{\nu}_0 a, \ldots, \bar{\nu}_0 f$ are bounded on $\Omega \times E_d$ and using (18) expressed in terms of the densities of $u^n_{x^ix^j}(dx)$ and dx with respect to a measure, we easily obtain

$$\bigwedge_{r\leq n}\Big[\bar{\nu}_0 a^{ij}(\omega_r)u^n_{x^ix^j}(dx) + (\bar{\nu}_0 b^i(\omega_r)u_{x^i} + \bar{\nu}_0 c(\omega_r)u^n + \bar{\nu}_0 f(\omega_r))\, dx\Big] = 0$$

in E_d. In this expression, let us pass to the limit as $n \to \infty$ by using

Thoerem 3.6.3. This theorem relates to parabolic operators; therefore, to use it, we first pass from the measures $u^n_{x^ix^j}(dx)$ and dx to the measures $u^n_{x^ix^j}(dt\,dx)$ and dt dx, obtain the corresponding equation in $(0, 1) \times E_d$, then add the zero summand $u^n_t dt\,dx$ to its left-hand side, and, finally, replace the infimum over $r \leqq n$ by the infimum over all r. Then, on applying Theorem 3.6.3 and coming back to the variables x, we obtain a function $u = u(x)$ satisfying assertions (a) and (b) and such that

$$\bigwedge_\omega \left[\bar{\nu}_0 a^{ij}(\omega) u_{x^ix^j}(dx) + (\bar{\nu}_0 b^i(\omega) u_{x^i} + \bar{\nu}_0 c(\omega) u + \bar{\nu}_0 f(\omega))\,dx\right] \leqq 0 \tag{20}$$

in E_d.

In fact the latter relation is an equality, since the measures under the infimum sign in (18) are positive for $r \leqq n$. Passing to the limit, one can see that this is true for all r whenever u^n is replaced by u, but then the measures in (20) are also positive as indefinite integrals of a *summable* function $\bar{\nu}_0 > 0$ with respect to positive measures.

We can now prove assertion (c) for the function u. Equality (20) is equivalent (cf. Section 3.6) to the set of two relations

$$\bar{\nu}_0 a^{ij}(\omega) u_{x^ix^j}(dx) << dx \text{ for all } \omega \in \Omega, \tag{21}$$

$$\inf_\omega \bar{\nu}_0(\omega)\left[a^{ij}(\omega) u^{(0)}_{x^ix^j} + b^i(\omega) u_{x^i} + c(\omega) u + f(\omega)\right] = 0$$

$$(\text{a.e.}\,E_d). \tag{22}$$

It follows from (21) that the measure $a^{ij}(\omega) \circ u_{x^ix^j}$ is absolutely continuous with respect to dx as an indefinite integral of a locally summable function $\bar{\nu}_0^{-1}(\omega)$ with respect to an absolutely continuous measure. To convert (22) to the form (16), we fix a point x_0, where (22) holds, and consider two cases: first, there exists an $\omega_0 \in \Omega$ such that $\operatorname{tr} \underline{a}(\omega_0, x_0) + |b(\omega_0, x_0)| + |c(\omega_0, x_0)| + |f(\omega_0, x_0)| = 0$; second, there is no such ω_0. In either case the coefficient of $\bar{\nu}_0$ in (22) is nonnegative at the point x_0 for any $\omega \in \Omega$, therefore, the left-hand side of (16) is nonnegative. In the first case, it is obviously equal to zero (cf. (1)). In the second case, we can prove that the left-hand side of (16) equals zero at x_0 using an assertion similar to Lemma 3 (see Lemmas 6.3.6 and 6.3.8 in Krylov [10]). The theorem is proved.

9. Remark. The uniqueness of the constructed function u can be studied by using Corollary 3.7.10.

The following fact which is quite similar to Theorem 7 is formulated without proof.

10. THEOREM. Let the conditions of Theorem 8 hold, let a domain D be contained in E_d, and let u be the function constructed in the proof of Theorem 8. For unit vectors $l \in E_d$, we introduce the functions $\bar{\mu}_0(l) = \bar{\mu}_0(l, x)$ and $\bar{\mu}_0 = \bar{\mu}_0(x)$ by formulas (14) and (15), where ν is replaced by ν_0 from (19). Then if $\bar{\mu}_0(l) \geqq \varepsilon$ in D for an $l \in E_d$ $|l| = 1$ with $\varepsilon > 0$, then the generalized derivative $u_{(l)(l)}$ is bounded in D, and if $\bar{\mu}_0 \geqq \varepsilon$ in D, then all generalized derivatives u_{xx} are bounded in D.

We now present one more example illustrating Theorems 1 and 7.

11. Example. Let σ, b, c, and f be independent of ω, bounded functions satisfying the conditions formulated at the outset of the section, let the functions satisfy inequalities (2.15), (2.16), (2), and (4), and let $c \leqq -\delta$, $a^{ij}\xi^i\xi^j \geqq \delta|\xi|^2$ on Q for all $\xi \in E_d$. Also, suppose that $f(T, x) = 0$. We assert that there exists a unique function $u \in C(\bar{Q})$ such that its Sobolev derivatives of the form u_x, u_t, and u_{xx} are bounded, $u(T, x) = 0$ on E_d, and

$$u_t \geqq 0, \quad a^{ij}u_{x^ix^j} + b^iu_{x^i} + cu + f \geqq 0,$$

$$u_t\left(a^{ij}u_{x^ix^j} + b^iu_{x^i} + cu + f\right) = 0 \text{ (a.e.}Q). \tag{23}$$

Indeed, the set of relations (23) is equivalent to the equation

$$\inf_{\omega=0,1,\ldots} \frac{1}{\omega+1}\left[u_t + \omega a^{ij}u_{x^ix^j} + \omega b^iu_{x^i} + \omega cu + \omega f\right] = 0 \text{ (a.e.}Q). \tag{24}$$

The latter equation can have only one solution in the class of functions under consideration, which easily follows from Theorem 3.4.6, where ψ is replaced by $\cosh(\gamma|x|)$ for a sufficiently small constant γ. Moreover, if we set $\Omega = \{0, 1, 2, \ldots\}$, $\sigma^k(\omega, z) = \sqrt{\omega}\,\sigma^k(z)$, $b(\omega, z) = \omega b(z)$, $c(\omega, z) = \omega c(z)$, $f(\omega, z) = \omega f(z)$, and $\nu(\omega) = (1 + \omega)^{-1}$ as dictated by (24), then, by Theorems 1 and 7, we obtain that equation (24) has a solution possessing the described properties.

In the context of this example, let us note without proof that if σ, b, c, and f are periodic in t with the period 1, $f(n, x) = 0$ for all integer $n = 1, 2, \ldots$, and $u^n(t, x)$ is the function associated with

$T = n$, then the functions $u^n(t, x)$ decreasing as $n \to \infty$ converge to a function $u(t)$ having bounded derivatives u_x, u_{xx}, and satisfying the equation

$$\inf_{t\in[0,1]} \left[a^{ij}(t, x)u_{x^i x^j}(x) + b^i(t, x)u_{x^i}(x) + \right.$$
$$\left. + c(t, x)u(x) + f(t, x)\right] = 0 \quad (\text{a.e.}E_d).$$

This enables us to solve the latter equation using less complex relationships (23) or (24). Note that the solvability of equation (24) were proved in Theorem 1 using passage to the limit in equations of the form

$$\inf_{\omega\leq m} \frac{1}{\omega+1}\left[u_t + \omega a^{ij}u_{x^i x^j} + \omega b^i u_{x^i} + \omega cu + \omega f\right] = 0,$$

which are equivalent to the equations

$$u_t = m(a^{ij}u_{x^i x^j} + b^i u_{x^i} + cu + f)_-.$$

In conclusion, it should be noted that a broad class of examples illustrating applications of these results will be shown in the first section of the next chapter.

7.5. An Example of One-Dimensional Degenerate Equations

In this chapter we study the necessity of condition (2.16) for Theorem 2.2 establishing an upper bound on $u_{(\xi)(\xi)}$, which is indepentent of T, and the necessity of this condition for Theorems 4.1, 4.7, 4.8, and 4.10 establishing the existence of a solution whose second derivatives are bounded from above in x.

The equation

$$2u'\tanh x - \lambda u - \frac{1}{\cosh x} = 0 \tag{1}$$

on $(-\infty, \infty)$ is a simple example illustrating the necessity. It is easy to see that, for any constant $\lambda > 0$, there exists a unique bounded function u having a Sobolev derivative and satisfying (1) almost everywhere. This function can be obtained in an explicit form (see (4)) which shows that $|u|$, $|u'|$, $|u''| \to 0$ as $|x| \to \infty$, moreover, if $\lambda > 4$, its second derivative is bounded, if $\lambda \in (2, 4]$, it is not true, and if $\lambda \in (0, 2]$, its first derivative is unbounded.

Equation (1) is suited for our theory since we have not assumed that $\sigma \not\equiv 0$ or Ω contains more than one point. For (1), condition (2.16) looks as

$$4\frac{1}{(\cosh x)^2} \leq (1 - \delta)\lambda, \tag{2}$$

and it holds with a $\delta \in (0, 1)$ if and only if $\lambda > 4$. Thus, by Theorem 4.8 with $\lambda > 4$, equation (1) must have a solution such that u" is bounded from above. Since equation (1) is linear, passing from u to (-u) we conclude that (-u") is also bounded from above, by Theorem 4.8. Theorem 4.8 thus ensures that there exists a solution of (1) with a bounded second derivative for $\lambda > 4$. By the above argument, equation (1) actually has a solution with a bounded second derivative only if $\lambda > 4$. In particular, we have shown that condition (2.16) is necessary for Theorem 4.8, and the coefficients of $(\xi, b_{(\xi)})$ and c in (2.16) are, generally speaking, best possible. An analysis of this example shows that condition (2.20) is necessary to estimate $|u'|$.

Equation (1) is of the first order and so the necessity of condition (2.16) may seem to follow from the fact that we did not distinguish between equations of the first and the second order. In fact, there are no conditions like (2.16) for uniformly nondegenerate equations considered in Chapter 6. Modifying equation (1), we shall later prove the necessity of condition (2.16) for 'weakly nondegenerate' equations among which we place equations with $\mu > 0$, where μ is taken from (3.31).

Let us consider the equation

$$\min_{\omega=0,1} [\omega u'' + 2u'\tanh x - \lambda u - (\cosh x)^{-1}] = 0. \tag{3}$$

For this equation, condition (2.16) is of the same form (2), and by Theorem 4.10, equation (3) with $\lambda > 4$ has a solution which is bounded together with its first and second derivatives. The analysis of this equation with $\lambda \in (0, 4]$ is presented as a number of assertions.

1. LEMMA. Let $-\infty \leqq r_1 < r_2 \leqq \infty$ and let functions u_1 and u_2 be defined, continuous, and bounded on the closure of the interval (r_1, r_2). Suppose that u_1' and u_2' are locally absolutely continuous on (r_1, r_2), the left-hand side of (3) with $u = u_1$ is greater than or equal to zero almost everywhere on (r_1, r_2) whereas for $u = u_2$ the opposite inequality holds. Finally, let $u_1(r_1) \leqq u_2(r_1)$, whenever $r_1 > -\infty$, and $u_1(r_2) \leqq u_2(r_2)$ for $r_2 < \infty$. Then $u_1 \leqq u_2$ on (r_1, r_2).

Proof. If $|r_1|, |r_2| < \infty$, then the lemma is an implication of Theorem 3.4.7. If $|r_1| + |r_2| = \infty$, then we denote by ν a positive root of the equation $\nu^2 + 2\nu - \lambda = 0$ and set $\psi(x) = \cosh \nu x$. Then $\omega\psi'' + 2\psi'\tanh x - \lambda\psi \leqq \psi'' + 2|\psi'| - \lambda\psi \leqq 0$, the left-hand side of (3) as computed on $u_2 + \delta_1\psi$ is negative for any $\delta_1 > 0$, and, by Theorem 3.4.7 with $x \in (r_1, r_2)$, $R > |x|$, we obtain

$$u_1(x) - (u_2(x) + \delta_1\psi(x)) \leqq \max_{y=(-R)\vee r_1, R\wedge r_2} (u_1(y) - (u_2(y) + \delta_1\psi(y)))_+.$$

Here the right-hand side tends to zero as $R \to \infty$, therefore, $u_1 \leqq u_2 + \delta_1\psi$ on (r_1, r_2) and, noting that δ_1 can be chosen arbitrarily small, we complete the proof.

2. LEMMA. Let $u \in C(E_1)$, let u' be a locally absolutely continuous function on E_1, and let u satisfy (3) (a.e.E_1). Then $u(-x) = u(x)$, $u(0) = -\lambda^{-1}$, and $u(x) \leqq w(|x|)$, where

$$w(x) := -(\sinh x)^{\lambda/2} \frac{1}{2} \int_x^\infty (\sinh y)^{-(\lambda/2+1)} dy, \quad x > 0. \tag{4}$$

Proof. Since equation (3) is invariant under the substitution $x \to -x$, it follows from Lemma 1 that $u(-x) \leqq u(x)$, i.e., $u(-x) = u(x)$. To prove the inequality $u(x) \geqq -\lambda^{-1}$, it suffices to put $u_1 = -\lambda^{-1}$ and $u_2 = u$ in Lemma 1. It follows that $u(0) \geqq -\lambda^{-1}$. To prove the opposite inequality, let us note that

$$2u'\tanh x - \lambda u - (\cosh x)^{-1} \geqq 0,$$

by virtue of (3). Moreover, straightforward consideration shows that the function $w(x)$ satisfies equation (1) for $x > 0$. Therefore, $2(u-w)' \times \tanh x \geqq \lambda(u-w)$ on $(0, \infty)$. It follows from the latter inequality that $(u-w)' \geqq 0$ whenever $u - w \geqq 0$. Specifically, for any $\varepsilon > 0$, the set $\{x > 0 : u(x) - w(x) > \varepsilon\}$ is of the form (r_ε, ∞), where $0 \leqq r_\varepsilon \leqq \infty$. We have $(u - w)' \geqq \lambda\varepsilon(2 \tanh x)^{-1}$ on this set, and since the latter function is nonintegrable at infinity ($\tanh x \to 1$ as $x \to \infty$) whereas the function $u - w$ is bounded, we therefore obtain $\{x > 0: u(x) - w(x) > \varepsilon\} = \emptyset$ and $u \leqq w$ for $x > 0$. Hence $u(0) \leqq w(0+)$. The value $w(0+)$ can easily be calculated e.g., by L'Hôpital's rule. We find that $w(0+) = -\lambda^{-1}$, which asserts the lemma.

3. LEMMA. Let $\lambda \in (0, 2]$. Then there is no function satisfying the hypotheses of Lemma 2.

Proof. Assume the contrary and prove that we then obtain $u(x) = \varphi(|x|)$, where

$$\varphi(x) := -\beta^{-2}(\cosh x)^{-1}(1 + \lambda^{-1}e^{-\beta x}), \quad \beta = \sqrt{1 + \lambda}. \tag{5}$$

It is obvious that $\varphi(|x|)$ has first derivative discontinuous at zero and the equality $u(x) = \varphi(|x|)$ yields the desired contradiction. By Lemma 2, it suffices for us to prove that $u(x) = \varphi(x)$ on $(0, \infty)$ and since $u(0) = \varphi(0) = -\lambda^{-1}$, it suffices, by Lemma 1, to verify that φ satisfies equation (3) on $(0, \infty)$. A simple calculation shows that

$$\varphi'' + 2\varphi'\tanh x - \lambda\varphi = (\cosh x)^{-1}$$

and thus it remains for us to show that $\varphi'' \leqq 0$ on $(0, \infty)$. We have

$$\begin{aligned}\psi(x) &:= \lambda(\lambda + 1)\cosh^3 x\varphi''(x) \\ &= -e^{-\beta x}\left[\beta^2\cosh^2 x + \beta \sinh 2x - 1 + \sinh^2 x\right] + \\ &\quad + \lambda(1 - \sinh^2 x),\end{aligned}$$

$$\psi'(x) = -e^{-\beta x}\left[\beta(2 - \lambda)\cosh^2 x + \sinh 2x\right] - \lambda \sinh 2x.$$

Hence $\psi(0) = 0$, $\psi'(x) \leqq 0$, $\psi(x) \leqq 0$, $\varphi''(x) \leqq 0$ on $(0, \infty)$. The lemma is proved.

We now consider the case when λ falls on the interval $(2, \infty)$. Let us fix a $\lambda \in (2, \infty)$ and define x_0 from the equation

$$\tanh x_0 = \frac{\sqrt{2\lambda^2 - 3\lambda + 1} - \sqrt{\lambda + 1}}{2\lambda}. \tag{6}$$

Since $\lambda > 2$, we obtain $x_0 > 0$. Also, let

$$N_1 = (1 - \tfrac{2}{\lambda})\frac{\cosh x_0}{\lambda \cosh^2 x_0 - 2}(\sinh x_0)^{-\lambda/2},$$

$$N_2 = \frac{1 - 2\tanh^2 x_0}{2\beta \tanh x_0 + 2\tanh^2 x_0 + \lambda}\cdot e^{\beta x_0},$$

$$v(x) = -(\sinh x)^{\lambda/2}\left[N_1 + \frac{1}{2}\int_x^{x_0} (\sinh y)^{-(\lambda/2+1)}\, dy\right]$$

$$\text{for } 0 < x \leqq x_0,$$

$$v(x) = -\beta^{-2}(\cosh x)^{-1}(1 + N_2 e^{-\beta x}) \text{ for } x > x_0. \tag{7}$$

4. LEMMA. Let $\lambda \in (2, \infty)$ and let u be a function satisfying the conditions of Lemma 2. Then $u(x) = v(|x|)$.

Proof. As in the previous proof, it suffices to prove that $v \in C^2_{loc}$ $(0, \infty)$ and v satisfies equation (3) on $(0, \infty)$. One can straightforwardly show that

$$2v'\tanh x - \lambda v = (\cosh x)^{-1} \text{ for } 0 < x < x_0, \tag{8}$$

$$v'' + 2v'\tanh x - \lambda v = (\cosh x)^{-1} \text{ for } x > x_0, \tag{9}$$

$$\begin{aligned} v''(x) &= -\frac{\lambda}{2}(\sinh x)^{\lambda/2} N_1 \left[1 + (\frac{\lambda}{2} - 1)\coth^2 x\right] + \\ &+ \frac{1}{2}(\frac{\lambda}{2} - 1)\frac{\cosh x}{\sinh^2 x} - \frac{\lambda}{2}(\sinh x)^{\lambda/2}\frac{1}{2}\int_x^{x_0}(\sinh y)^{-\lambda/2 - 1}\, dy \times \\ &\times \left[1 + (\frac{\lambda}{2} - 1)\coth^2 x\right] \text{ for } 0 < x < x_0, \end{aligned} \tag{10}$$

$$\begin{aligned} (\lambda + 1)\cosh^3 x\, v''(x) &= -N_2 e^{-\beta x}(\lambda \cosh^2 x + 2\sinh^2 x + \\ &+ \beta \sinh 2x) + 1 - \sinh^2 x \text{ for } x > x_0. \end{aligned} \tag{11}$$

This and the definition of N_1 and N_2 imply that $v''(x_0 +) = v''(x_0 -) = 0$. The equality $v(x_0 +) = v(x_0 -)$ is also true and, owing to some arithmetical coincidence, it turns out equivalent to the equation $2\lambda(\tanh x_0)^2 + 2\beta \tanh x_0 - \lambda + 2 = 0$, while (6) actually defines a root of this equaiton. Relations (8) and (9) now automatically imply that $v'(x_0 +) = v'(x_0 -)$. Thus v is twice continuously differentiable on $(0, \infty)$.

In order to prove that v satisfies equation (3) on $(0, x_0]$, we, in view of (8), have to prove that $v'' \geqq 0$ on $(0, x_0]$. Let us consider a function $h := v''\chi$, where

$$\chi(x) := \left\{\frac{\lambda}{2}(\sinh x)^{\lambda/2}\left[1 + (\frac{\lambda}{2} - 1)\coth^2 x\right]\right\}^{-1}$$

on $(0, x_0]$. It is clear that $\chi > 0$ and $h(x_0) = 0$. Therefore, if we prove that $h' \leqq 0$ on $(0, x_0]$, then we shall obtain $h \geqq 0$ and $v'' \geqq 0$ on $(0, x_0]$. It follows from (10) that

$$\begin{aligned} h' &= \frac{1}{2\lambda}(\sinh x)^{-\lambda/2 - 1}(\frac{\lambda}{2}\cosh^2 x - 1)^{-2}(\lambda \cosh^4 x + \\ &+ (2 - 3\lambda)\cosh^2 x + 2\lambda - 2). \end{aligned}$$

The sign is completely determined by the third multiplier which is denoted by g. It is a quadratic function of $\cosh^2 x$, whose roots are $\cosh x_1 = 1$ $(x_1 = 0)$, $\cosh^2 x_2 = 2\lambda^{-1}(\lambda - 1)$. Hence, for $x \in (0, x_2)$, we obtain $g \leqq 0$ and $h' \leqq 0$. If we show that $x_2 > x_0$, i.e., $\tanh x_0 < \tanh x_2$, then it will be proved that $h' \leqq 0$ on $(0, x_0]$. The inequality $\tanh x_0 <$ $\tanh x$ can be proved as follows:

$$\tanh x_0 = \frac{\lambda - 2}{\sqrt{(\lambda - 1)(2\lambda - 1)} + \sqrt{\lambda + 1}}$$
$$< \frac{\lambda - 2}{\sqrt{(\lambda - 1)(2\lambda - 4)}} = \tanh x_2.$$

In order to prove that v satisfies equation (3) on (x_0, ∞), by virtue of (9), we need to show that $v'' \leqq 0$ on (x_0, ∞). The inequality $v'' \leqq 0$ on (x_0, ∞), by (11), is equivalent to the fact that the function

$$\frac{1 - \sinh^2 x}{\cosh^2 x + 2\sinh^2 x + \beta \sinh 2x} e^{\beta x} \tag{12}$$

takes its greatest value on $[x_0, \infty)$ at the point x_0. It can easily be shown that the derivative of (12) has the same sign as $(\lambda - 2) - 2\beta \tanh x - 2\lambda \tanh^2 x$, while the latter quadratic function of tanh x is negative for $\tanh x \geqq \tanh x_0$ since $\tanh x_0$ is its root. Therefore, $v'' \leqq 0$ on (x_0, ∞) and the lemma is proved.

5. THEOREM. Let $\lambda \in (2, 4]$, and let u be a function satisfying the conditions of Lemma 2. Then $u \in C^2_{loc}(E_1 \setminus \{0\})$ and $u''(x) \to \infty$ as $x \to 0$.

Proof. By Lemma 4 for $x > 0$, on integrating by parts in (10), we obtain

$$u''(x) = -\left(\frac{\lambda}{2}\right)^2 (\sinh x)^{\lambda/2}\left[N_1 + \frac{1}{2}\int_x^{x_0} (\sinh y)^{-\lambda/2 - 1}\, dy\right] -$$
$$- \frac{1}{2}\left(\frac{\lambda}{2} - 1\right)(\sinh x)^{\lambda/2-2}\Big\{\lambda N_1 - (\sinh x_0)^{-\lambda/2}\cosh x_0 -$$
$$- \left(\frac{\lambda}{2} - 1\right)\int_x^{x_0} (\sinh y)^{-\lambda/2 + 1}\, dy\Big\}.$$

The first summand has a finite limit as $x \downarrow 0$, as has already been mentioned. Moreover, $\lambda N_1 - (\sinh x_0)^{-\lambda/2}\cosh x_0 < 0$, consequently, the expression in the braces is negative and $u''(x) \to \infty$ as $x \downarrow 0$. Finally, $u''(-x) = u''(x)$, therefore, generally $u''(x) \to \infty$ as $x \to 0$. The theorem is proved.

This theorem demonstrates the necessity of condition (2.16) for Theorems 4.8 and 4.10. Slightly modifying equation (3), we can also show the necessity of condition (2.16) for Theorem 2.2. Indeed, let us take the function $\psi_p(g^1, g^2)$ from the proof of Lemma 3.2 with $r = 2$, $p \geqq 1$ and consider Cauchy's problem

$$u_t + \psi_p(\frac{1}{p} u_{xx} + 2u_x \tanh x - \lambda u - (\cosh x)^{-1},$$
$$(1 + \frac{1}{p})u_{xx} + 2u_x \tanh x - \lambda u - (\cosh x)^{-1}) = 0 \tag{13}$$

on $(0, T) \times E_1$ with zero condition at $t = T$. It is easy to derive from relationships (3.16) and (3.17) and the results of previous chapters that this problem has an infinitely differentiable solution $u^{p,T}$ such that all derivatives of $u^{p,T}$ vanish as $|x| \to \infty$. Having written equation (13) in the form (3.19), one can see that condition (2.16) is equivalent to the condition $\lambda > 4$. If $\lambda \leqq 4$, then Theorem 2.2 cannot be true since, if otherwise, on letting first $p \to \infty$ and then $T \to \infty$ we, by Theorem 4.8, would obtain a bounded solution to equation (3) whose second derivative is bounded from above and, by Theorem 4.10, it is also bounded from below.

Finally, let us discuss Theorems 4.1 and 4.7. Let $\Omega = \{0, 1, 2, \ldots\}$

$$L(\omega)u = \omega(2u_x \tanh x - \lambda u), \text{ if } \omega \text{ is even},$$

$$L(\omega)u = \omega(u_{xx} + 2u_x \tanh x - \lambda u), \text{ if } \omega \text{ is odd},$$

$$f(\omega) = -\omega(T - t)(\cosh x)^{-1}, \quad \nu(\omega) = (1 + \omega)^{-1}.$$

Considering the function $\psi = (N + |x|^2)\exp N(T - t)$ with a sufficiently large N and applying Corollary 3.7.8, we obtain that there can exist only one function u such that $u \in C([0, T] \times E)$, u_t and u_x are bounded, Sobolev's derivative u_{xx} exists and is bounded from above, and

$$u(T, x) = 0, \quad \inf_{\omega} \frac{1}{1+\omega}[u_t + L(\omega)u + f(\omega)] = 0 \tag{14}$$

almost everywhere in $(0, T) \times E_d$. (To avoid confusion, note that $L(\omega)$ here and in Corollary 3.7.8 are different.) We now show that if $\lambda \leqq 4$, i.e., if condition (2.16) is violated for equation (14), then the function with the above properties cannot exist. This will show the necessity of condition (2.16) for Theorems 4.1 and 4.7.

Suppose that $\lambda \leq 4$ and the function u does exist. It is almost obvious that in such a case, $\xi^{-1}u(T - \xi(T - t), x)$ for any $\xi > 0$ also satisfies equation (14) in $(T(1 - \xi^{-1}), T) \times E_1$ and, by virtue of the above uniqueness, we have $u(t, x) = \xi^{-1}u(T - \xi(T - t), x)$ for $\xi \leq 1$, $t \in [0, T]$. Hence $\xi u(0, x) = u(T(1 - \xi), x)$, $u(t, x) = (T - t)u(x)$. Thus defined, function $u(x)$, on substituting $u(t, x) = (T - t)u(x)$ into (14), turns out to satisfy the equation

$$\min(-u,\ 2u_x \tanh x - \lambda u - (\cosh x)^{-1},$$

$$u_{xx} + 2u_x \tanh x - \lambda u - (\cosh x)^{-1}) = 0.$$

This equation can be studied by literally repeating the analysis of equation (3) and it follows either that, by Lemma 3, there is no solution to this equation, or that, by Theorem 5, the second derivative of the solution with respect to x is unbounded from above. The latter yields the desired contradiction.

6. Remark. Let us add $\varepsilon u''$ to the left-hand side of (3) with $\varepsilon > 0$ and denote by u^{ε} the bounded solution to the resulting equation. It can be shown that $u^{\varepsilon}(x) \to \varphi(|x|)$ (see (5)) as $\varepsilon \downarrow 0$, whenever $\lambda \in (0, 2]$ and $u^{\varepsilon}(x) \to v(|x|)$ (see (7)) as $\varepsilon \downarrow 0$ for $\lambda > 2$.

7.6. Comments

Section 1. The formalism associated with (1), (2), and (4) resembles the Ito formalism, well-known in the theory of Ito stochastic equations. It is applicable if the matrix (a^{ij}) can be represented as $\frac{1}{2}\sigma\sigma^*$ where σ is a smooth matrix. The possibility of such a representation is quite a profound problem, whose partial solution in the case $(a^{ij}) \in C^2$, $(a^{ij}) \geq 0$ has been given in Freidlin [2] and Phillips and Sarason [1], where it is proved that $(a^{ij})^{1/2}$ satisfies a Lipschitz condition with respect to x.

Section 2. Condition (16) first appeared in Krylov [6]. Theorems 2 and 6 formulated in different terms were proved by probabilistic means in Krylov [10]. A probabilistic counterpart of Theorem 2 for elliptic equations is discussed in Krylov [4, 6] (see also Lions [1, 3] and Krylov [10]).

Section 3. The probabilistic proofs of Theorems 1, 8, 9 and their extensions are given in Krylov [14]. A result concerning elliptic equations which corresponds to Theorems 1 and 9 in somewhat different terms can be found in Lions [1, 3]. Bellman integral/differential equations were analysed in Pragarauskas [1, 2]. The proof of Lemma 2 and formula (20) are taken from Krylov [23]. The reader who is interested in the case $a^{ij} \equiv 0$ is referred to Kruzhkov [1, 2] and Crandall and Lions [1].

Section 4. The normalized Bellman equation was introduced in Krylov [3, 5, 6]. Results of this section can be obtained by combining those in Krylov [16, 10].

Section 5. The example involving equation (3) is taken from Genis and Krylov [1].

Chapter 8

DEGENERATE NONLINEAR EQUATIONS IN A DOMAIN

In this chapter we continue our study of Bellman equations. In Section 1, a formal technique, which will be thoroughly discussed in Section 3, is used to derive the solvability of some Bellman equations in a ball and a circular cylinder from the results of the previous chapter concerning the solvability of Bellman equations in the whole space. The subclass of Bellman equations in question is not as narrow as might appear at first glance, and Section 2 presents some illustrative examples. In Sections 4 to 8, we consider general Bellman equations in smooth domains.

8.1. Equations with Constant 'Coefficients' in a Ball and a Circular Cylinder

Given constants $K \geqq \varepsilon > 0$, $D = \{x \in E_d : |x| < 1\}$, and a set Ω of indices, let a symmetric nonnegative $d \times d$ matrix $a(\omega) = (a^{ij}(\omega))$ and a real-valued function $f(\omega) = f(\omega, x)$ on E_d be defined for every $\omega \in \Omega$. Note that a is independent of x. Let $\psi(x) = 1 - |x|^2$. The first objective of this section is to prove the following theorem.

1. THEOREM. For all $\omega \in \Omega$, let $f(\omega) \in C^2(E_d)$ and let the norms of $f(\omega)$ in $C^2(E_d)$ do not exceed K. (E_d can be replaced by D (see Remark 8.5)). Also, let $\varepsilon \leqq \operatorname{tr} a(\omega) \leqq K$ on Ω. Then there exists a function u on $\bar{D}$ enjoying the following properties:

(a) $u \in C(\bar{D})$, $|u| \leqq N(K, \varepsilon)\psi$ on $\bar{D}$;

(b) $u - N(K, \varepsilon)|x|^2$ is concave on D and on D there exist Sobolev derivatives u_x and measures u_{xx};

(c) $a^{ij}(\omega)u_{x^ix^j}(dx) << dx$ for every $\omega \in \Omega$

$$\inf_{\omega\in\Omega}\left[a^{ij}(\omega)u^{(0)}_{x^ix^j} + f(\omega)\right] = 0 \quad \text{a.e.} D; \tag{1}$$

(d) for a unit vector $\xi \in E_d$, if

$$\inf_{\eta:(\eta,\xi)=1}\ \sup_{\omega\in\Omega} a^{ij}(\omega)\eta^i\eta^j > 0, \tag{2}$$

then the Sobolev derivative $u_{(\xi)(\xi)}$ is bounded on D, but if

$$\sup_{\omega\in\Omega} a^{ij}(\omega)\xi^i\xi^j > 0, \tag{3}$$

for all $\xi \in E_d$, $\xi \neq 0$, then all Sobolev derivatives u_{xx} are bounded on D.

Moreover, if a function $v \in C(\bar{D})$ is such that $v = 0$ on ∂D and v enjoys properties (b) and (c), then $v = u$ on $\bar{D}$.

Proof. To prove the uniqueness, it suffices, by Corollary 3.7.9, to add u_t to the left-hand side of equation (1), to consider this equation in $Q = (0, \infty) \times D$, and to take note that

$$e^{-\varepsilon t}\left[(\psi e^{\varepsilon t})_t + a^{ij}(\psi e^{\varepsilon t})_{x^i x^j}\right] = \varepsilon\psi - 2 \operatorname{tr} a \leqq 0$$

in Q. Assertion (d) follows from (c) as in Theorem 2.2.6 (see, also, Lemma 3.6.2). To prove the existence of a function u possessing properties (a) — (c), we use a technique which will be explained in Section 3 below.

The original space E_d will be thought of as a subspace of E_{d+4} namely as $\{x \in E_{d+4} : x^{d+1} = \dots = x^{d+4} = 0\}$. It is also assumed that *Greek superscripts* take on integer values from $d + 1$ to $d + 4$ and repeated Greek superscripts symbolize summation over this range. Repeated italic superscripts symbolize summation over the range from 1 to d unless a special stipulation is made. It is clear that all functions initially defined on E_d are also defined on E_{d+4} independent of x^i for $i > d$. Let us denote

$$[x] = (x^\nu x^\nu)^{1/2} = \left(\sum_{\nu=d+1}^{d+4} (x^\nu)^2\right)^{1/2} \tag{4}$$

and, for i, j = 1, ..., d, define the operators

$$L_{ij}v := \left(x^\nu \frac{\partial}{\partial x^i} - x^i \frac{\partial}{\partial x^\nu}\right)\left(x^\nu \frac{\partial}{\partial x^j} - x^j \frac{\partial}{\partial x^\nu}\right)v - 2\delta^{ij}v, \tag{5}$$

on functions v defined in E_{d+4}, where summation is carried out over $\nu = d + 1, \dots, d + 4$. Consider the equation

$$\inf_{\omega\in\Omega}\left[\gamma\delta^{ij}v_{x^i x^j} + \gamma\delta^{\mu\nu}v_{x^\mu x^\nu} + a^{ij}(\omega)L_{ij}v + f(\omega)\right] = 0 \tag{6}$$

on E_{d+4} for $\gamma > 0$.

To prove the solvability of this equation, we would like to apply Theorem 7.4.10 (with $\nu(\omega, z) \equiv 1$ in (7.4.1)). The fact that the countability of Ω is not assumed presents no difficulties since we can replace Ω by its countable subset without changing the infimum in (1) in view of the continuity of f in x which is uniform with respect to ω (cf. the proof of Theorem 6.1.5 (d)). According to the plan adopted in Chapter 7, we further should define vectors $\bar{\sigma}$ and $\bar{b}$ and functions $\bar{c}$ and $\bar{f}$ on E_{d+4} such that equation (6) assume the appropriate form. We denote by $\sigma^k(\omega)$ the kth column of the matrix $\sqrt{2a(\omega)}$ and, for i, k = 1, ..., d, μ, ν = d + 1, ..., d + 4, set

$$\bar{f}(\omega) = f(\omega), \qquad \bar{c}(\omega) = -2 \operatorname{tr} a(\omega)$$

$$\bar{b}^i(\omega) = -4a^{ij}(\omega)x^j, \qquad \bar{b}^\nu(\omega) = -x^\nu \operatorname{tr} a(\omega),$$

$$\bar{\sigma}^{i\nu k}(\omega) = x^\nu \sigma^{ik}(\omega), \qquad \bar{\sigma}^{\mu\nu k}(\omega) = -\delta^{\mu\nu}\sigma^{jk}(\omega)x^j,$$

$$\tilde{\sigma}^{ik}(\omega) = \delta^{ik}\sqrt{2\gamma}, \qquad \tilde{\sigma}^{\mu k}(\omega) = 0,$$

$$\tilde{\sigma}^{i\mu}(\omega) = 0, \qquad \tilde{\sigma}^{\nu\mu}(\omega) = \delta^{\mu\nu}\sqrt{2\gamma}. \tag{7}$$

These formulas define a vector $\bar{b} \in E_{d+4}$ and d_1 = 5d + 4 vectors $\bar{\sigma}^{\nu k}$, $\tilde{\sigma}^k$, $\tilde{\sigma}^\mu \in E_{d+4}$. Also, let

$$\bar{a}^{ij} = \frac{1}{2}\sum_{k=1}^{d}\sum_{\nu=d+1}^{d+4} \bar{\sigma}^{i\nu k}\bar{\sigma}^{j\nu k} + \frac{1}{2}\sum_{k=1}^{d+4} \tilde{\sigma}^{ik}\tilde{\sigma}^{jk},$$

i, j = 1, ..., d + 4.

A simple calculation, which is omitted, shows that the expression in the square brackets in (6) coincides with

$$\sum_{i,j=1}^{d+4} \bar{a}^{ij}(\omega)v_{x^ix^j} + \sum_{i=1}^{d+1} \bar{b}^i(\omega)v_{x^i} + \bar{c}(\omega)v + \bar{f}(\omega).$$

Consequently, equation (6) falls within the scheme considered in Section 7.4. Verification of conditions (7.2.15) and (7.4.4) presents no difficulties. The left-hand side of (7.2.16), for the introduced $\bar{b}$ and $\bar{\sigma}$ and unit vectors $\xi \in E_{d+4}$, is equal to

$$2\delta[\xi]^2 \operatorname{tr} a + 8\delta a^{ij}\xi^i\xi^j \leq 8\delta \operatorname{tr} a = -4\delta\bar{c}. \tag{8}$$

The latter is less than $[-(1 - \delta)\bar{c}]$ whenever $\delta < 1/5$. Therefore, condition (7.2.16) holds. Finally, for unit vectors $\xi \in E_{d+4}$,

$$\sum_{i,j=1}^{d+4} \bar{a}^{ij}\xi^i\xi^j \geq \gamma. \tag{9}$$

By Theorem 7.4.10 and Theorem 7.4.8, equation (6) as considered almost everywhere has a bounded solution v^γ, whose Sobolev derivatives $v^\gamma_{x^i}$, $v^\gamma_{x^ix^j}$ (i, j = 1, ..., d + 4) are also bounded. Moreover, for all unit vectors $\xi \in E_{d+4}$,

$$|v^\gamma|, \ |v^\gamma_x|, \ \sum_{i,j=1}^{d+4} v^\gamma_{x^ix^j} \xi^i\xi^j \leqq N(K, \varepsilon) \tag{10}$$

almost everywhere on E_{d+4}. This is a unique solution. Indeed, note that the operator

$$\gamma\delta^{ij} \frac{\partial^2}{\partial x^i \partial x^j} + \gamma\delta^{\mu\nu} \frac{\partial^2}{\partial x^\mu \partial x^\nu} + a^{ij}(\omega)L_{ij},$$

as applied to $\cosh \delta|x|$, where $|x|$ is the length of $x \in E_{d+4}$, yields

$$\gamma\delta^2 \cosh \delta|x| + (d + 3)\gamma\delta^2 \frac{\sinh \delta|x|}{\delta|x|} - 2 \operatorname{tr} a \cosh \delta|x|$$

$$\leqq ((d + 4)\gamma\delta^2 - 2\varepsilon)\cosh \delta|x|.$$

The latter expression is negative for suffiently small δ, and $\cosh \delta|x| \to \infty$ as $|x| \to \infty$. Therefore, the uniqueness follows from Theorem 3.4.7 and the method underlying the proof of Theorem 3.4.6.

In turn, it follows from the uniqueness and the easily verifiable invariance of equation (6) under orthogonal transformations of the variables $x^{d+1}, \ldots, x^{d+4}$ that v^γ, as a function of $x \in E_{d+4}$, depends only on $x^1, \ldots, x^d$, [x]. For an R 0, we now consider equation (6) in the ball $\{x \in E_{d+4} : |x| < R\}$ with the boundary condition $v = v^\gamma$. Of course, v^γ is its unique solution in the class of continuous functions with locally bounded derivatives with respect to x^i and x^ix^j (see Theorem 3.4.7). On the other hand, by virtue of (9) and in view of Theorem 6.4.5 (where M_0^F is an arbitrary positive number; see also Theorem 6.1.5), this problem has a continuous solution whose second derivatives are continuous in $\{x \in E_{d+4} : |x| < R\}$. Thus $v^\gamma \in C^2_{loc}(E_{d+4})$.

On letting $\gamma \downarrow 0$, we next want to obtain from v^γ a function $v(x^1, \ldots, x^d, [x])$, to set

$$u(x^1, \ldots, x^d) = \psi(x^1, \ldots, x^d)v(x^1, \ldots, x^d, \sqrt{\psi(x^1, \ldots, x^d)})$$

and to show that u is the desired function. First of all, let us consider three straightforwardly verifiable formulas. It turns out that if $g \in C^2_{loc}(E_{d+4})$, $g = g(x^1, \dots, x^d, [x])$, and a function $h(x, R)$ is defined on the closure of the cone

$$\mathbb{A} := \{(x, R) : x \in E_d, \quad 0 < |x| < R\}$$

as

$$h(x, R) = (R^2 - |x|^2) g(x, \sqrt{R^2 - |x|^2}),$$

then, for $i, j = 1, \dots, d$, we obtain

$$h_{x^i x^j} = \frac{1}{2} [(L_{ij} + L_{ji}) g](x, \sqrt{R^2 - |x|^2}), \tag{11}$$

$$(\delta^{kr} g_{x^k x^r})(x, \sqrt{R^2 - |x|^2}) = (R^2 - |x|^2)^{-1} \left(\frac{\partial}{\partial x^k} + \frac{x^k}{R} \frac{\partial}{\partial R}\right)\left(\frac{\partial}{\partial x^k} + \frac{x^k}{R} \frac{\partial}{\partial R}\right) h, \tag{12}$$

$$(\delta^{\mu\nu} g_{x^\mu x^\nu})(x, \sqrt{R^2 - |x|^2}) = R^{-2} h_{RR} - R^{-3} h_R \tag{13}$$

in $\mathbb{A}$.

Accordingly, let us denote

$$u^\gamma(x, R) = (R^2 - |x|^2) v^\gamma(x, \sqrt{R^2 - |x|^2}) \tag{14}$$

on $\bar{\mathbb{A}}$.

Then by (6), (11) to (13), we obtain

$$\inf_{\omega \in \Omega} \left[\gamma P u^\gamma + a^{ij}(\omega) u^\gamma_{x^i x^j} + f(\omega)\right] = 0 \tag{15}$$

on $\mathbb{A}$, where P is defined as the sum of the operators on the right of (12) and (13). Now, let $\gamma \to 0$. Estimates (10) make it possible to select a subsequence $\gamma_n \downarrow 0$ (see, however, Remark 3) such that the function v^γ uniformly converge to a function v on every compactum of E_{d+4}. The corresponding subsequence of function u^γ uniformly converges on every compactum lying in $\bar{\mathbb{A}}$. We denote its limit by u. It follows from (10) and (14) that the first derivatives of u^γ with respect to (x, R) are bounded uniformly in γ in every compactum lying in $\bar{\mathbb{A}}$, whereas the second derivatives of u^γ with respect to any vector lying in the space of (x, R) are bounded from above uniformly in γ in every compactum lying in $\mathbb{A}$. Indeed, the first assertion is clear, and to prove the second one, it suffices to note that the second derivative of u^γ along a vector is expressed in terms of the first derivatives of v^γ and the second derivative of v^γ along some other vector multiplied by $R^2 - |x|^2$. Therefore, $(-u^\gamma)$, $(-u) \in \tilde{W}^{bv,+}_{loc}((0, 1) \times \mathbb{A})$ (note that $u^\gamma(t, x, R) := u^\gamma(x, R))$.

Furthermore, in Chapter 7, we often passed to the limit in situations like that for equations of type (15) (see, for exmaple, the proof of Lemma 7.3.4). It therefore follows from (15) that

$$\bigwedge_{\omega\in\Omega}\left[a^{ij}(\omega)\bar{u}_{x^ix^j} + f(\omega)\circ\bar{\Lambda}\right] = 0, \tag{16}$$

in $(0, 1) \times \mathbb{A}$ as $\gamma = \gamma_n \downarrow 0$, where $\bar{\Lambda}$ is the Lebesgue measure on $(0, 1) \times \mathbb{A}$ and $\bar{u}_{x^ix^j}$ are measures in $(0, 1) \times \mathbb{A}$ which are generalized derivatives of the function $u(t, x, R) = u(x, R)$.

Also, note that at every point $x \in E_{d+4}$ the operator L_{ij} in (5) can obviously be represented as a sum of a zero-order operator, a first-order operator, and a double-differentiation operator along a vector (changing with x and, by the way, orthogonal to x). It follows from (10) and (11) that $\xi^i\xi^j u^\gamma_{x^ix^j} \leqq N(R, K, \varepsilon)$ on $\mathbb{A}$ for all unit vectors $\xi \in E_d$ and the function $u^\gamma - N(R, K, \varepsilon)|x|^2$, and consequently the function $u - N(R, K, \varepsilon)|x|^2$ are concave in x on

$$S_R = \{x \in E_d : |x| < R\}$$

for every $R > 0$.

Specifically, for every $R > 0$ there exist measures on S_R whose differentials $u_{x^ix^j}(dx, R)$ are generalized derivatives of $u(\cdot, R)$ in S_R. In S_R for any $\omega \in \Omega$ we have

$$a^{ij}(\omega)u_{x^ix^j}(dx, R) + f(\omega)dx \geqq 0. \tag{17}$$

Indeed, otherwise there would exist an $R > 0$, an $\omega \in \Omega$, and a non-negative function $\zeta \in C_0^\infty(S_R)$ such that the integral of ζ over S_R with respect to the left-hand side of (17) is strictly negative. On integrating by parts this integral, multiplying by an appropriate function $\eta(t, R)$ such that $\eta\zeta \in C_0^\infty((0, 1) \times \mathbb{A})$, and further integrating with respect to dt dR, we would obtain a strictly negative integral of $\eta\zeta$

with respect to the measure under the infimum in (16), which is impossible in view of (16). It follows from (17) and the estimate $u_{x^ix^j}(dx, R)\xi^i\xi^j \leqq N(R, K, \varepsilon)|\xi|^2\, dx$ that for any $\omega \in \Omega$

$$a^{ij}(\omega)u_{x^ix^j}(dx, R) \leqq N\, dx, \quad |a^{ij}(\omega)u_{x^ix^j}(dx, R)| \leqq N\, dx,$$

$$a^{ij}(\omega)u_{x^ix^j}(dx, R) \ll dx \tag{18}$$

on S_R.

By Lemma 3.6.2, it follows from (16) that

$$\inf_{\omega\in\Omega}\left[a^{ij}(\omega)\bar{u}^{(0)}_{x^ix^j} + f(\omega)\right] = 0 \text{ a.e. on } (0, 1) \times \mathbb{A}. \tag{19}$$

Here, by the Aleksandrov–Busemann–Feller theorem, (see Section 1.2), $\bar{u}^{(0)}_{x^ix^j}(t_0, x_0, R_0)$ for almost all (t_0, x_0, R_0) can be taken as the coefficients in the second-order Taylor series expansion of $u(x, R_0)$ at the point x_0. By the same theorem, this is also true for $u^{(0)}_{x^ix^j}(x_0, R_0)$. Theorefore,

$$\bar{u}^{(0)}_{x^ix^j}(t_0, x_0, R_0) = u^{(0)}_{x^ix^j}(x_0, R_0) \text{ a.e. on } (0, 1) \times \mathbb{A}.$$

Using this fact, we convert (19) to the form

$$\inf_{\omega\in\Omega}\left[a^{ij}(\omega)u^{(0)}_{x^ix^j}(x, R) + f(\omega, x)\right] = 0 \text{ a.e. on } \mathbb{A}. \tag{20}$$

By Fubini's theorem, we conclude that for almost all R the equality in (20) holds for almost all $x \in S_R$. This together with (18) and Lemma 3.6.2 yields

$$\bigwedge_{\omega\in\Omega} (a^{ij}(\omega)u_{x^ix^j}(dx, R) + f(\omega, x)\, dx) = 0 \tag{21}$$

for almost all R on S_R. It is easy to pass to the limit in the latter equality, therefore, it holds on S_R for all rather than almost all R. To complete the proof of the theorem, it remains to take $u(x, 1)$ as a desired function and to take note that the inequality $|u| \leqq N(K, \varepsilon)\psi$ follows from (10) and (14). The theorem is proved.

2. Remark. By Remark 2.2.8, all second derivatives that are needed to compute the left-hand side of (1) are Sobolev derivatives.

3. Remark. $|v - v^{\gamma}| \leq N(K, \varepsilon)\sqrt{\gamma d}$. Indeed, for a fixed $\kappa > 0$, the function $v^{\kappa\gamma^2}(x)$, as a function of $\gamma \in (-\infty, \infty)$ and $x \in E_{d+4}$, satisfies equation (6), where γ should be replaced by $\kappa\gamma^2$. (The fact that no derivatives with resepct to γ come into (6) makes no difference.) Regarding this equation in variables (γ, x) and applying, as in the proof of Theorem 1, Theorem 7.4.8, we can see that inequality (8) has to be checked, where the expression

$$(1 + \delta)2\kappa(d + 4)(\xi_0)^2 + 4\kappa|\xi|^2,$$

appearing due to differentiation of $\bar{b}$, $\bar{\sigma}^{\nu k}$, $\tilde{\sigma}^k$, and $\tilde{\sigma}^{\mu}$ with respect to the new variable should be added to the left-hand side. The modified inequality (8) should be checked for $\xi_0 \in (-\infty, \infty)$, $\xi \in E_{d+4}$, and $(\xi_0)^2 + |\xi|^2 = 1$. It is clear that the inequality holds for $\delta \leq 0.1$ and a sufficiently small κ of the form $d^{-1}\tau(\varepsilon)$. For such a κ, the derivative of $v^{\kappa\gamma^2}$ with respect to $\gamma \in (-\infty, \infty)$ is bounded in view of Theorem 7.4.8, hence our assertion is proved.

We now turn to parabolic equations. Let us fix $T \in (0, \infty)$ and take the objects introduced in the beginning of this section, but now assume furhter that f depends on t so that $f(\omega) = f(\omega, t, x)$, $t > 0$, and $x \in E_d$. Also, let $Q = (0, T) \times D$.

4. THEOREM. Let $f(\omega) \in C^2((0, T) \times E_d)$, $|f(\omega)| + |f_t(\omega)| + |f_x(\omega)| + (f_{(l)(l)}(\omega))_+ \leq K(1 + \operatorname{tr} a(\omega))$ on $(0, T) \times E_d$ for all $\omega \in \Omega$ and unit vectors $l \in E_d$. Also, let $a(\omega)$ be bounded on Ω and let $|f(\omega, T, x)| \leq K$ for all $\omega \in \Omega$ and $x \in E_d$. Then there exists a function on $\bar{Q}$ that posesses the following properties:

(a) $u \in C(\bar{Q})$, $|u| \leq N(K, T)\psi$, and $u(T, x) = 0$;

(b) $u - N(K, T)|x|^2$ is concave in x on Q, and on Q there exist Sobolev derivatives u_x, u_t, and measures u_{xx};

(c) $a^{ij}(\omega)u_{x^ix^j}(dt\,dx) \ll dt\,dx$ for any $\omega \in \Omega$ and

$$\inf_{\omega\in\Omega}\left[u_t + \psi a^{ij}(\omega)u^{(0)}_{x^ix^j} + \psi f(\omega)\right] = 0 \text{ a.e. } Q; \tag{22}$$

(d) if (2) holds for a unit $\xi \in E_d$, then the Sobolev derivative $u_{(\xi)(\xi)}$ is bounded on Q, but if inequality (3) holds for all $\xi \in E_d$, $\xi \neq 0$, then all Sobolev derivatives u_{xx} are bounded on Q.

Moreover, if a function $v \in C(\bar{Q})$ is such that $v = 0$ on $\partial' Q$ and possesses properties (b) and (c), then $u = v$ on $\bar{Q}$.

Proof. We follow the proof of Theorem 1 closely. As in that proof, we replace Ω by its countable subset such that the infimum in (22) remains unchanged for any (t, x) and $u^{(0)}_{x^i x^j}$. The uniqueness now follows from Corollary 3.7.2. To prove existence, we make use of the denotation introduced in the proof of Theorem 1 and consider the equation

$$\inf_{\omega \in \Omega}\left[v_t + \gamma \delta^{ij} v_{x^i x^j} + \gamma \delta^{\mu\nu} v_{x^\mu x^\nu} + a^{ij}(\omega) L_{ij} v - v + f(\omega) e^t\right] = 0 \tag{23}$$

in $(0, T) \times E_{d+4}$ with $\gamma > 0$ and the boundary condition $v(T, x) = 0$. Using (7) and checking the hypothesis of Theorem 7.4.1, we obtain, by Theorems 7.4.1 and 7.4.7, the function v^γ, which is a solution to this problem such that the Sobolev derivatives $v^\gamma_{x^i x^j}$ $(i, j = 1, \ldots, d + 4)$ are bounded and

$$|v^\gamma|,\ |v^\gamma_t|,\ |v^\gamma_x|,\ \sum_{i,j=1}^{d+4} v^\gamma_{x^i x^j} \xi^i \xi^j \leqq N(K, T)$$

almost everywhere on $(0, T) \times E_{d+4}$ for all unit vectors $\xi \in E_{d+4}$. By Theorem 3.4.6 (again setting $\psi = \cosh \delta|x|$) the function v^γ is uniquely defined. It follows, as in the proof of Theorem 1, that v^γ depends only on $t, x^1, \ldots, x^d, [x]$. On applying Theorem 6.4.4, we can show that $v^\gamma \in C^2_{loc}([0, T] \times E_{d+4})$.

Now, we again take the cone $\mathbf{A}$, set $u^\gamma(t, x, R) = (R^2 - |x|^2) v^\gamma(t, x, \sqrt{R^2 - |x|^2}) e^{-t}$ in $[0, T] \times \bar{\mathbf{A}}$, and find a sequence $\gamma_n \downarrow 0$ such that v^{γ_n} converges to a function v. We denote by u the corresponding limit of u^{γ_n}. Formulas (11) to (13) and equation (23), as $\gamma = \gamma_n \downarrow 0$, yield

$$\bigwedge_\omega ((R^2 - |x|^2)^{-1} u_t \circ \bar{\Lambda} + a^{ij}(\omega) \bar{u}_{x^i x^j} + f(\omega) \cdot \bar{\Lambda}) = 0$$

in $(0, T) \times \mathbf{A}$ (cf. (16)). Furthermore, using suitably modified arguments of the proof of Theorem 1, we can convert this relationship to the form

$$\bigwedge_\omega ((R^2 - |x|^2)^{-1} u_t \, dt\, dx + a^{ij}(\omega) u_{x^i x^j}(dt\, dx, R) + $$
$$+ f(\omega) dt\, dx) = 0, \tag{24}$$

and the latter equality is true for all R in $(0, T) \times S_R$ (cf. (21)). It is also proved in passing that $u(t, x, R) - N(R, K, T)|x|^2$ is concave in x on $C_{T,R}$ for every R, $|u_t| \leq (R^2 - |x|^2)N(K, T)$, and $|u_x| \leq N(R, K, T)$. It remains to set $u(t, x) = u(t, x, 1)$ and to take note that assertion (d) follows from (c) (and from the inequality $|u_t| \leq N\psi$) as in Theorem 2.2.6, and (24), by Lemma 3.6.2, implies

$$a^{ij}(\omega)u_{x^ix^j}(dt\ dx) << dt\ dx, \ \omega \in \Omega,$$

$$\inf_{\omega} \psi^{-1}\left[u_t + \psi a^{ij}(\omega)u^{(0)}_{x^ix^j} + \psi f(\omega)\right] = 0 \text{ a.e. } Q.$$

Note that the latter is equivalent to (22) since $\psi > 0$ on Q. The theorem is proved.

We complete the section by extending Theorem 4 to the case of unbounded $a(\omega)$.

5. THEOREM. Let all assumptions of Theorem 4 be true except the assumption concerning the boundedness of $a(\omega)$ on Ω. Then there exists a function u on $\bar{Q}$ which possesses properties (a), (b), and (d) of Theorem 4 and such that $a^{ij}(\omega)u_{x^ix^j}(dt\ dx) << dt\ dx$ for any $\omega \in \Omega$ and

$$\inf_{\omega\in\Omega} \nu(\omega)\left[u_t + \psi a^{ij}(\omega)u^{(0)}_{x^ix^j} + \psi f(\omega)\right] = 0 \text{ a.e. } Q. \tag{25}$$

for any nonnegative bounded function $\nu(\omega, z)$ defined on $\Omega \times Q$ such that νa is bounded. Moreover, $|u_t| \leq N(K, T)\psi$. Finally, the function possessing the above properties is unique.

Proof. Given a function u, for any ν relationship (25) follows from the same relationship for $\nu = \nu_0 := (1 + \text{tr } a)^{-1}$ (see Lemma 7.4.3). When $\nu = \nu_0$ we can assume as above that Ω is countable. Let $\Omega = \{\omega_1, \omega_2, \ldots\}$ and $\Omega_n = \{\omega_1, \ldots, \omega_n\}$. Let us substitute Ω_n for Ω in Theorem 4 and denote the respective function u by u^n. It is clear that

$$\bigwedge_{\omega\in\Omega_n} \nu_0(\omega)\left[u^n_t\ dt\ dx + \psi a^{ij}(\omega)u^n_{x^ix^j}(dt\ dx) + \right.$$

$$\left. + \psi f(\omega)\ dt\ dx\right] = 0. \tag{26}$$

on Q. Moreover, we could see in Theorem 4 that $|u^n|$, $|u^n_x| \leq N(K, T)$,

$|u^n_t| \leq \psi N(K, T)$, and $u^n - N(K, T)|x|^2$ is concave in x on Q. This enables us easily to pass to the limit in (26) and obtain the desired function (cf. the proof of Theorem 7.4.1).

The uniqueness immediately follows from Corollary 3.7.6 (the inequality $|u_t| \leq N\psi$) and the fact that the operator

$$\nu_0\left[\frac{\partial}{\partial t} + \psi a^{ij}\frac{\partial^2}{\partial x^i \partial x^j}\right],$$

applied to $N_1\psi e^{-2t}$, yields $(-2)N_1\psi e^{-2t}$. The theorem is proved.

8.2. Examples of Equations with Monge—Ampère Operators, and Other Examples

In this section we consider examples of nonlinear equations admitting the application of the results established in Section 1. All the examples are based on explicit calculation of infima in (1.1) and (1.25) or equation (1.1) and (1.25) rewritten in a more 'explicit' form. Let us fix $T > 0$ and denote $D = \{x \in E_d : |x| < 1\}$ and $Q = (0, T) \times D$.

1. Example. Consider the equation

$$u_{x^1x^1} - (u_{x^2x^2})_- - \dots - (u_{x^dx^d})_- = f \text{ (a.e.).} \qquad (1)$$

in D. Assume that $f \in C^2(E_d)$. Then it turns out that this equation with the boundary data $u = 0$ on ∂D has a unique solution in the class of functions which are continuous on $\bar{D}$ and possess bounded first and second generalized derivatives.

Indeed, equation (1) is of the form

$$\inf(\omega^i u_{x^ix^i} - f) = 0 \quad \text{(a.e.)},$$

where the infimum is taken over all $\omega = (\omega^i) \in E_d$ such that $\omega^1 = 1$ and $0 \leq \omega^i \leq 1$ for $i \geq 2$. Therefore, Theorem 1.1 is applicable to equation (1) and it remains to note that the left-hand side of (1.3) is equal to $|\xi|^2$.

2. Example. Given the same function f, with $d \geq 2$ in D let us consider the so-called simplest Monge—Ampère equation

$$\det(u_{x^ix^j}) = (f_+)^d \quad \text{(a.e.).} \qquad (2)$$

This equation with zero boundary data has a unique solution in the class of functions which are convex and possess the same properties as in the first example. Moreover, if $f > 0$ on $\bar{D}$, then $u \in C^{3+\alpha}(\bar{D})$ for any $\alpha \in (0, 1)$.

Indeed, by Lemma 3.2.2, equation (2) with the condition $(u_{x^i x^j}) \geqq 0$ is equivalent to the equation

$$\inf\left(\omega^{ij}u_{x^i x^j} - \sqrt[d]{\det \omega}\, fd\right) = 0 \quad (\text{a.e.}), \tag{3}$$

where the infimum is taken over the set of all positive $d \times d$ matrices $\omega = (\omega^{ij})$ such that $\omega = \omega^*$ and $\operatorname{tr} \omega = 1$. Therefore, Theorem 1.1 is applicable to equation (2) and the left-hand side of (1.3) is equal to $d^{-1}|\xi|^2$.

The smoothness can be improved in the following manner. If $f > 0$ and a solution u with bounded second derivatives is available, then it follows from (2) that the eigenvalues of the matrix $(u_{x^i x^j})$ are bounded from below (a.e.) by a known strictly positive constant. Then it is pointless to take the infimum in (3) over degenerate ω and, moreover, we can take the infimum under the additional condition $\omega^{ij}\xi^i\xi^j \geqq \varepsilon|\xi|^2$ for all $\xi \in E_d$, where ε is a sufficiently small positive constant. It can now be seen that all conditions in Example 6.1.8 except (6.1.8) hold for the operator on the left in (3). Condition (6.1.8) can easily be satisfied if we note that the function $v := u(2 - |x|^2)^{-1}$ satisfies the equation

$$\inf\left[(2 - |x|^2)\omega^{ij}v_{x^i x^j} - 4\omega^{ij}x^i v_{x^j} - 2v - \sqrt[d]{\det \omega}\, fd\right] = 0 \quad (\text{a.e.}), \tag{4}$$

where the infimum is taken over $\omega = \omega^* \geqq \varepsilon(\delta^{ij})$, $\operatorname{tr} \omega = 1$ as above. By Theorem 6.4.2, equation (4) with zero boundary condition has a $C^{2+\beta}(\bar{D})$-solution ∂D with $\beta \in (0, 1)$. By the uniqueness Theorem 3.4.7, this solution coincides with v. Consequently, $v, u \in C^{2+\beta}(\bar{D})$. Finally, note that if $F(u_{ij}) := \det(u_{ij})$, then the matrix $(F_{u_{ij}}(u_{x^k x^r}))$ has entries belonging to $C^{2+\beta}(\bar{D})$ and it is uniformly nondegenerate in $\bar{D}$ as coinciding with the inverse of $(u_{x^i x^j})$ multiplied by $\det(u_{x^i x^j})$. By well-known regularity theorems (see, for example, Agmon, Duglis and Nirinberg [1], Theorem 11.1) applied to (2), we concude that $u \in C^{3+\alpha}(\bar{D})$ for any $\alpha \in (0, 1)$.

3. Example. If $\det(u_{x^i x^j})$ on the left of (2) is replaced by $u_{x^1 x^1} \cdots u_{x^d x^d}$, then the resulting equaitons admit the same assertions where

'convex' should be replaced by 'convex in every variable x^i'. This fact can be proved by repeating the previous arguments with ω^{ij} substituted by $\delta^{ij}\omega^{ii}$ in (3) and (4) and applying Lemma 3.2.2 to diagonal matrices.

4. Example. The assertions in Example 2 can easily be extended to the equation

$$\det\left(u_{x^ix^j} + w_{ij}\right) = (f_+)^d,$$

where $w_{ij} = w_{ji}$ and f are known functions of the class $C^2(E_d)$. Note that here we consider functions such that $(u_{x^ix^j} + w_{ij}) \geqq 0$ rather than convex functions. This example obviously relates to equation (2) with nonzero boundary data.

5. Example. Let $w \in C^4(E_d)$, $w = 0$ on ∂D, and the function w should be expressed on $\bar{D}$ as a difference between two functions such that they are zero on ∂D and convex, i.e., $w = u - (u - w)$. This can be done in many different ways. However, if the condition is imposed that the integral of $|u|$ over D should be as small as possible, then it can be shown that there appears the equation

$$\det\left(u_{x^ix^j}\right)\det\left(u_{x^ix^j} - w_{x^ix^j}\right) = 0 \text{ (a.e.)} \tag{5}$$

for u and we would like to obtain a solution such that the matrices $(u_{x^ix^j})$ and $(u_{x^ix^j} - w_{x^ix^j})$ are nonnegative and this solution equals zero on ∂D. It turns out that the desired solution always exists and it is unique in the class of functions whose first and second generalized derivatives are bounded. Indeed, equation (5) under the additional condition on the matrices involved is equivalent to the equation

$$\inf\left[(\omega_1^{ij} + \omega_2^{ij})u_{x^ix^j} - \omega_2^{ij}w_{x^ix^j}\right] = 0 \text{ (a.e.)},$$

where the infimum is taken subject to $\omega_1 = \omega_1^* \geqq 0$, $\omega_2 = \omega_2^* \geqq 0$, and $\operatorname{tr}\omega_1 + \operatorname{tr}\omega_2 = 1$. Theorem 1.1 can be applied to the latter equation, the left-hand side of (1.3) being equal to $d^{-1}|\xi|^2$.

6. Example. Consider the strongly elliptic Monge—Ampère equation

$$\det\left(u_{x^ix^j}\right) = A^{ij}u_{x^ix^j} + f, \tag{6}$$

in D, where $d \geqq 2$, $f = f(x)$ and $A^{ij} = A^{ij}(x)$ are given functions,

$f \in C^2(E_d)$, $A = (A^{ij}) \in C^2(E_d)$ (it suffices that f and A belong to $C^2(D)$ (see Remark 8.5)), $f \geq \delta$, $A \geq \delta$, $A = A^*$, and $\delta > 0$ is a constant. It turns out that under these conditions, equation (6) defined in D with zero boundary data on ∂D has a unique solution which is convex and belongs to $C^{3+\alpha}(\bar{D})$ for any $\alpha \in (0, 1)$.

To derive this fact from Theorem 1.1, we make use of the following lemma (cf. Krylov [23]).

7. LEMMA. Let Ω be the set of all strictly positive symmetric $d \times d$ matrices with unit trace, let $\omega \in \Omega$, let $B = (B^{ij})$ and $u = (u_{ij})$ be $d \times d$ symmetric matrices and $B > 0$, and let g be a strictly positive scalar. Then:

(a) the set of equations in $s, \gamma \in (0, \infty)$

$$s = \det(\gamma\omega + sB), \tag{7}$$

$$1 - gs = [\det(\gamma\omega + sB)]'_s \tag{8}$$

has a unique solution;

(b) the function $(d-1+gs)\gamma^{-1}$ and its first and second derivatives with respect to B and g are continuous and bounded on every set $\{(\omega, B, g) : \omega \in \Omega, \varepsilon \leq B \leq N, g \geq \varepsilon\}$, where $N < \infty$ and $\varepsilon > 0$;

(c) the relationships

$$(u_{ij}) \geq 0, \quad \det u = \operatorname{tr} Bu + g \tag{9}$$

hold if and only if

$$\inf_{\omega\in\Omega}[\operatorname{tr} \omega u - (d - 1 + gs)\gamma^{-1}] = 0, \tag{10}$$

where (s, γ) is a solution to (7) and (8);

(d) if relationships (9) (or (10)) hold, then $u > 0$, $\sigma := u^{-1} - B \times \det u^{-1} \geq gu^{-1} \det u^{-1}$ and the infimum in (10) is attained at

$$\omega = \sigma(\operatorname{tr} \sigma)^{-1} \geq gu^{-1}(\operatorname{tr} u^{-1})^{-1} \det u^{-1}.$$

After formulating this lemma without proof, we next turn to an examination of equation (6). Let $s = s(\omega, B, g)$, $\gamma = \gamma(\omega, B, g)$ be the solution to system (7) and (8). For $\omega \in \Omega$ and $x \in E_d$, we denote

$$h(\omega) = h(\omega, x)$$

$$= [d - 1 + f(x)s(\omega, A(x), f(x))]\gamma^{-1}(\omega, A(x), f(x))$$

and consider the equation

$$\inf_{\omega\in\Omega}\left[\omega^{ij}u_{x^ix^j} - h(\omega)\right] = 0 \tag{11}$$

in D.

By Theorem 1.1, on D there exists a unique function u which is continuous in $\bar{D}$, equal to zero on ∂D, possessing bounded generalized second derivatives, and satisfying (11) almost everywhere. By Lemma 7(c), this function is convex, and since $A^{ij}u_{x^ix^j} + f \geq f \geq \delta$, it follows that $\det(u_{x^ix^j}) \geq \delta$ on D (a.e.). Therefore, the matrix $(u_{x^ix^j})$ is uniformly positive and, by Lemma 7(d), the function u in D (a.e.) satisfies the equation resulting from (11) if Ω is replaced by $\Omega \cap \{\omega \geq (\varepsilon^{ij})\}$ with an $\varepsilon > 0$. After such a replacement we obtain, as in Example 2, that $u \in C^{2+\beta}(\bar{D})$ for a $\beta \in (0, 1)$. Finally, if $F(u_{ij}, x) := \det(u_{ij}) - A^{ij}u_{ij}$, then the matrix $(F_{u_{ij}}(u_{x^kx^r}))$ coincides with

$$\left(u_{x^ix^j}\right)^{-1}\det\left(u_{x^kx^r}\right) - A^{ij},$$

and, by Lemma 7(d), the latter is uniformly nondegenerate so that the inclusion $u \in C^{3+\alpha}(\bar{D})$ is obtained for any $\alpha \in (0, 1)$ as in Example 2.

8. <u>Example</u>. We now apply Theorem 1.5 to the parabolic Monge—Ampère equation in Q

$$u_t \det\left(u_{x^ix^j}\right) = \psi(f_+)^{d+1} \quad \text{(a.e.)}, \tag{12}$$

where $\psi = 1 - |x|^2$, $f = f(t, x) \in C^2([0, T] \times E_d)$, and $f(T, x) = 0$ on E_d. We look for a solution to (12) in the class of functions which are continuous in Q, equal to zero on $\partial'Q$, having bounded generalized derivatives u_t, u_x, and u_{xx}, increasing in t, and convex in x. The two latter properties together with equation (12) are equivalent to

$$\inf_{\omega\in\Omega_0}\Big(r(\omega)u_t + \omega^{ij}u_{x^ix^j} -$$

$$- \sqrt[d+1]{\psi r(\omega)\det\omega}\, f(d + 1)\Big) = 0 \quad \text{(a.e.)}, \tag{13}$$

as follows from Lemma 3.2.2. In (13), Ω_0 is the set of all symmetric $d \times d$ matrices ω satisfying the conditions $\omega \geq 0$, $\operatorname{tr}\omega < 1$, and $r(\omega) =$

$1 - \operatorname{tr}\omega$.

Furthermore, denote by Ω the set of all pairs (r, ω), where $r > 0$, ω is a symmetric nonnegative $d \times d$ matrix. For $(r, \omega) \in \Omega$, it is obvious that $(r + \operatorname{tr}\omega)^{-1}\omega \in \Omega_0$. Moreover, $(r(\omega), \omega) \in \Omega$, whenever $\omega \in \Omega_0$. Therefore, equation (13) is equivalent to

$$\inf_{(r,\omega)\in\Omega} (r + \operatorname{tr}\omega)^{-1}\Big(ru_t + \omega^{ij}u_{x^ix^j} - \\ - \sqrt[d+1]{\psi r \det \omega\,(d + 1)f}\Big) = 0 \text{ (a.e.)}. \tag{14}$$

The set Ω is a cone with respect to ω, so that passing from (r, ω) in (14) to $(r, \psi\omega)$, we obtain the equation

$$\inf_{(r,\omega)\in\Omega} r(r + \psi \operatorname{tr}\omega)^{-1}\Big(u_t + \psi a^{ij}(r, \omega)u_{x^ix^j} - \\ -\psi \sqrt[d+1]{\det a(r, \omega)(d + 1)f}\Big) = 0 \text{ (a.e.)},$$

where $a^{ij}(r, \omega) = r^{-1}\omega^{ij}$. By Theorem 1.5, the latter equation, and hence equation (13) too, has a solution which is continuous in $\bar{Q}$, equal to zero on $\partial'D$, and possesses bounded derivatives u_t, u_x, and u_{xx}. For this solution, we also have $|u_t| \leq N\psi$; this, together with the previous properties, uniquely characterizes the function u.

9. Example. Let us take ψ and f from the previous example and show that there exists a continuous function in $\bar{Q}$ which is equal to zero on $\partial'Q$, possesses bounded generalized derivatives u_t, u_x, and u_{xx} and is such that

$$u_t + \psi f \geq 0, \quad \Delta u + f \geq 0, \quad (u_t + \psi f)(\Delta u + f) = 0 \quad \text{a.e. } Q. \tag{15}$$

To this end, it suffices to apply Theorem 1.5 and take note that the set of three relationships (15) is equivalent to the equation

$$\inf_{\omega\in(0,1]} \omega\left[u_t + \psi\,\frac{1-\omega}{\omega}\,\Delta u + \psi\,\frac{1}{\omega}\,f\right] = 0 \text{ a.e.}Q.$$

8.3. Interrelation Between the Equation in a Domain of a Euclidean Space and the Equation on a Manifold

In this section we outline the method of research into degenerate equations in a domain that will be used in the remaining part of the book. Let $\psi \in C^2(E_d)$ and let $D := \{x \in E_d : \psi(x) > 0\}$ be a nonempty domain in E_d. In order to explain our basic idea, we consider the simplest problem

$$\Delta u + f = 0 \text{ in } D, \tag{1}$$

$$u = 0 \text{ on } \partial D, \tag{2}$$

where f is a given function. In view of condition (2), it is natural to express the unknown function as $u = v\psi$. Then (1) yields the following equation for v:

$$\psi\Delta v + 2\psi_{x^i} v_{x^i} + v\Delta\psi + f = 0. \tag{3}$$

It seems that a boundary condition on v should be added to this equation. However, this is superfluous since equation (3) can have only one bounded solution in D. Indeed, the difference between two solutions satisfies the equation $\psi\Delta w + 2\psi_{x^i} w_{x^j} + w\Delta\psi = 0$, i.e., $\Delta(\psi w) = 0$ in D and $\psi w = 0$ on ∂D, hence $\psi w = 0$ and $w = 0$ in D. Thus, equation (3) can be considered without boundary conditions so that it is possible to treat this equation exactly as we treated equations in the whole space in Section 7.2.

We encounter only one difficulty here; this is that the matrix $\psi(\delta^{ij})$ composed of the efficients of the second derivatives in equation (3) cannot, generally speaking, be represented as $(\sigma^{ik}\sigma^{jk})$, where σ^{ik} are smooth functions. In order to obtain such a representation, it is required that the function ψ in D can be expressed in terms of a sum of squared functions possessing bounded derivatives. But then, as the argument approaches D, the function ψ as a squared distance between the argument and ∂D would tend to zero; the same would be true for the solution to problem (1) and (2), and this would impose a highly complex condition on f. This difficulty can be surmounted, for example, by multiplying (3) by ψ, but in such a case conditions (7.2.15) and (7.2.16) become absolutely impractable. Therefore, we follow another path. A surface is sought for 'above' the domain D, the function ψ is 'lifted' onto this surface and it is required that the 'lifted' function be the square of a smooth function.

For example, we can take a surface in E_{d+1} specified by the equation $r^2 = \psi(x)$, $r \in [0, \infty)$, $x \in E_d$. It is clear that both the function v and equation (3) must be 'lifted' onto this surface. In other words we have to find a function $w(x, r)$ such that $v(x) = w(x, \sqrt{\psi(x)})$. The latter expression makes it possible to represent the derivatives of v in terms of w. For example, if $r^2 = \psi(x)$, we have

$$v_{x^i} = \left(\frac{\partial}{\partial x^i} + \frac{1}{2\sqrt{\psi}}\psi_{x^i}\frac{\partial}{\partial r}\right)w = w_{x^i} + \frac{1}{2r}\psi_{x^i}w_r, \tag{4}$$

$$v_{x^ix^j} = \left(\frac{\partial}{\partial x^i} + \frac{1}{2\sqrt{\psi}}\psi_{x^i}\frac{\partial}{\partial r}\right)\left(\frac{\partial}{\partial x^j} + \frac{1}{2\sqrt{\psi}}\psi_{x^j}\frac{\partial}{\partial r}\right)w =$$

$$= w_{x^ix^j} + \frac{1}{2r}\psi_{x^i}w_{x^jr} + \frac{1}{2r}\psi_{x^j}w_{x^ir} +$$

$$+ \frac{1}{4r^2}\psi_{x^i}\psi_{x^j}\left(w_{rr} - \frac{1}{r}w_r\right) + \frac{1}{2r}\psi_{x^ix^j}w_r.$$

By (3), these formulas enable us to write an equation such that w has to be its solution on the surface $r^2 = \psi(x)$. It turns out that although this equation involves various derivatives with respect to x and r it is exactly the equation on the surface $r^2 = \psi(x)$, i.e., the function satisfying the equation on the surface $r^2 = \psi(x)$ can be uniquely reproduced on this surface. Not all equations possess this property. For example, let $d = 1$ and let $\varphi''(x) = 0$ only for $x = \pm 1$: then nothing could be said about the values of φ for $x = \pm 1$. Of course, this property follows from the fact that, if w satisfies our equation on the set $r^2 = \psi(x)$, then $w(x, \sqrt{\psi(x)})$ satisfies (3) in D.

An unpleasant fact concerning the equation to which w is a solution is that the coefficient of w_r in it is unbounded on the set $r^2 = \psi(x)$. However, we managed to note that substituting (4) into (3) and combining the term involving w_{rr} with the terms involving $\frac{1}{r}w_r$, we obtain $\frac{1}{4}|\psi_x|^2 \times$ $\times (w_{rr} + \frac{3}{r}w_r)$. Here $w_{rr} + \frac{3}{r}w_r$ is the radial component of the four-dimensional Laplace operator. Therefore, if r is interpreted as $((x^{d+1})^2 + \dots + (x^{d+4})^2)^{1/2}$, then

$$w_{rr} + \frac{3}{r}w_r = \sum_{\nu=d+1}^{d+4} w_{x^\nu x^\nu}.$$

In case of such an interpretation, $r^2 = \psi(x)$ describes a surface in E_{d+4}, the equation for the function w, which is spherically symmetric with respect to $x^{d+1}, \dots, x^{d+4}$, is written for this new surface, and it turns out that the coefficients of this equation are smooth and bounded. On carrying out the above transformations, we may seem to have lost the target of the first step, i.e., the representation of the matrix composed of the coefficients of second derivatives in the equation for w in the form $(\sigma^{ij}\sigma^{jk})$ with smooth σ^{ik}. However, in fact we have not lost it. Indeed, the desired representation merely means that the principal part of the operator can be represented as a sum of pure second derivatives along some vectors depending smoothly on their arguments. On the other hand, by (4), replacing the terms involving first derivatives by dots, we obtain

$$\psi v_{x^i x^j} = r^2 v_{x^i x^j} = x^\nu x^\nu w_{x^i x^j} + \frac{1}{2}\psi_{x^i} x^\nu w_{x^j x^\nu} +$$

$$+ \frac{1}{2}\psi_{x^j} x^\nu w_{x^i x^\nu} + \frac{1}{4}\psi_{x^i}\psi_{x^j}\delta^{\mu\nu} w_{x^\mu x^\nu} + \ldots =$$

$$= \left(x^\nu \frac{\partial}{\partial x^i} + \frac{1}{2}\psi_{x^i}\frac{\partial}{\partial x^\nu}\right)\left(x^\nu \frac{\partial}{\partial x^j} + \frac{1}{2}\psi_{x^j}\frac{\partial}{\partial x^\nu}\right)w + \ldots,$$

on the surface $r^2 = \psi(x)$, where the sum is taken over ν and μ within the limits from $d + 1$ to $d + 4$. Obviously, this calculation shows that the desired representation does exist.

Once the equation for w is written in the coordinates $x^1, \ldots, x^{d+4}$, we can forget that it is an equation on the surface and we can study it as an equation in the whole space E_{d+4} using the methods of the previous chapter. That is the approach we have used in Section 1. In the remainder of this book, we shall essentially use the fact that the equation for w can be considered only on the surface.

Let us now formulate the rule of reducing an equation in a domain to an equation on an auxiliary surface. We will follow the stipulation in the proof of Theorem 1.1. Recall that E_d is thought of as a subspace $\{x \in E_{d+4} : x^{d+1} = \ldots = x^{d+4} = 0\}$ of the space E_{d+4}, all functions of t and x, where $x \in E_d$, are also assumed to be functions of $t, x^1, \ldots, x^{d+4}$ which are independent of $x^{d+1}, \ldots, x^{d+4}$, the Greek superscripts take on integer values from $d + 1$ to $d + 4$, and the summation is carried out over repeated Greek superscripts within these limits. Repeated italic superscripts symbolize summation over the range from 1 to d unless a special stipulation is made. Let us define $[x]$ on E_{d+4} by formula (1.4). Let us also set

$$L_0 w = [x]^2 w, \tag{5}$$

$$L_i w = [x]^2 w_{x^i} + \frac{1}{2}\psi_{x^i} x^\nu w_{x^\nu} + \psi_{x^i} w$$

$$= x^\nu\left(x^\nu \frac{\partial}{\partial x^i} + \frac{1}{2}\psi_{x^i}\frac{\partial}{\partial x^\nu}\right)w + \psi_{x^i} w, \tag{6}$$

$$L_{ij} w = \left(x^\nu \frac{\partial}{\partial x^i} + \frac{1}{2}\psi_{x^i}\frac{\partial}{\partial x^\nu}\right)\left(x^\nu \frac{\partial}{\partial x^j} + \frac{1}{2}\psi_{x^j}\frac{\partial}{\partial x^\nu}\right)w + \psi_{x^i x^j} w. \tag{7}$$

Finally, let us take any $T \in (0, \infty)$ and denote $Q = (0, T) \times D$, $U = \{x \in E_{d+4} : 0 < [x]^2 = \psi(x)\}$, and $U_T = (0, T) \times U$.

1. THEOREM. Let $|\psi_x| > 0$ on ∂D, let D be a bounded domain, let $u \in C^2(Q) \cap C(\bar{Q})$, and let $u = 0$ on $\partial_x Q$. Let a function $F(u_{ij}, u_i, u, z)$ be defined for $z \in Q$ and real-valued u_{ij}, u_i, u, z. Also, suppose that

$$u_t + F\left(u_{x^i x^j}, u_{x^i}, u, z\right) = 0 \tag{8}$$

in Q and introduce the function

$$w(t, x) = u(t, x)\psi^{-1}(x). \tag{9}$$

Then $w_x \in C(Q)$ and

$$[x]^2 w_t + F(\tfrac{1}{2}(L_{ij} + L_{ji})w, L_i w, \psi w, t, x) = 0. \tag{10}$$

on U_T. Moreover, if $u_x \in C^2(Q)$ and $\psi \in C^3(E_d)$, then $w_t, w_x, w_{xx} \in C(Q)$..

Proof. Equality (10) on U_T can be derived from (8) in Q using elementary calculations. The continuity of w_x in Q is clear. To prove the boundedness of w_x near $\partial_x Q$, we consider a point $x_0 \in \partial D$ and, using a transformation of the class C^2, straighten the boundary of D in a neighbourhood of this point so that in the neighbourhood $\psi(x)$ is sent to x^1 whereas the straightened part of the boundary lies in the plane $x^1 = 0$. Then for x lying in the neighbourhood of the straightened part of the boundary formula (9), in the new coordinates, is of the form

$$\begin{aligned} w(t, x) &= (x^1)^{-1} u(t, x) \\ &= \int_0^1 u_{x^1}(t, rx^1, x^2, \ldots, x^d)\,dr, \end{aligned}$$

whence one can see that the gradient of w with respect to x is bounded in the new coordinates and, therefore, in the old coordinates when x lies near x_0. Similarly we can prove the last assertion of the theorem wherein the condition $\psi \in C^3(E_d)$ is imposed to preserve the boundedness of u_x, u_{tx}, u_{xx} and u_{xxx} under straightening the boundary based on a transformation of the class C^3. The theorem is proved.

8.4. Permutation of a Differential Operator with an Elliptic Operator on a Manifold

As we could see in Section 3, (3.10) is an equation in w on the set U_T. Here we partly formalize this fact and make, as in Section 7.1, the first step towards estimating first and second derivatives of w on U_T as in Section 7.2.

1. Definition. Let $V \in E_d$ be a domain and let real-valued functions $a^{ij}(x) = a^{ji}(x)$, $b^i(x)$, $c(x)$ be defined for $x \in V$, $i, j = 1, \dots, d$. Let a set U be contained in V. The operator

$$Lu := a^{ij}u_{x^ix^j} + b^iu_{x^i} + cu \tag{1}$$

is said to act on the set U and we write $L \in \mathcal{L}(U)$ if $Lu = 0$ on U for any function $u \in C^2_{loc}(V)$ such that $u = 0$ on U.

2. LEMMA. Let $\psi^1, \dots, \psi^k \in C^2_{loc}(E_d)$, $M = \{x : \psi^1(x) = \dots = \psi^k(x) = 0\}$, let U be a (nonempty) open, in the relative topology on M, subset of M, rank $(\psi^i_{x^j}; i = 1, :\dots, k, J = 1, \dots, d) = k$ on u, let a^{ij}, b^i, and c be defined in a neighbourhood of the set U such that $(a^{ij}) = (a^{ij})^* \geq 0$ on U, and let L be an operator in (1) and $L\psi^r = L(\psi^r)^2 = 0$ on U for $r = 1, \dots, k$. Then $L \in \mathcal{L}(U)$. Moreover, if a function u of the class C^2 defined in a neighbourhood of U attains its local maximum in the relative topology on U at a point $x_0 \in U$, then $Lu(x_0) - c(x_0)u(x_0) \leq 0$.

Proof. We fix a point $x_0 \in U$ and its sufficiently small neighbourhood V (in E_d). Without loss of generality we can assume that the matrix $(\psi^i_{x^j}(x_0); i, j \leq k)$ is nonsingular and the transformation $x \to \tilde{x} = (\psi^1(x), \dots, \psi^k(x), x^{k+1}, \dots, x^d)$ in V has an inverse and the inverse transformation is twice continuously differentiable. Then any function u of the class $C^2(V)$ can be written as $\tilde{u}(\psi^1(x), \dots, \psi^k(x), x^{k+1}, \dots, x^d)$ and

$$Lu = \tilde{u}_{\tilde{x}^r\tilde{x}^s} a^{ij}\psi^r_{x^i}\psi^s_{x^j} + 2\tilde{u}_{\tilde{x}^r\tilde{x}^q} a^{qi}\psi^r_{x^i} +$$

$$+ \tilde{u}_{\tilde{x}^r}(L - c)\psi^r + a^{pq}\tilde{u}_{\tilde{x}^p\tilde{x}^q} + b^p\tilde{u}_{\tilde{x}^p} + c\tilde{u} \tag{2}$$

at the respective points where the sum is taken over the range r, s = 1,

$\ldots, k, p, q = k+1, \ldots, d, i, j = 1, \ldots, d$. If $u = 0$ on $V \cap U$, then the last four terms disappear since $\tilde{u}(0, \ldots, 0, x^{k+1}, \ldots, x^d) = 0$ and $(L - c)\psi^r = -c\psi^r = 0$. Moreover,

$$L(\psi^r)^2 = 2\psi^r L\psi^r + 2a^{ij}\psi^r_{x^i}\psi^r_{x^j} - c(\psi^r)^2, \tag{3}$$

where no summation over r is assumed, and the equalities $L\psi^r = L(\psi^r)^2 = 0$ on U imply that $a^{ij}\psi^r_{x^i}\psi^r_{x^j} = 0$ on U. Since $(a^{ij}) \geqq 0$, $a^{ij} = a^{ji}$, we obtain $a^{ij}\psi^r_{x^i} = 0$ on U and therefore, the right-hand side of (2) is zero on $U \cap V$.

We have proved the first assertion and the fact that Lu always coincides with the sum of the last three terms in (2) on $U \cap V$. Applying the ordinary maximum principle we obtain the second assertion of the lemma. The lemma is proved.

3. Remark. The first assertion of the lemma follows from its second assertion and the conditions $L\psi^r = L(\psi^r)^2 = 0$ on U are needed to guarantee $L \in \mathcal{L}(U)$.

4. Remark. A discussion of (3) shows that the conditions $L\psi^1 = L(\psi^1)^2 = 0$ and $L\psi^1 = a^{ij}\psi^1_{x^j} = 0$, $i = 1, \ldots, d$ are equivalent on U.

For the rest of the section we fix a function $\psi \in C^2_{loc}(E_d)$, denote $M = \{x : \psi(x) = 0\}$, and suppose that U is a (nonempty) subset of M open in the relative topology on M and $|\psi_x| > 0$ on U. We also fix a neighbourhood (in E_d) V of the set U. Furthermore, we denote by $\tilde{\xi}$ and $\tilde{\eta}$, as in Section 7.1, the pairs (ξ, ξ^{d+1}) and (η, η^{d+1}), where $\xi, \eta \in E_d$ and $\xi^{d+1}, \eta^{d+1} \in E_1$. We put

$$U' = \{(x, \tilde{\xi}) \in E_{2d+1} : x \in U, \psi_{(\xi)}(x) = 0\}, \tag{4}$$

$$U'' = \{(x, \tilde{\xi}, \tilde{\eta}) \in E_{3d+2} : x \in U, \psi_{(\xi)}(x) = \\ = \psi_{(\xi)(\xi)}(x) + \psi_{(\eta)}(x) = 0\}. \tag{5}$$

Note that U' is the tangent manifold to U and U" is a part of the second tangent manifold to U. We also need the following objects, two of which we have already met in Section 7.1 (see (7.1.5) and (7.1.14)):

$$\partial(\tilde{\xi}) = \xi^1 \frac{\partial}{\partial x^1} + \ldots + \xi^d \frac{\partial}{\partial x^d} + \xi^{d+1}, \quad \partial(\tilde{\xi}, \tilde{\eta}) = \partial^2(\tilde{\xi}) + \partial(\tilde{\eta}),$$

$$\partial(\xi) = \xi^1 \frac{\partial}{\partial x^1} + \ldots + \xi^d \frac{\partial}{\partial x^d}, \quad \partial(\xi, \eta) = \partial^2(\xi) + \partial(\eta),$$

$$n(x) = |\psi_x(x)|^{-1}\psi_x(x), \quad \tilde{\partial}_i = \frac{\partial}{\partial x^i} - n^i(x)n^j(x)\frac{\partial}{\partial x^j},$$

$$\tilde{\partial}(\tilde{\xi}) = \xi^1\tilde{\partial}_1 + \ldots + \xi^d\tilde{\partial}_d + \xi^{d+1}, \quad \tilde{\partial}(\tilde{\xi}, \tilde{\eta}) = \tilde{\partial}(\tilde{\xi})\tilde{\partial}(\tilde{\xi}) + \tilde{\partial}(\tilde{\eta}),$$

$$\tilde{\partial}(\xi) = \xi^1\tilde{\partial}_1 + \ldots + \xi^d\tilde{\partial}_d, \quad \tilde{\partial}(\xi, \eta) = \tilde{\partial}(\xi)\tilde{\partial}(\xi) + \tilde{\partial}(\eta). \tag{6}$$

It is clear that, for $x \in U$, $n(x)$ is a normal vector to U at the point x and $\tilde{\partial}_i$ is the operator of differentiation along the vector which results from projecting the ith basic vector on the tangent plane to U at the point x. It is clear that any operator expressed in terms of $\tilde{\partial}_i$ acts on U. The inverse assertion is stated in the following lemma which can be proved using a straightforward calculation and applying Remark 4.

5. LEMMA. Let the operator L in (1) belong to $\mathcal{L}(U)$. Then $Lu = a^{ij}\tilde{\partial}_i\tilde{\partial}_j u + b^i\tilde{\partial}_i u + cu$ on U for any function $u \in C^2_{loc}(V)$.

It is clear that the manifold U' is entirely defined by U. The same turns out to be true for U'' which immediately follows from the following lemma.

6. LEMMA. (a) If $(x, \tilde{\xi}, \tilde{\eta}) \in U''$, we have

$$\partial(\tilde{\xi})u(x) = \tilde{\partial}(\tilde{\xi})u(x), \quad \partial(\tilde{\xi}, \tilde{\eta})u(x) = \tilde{\partial}(\tilde{\xi}, \tilde{\eta})u(x)$$

for any $u \in C^2_{loc}(V)$.

(b) Let $x \in U$ and let $\tilde{\xi}, \tilde{\eta} \in E_{d+1}$, then $(x, \tilde{\xi}, \tilde{\eta})$ belongs to U'' iff

$$\partial(\tilde{\xi}, \tilde{\eta})u(x) = \partial(\tilde{\xi})u(x) = 0. \tag{7}$$

for all $u \in C^2_{loc}(V)$ which are equal to zero on U.

Proof. Assertion (a) can be proved by elementary calculations. In order to prove the 'if' part of (b), we should set $u = \psi$. To prove the 'only if' part, we take note that $\tilde{\partial}(\tilde{\zeta})u = 0$ on U for any $\tilde{\zeta} \in E_{d+1}$ and any function u which is zero on U, in view of the argument preceding

Lemma 5. It immediately follows that (7) holds for any $\tilde{\xi}$, $\tilde{\eta} \in E_{d+1}$ whenever ∂ is replaced by $\tilde{\partial}$. By the first assertion, there is no need for such replacement if $(x, \tilde{\xi}, \tilde{\eta}) \in U''$. The lemma is proved.

After becoming acquainted with the manifolds U, U', and U" and the properties of operators acting on U we pass to the permutability of the operators $\partial(\tilde{\xi})$ and $\partial(\tilde{\xi}, \tilde{\eta})$ with the operator L.

7. THEOREM. Let d_1 be an integer more than or equal to 1 and, for i = 1, ..., d, k = 1, ..., d_1, let functions $\sigma^{ik}(x)$, $b^i(x)$, and $c(x)$ continuously differentiable with respect to x be defined in V. Let us set $a^{ij} = \frac{1}{2}\sigma^{ik}\sigma^{jk}$ and take an arbitrary function $\pi_1(x, \tilde{\xi})$ with values in E_{d_1}, and introduce the operators L and $\tilde{L}$ using formulas (7.1.2) and (7.1.4). Then:

(a) for $u \in C^2_{loc}(V)$ we have $Lu = \tilde{L}u$ on $V \times E_{d+1}$;

(b) $\partial(\tilde{\xi})Lu = \tilde{L}\partial(\tilde{\xi})u$ on $V \times E_{d+1}$ if $u \in C^3_{loc}(V)$;

(c) if $L\psi = L\psi^2 = 0$ on U and $\psi \in C^3_{loc}(V)$, then $\tilde{L} \in L(U')$, and if a function $v(x, \tilde{\xi})$ of the class $C^2_{loc}(V \times E_{d+1})$ attains its local, in the relative topology on U', maximum at a point $(x_0, \tilde{\xi}_0)$, then $\tilde{L}v - cv \leq 0$ at the point $(x_0, \tilde{\xi}_0)$.

Proof. Assertions (a) and (b) are the assertions (a) and (c) of Lemma 7.1.1 respectively. They do not relate to any structure of the manifolds U and U'. To prove (c), it suffices, by Lemma 2, to check whether the gradients of the functions $\psi(x)$ and $\psi_{(\xi)}(x)$ depending on $(x, \tilde{\xi})$ are linearly independent on U' and $\tilde{L}\psi = \tilde{L}\psi^2 = \tilde{L}\partial(\tilde{\xi})\psi = \tilde{L}[\partial(\tilde{\xi})\psi]^2 = 0$ on U'. The independence of the gradients is clearly ensured by the assumption $|\psi_x| > 0$ on U, the equalities $\tilde{L}\psi = \tilde{L}\psi^2 = 0$ on U' hold by hypothesis and assertion (a). Moreover, assertion (b) implies $\tilde{L}\partial(\tilde{\xi})\psi = \partial(\tilde{\xi})L\psi$, the latter being equal to zero on U' since $L\psi = 0$ on U and ξ is tangent to U at x if $(x, \tilde{\xi}) \in U'$. Finally, a formula of the form (3) shows that

$$\tilde{L}[\partial(\tilde{\xi})\psi]^2 = (\partial(\tilde{\xi})\psi)_{(\tilde{\sigma}^k)}(\partial(\tilde{\xi})\psi)_{(\tilde{\sigma}^k)}$$

on U'. Straightforward calculations can show that for $x \in U$ and $\tilde{\xi} \in E_{d+1}$ we have

$$(\partial(\tilde{\xi})\psi)_{(\tilde{\sigma}^k)} = \partial(\tilde{\xi})\left(\psi_{(\sigma^k)}\right)$$

and the latter is equal to zero at the points $(x, \tilde{\xi}) \in U'$ since, by

Remark 4, $2a^{ij}\psi_{x^i}\psi_{x^j} = \psi_{(\sigma^k)}\psi_{(\sigma^k)} = 0$ and $\psi_{(\sigma^k)} = 0$ on U, and the vector ξ is tangent ot U at x. The theorem is proved.

8. THEOREM. Let d_1 be an integer greater than or equal to 1 and for $i = 1, \ldots, d$, $k = 1, \ldots, d_1$, let functions $\sigma^{ik}(x)$, $b^i(x)$, and $c(x)$ twice continuously differentiable with respect to x be defined in V. Let $a^{ij} = \frac{1}{2}\sigma^{ik}\sigma^{jk}$, let $\pi_1(x, \tilde{\xi}, \tilde{\eta})$ and $\pi_2(x, \tilde{\xi}, \tilde{\eta})$ be arbitrary functions with values in E_{d_1} and let us define the operators L and $\tilde{L}$ using formulas (1), (7.1.3), (7.1.13), and formula (7.1.4) wherein $\tilde{\sigma}^k(x, \tilde{\xi})$ $\tilde{b}(x, \tilde{\xi})$ are replaced by $\tilde{\sigma}^k(x, \tilde{\xi}, \tilde{\eta})$ and $\tilde{b}(x, \tilde{\xi}, \tilde{\eta})$ respectively. Then:

(a) the operator $\tilde{L}$ applied to smooth functions $v(x, \tilde{\xi})$ yields the same result as the operator $\tilde{L}$ in Theorem 7;

(b) $\partial(\tilde{\xi}, \tilde{\eta})Lu = \tilde{L}\partial(\tilde{\xi}, \tilde{\eta})u$ on $V \times E_{2d+2}$ if $u \in C^4_{loc}(V)$;

(c) if $L\psi = L\psi^2 = 0$ on U and $\psi \in C^4_{loc}(V)$, then $\tilde{L} \in \mathcal{L}(U'')$ and else if a function $v(x, \tilde{\xi}, \tilde{\eta})$ of the class $C^2_{loc}(V \times E_{2d+2})$ attains its local, in the relative topology on U", maximum at $(x_0, \tilde{\xi}_0, \tilde{\eta}_0)$, then $\tilde{L}v - cv \leqq 0$ at $(x_0, \tilde{\xi}_0, \tilde{\eta}_0)$.

Proof. Again, assertions (a) and (b) are known from Section 7.1. To prove (c), it suffices, by Lemma 2, to check that the gradients of ψ, $\partial(\tilde{\xi})\psi$, and $\partial(\tilde{\xi}, \tilde{\eta})\psi$ are independent on U" as functions of $(x, \tilde{\xi}, \tilde{\eta})$ and to prove that $\tilde{L}\psi_i = \tilde{L}\psi_i^2 = 0$ for $i = 1, 2, 3$ on U", where $\psi_1 = \psi$, $\psi_2 = \partial(\tilde{\xi})\psi$, $\psi_3 = \partial(\tilde{\xi}, \tilde{\eta})\psi$. The independence of the gradients immediately follows from the condition $|\psi_x| > 0$ on U. By assertion (a) and Theorem 7 it remains for us to consider only those equalities among the desired ones that involve ψ_3. The equality $\tilde{L}\psi_3 = 0$ on U", by Lemma 6(b) and assertion (b) of the current theorem, follows from the fact that $\tilde{L}\psi_3 = \partial(\tilde{\xi}, \tilde{\eta})L\psi$ and the equality $\tilde{L}\psi_3^2 = 0$ on U" follows from the formula

$$\tilde{L}\psi_3^2 = (\partial(\tilde{\xi}, \tilde{\eta})\psi)_{(\tilde{\sigma}^k)}(\partial(\tilde{\xi}, \tilde{\eta})\psi)_{(\tilde{\sigma}^k)},$$

which holds on U" by virtue of the equality $\tilde{L}\psi_3 = 0$, from the formula

$$(\partial(\tilde{\xi}, \tilde{\eta})\psi)_{(\tilde{\sigma}^k)} = \partial(\tilde{\xi}, \tilde{\eta})\left(\psi_{(\sigma^k)}\right),$$

which can directly be checked for $(x, \tilde{\xi}) \in U'$ and $\tilde{\eta} \in E_{d+1}$, from Lemma

6(b), and from the equality $\psi_{(\sigma^k)} = 0$ on U established in the previous proof. The theorem is proved.

8.5. The Estimates of the Derivatives of a Solution to a Non-Linear Equation on a Manifold

Let $\delta \in (0, 1)$, let K and K_1 be positive constants, let $\psi \in C^2_{loc}(E_d)$, let a domain V be contained in E_d, let $|\psi_x| > 0$ on V, and let d_1 be an integer greater than or equal to 1. Suppose that the set $U := V \cap \{x : \psi(x) = 0\}$ is nonempty. We also fix an abstract set Ω and for all $\omega \in \Omega$, $i = 1, \ldots, d$, and $k = 1, \ldots, d_1$ let real-valued functions $\sigma^{ik}(\omega, x)$, $b^i(\omega, x)$, $c(\omega, x)$, $f(\omega, x)$ be defined in V, which are continuously differentiable with respect to x and bounded in ω for every $x \in V$. We will follow the arguments in Section 7.2 and, therefore, use the notation in (7.2.1) to (7.2.3) wherein z is replaced by x, and assume that, for any $x \in U$, u_{ij}, u_i, and u, the infimum in (7.2.3) is attained for an $\omega_0 = \omega_0(u_{ij}, u_i, u, x)$.

In Section 7.2 we estimated the expression of $u_{(\xi)}$ in terms of the Euclidean norm of ξ and, therefore, we estimated the Euclidean norm of the gradient of u. This was quite a natural approach in case of the equation in a Euclidean space. But as for an equation on a manifold, it is natural to introduce an appropriate Riemann metric and to estimate the norm of the gradient of u in accordance with this metric. We are not going to develop our subsequent approach from the point of view of Riemann spaces; however, keeping in mind that such a viewpoint exists, we introduce a $d \times d$ nonnegative symmetric matrix $B(x) = (B^{ij}(x))$ in V and suppose that $B \in C^2_{loc}(V)$. Throughout this section it is also assumed that

$$L(\omega)\psi = L(\omega)\psi^2 = 0$$

on U for all $\omega \in \Omega$. As always, we put

$$L^0(\omega) = L(\omega) - c(\omega),$$

and use the notaiton U', $\partial(\tilde{\xi})$, etc. of Section 4 with no further explanation. We also introduce

$$\Gamma_1(\xi) := -L^0(B\xi, \xi) - 2\left(B_{(\sigma^k)}\xi, \sigma^k_{(\xi)}\right) - \left(B\sigma^k_{(\xi)}, \sigma^k_{(\xi)}\right) - $$

$$- 2(B\xi, b_{(\xi)}) - (2 - \delta)c(B\xi, \xi) + K \sum_{k=1}^{d_1} (B\xi, \sigma^k)^2. \quad (1)$$

1. LEMMA. For all $\omega \in \Omega$, $x \in U$, $\xi \perp \psi_x(x)$, $\xi \neq 0$, let the inequalities

$$\Gamma_1(\xi) > 0, \quad (B\xi, \xi)^{1/2}(|f_{(\xi)}| + |c_{(\xi)}|) \leqq K\Gamma_1(\xi),$$

$$(B\xi, \xi)|f|, \quad (B\xi, \xi)c \leqq K\Gamma_1(\xi)$$

hold.

Then there exists a constant $\lambda = \lambda(K, \delta) \geqq 1$ such that the function

$$\varphi(x, \tilde{\xi}) := [\lambda(B(x)\xi, \xi) + |\xi^{d+1}|^2 + 1]^\gamma,$$

where $\gamma = (2 - \delta)^{-1}$ as considered on the set

$$U' \cap \{(B\xi, \xi) \geqq |\xi^{d+1}|^2 + 1\} \tag{2}$$

satisfies the inequality

$$\tilde{L}(\omega)\varphi + \partial(\tilde{\xi})f(\omega) < 0, \tag{3}$$

for all $\omega \in \Omega$, where $L(\omega)$ is constructed from $\pi_1^k = \lambda(B\xi, \sigma^k(\omega))$, $\sigma^k(\omega)$, $b(\omega)$, and $c(\omega)$ for every ω as in Theorem 4.7.

Proof. As in (7.2.18), we have

$$\gamma^{-1}\varphi^{1/\gamma-1}\tilde{L}\varphi \leqq \tilde{L}^0\varphi^{1/\gamma} + \gamma^{-1}c\varphi^{1/\gamma}. \tag{4}$$

Straightforward calculations based on (7.1.9) and (7.1.11) show that the right-hand side of (4) is equal to

$$-\lambda\Gamma_1(\xi) + \lambda(K - \lambda)\Sigma(B\xi, \sigma^k)^2 + 2\xi^{d+1}c_{(\xi)} +$$
$$+ (2 - \delta)c(|\xi^{d+1}|^2 + 1).$$

For $\lambda \geqq K$ this expression, by our hypothesis, does not exceed

$$-\lambda\Gamma_1(\xi) + 2(B\xi, \xi)^{1/2}|c_{(\xi)}| + 2(B\xi, \xi)c_+ \leqq (4K - \lambda)\Gamma_1(\xi)$$

on set (2).

On the other hand, we have (supposing that $\lambda \geqq 1$)

$$\varphi \leqq 2\lambda^\gamma(B\xi, \xi)^\gamma, \quad \varphi^{1/\gamma-1} \leqq 4\lambda^{1-\gamma}(B\xi, \xi)^{1/2},$$

$$\gamma^{-1}\varphi^{1/\gamma-1}\partial(\tilde{\xi})f \leqq 8\lambda^{1-\gamma}[(B\xi, \xi)^{1/2}|f_{(\xi)}| +$$
$$+ (B\xi, \xi)|f|] \leqq 16K\lambda^{1-\gamma}\Gamma_1(\tilde{\xi})$$

on (2). It follows that the left-hand side of (3) multiplied by $\gamma^{-1}\varphi^{1/\gamma-1}$ does not exceed $(4K - \lambda + 16K\lambda^{1-\gamma})\Gamma_1$, while the latter is negative for an appropriate $\lambda(K, \delta) \geq 1$. The lemma is proved.

The function φ in Lemma 1 will act as a barrier in estimation of the first derivatives of a solution to a nonlinear equation on U. To estimate the second derivatives, we need the following notions. Let us put (see (1))

$$H = H(x) = (\delta^{ij} - n^i(x)n^j(x)), \quad \hat{B} = HBH,$$

$$\hat{\Gamma}_1(\xi) = \Gamma_1(\xi) - \delta(\hat{B}\sigma^k_{(\xi)}, \sigma^k_{(\xi)}),$$

$$\Gamma_2(\xi) = (B\xi, \xi)\left[\hat{\Gamma}_1(\xi) + (1 - \frac{\delta}{2})c(B\xi, \xi)\right] -$$

$$- \frac{1}{2}\sum_{k=1}^{d_1}(B_{(\sigma^k)}\xi + 2B\sigma^k_{(\xi)}, \xi)^2. \tag{5}$$

It is clear that $H(x)$ with $x \in U$ is a projection operator on the tangent plane to U at x. Therefore, for example,

$$(\hat{B}\xi, \xi) = (B\xi, \xi) \text{ on } U', \quad \hat{B}n = 0,$$

$$\hat{B}_{(\sigma^k)}n = -\hat{B}n_{(\sigma^k)} = -HBn_{(\sigma^k)}, \tag{6}$$

$$\left(\hat{B}_{(\sigma^k)}n, n_{(\sigma^k)}\right) = -\left(Bn_{(\sigma^k)}, n_{(\sigma^k)}\right) \text{ on } U.$$

Also, note that $H\sigma^k = \sigma^k$ on U,

$$|\psi_x|^{-1}\psi_{(\xi)(\xi)} = (n_{(\xi)}, \xi) \text{ on } U', \tag{7}$$

and let us single out the following.

2. Remark. If $\psi \in C^3_{loc}(V)$, then the right-hand side of (1) does not change on U' if B is replaced by $\hat{B}$.

Indeed, the two latter summands in (1) do not change since $H\xi = \xi$ and $H\sigma^k = \sigma^k$, while the sum of the remaining terms, by (7.1.11), coincides with $-\tilde{L}^0(B\xi, \xi)$ if $\pi_1 = 0$. It is easily seen that after replacing B by $\hat{B}$ in the terms in question their sum is $-\tilde{L}^0(\hat{B}\xi, \xi)$. Finally, $\tilde{L}(B\xi, \xi) = \tilde{L}(\hat{B}\xi, \xi)$ on U' by Theorem 4.7(c) since $(B\xi, \xi) = (\hat{B}\xi, \xi)$ on U'.

3. LEMMA. Let functions σ^k, b, c, and f be twice continuously differentiable with respect to x on V, let $\psi \in C^3_{loc}(V)$, let $\chi(\omega, x)$ be a nonnegative function defined on $\Omega \times U$, and for all $\omega \in \Omega$, $(x, \tilde{\xi})$, $(x, \tilde{\xi}_1) \in U'$ such that $\xi \neq 0$, let the following inequalities hold:

$$0 < \Gamma_2(\xi), \quad \chi(B\xi, \xi)^2 \leqq K\Gamma_2(\xi), \quad c \leqq 0, \quad |f| \leqq K\chi, \tag{8}$$

$$((L^0\hat{B})n, n)^2 + \left(Bn_{(\sigma^k)}, n_{(\sigma^k)}\right) \leqq K, \quad (L^0\hat{B}\xi, n)^2 \leqq K\hat{\Gamma}_1(\xi), \tag{9}$$

$$(n_{(\xi)}, \xi)^2 + (\hat{B}\tilde{\partial}^2(\xi)\sigma^k, \tilde{\partial}^2(\xi)\sigma^k) \leqq K\Gamma_2(\xi), \tag{10}$$

$$(\hat{B}\tilde{\partial}^2(\xi)b, \tilde{\partial}^2(\xi)b) \leqq K\chi\Gamma_2(\xi), \tag{11}$$

$$(B\xi, \xi)|\tilde{\partial}^2(\xi)c|, \quad (B\xi, \xi)\tilde{\partial}^2(\xi)f \leqq K\Gamma_2(\xi), \tag{12}$$

$$(B\xi, \xi)|c_{(\xi)}|^2 + (B\xi, \xi)|f_{(\xi)}|^2 \leqq K\chi\Gamma_2(\xi), \tag{13}$$

$$\left(\hat{B}_{(\sigma^k)}\xi_1, \tilde{\partial}^2(\xi)\sigma^k\right) \leqq K\hat{\Gamma}_1^{1/2}(\xi_1)\Gamma_2^{1/2}(\xi). \tag{14}$$

Then there exists a constant $\lambda = \lambda(K, \delta) \geqq 1$ such that the function

$$\varphi(x, \tilde{\xi}, \tilde{\eta}) := [\lambda^2(B(x)\xi, \xi)^2 + \lambda(\hat{B}(x)\eta, \eta) + \lambda|\xi^{d+1}|^4 + |\eta^{d+1}|^2 + 1]^\gamma,$$

where $\gamma = (2 - \delta)^{-1}$ considered on the set

$$U'' \cap \{(B\xi, \xi)^2 \geqq (\hat{B}\eta, \eta) + |\xi^{d+1}|^4 + |\eta^{d+1}|^2 + 1\} \tag{15}$$

satisfies the inequality

$$\tilde{L}(\omega)\varphi + \partial(\tilde{\xi}, \tilde{\eta})f(\omega) < 0, \tag{16}$$

for all $\omega \in \Omega$, where $\tilde{L}(\omega)$ is constructed as in Theorem 4.8 and π_1, π_2 are chosen according to a certain rule.

<u>Proof</u>. From (7.1.10), (7.1.12), (7.1.15), and (7.1.16) we obtain that $\tilde{L}^0\varphi^{1/\gamma}$ coincides with $I_1 + I_2 + I_3 + I_4$, where

$$I_1 = -2\lambda^2\Gamma_2(\xi) - (2 - \delta)\lambda^2 c(B\xi, \xi)^2 + 2\lambda^2 K(B\xi, \xi)\sum_{k=1}^{d_1}(B\xi, \sigma^k)^2 - 2\lambda^2\delta(\hat{B}\sigma^k_{(\xi)}, \sigma^k_{(\xi)})(B\xi, \xi),$$

$$I_2 = \lambda L^0(\hat{B}\eta, \eta) + 2\lambda(\hat{B}\eta, \partial(\xi, \eta)b) + $$
$$+ \lambda(\hat{B}\partial(\xi, \eta)\sigma^k, \partial(\xi, \eta)\sigma^k) + 2\lambda\left(\hat{B}_{(\sigma^k)}\eta, \partial(\xi, \eta)\sigma^k\right),$$

$$I_3 = -4\lambda^2(B\xi, \xi)\pi_1^k(B\xi, \sigma^k) - 4\lambda\pi_1^k(\hat{B}\eta, \sigma^k_{(\xi)}) -$$
$$- 2\eta^{d+1}|\pi_1|^2 + 6\lambda|\xi^{d+1}|^2|\pi_1|^2 - 2\lambda\pi_2^k(\hat{B}\eta, \sigma^k) + |\pi_2|^2,$$

$$I_4 = 4\lambda(\xi^{d+1})^3 c_{(\xi)} + 2\eta^{d+1}\partial(\xi, \eta)c.$$

The following computations are conducted for fixed $\lambda \geqq 1$ and $(x, \tilde{\xi}, \tilde{\eta})$ taken from the set (15) and for any $\varepsilon \in (0, \delta]$. Let us denote by N constants depending only on K and δ, and set (cf. (7), (4.5))

$$\xi_1 := \eta + (n_{(\xi)}, \xi)n = H\eta.$$

Note that $(x, \xi_1, 0) \in U'$ and $\partial(\xi, \eta) = \tilde{\partial}^2(\xi) + \partial(\xi_1)$ by Lemma 4.6(a). Hence using (6), (9), (10), (14) and the Cauchy—Buniakovski inequality, we obtain

$$\left(\hat{B}_{(\sigma^k)}\eta, \partial(\xi, \eta)\sigma^k\right) = \left(\hat{B}_{(\sigma^k)}\xi_1, \sigma^k_{(\xi_1)}\right) +$$
$$+ (n_{(\xi)}, \xi)\left(\hat{B}n_{(\sigma^k)}, \sigma^k_{(\xi_1)}\right) +$$
$$+ (n_{(\xi)}, \xi)\left(\hat{B}n_{(\sigma^k)}, \tilde{\partial}^2(\xi)\sigma^k\right) +$$
$$+ \left(\hat{B}_{(\sigma^k)}\xi_1, \tilde{\partial}^2(\xi)\sigma^k\right)$$
$$\leqq \left(\hat{B}_{(\sigma^k)}\xi_1, \sigma^k_{(\xi_1)}\right) + \varepsilon\left(\hat{B}\sigma^k_{(\xi_1)}, \sigma^k_{(\xi_1)}\right) +$$
$$+ N\varepsilon^{-1}\Gamma_2(\xi) + \varepsilon\hat{\Gamma}_1(\xi_1).$$

Let us show how it is possible to estimate one more term entering I_2:

$$(\hat{B}\partial(\xi, \eta)\sigma^k, \partial(\xi, \eta)\sigma^k)$$
$$\leqq (1 + \varepsilon)\left(\hat{B}\sigma^k_{(\xi_1)}, \sigma^k_{(\xi_1)}\right) + \left(1 + \frac{1}{\varepsilon}\right)(\hat{B}\tilde{\partial}^2(\xi)\sigma^k, \tilde{\partial}^2(\xi)\sigma^k)$$
$$\leqq (1 + \varepsilon)\left(\hat{B}\sigma^k_{(\xi_1)}, \sigma^k_{(\xi_1)}\right) + N\varepsilon^{-1}\Gamma_2(\xi$$

Similarly, using (8) and (11) (and also the obvious inequality $\Gamma_2 \leqq (B\xi,\ \xi)\hat{\Gamma}_1$ and $\chi(B\xi,\ \xi) \leqq K\hat{\Gamma}_1$, the latter holding by virtue of (8)) we can estimate other summands in I_2 so that

$$\lambda^{-1}I_2 \leqq \left[((L^0\hat{B})\xi_1,\ \xi_1) + 2\left(\hat{B}_{(\sigma^k)}\xi_1,\ \sigma^k_{(\xi_1)}\right) + \left(\hat{B}\sigma^k_{(\xi_1)},\ \sigma^k_{(\xi_1)}\right) + 2\left(\hat{B}b_{(\xi_1)},\ \xi_1\right)\right] + \varepsilon\left(\hat{B}\sigma^k_{(\xi_1)},\ \sigma^k_{(\xi_1)}\right) + N\varepsilon^{-1}\Gamma_2(\xi) + \varepsilon\hat{\Gamma}_1(\xi_1).$$

By Remark 2, we can replace $\hat{B}$ in the square brackets by B hence, by (1) and (5), we conclude that for $\varepsilon \leqq \delta$

$$\lambda^{-1}I_2 \leqq -(1 - \varepsilon)\hat{\Gamma}_1(\xi_1) - (2 - \delta)c(B\xi_1,\ \xi_1) + $$
$$+ K\Sigma(B\xi_1,\ \sigma^k)^2 + N\varepsilon^{-1}\Gamma_2(\xi).$$

Prior to estimating I_3, we fix the values of π_1 and π_2:

$$\pi_1^k = \lambda\varepsilon(B\sigma^k,\ \xi),\quad \pi_2^k = \lambda(\hat{B}\sigma^k,\ \eta) = \lambda(B\sigma^k,\ \xi_1).$$

It has been assumed that $\lambda \geqq 1$ and $(x,\ \tilde{\xi},\ \tilde{\eta})$ belongs to set (15). Therefore,

$$I_3 = \varepsilon^2(6\lambda|\xi^{d+1}|^2 - 2\eta^{d+1})\lambda^2\Sigma(B\sigma^k,\ \xi)^2 -$$
$$- 4\varepsilon\lambda^3(B\xi,\ \xi)\Sigma(B\sigma^k,\ \xi)^2 - 4\varepsilon\lambda^2(B\sigma^k,\ \xi)(\hat{B}\eta\ ,\ \sigma^k_{(\xi)}) -$$
$$- \lambda^2\Sigma(\hat{B}\sigma^k,\ \xi_1)^2$$
$$\leqq -(4\varepsilon\lambda^3 - 6\lambda^3\varepsilon^2 - 2\lambda^2\varepsilon^2)(B\xi,\ \xi)\Sigma(B\sigma^k,\ \xi)^2 +$$
$$+ 4\varepsilon\lambda^2(B\xi,\ \xi)\Sigma|(B\sigma^k,\ \xi)|(\hat{B}\sigma^k_{(\xi)},\ \sigma^k_{(\xi)})^{1/2} - \lambda^2\Sigma(\hat{B}\sigma^k,\ \xi_1)^2$$
$$\leqq -\lambda^3(2\varepsilon - 8\varepsilon^2)(B\xi,\ \xi)\Sigma(B\sigma^k,\ \xi)^2 +$$
$$+ 2\varepsilon\lambda^2(B\xi,\ \xi)(\hat{B}\sigma^k_{(\xi)},\ \sigma^k_{(\xi)}) - \lambda^2\Sigma(\hat{B}\sigma^k,\ \xi_1)^2.$$

Similarly, using the latter calculation and inequalities (12), (13) and (8), we can show that

$$I_4 \leqq 4\lambda(B\xi,\ \xi)^{3/2}|c_{(\xi)}| + 2(B\xi,\ \xi)|c_{(\xi_1)}| + 2(B\xi,\ \xi)|\tilde{\partial}^2(\xi)c|$$

$$\leqq N(\lambda + 1)\Gamma_2(\xi) + 2(B\xi, \xi)(K\chi\hat{\Gamma}_1(\xi_1))^{1/2}$$

$$\leqq N\lambda\Gamma_2(\xi) + N\hat{\Gamma}_1(\xi_1),$$

$$\gamma^{-1}c\varphi^{1/\gamma} \leqq (2-\delta)c\lambda^2(B\xi, \xi)^2 + (2-\delta)c\lambda(\hat{B}\xi_1, \xi_1),$$

$$\gamma^{-1}\varphi^{1/\gamma-1}\partial(\tilde{\xi}, \tilde{\eta})f \leqq N\lambda^{2-2\gamma}(B\xi, \xi)[(\tilde{\partial}^2(\xi)f)_+ + 2|\xi^{d+1}f_{(\xi)}| + |f_{(\xi_1)}| + (|\xi^{d+1}|^2 + |\eta^{d+1}|)|f|]$$

$$\leqq N\lambda^{2-2\gamma}\Gamma_2(\xi) + N\lambda^{2-2\gamma}(B\xi, \xi)|f_{(\xi_1)}|$$

$$\leqq N\lambda^{2-2\gamma}\Gamma_2(\xi) + N\lambda^{2-2\gamma}(B\xi,\xi)(K\chi\hat{\Gamma}_1(\xi_1))^{1/2}$$

$$\leqq N\lambda^{2-2\gamma}(\hat{\Gamma}_1(\xi_1) + \Gamma_2(\xi)).$$

Finally, using inequality (4), we obtain that the left-hand side of (16) multiplied by $\gamma^{-1}\varphi^{1/\gamma-1}$ does not exceed

$$(N\varepsilon^{-1}\lambda - 2\lambda^2)\Gamma_2(\xi) + (2\lambda^2K - \lambda^3(2\varepsilon - 8\varepsilon^2))(B\xi, \xi)\Sigma(B\sigma^k, \xi)^2 + (N\lambda^{2-2\gamma} - \lambda(1-\varepsilon))\hat{\Gamma}_1(\xi_1) + (K\lambda - \lambda^2)\Sigma(B\sigma^k, \xi_1)^2.$$

It follows that for $\varepsilon = 8^{-1} \wedge \delta$ we can choose a large $\lambda = \lambda(K, \delta) \geqq 1$ such that the latter expression become negative. The lemma is proved.

The conditions of Lemmas 1 and 3 assocaited with the functions Γ_1, $\hat{\Gamma}_1$, and Γ_2 look rather cumbersome. We therefore state below somewhat simpler conditions. The remarks to come demonstrate two basic cases: when the functions Γ_1, $\hat{\Gamma}_1$, and Γ_2 are positive in view of the fact that c is negative, and when the first summand in (1) ensures these functions being positive if $c \equiv 0$.

4. Remark. Suppose that for all $\omega \in \Omega$

$$L^0(B\xi, \xi) + 2(B_{(\sigma^k)}\xi, \sigma^k_{(\xi)}) + (B\sigma^k_{(\xi)}, \sigma^k_{(\xi)}) + 2(B\xi, b_{(\xi)}) < -2(1-\delta)c(B\xi, \xi) + K\sum_{k=1}^{d_1}(B\xi, \sigma^k)^2,$$

$$(B\xi, \xi)^{1/2}(|f_{(\xi)}| + |c_{(\xi)}|) + (B\xi, \xi)(|f| + c_+)$$

$$\leq K\left(-c(B\xi,\ \xi) + K\sum_{k=1}^{d_1}(B\xi,\ \sigma^k)^2\right)$$

on U' with $\xi \neq 0$. Then the hypotheses of Lemma 1 hold true with a constant $N(K, \delta)$ instead of K.

This remark immediately follows from the inequality $\Gamma_1(\xi) > \delta(-c\ \ (B\xi,\ \xi) + K\Sigma(B\xi,\ \sigma^k)^2)$, which coincides with (17) if $(1 + \delta)K$ in the definition (1) of the function Γ_1 is substituted for K.

5. Remark. Suppose that there exist a constant $\delta_1 \in (0, 1)$ and a function $\tilde{u} \in C^2_{loc}(V)$ such that $\tilde{u} \geqq 0$ on V and for all $\omega \in \Omega$ the inequalities

$$\tilde{u}\left[(\sigma^k_{(\xi)},\ \sigma^k_{(\xi)}) + 2(\xi,\ b_{(\xi)})\right] + 2\tilde{u}_{(\sigma^k)}(\sigma^k_{(\xi)},\ \xi)$$
$$< -(1 - \delta_1)[L^{\Theta}\tilde{u} + (2 - \delta)c\tilde{u}] + K\tilde{u}^2\sum_{k=1}^{d_1}(\sigma^k,\ \xi)^2, \tag{19}$$

$$\tilde{u}^{1/2}(|f_{(\xi)}| + |c_{(\xi)}|) + \tilde{u}(|f| + c_+)$$
$$\leqq K\left(-L^0\tilde{u} - (2 - \delta)c\tilde{u} + K\tilde{u}^2\sum_{k=1}^{d_1}(\sigma^k,\ \xi)^2\right) \tag{20}$$

hold on U' with $|\xi| = 1$. Then for $(B\xi,\ \xi) := \tilde{u}|\xi|^2$ the hypothesis of Lemma 1 holds true with a constant $N(K, \delta_1)$ instead of K.

Here inequality (19) with $|\xi| = 1$ is equivalent to the fact that if K in the definition of Γ_1 is replaced by $(1 + \delta_1)K$, then Γ_1 turns out to be strictly greater than δ_1 multiplied by the expression between the outer brackets on the right of (20). But in view of homogeneity, it suffices to verify the conditions of Lemma 1 only when $|\xi| = 1$.

6. Remark. Let σ, b, c, and f be twice continuously differentiable with respect to x on V, let $\psi \in C^3_{loc}(V)$, and let the sum of squared σ^{ik}, b^i, f, B^{ij}, ψ_{x^i}, all their squared first and second derivatives, and all squared first and second derivatives of c on U not exceed K. Also, let

$$(B\xi,\ \xi)\Big[L^0(B\xi,\ \xi) + 2\left(B_{(\sigma^k)}\xi,\ \sigma^k_{(\xi)}\right) + (B\sigma^k_{(\xi)},\ \sigma^k_{(\xi)}) +$$
$$+ 2(Bb_{(\xi)},\ \xi)\Big] + \frac{1}{2}\sum_{k=1}^{d_1}\left(B_{(\sigma^k)}\xi + 2B\sigma^k_{(\xi)},\ \xi\right)^2$$

$$\leq (B\xi, \xi)\left[-(1-\delta)c(B\xi, \xi) + K\sum_{k=1}^{d_1}(B\sigma^k, \xi)^2\right], \tag{21}$$

$$(B\xi,\xi) \geq \delta_1|\xi|^2, \quad |\psi_x| \geq \delta_1, \quad c \leq -\delta_1, \tag{22}$$

for all $\omega \in \Omega$ on U', where δ_1 is a strictly positive constant. Under these conditions, the conditions of Lemma 3 also hold if K, δ, χ in (1), (5), and Lemma 3 are replaced by N, δ_2, 1, where N and δ_2 depend only on K, δ, δ_1, and d and $\delta_2 \in (0, 1)$.

Indeed, it follows from the first assumption and (22) that inequalities (9) — (14) hold if χ, K, $\hat{\Gamma}_1(\xi)$, and $\Gamma_2(\xi)$ in them are replaced by 1, $N(K, \delta, \delta_1, d)$, $|\xi|^2$, $|\xi|^4$. Moreover, let us assume that (1) and (5) define Γ_1, $\hat{\Gamma}_1$, Γ_2 as functions of K, δ, ξ. Then inequality (21) acquire the form

$$\Gamma_2(K, \delta_2, \xi) \geq \left[\left(\frac{1}{2}\delta_2 - \delta\right)c(B\xi, \xi) - \delta_2(\hat{B}\sigma^k_{(\xi)}, \sigma^k_{(\xi)})\right](B\xi, \xi).$$

When $\delta_2 < 2\delta$, the latter expression is greater than

$$|\xi|^4\delta_1\left[-\left(\frac{1}{2}\delta_2 - \delta\right)\delta_1^2 - \delta_2 N(K, \delta_1, d)\right],$$

by choosing a sufficiently small δ_2 we can make it greater than $\delta_2|\xi|^4$, and then $\Gamma_2(K, \delta_2, \xi) \geq \delta_2|\xi|^4$. Finally, for any B it follows from the inequality $c \leq 0$ that $\Gamma_2 \leq (B\xi, \xi)\hat{\Gamma}_1$, hence we obtain $N\Gamma_2(K, \delta_2, \xi) \geq (B\xi, \xi)^2$, $N\hat{\Gamma}_1(K, \delta_2, \xi) \geq |\xi|^2$. This proves the remark.

7. Remark. Let σ, b, c, f be twice continuously differentiable with respect to x on V, let $\psi \in C^3(V)$, let $\chi(\omega, x)$ be a nonnegative function defined on $\Omega \times U$, let $\delta_1 \in (0, 1)$, and let $\tilde{u}$ be a function of $C^2(V)$. For $x \in U$, $\xi \in E_d$, and $|\xi| = 1$, let

$$M(\omega, x, \xi) = -\frac{1-\delta_1}{2}[L^0\tilde{u} + (2-\delta)c\tilde{u}] + $$
$$+ K\tilde{u}^{\frac{3}{2}}\sum_{k=1}^{d_1}(\sigma^k, \xi)^2$$

and suppose that the inequalities

$$\tilde{u}\left[(1+\delta)(\sigma^k_{(\xi)}, \sigma^k_{(\xi)}) + 2(\xi, b_{(\xi)}) + 2\sum_{k=1}^{d_1}(\sigma^k_{(\xi)}, \xi)^2\right] +$$

$$+ 2\tilde{u}_{(\sigma^k)}(\sigma^k_{(\xi)}, \xi) \leq M, \tag{23}$$

$$|\psi_x| \geq \delta_1, \quad M \geq \delta_1, \quad \chi > 0, \quad \tilde{u} > 0,$$

$$\chi\tilde{u} \leq KM, \quad c \leq 0, \quad |f| \leq K\chi, \quad |b| + \text{tr } a \leq K,$$

$$\chi(\tilde{\partial}^2(\xi)\sigma^k, \tilde{\partial}^2(\xi)\sigma^k) + |\tilde{\partial}^2(\xi)b|^2 + \chi|\tilde{\partial}^2(\xi)c| + \tag{24}$$

$$+ \chi(\tilde{\partial}^2(\xi)f)_+ + |c_{(\xi)}|^2 + |f_{(\xi)}|^2 \leq K\chi M$$

hold for all $\omega \in \Omega$, $x \in U$, and unit $\xi \perp \psi_x(x)$. Also, suppose that the norm of $\tilde{u}^{1/2}$ in $C^2(V)$ does not exceed K. Then the hypothesis of Lemma 3 holds true with $B := \tilde{u}^{1/2}(\delta^{ij})$ and a constant N in place of K, the constant N depending only on K, δ_1, d, and the norm of ψ in $C^3(V)$.

Indeed, it immediately follows from our conditions that inequalities (8) — (14) hold for $|\xi| = |\xi_1| = 1$ if K, $\Gamma_2(\xi)$, $\hat{\Gamma}_1(\xi)$, and $\hat{\Gamma}_1(\xi_1)$ on the right are replaced by N, $M(\xi)$, $M(\xi)$, $M(\xi_1)$. Moreover, if K in the definition of Γ_1, $\hat{\Gamma}_1$, and Γ_2 is replaced by $(1 - \delta_1)^{-1}K$, then simple calculations show that inequality (23) is equivalent to the inequality

$$\Gamma_2(\xi) \geq \frac{\delta_1}{1-\delta_1} M + \delta\tilde{u}\,[(\sigma^k_{(\xi)}, \sigma^k_{(\xi)}) - (H\sigma^k_{(\xi)}, \sigma^k_{(\xi)})].$$

The second summand on the right is positive and $\Gamma_2(\xi) \geq \delta_1 M/(1 - \delta_1)$. Finally, $\Gamma_2 \leq (B\xi, \xi)\hat{\Gamma}_1$, $\hat{\Gamma}_1(\xi)\tilde{u}^{1/2} \geq \delta_1 M/(1 - \delta_1)$, while it suffices to verify (8) — (14) only for unit vectors ξ and ξ_1.

We have shown several cases when the sign of $\Gamma_1(\xi)$ is defined by two summands in (1). There also are a number of important cases when $(B\xi, b_{(\xi)})$ plays a dominant role. For the sake of brevity, we do not, however, consider them here and, using Lemmas 1 and 3, turn to estimating the derivatives of solutions to equations.

8. THEOREM. Let ψ, $u \in C^3_{loc}(V)$, let the conditions of Lemma 1 hold (cf. Remarks 4 and 5), let $c \leq 0$, $|u| \leq K$, and $|u_{(\xi)}|^2 \leq M^2(B\xi, \xi)$ on U' for a constant $M \geq 0$, and let $(B\xi, \xi) > 0$ with $\xi \neq 0$ on U'. Let

$$F\left(u_{x^ix^j}(x),\ u_{x^i}(x),\ u(x),\ x\right) = 0 \tag{25}$$

on U. Also, suppose that $\psi_1 \in C^2_{loc}(V)$ is a nonnegative function such that $L(\omega)\psi_1 \leqq 0$ on U for all $\omega \in \Omega$. For m = 1, 2, ..., let

$$U_m = \{x \in U : \psi_1(x) < m\}. \tag{26}$$

Suppose that U_m is bounded, u, $\tilde{\partial}_i u$, and B can be extended to continuous functions in the relative topology on $\bar{U}_m$ for any m, and

$$|\xi^i\tilde{\partial}_i u(x) + \xi^{d+1}u(x)|^2 \leqq K_1^2[\lambda(B(x)\xi,\ \xi) + |\xi^{d+1}|^2],$$

$$(B(x)\xi,\ \xi) > 0$$

for all $x \in \bar{U}_m \setminus U$, $\xi \perp \psi_x(x)$, $\xi \neq 0$, $\xi^{d+1} \in E_1$, $m \geqq 1$, where λ is taken from Lemma 1. Then

$$|\partial(\tilde{\xi})u|^2 \leqq (K_1^2 \vee N_1)[\lambda(B\xi,\ \xi) + |\xi^{d+1}|^2] \tag{27}$$

on U', where $N_1 = N_1(K,\ \delta)$. Specifically, $|u_{(\xi)}|^2 \leqq \lambda(K_1^2 \vee N_1)\cdot(B\xi,\ \xi)$ on U'.

Proof. We fix an $\varepsilon > 0$, take the function φ from Lemma 1, and consider the function

$$v(x,\ \tilde{\xi}) := \partial(\tilde{\xi})u(x) - \varphi(x,\ \tilde{\xi}) - \varepsilon\psi_1(x)$$

on U'.

There are three possible cases: (i) $v \leqq 0$ on U'; (ii) $v > 0$ at some points of U', but v does not attain its supremum on U'; (iii) there exists a point $(x_0,\ \xi_0) \in U'$ at which v attains its maximum value on $v(x_0,\ \tilde{\xi}_0) > 0$.

We first consider the third case. Let us show that

$$(B(x_0)\xi_0,\ \xi_0) < |\xi_0^{d+1}|^2 + 1. \tag{28}$$

Indeed, otherwise, using the appropriate $\omega \in \Omega$ and applying Theorem 4.7 exactly as in the proof of Lemma 7.2.1 (cf. (7.2.12) and (7.2.14)) we would obtain at $(x_0,\ \tilde{\xi}_0)$

$$0 \geqq \tilde{L}(\omega)v - cv,\quad 0 \geqq \tilde{L}(\omega)v,\quad \varepsilon L(\omega)\psi_1 + \tilde{L}(\omega)\varphi \geqq \tilde{L}\partial(\tilde{\xi})u,$$

$$0 = \tilde{L}(\omega)\partial(\tilde{\xi})u + \partial(\tilde{\xi})f,\quad \varepsilon L(\omega)\psi_1 + \tilde{L}(\omega)\varphi + \partial(\tilde{\xi})f(\omega) \geqq 0;$$

the latter is, however, impossible at the points of set (2) in view of (3).

Furthermore, by differentiation of v with respect to $\tilde{\xi}$ along U', we obtain

$$u_{(\zeta)}(x_0) = 2\lambda\gamma\varphi(x_0, \tilde{\xi}_0)^{1-1/\gamma}(B(x_0)\xi_0, \zeta),$$

$$u(x_0) = 2\gamma\varphi(x_0, \tilde{\xi}_0)^{1-1/\gamma}\xi_0^{d+1} \tag{29}$$

for any $\zeta \perp \psi_x(x_0)$.

Of course, it follows that

$$v(x_0, \tilde{\xi}_0) = 2\gamma\varphi(x_0, \tilde{\xi}_0)^{1-1/\gamma}\Big[\lambda(B(x_0)\xi_0, \xi_0) + |\xi_0^{d+1}|^2\Big] - $$

$$- \varphi(x_0, \tilde{\xi}_0) - \varepsilon\psi_1(x_0) \leqq (2\gamma - 1)\varphi(x_0, \tilde{\xi}_0).$$

It follows from (28) and (29) that

$$\varphi^{1/\gamma} \leqq (\lambda + 1)(|\xi_0^{d+1}|^2 + 1),$$

$$K \geqq 2\gamma|\xi_0^{d+1}|\Big[(\lambda + 1)(|\xi_0^{d+1}|^2 + 1)\Big]^{\gamma-1}, \quad |\xi_0^{d+1}| \leqq N(K, \delta).$$

Therefore,

$$v(x, \tilde{\xi}) \leqq v(x_0, \tilde{\xi}_0) \leqq (2\gamma - 1)(\lambda + 1)^{\gamma}\Big[|\xi_0^{d+1}|^2 + 1\Big]^{\gamma} \leqq N, \tag{30}$$

on U', where $N = N(K, \delta)$. If we assume that $N \geqq 0$, then inequality (30) holds in the first case as well.

Let us see what happens in the second case. Let $(x_n, \tilde{\xi}_n)$ be an arbitrary sequence of points of U' which maximizes v on U'. Note that $\partial(\tilde{\xi})u \leqq M(B\xi, \xi)^{1/2} + K|\xi^{d+1}|$ and $\gamma > 1/2$. Therefore, it follows from the condition $v(x_n, \tilde{\xi}_n) \nrightarrow -\infty$ that $(B(x_n)\xi_n, \xi_n)$, $|\xi_n^{d+1}|$, and $\partial(\tilde{\xi}_n) \times u(x_n)$ are bounded. Moreover, $v(x_n, \tilde{\xi}_n) \geqq 0$ and $\partial(\tilde{\xi}_n)u(x_n) \geqq \varepsilon\psi_1(x_n)$ for sufficiently large n, $\psi_1(x_n)$ is bounded, and $x_n \in U_m$ for an m and all n. By hypothesis, $(B(x)\xi, \xi) > 0$ for $x \in \bar{U}_m$, $\xi \perp \psi_x(x)$, $\xi \neq 0$. Since $\bar{U}_m$ is compact, we have $(B(x)\xi, \xi) \geqq \delta_m|\xi|^2$ for $x \in U_m$, $\xi \perp \psi_x(x)$, and a strictly positive constant δ_m, while the boundedness of $(B(x_n)\xi_n, \xi_n)$ implies the boundedness of $|\xi_n|$. Now without loss of generality we can assume that the sequence $(x_n, \tilde{\xi}_n)$ has a limit, say, $(x_0, \tilde{\xi}_0)$. It is clear that $(x_0, \tilde{\xi}_0) \notin U'$ (the second case), $\psi_{(\xi_0)}(x_0) = 0$, $x_0 \notin U$,

$x_0 \in \bar{U}_m \setminus U$, and on U'

$$v(x, \tilde{\xi}) \leq \lim_{n\to\infty} v(x_n, \tilde{\xi}_n) \leq I(K_1) := \sup_{t\geq 0}[K_1 t - (t^2 + 1)^{\gamma}]. \quad (31)$$

In the derivation of the second inequality, $\lambda(B(x_0)\xi_0, \xi_0) + |\xi_0^{d+1}|^2$ is interpreted as t^2.

Our consideration of the three cases imply that either (30) or (31) holds at every point $(x, \tilde{\xi}) \in U'$, i.e., $v \leq N \vee I(K_1)$. Since ε is arbitrary, we obtain $\partial(\tilde{\xi})u - \varphi \leq N \vee I(K_1)$ on U'. The latter inequality can be rewritten equivalently as $Pt - (t^2 + 1)^{\gamma} \leq N \vee I(K_1)$, where $t \geq 0$, $t^2 = \lambda(B\xi, \xi) + |\xi^{d+1}|^2$, $P = P(x, \tilde{\xi}) = |\partial(\tilde{\xi})u|t^{-1}$. Of course, it follows that $I(P) \leq N \vee I(K_1)$. To complete the proof of inequality (27) which is equivalent to the inequality $P \leq K_1 \vee N_1^{1/2}$, it remains to note that the function $I(K_1)$ is, by the strict convexity of $(t^2 + 1)^{\gamma}$, strictly increasing in K_1 with $K_1 > K_0$, where K_0 is found from the condition that the straight line $K_0 t$ is tangent to $(t^2 + 1)^{\gamma}$. The theorem is proved.

We now come to estimating the second derivatives. Here we establish a rougher estimate than (27).

9. THEOREM. Let ψ, $u \in C^4_{loc}(V)$, let u satisfy equation (25) on U, let the conditions of Lemma 3 hold (cf. Remarks 6 and 7), let $|u| \leq K$, let $|u_{(\xi)}|^2$, $\tilde{\partial}^2(\xi)u \leq M(B\xi, \xi)$ on U' for a constant M, and let $(B\xi, \xi) > 0$ for $\xi \neq 0$ on U'. Let $\psi_1 \in C^2_{loc}(V)$, $\psi_1 \geq 0$, and $L(\omega)\psi_1 \leq 0$ on U for all $\omega \in \Omega$. We take U_m from (26) and suppose that U_m is bounded, u, $\tilde{\partial}_i u$, $\tilde{\partial}_i\tilde{\partial}_j u$, and B for every m can be extended to functions which are continuous in the relative topology on $\bar{U}_m$, and for every $m \geq 1$,

$$|\xi^i\tilde{\partial}_i u(x)|^2, \quad \xi^i\xi^j\tilde{\partial}_i\tilde{\partial}_j u(x) \leq K(B(x)\xi, \xi), \quad (B(x)\xi, \xi) > 0,$$

whenever $x \in \bar{U}_m \setminus U$, $\xi \perp \psi_x(x)$, and $\xi \neq 0$. Then

$$|u_{(\xi)}|^2, \quad \tilde{\partial}^2(\xi)u(x) \leq N(K, \delta)(B(x)\xi, \xi) \quad (32)$$

on U'.

<u>Proof</u>. First, let us note that (from the parenthetical remark on p. 391) if the conditions of Lemma 3 hold, then so do the conditions of Lemma 1 and, therefore, the first estimate in (32) follows from the previous theorem. We further fix $\varepsilon > 0$, take the function φ from Lemma 3, and consider the function

$$v(x, \tilde{\xi}, \tilde{\eta}) := \tilde{\partial}(\tilde{\xi}, \tilde{\eta})u - \varphi(x, \tilde{\xi}, \tilde{\eta}) - \varepsilon\psi_1(x)$$

on U".

Again, as above, only one of the three cases can occur, and we start by studying the third case. Let us show that

$$(B(x_0)\xi_0, \xi_0)^2 < (\hat{B}(x_0)\eta_0, \eta_0) + |\xi_0^{d+1}|^4 + |\eta_0^{d+1}|^2 + 1. \quad (33)$$

Indeed, otherwise the point $(x_0, \tilde{\xi}_0, \tilde{\eta}_0)$ would lie in the set (15) and, by Lemma 4.6, Theorem 4.8, and Lemma 3,

$$0 \geqq \tilde{L}(\omega)v - c(\omega)v, \quad 0 \geqq \tilde{L}(\omega)v,$$

$$\varepsilon L(\omega)\psi_1 + \tilde{L}(\omega)\varphi \geqq \tilde{L}(\omega)\tilde{\partial}(\tilde{\xi}, \tilde{\eta})u = \tilde{\partial}(\tilde{\xi}, \tilde{\eta})L(\omega)u,$$

$$\varepsilon L(\omega)\psi_1 > \tilde{\partial}(\tilde{\xi}, \tilde{\eta})[L(\omega)u + f(\omega)] \quad (34)$$

at this point for any $\omega \in \Omega$. If $\omega = \omega_0 = \omega_0(u_{x^ix^j}(x_0), u_{x^i}(x_0), u(x_0), x_0)$, then inequality (34) is impossible by Lemma 4.2 since clearly $\tilde{\partial}(\tilde{\xi}, \tilde{\eta})(\psi^k) = 0$ on U for all $k = 1, 2, \ldots$, $\tilde{\xi}, \tilde{\eta} \in E_{d+1}$ and, in addition, the function $L(\omega_0)u(x) + f(\omega_0, x)$ attains its minimum value on U, which is equal to zero, at x_0.

On proving (33), we further note that $v(x_0, \tilde{\xi}_0, \tilde{\eta}_0)$ is the greatest value of $v(x_0, \tilde{\xi}, \tilde{\eta})$ not only over $(x_0, \tilde{\xi}, \tilde{\eta}) \in U''$, but also over all $\tilde{\xi}, \tilde{\eta}$ such that only the condition $\xi \perp \psi_x(x_0)$ is taken into account. This follows from the obvious equality

$$v(x, \xi, \xi^{d+1}, \eta, \eta^{d+1}) = v(x, \xi, \xi^{d+1}, H(x)\eta, \eta^{d+1}),$$

$$x \in U \quad (35)$$

(H was introduced in (5)) and the fact that for any $\xi \perp \psi_x(x)$ and $\eta \in E_d$ the vector $\eta_1 = H(x)\eta - \psi_{(\xi)(\xi)}(x)|\psi_x|^{-2}\psi_x(x)$ is such that $(x, \xi, \xi^{d+1}, \eta_1, \eta^{d+1}) \in U''$ and $H\eta_1 = H\eta$. Hence using differentiation of v with respect to $\tilde{\xi}, \tilde{\eta}$ at the maximum point for any $\zeta \perp \psi_x(x_0)$, $\zeta_1 \in E_d$, we obtain

$$u(x_0) = 2\gamma\varphi(x_0, \tilde{\xi}_0, \tilde{\eta}_0)^{1-1/\gamma}\eta_0^{d+1} =: \rho\eta_0^{d+1},$$

$$2\xi_0^{d+1}u(x_0) + 2\tilde{\partial}(\xi_0)u(x_0) = \rho 2\lambda(\xi_0^{d+1})^3, \quad (36)$$

$$\tilde{\partial}(\zeta_1)u(x_0) = \rho\lambda(\hat{B}(x_0)\eta_0, \zeta_1),$$

$$(\zeta^i\xi_0^j + \xi_0^i\zeta^j)\tilde{\partial}_i\tilde{\partial}_j u(x_0) + 2\xi_0^{d+1}\tilde{\partial}(\zeta)u(x_0) \tag{36}$$

$$= \rho 2\lambda^2(B(x_0)\xi_0, \xi_0)(B(x_0)\xi_0, \zeta).$$

On substituting $\zeta_1 = \eta_0$ and $\zeta = \xi_0$ into the third and the fourth inequality respectively, we obtain

$$\tilde{\partial}(\tilde{\xi}_0, \tilde{\eta}_0)u(x_0) = \rho(\varphi^{1/\gamma} - 1) \leqq 2\gamma\varphi(x_0, \tilde{\xi}_0, \tilde{\eta}_0),$$

$$v(x_0, \tilde{\xi}_0, \tilde{\eta}_0) \leqq (2\gamma - 1)\varphi(x_0, \tilde{\xi}_0, \tilde{\eta}_0). \tag{37}$$

Moreover, as mentioned above, the estimate of $\tilde{\partial}_i u$ had already been obtained; it therefore follows from (33) and (36) that

$$|\eta_0^{d+1}| \leqq N\rho^{-1}, \quad (\hat{B}(x_0)\eta_0, \eta_0) \leqq N\rho^{-1}|\tilde{\partial}(\eta_0)u(x_0)|$$

$$\leqq N\rho^{-1}(\hat{B}(x_0)\eta_0, \eta_0)^{1/2}, \quad |\tilde{\partial}(\xi_0)u(x_0)|$$

$$\leqq N(B(x_0)\xi_0, \xi_0)^{1/2}$$

$$\leqq N(\rho^{-1/2} + |\xi_0^{d+1}| + 1), \quad \rho|\xi_0^{d+1}|^3$$

$$\leqq N(|\xi_0^{d+1}| + \rho^{-1/2} + 1), \quad (\rho^{1/2}|\xi_0^{d+1}|)^3$$

$$\leqq N(\rho^{1/2}|\xi_0^{d+1}| + 1), \quad |\xi_0^{d+1}| \leqq N\rho^{-1/2},$$

$$\varphi^{1/\gamma} \leqq N(\rho^{-2} + 1), \quad (2\gamma)^2\varphi^{2-1/\gamma} = \rho^2\varphi^{1/\gamma} \leqq N,$$

$$2 - \frac{1}{\gamma} = \delta, \quad \varphi \leqq N,$$

where $N = N(K, \delta)$, $\varphi = \varphi(x_0, \tilde{\xi}_0, \tilde{\eta}_0)$. (Note that $1 - \gamma^{-1} < 0$, $\varphi \geqq 1$, therefore $1 + \sqrt{\rho} \leqq 1 + \sqrt{2\gamma}$.) This together with (37) yields

$$\tilde{\partial}(\tilde{\xi}, \tilde{\eta})u(x) \leqq \varphi(x, \tilde{\xi}, \tilde{\eta}) + \varepsilon\psi_1(x) + N$$

for all $(x, \tilde{\xi}) \in U'$ and $\tilde{\eta} \in E_{d+1}$. This inequality is also clear in the first case and it is easy to prove it in the second case using (35) as in the previous proof (cf. (31)). Letting $\varepsilon \downarrow 0$ in this inequality and putting $\tilde{\eta} = 0$, $\xi^{d+1} = 0$, and $(B\xi, \xi) = 1$, we can see that the second inequality in (32) holds whenever $(B\xi, \xi) = 1$. In view of homogeneity, it also holds for all values of $(B\xi, \xi)$. The theorem is proved.

Theorems 8 and 9 on the estimates of the derivatives of a solution to equation (25) on the manifold will be used in forthcoming sections. Theorem 8 will be used in Section 6, Theorem 9 in Section 8. In addition, the material of Section 7 will call for some modifications of Theorems 8 and 9 where Remarks 5 and 7 are used for checking the conditions of Lemmas 1 and 3 and the set U is an open subset of E_d. The case when U is a domain of E_d can quite formally be considered using the scheme of this section. Indeed, it suffices to consider E_d as a hyperplane $\{\psi(x^1, \ldots, x^{d+1}) = 0\}$ in E_{d+1}, where $\psi := x^{d+1}$, and to take $V = U \times (-\infty, \infty)$. Note that this interpretation makes checking conditions (9) in Lemma 3 superfluous since the left-hand sides of these inequalities are equal to zero. Therefore, we can free Remark 7 of the requirement that b (see (24)) and the second derivatives of $\tilde{u}^{1/2}$ are bounded. Also, note that if $\tilde{u}$ were constant, then for $B = \tilde{u}^{1/2}(\delta^{ij})$ the left-hand side of (14) would be zero and we would need neither the boundedness of tr a, nor the inequality $M \geqq \delta_1$ (see (24)).

We have discussed Remark 7. The case of Remark 5 is much simpler so, summing up the above arguments, we can obtain the following results from Theorems 8 and 9 and Remarks 5 and 7.

10. THEOREM. Let D be a domain of E_d and for all $\omega \in \Omega$, $i = 1, \ldots, d$, $k = 1, \ldots, d_1$, let real-valued functions σ^{ik}, b^i, c, and f be defined in D, which are continuously differentiable with respect to x and bounded in ω for every x. Let us adopt notations (7.2.1) — (7.2.3) and suppose that for any $x \in D$, u_{ij}, u_i, and u the infimum in (7.2.3) is attained at an ω after replacing z by x. Also, let functions $\tilde{u}$, $\psi_1 \in C^2_{loc}(D)$, $\tilde{u} \in C(\bar{D})$ be defined. Suppose that for all $\omega \in \Omega$, $x \in D$, $\xi \in E_d$ $\xi \neq 0$ inequalities (19) and (20) hold and, moreover, $K \geqq \tilde{u} \geqq \delta_1$, $c \leqq 0$, $\psi_1 \geqq 0$, $L\psi_1 \leqq 0$, where $\delta_1 \in (0, 1)$ is a constant. Finally, suppose that $u \in C^1(\bar{D}) \cap C^3_{loc}(D)$, u satisfies equation (25) on D, $|u| \leqq K$ in D, every set $D_m := D \cap \{\psi_1(x) < m\}$ is bounded, $|u_x| \leqq K$ in $\bar{D}_m \setminus D$.

Then $|u_x| \leqq N(K, \delta, \delta_1)$ in D.

11. THEOREM. Let the assumptions of Theorem 10 hold true, and let σ, b, c, and f be twice continuously differentiable on D. Let us take the function $M(\omega, x, \xi)$ from Remark 7 and require that for a nonnegative function $\chi(\omega, x)$ and all $\omega \in \Omega$, $x \in D$, and unit vectors $\xi \in E_d$ in-

equality (23) hold and, moreover,

$$\chi > 0, \quad \chi\tilde{u} \leq KM, \quad |f| \leq K\chi, \tag{38}$$

$$\chi(a\tilde{u}_x, \tilde{u}_x) + \chi(\sigma^k_{(\xi)(\xi)}, \sigma^k_{(\xi)(\xi)}) + |b_{(\xi)(\xi)}|^2 + + \chi|c_{(\xi)(\xi)}| + \chi(f_{(\xi)(\xi)})_+ + |c_{(\xi)}|^2 + + |f_{(\xi)}|^2 \leq K\chi M. \tag{39}$$

(It is clear that if the inequalities (38), (39), $\tilde{u} \geq \delta_1$, and (23) hold, then (19) and (20) also hold with a different constant K.) Finally, suppose that $u \in C^2(\bar{D}) \cap C^4_{loc}(D)$ and $u_{(\xi)(\xi)} \leq K$ on $\bar{D}_m \setminus D$ for all unit vectors $\xi \in E_d$, $m \geq 1$.

Then $|u_x|$, $u_{(\xi)(\xi)} \leq N(K, \delta, \delta_1)$ on D for all unit vectors $\xi \in E_d$.

12. <u>Remark</u>. We obtain a typical example of the application of Theorem 11 in the case when D is bounded and $\psi_1 \equiv 0$. Then $|u_x|$ and $u_{(\xi)(\xi)}$ in Theorem 11 are required to be *a priori* bounded on ∂D. Sometimes it is not necessary that they should be bounded on the whole of ∂D. For example, if the operators L are of the form of the sum $\partial/\partial x^1$ and an operator which is free of the derivatives with respect to x^1, then we deal with a parabolic equation. It is known from the theory of parabolic equations that if $D = (0, 1) \times \{|x^i| < 1, i \geq 2\}$ we should not require anything on $\partial D \cap \{x^1 = 0\}$. Using the function $\psi_1 = (x^1)^{-1}$, we can treat this case in the framework of the general scheme.

13. <u>Remark</u>. The conditions concerning inequalities (23), (38), and (39) in Theorem 11 can be replaced by the requirement that

$$(1 + \delta)\sum_{k=1}^{d_1} |\sigma^k_{(\xi)}|^2 + 2(b_{(\xi)}, \xi) + 2\sum_{k=1}^{d_1} (\xi, \sigma^k_{(\xi)})^2 < -(1 - \delta)c + K(a\xi, \xi), \tag{40}$$

$$\sum_k |\sigma^k_{(\xi)(\xi)}|^2 + |c_{(\xi)(\xi)}| + (f_{(\xi)(\xi)})_+ \leq K(|c| + (a\xi, \xi)), \tag{41}$$

$$|b_{(\xi)(\xi)}|^2 + |c_{(\xi)}|^2 + |f_{(\xi)}|^2 \leq K|c|(|c| + (a\xi, \xi)),$$

$$|f| \leq -Kc. \tag{42}$$

for all $\omega \in \Omega$, $x \in D$, and unit vectors $\xi \in E_d$.

This remark follows readily from Theorem 9, if we reduce the argument of Remark 6 to the case when B is a unit matrix, $\delta_2 = \delta$, $\chi = -c$, and U is a domain of E_d. Also, note that conditions (40) — (42) are weaker than (7.2.15) and (7.2.16) particularly if $(a\xi, \xi) \geqq \delta|\xi|^2$ for all ξ. Moreover, the inequality in (40) must be strict as can easily be seen when σ, b, c, f are zero.

8.6. The Estimates of the Second Mixed Derivatives on the Boundary of a Domain.

Suppose that γ, κ, $\delta \in (0, 1)$, $K \geqq 1$, d and d_1 are integers greater than or equal to 1, $\psi \in C^2_{loc}(E_d)$, $D := \{x \in E_d : \psi(x) > 0\}$ is a non-empty bounded domain, $\psi \in C^3_{loc}(D)$, $|\psi_x| > 0$ on ∂D, Ω is a set, and for $\omega \in \Omega$, $i = 1, \ldots, d$, $k = 1, \ldots, d_1$, $x \in D$ real-valued functions $\sigma^{ik}(\omega, x)$, $b^i(\omega, x)$, $c(\omega, x)$, and $f(\omega, x)$ are defined and bounded in ω for every $x \in D$. We adopt the notation of (7.2.1) — (7.2.3) and assume that for all $x \in D$, u_{ij}, u_i, and u the infimum in (7.2.3) is attained (of course z in (7.2.1) — (7.2.3) is replaced by x).

We will also need a matrix function $B(x) = (B^{ij}(x); i, j = 1, \ldots, d)$ such that $B = B^* \geqq 0$, $B \in C^2_{loc}(D)$. Denote

$$\Delta_\kappa D = \{x : 0 < \psi(x) < \kappa\}.$$

The main results of this section arise in connection with the following theorem.

1. THEOREM. Let $u \in C^2(\Delta_\kappa D) \cap C^3_{loc}(\Delta_\kappa D) \cap C(\overline{\Delta_\kappa D})$, let $|u| \leqq K\psi$ in $\Delta_\kappa D$, and let

$$F\left(u_{x^i x^j}, u_{x^i}, u, x\right) = 0 \tag{1}$$

in $\Delta_\kappa D$. Suppose that the following conditions hold:

(a) σ, b, c, and f are continuously differentiable with respect to x in $\Delta_\kappa D$;

(b) for all $x \in \Delta_\kappa D$, $\xi \in E_d$, and $\omega \in \Omega$

$$0 < |\psi_x| \leqq K, \quad |B\xi| \leqq K|\xi|, \quad \bar{c} := L\psi < 0, \tag{2}$$

$$\operatorname{tr} a + |b| + |c| + |f| \leqq K|\bar{c}|, \tag{3}$$

$$(\sigma^k_{(\xi)}, \sigma^k_{(\xi)}) + \psi^\gamma |L^0(B\xi, \xi)|$$

$$\leq K|\bar{c}|[(B\xi, \xi) + |\psi_{(\xi)}|^2], \quad (L^0 = L - c) \tag{4}$$

$$|(B\xi, \xi)_x| \leq K[(B\xi, \xi) + |\psi_{(\xi)}|^2], \tag{5}$$

$$|b_{(\xi)}| + |c_{(\xi)}| + |f_{(\xi)}| + |\bar{c}_{(\xi)}|$$

$$\leq K|\bar{c}|[(B\xi, \xi) + |\psi_{(\xi)}|^2]^{1/2}, \tag{6}$$

$$2 \operatorname{tr} aB + L^0\psi \leq 2(1 - \delta)|\bar{c}| + K(a\psi_x, \psi_x), \tag{7}$$

$$\frac{1}{2}\sum_k |(\sigma^k, \psi_x)_{(\xi)}|^2 + 2(B\xi, (a\psi_x)_{(\xi)}) + (a\psi_x, (B\xi, \xi)_x)$$

$$\leq (1 - \delta)|\bar{c}|[(B\xi, \xi) + K|\psi_{(\xi)}|^2]; \tag{8}$$

(c) in $\Delta_\kappa D$, the matrix $B + \frac{1}{4}\psi^{-1}\psi_x\psi_x^*$ has the inverse C which is bounded in $\Delta_\kappa D$.

Then there exist constants $\rho_0 \in (0, \kappa)$, N_1, $N_2 \geq 0$ depending only on γ, κ, δ, K such that for $\rho \geq \rho_0$

$$\left(C\left(\frac{u}{\psi}\right)_x, \left(\frac{u}{\psi}\right)_x\right) + N_1\left(\frac{u}{\psi}\right)^2$$

$$\leq \max_{x:\psi(x)=\rho}\left[\left(C\left(\frac{u}{\psi}\right)_x, \left(\frac{u}{\psi}\right)_x\right) + N_1\left(\frac{u}{\psi}\right)^2\right] \vee N_2 \tag{9}$$

in $\Delta_\rho D$.

Before going on to the proof of this theorem we discuss its assertions and hypotheses.

2. Remark. The multiplier ψ^γ of the second summand on the left of (4) is introduced in order that we can consider functions B whose second derivatives tend to infinity as $x \to \partial D$.

3. Remark. Let all conditions of Theorem 1 except inequality (8) hold and let this inequality be replaced by

$$\frac{1}{2}(1 + \delta)\sum_k |(\sigma^k, \psi_x)_{(\xi)}|^2 + 2(B\xi, (a\psi_x)_{(\xi)})$$

$$\leq (1 - \delta)|\bar{c}|[(B\xi, \xi) + K|\psi_{(\xi)}|^2]. \tag{10}$$

Then the asserion of Theorem 1 remains valid; we need only replace C by the inverse of the matrix

$$(1 - N_3\psi)B + \frac{1}{4}\psi^{-1}\psi_x\psi_x^*,$$

where $N_3 = N_3(\delta, K)$ and ρ_0 is chosen so that $1 - N_3\rho_0 \geqq 1/2$.

In order to verify this, it suffices to show that if B in (2) — (8) is replaced by $(1 - N_3\psi)B$, then these conditions with the appropriate $N_3(\delta, K) > 0$ hold on $\Delta_{\kappa_1}D$ with a constant which replaces K and depends only on κ, δ, K just as κ_1 does, and with δ replaced by $\frac{1}{2}\delta$.

It follows from (2), (3), and (5) that for $\kappa_1 \leqq \kappa \wedge N_3^{-1}$, $N_3 = \delta^{-1}K^3$

$$\begin{aligned}
&(a\psi_x, [(1 - N_3\psi)(B\xi, \xi)]_x) \\
&= (1 - N_3\psi)(a\psi_x, (B\xi, \xi)_x) - N_3(a\psi_x, \psi_x)(B\xi, \xi) \\
&\leqq (a\psi_x, \psi_x)^{1/2}K^{3/2}|\bar{c}|^{1/2}[(B\xi, \xi) + |\psi_{(\xi)}|^2] - \\
&\quad - N_3(a\psi_x, \psi_x)(B\xi, \xi) \\
&\leqq \frac{\delta}{2}|\bar{c}|[(B\xi, \xi) + |\psi_{(\xi)}|^2] + \delta^{-1}K^3(a\psi_x, \psi_x)[(B\xi, \xi) + \\
&\quad + |\psi_{(\xi)}|^2] - N_3(a\psi_x, \psi_x)(B\xi, \xi) \leqq \frac{\delta}{2}|\bar{c}|[(B\xi, \xi) + \\
&\quad + (1 + 2\delta^{-2}K^6)|\psi_{(\xi)}|^2]
\end{aligned}$$

in $\Delta_{\kappa_1}D$.

This and (10) with κ_1 satisfying the inequality $(1 - N_3\kappa_1)^{-1} \leqq 1 + \delta$ imply that in $\Delta_{\kappa_1}D$ the left-hand side of (8) with $(1 - N_3\psi)B$ in place of B does not exceed

$$\begin{aligned}
&(1 - N_3\psi)(1 - \delta)|\bar{c}|[(B\xi, \xi) + K|\psi_{(\xi)}|^2] + \\
&\quad + \frac{\delta}{2}|\bar{c}|[(B\xi, \xi) + N_4|\psi_{(\xi)}|^2] \\
&\leqq \left(1 - \frac{\delta}{2}\right)|\bar{c}|[(B\xi, \xi) + N_5|\psi_{(\xi)}|^2].
\end{aligned}$$

Furthermore, if κ_1 satisfies the inequality $(1 - \delta/2)(1 - N_3\kappa_1)^{-1} \leqq$ $\leqq 1 - \delta/4$ then in $\Delta_{\kappa_1}D$ the latter is bounded above by

$$\left(1 - \frac{\delta}{4}\right)|\bar{c}|\,[(1 - N_3\psi)(B\xi, \xi) + N_5|\psi_{(\xi)}|^2].$$

On fixing $N_3 = \delta^{-1}K^3$ and the greatest κ_1 such that $\kappa_1 \leqq \kappa \wedge N_3^{-1}$, $(1 - N_3\kappa_1)^{-1} \leqq 1 + \delta$, $(1 - \delta/2)(1 - N_3\kappa_1)^{-1} \leqq 1 - \delta/4$, we have thus checked condition (8) in $\Delta_{\kappa_1}D$ for $(1 - N_3\psi)(B\xi, \xi)$ with $\delta/4$ instead of δ and $N_5 = N_5(\delta, K)$ instead of K. The reader can easily verify the possibility of such replacement in the rest of condition (2) — (7) if, on considering (4), it is noted that

$$L^0(1 - N_3\psi)(B\xi, \xi)$$

$$= (1 - N_3\psi)L^0(B\xi, \xi) - 2N_3(a\psi_x, (B\xi, \xi)_x) -$$

$$- N_3(B\xi, \xi)(\bar{c} - c\psi).$$

We now go on to discuss the conditions of the theorem. Inequalities (7) and (8) are by far the most rigid conditions. Therefore, note that sometimes it is possible to furnish (7) by increasing the constant K. It turns out that the 'assisting' expression $(a\psi_x, \psi_x)$ can be introduced into other inequalities as well.

4. THEOREM. Let $K_1 > 0$ be a constant, let all conditions of Theorem 1 hold except for condition (b), and in place of that for all $x \in \Delta_K D$, $\xi \in E_d$, $\omega \in \Omega$, let

$$0 < |\psi_x| \leqq K, \quad |B\xi| \leqq K|\xi|, \quad \tilde{c} := L\psi - 2K_1(a\psi_x, \psi_x) < 0, \tag{11}$$

$$\operatorname{tr} a + |b| + |c| + |f| \leqq K|\tilde{c}|, \tag{12}$$

$$(\sigma^k_{(\xi)}, \sigma^k_{(\xi)}) + \psi^\gamma|L^0(B\xi, \xi)| \leqq K|\tilde{c}|\,[(B\xi, \xi) + |\psi_{(\xi)}|^2], \tag{13}$$

$$|(B\xi, \xi)_x| \leqq K[(B\xi, \xi) + |\psi_{(\xi)}|^2], \tag{14}$$

$$|b_{(\xi)}| + |c_{(\xi)}| + |f_{(\xi)}| + |(L(\psi - K_1\psi^2))_{(\xi)}|$$
$$\leqq K|\tilde{c}|\,[(B\xi, \xi) + |\psi_{(\xi)}|^2]^{1/2}, \tag{15}$$

$$\operatorname{tr} aB + \frac{1}{2}L^0\psi \leqq (1 - \delta)|\tilde{c}| + K(a\psi_x, \psi_x), \tag{16}$$

$$(1 + \delta)\,\frac{1}{2}\sum_k|(\sigma^k, \psi_x)_{(\xi)}|^2 + 2(B\xi, (a\psi_x)_{(\xi)})$$
$$\leqq (1 - \delta)|\tilde{c}|\,[(B\xi, \xi) + K|\psi_{(\xi)}|^2]. \tag{17}$$

Then there exist constants $\rho_0 \in (0, \kappa)$, N_1, N_2, N_3, depending only on γ, κ, δ, K, K_1, such that $1 - 2K_1\rho_0 \geqq 1/2$, $1 - N_3\rho_0 \geqq 1/2$, and for $\rho \leqq \rho_0$ inequality (9) holds in $\Delta_\rho D$, where C is replaced by the inverse matrix to $(1 - N_3\tilde{\psi})B + \frac{1}{4}\tilde{\psi}^{-1}\tilde{\psi}_x\tilde{\psi}_x^*$ where

$$\tilde{\psi} := \psi - K_1\psi^2.$$

Proof. We first find $\kappa_1 = \kappa_1(\kappa, \delta, K, K_1)$ such that the function $\tilde{\psi}$ replacing ψ satisfies inequalities (2) — (7) and (10) in $\Delta_{\kappa_1} D$ with $\delta/4$ instead of δ and $N(\delta, K, K)$ instead of K. To this end, let us note that

$$L\tilde{\psi} = \tilde{c} - 2K_1\psi L\psi + K_1\psi^2 c = \tilde{c}(1 - 2K_1\psi) - 4K_1^2\psi(a\psi_x, \psi_x) + K_1\psi^2 c,$$

$$\tilde{\psi}_{(\xi)} = (1 - 2K_1\psi)\psi_{(\xi)}, \quad (a\tilde{\psi}_x, \tilde{\psi}_x) = (1 - 2K_1\psi)^2(a\psi_x, \psi_x).$$

This and (12) show that for every $\varepsilon \in (1/2, 1)$ we can take $\kappa_1 = \kappa_1(\kappa, \delta, K, K_1, \varepsilon) \in (0, \kappa]$ such that $1 - 2K_1\kappa_1 \geqq 3/4$ and the inequalities

$$L\tilde{\psi} \leqq \varepsilon\tilde{c}, \quad |\psi_{(\xi)}| \leqq \frac{4}{3}|\tilde{\psi}_{(\xi)}|, \quad (a\psi_x, \psi_x) \leqq 2(a\tilde{\psi}_x, \tilde{\psi}_x) \tag{18}$$

hold in $\Delta_{\kappa_1} D$.

It follows that inequalities (12) — (15) in $\Delta_{\kappa_1} D$ remain in force if $\tilde{c}$ and K in them are replaced by $L\tilde{\psi}$ and $2K$, respectively, and $L\tilde{\psi} < 0$. Specifically, inequalities (2) — (6) hold in $\Delta_{\kappa_1} D$ for the function $\tilde{\psi}$ instead of ψ and the constant $4K$ instead of K. Moreover, it follows from (12) and (18) that

$$\begin{aligned} L^0\tilde{\psi} &= L^0\psi + L\tilde{\psi} - \tilde{c} + K_1\psi^2 c - 2K_1(a\psi_x, \psi_x) \\ &\leqq L^0\psi + (1 - \varepsilon)|\tilde{c}| + KK_1\psi^2|\tilde{c}| \\ &\leqq L^0\psi + (1 - \varepsilon + KK_1\kappa_1^2)|\tilde{c}|. \end{aligned}$$

Therefore, taking (16) and (18) into account, we obtain

$$\operatorname{tr} aB + \frac{1}{2} L^0\tilde{\psi} \leqq (1 - \delta + 1 - \varepsilon + KK_1\kappa_1^2)\varepsilon^{-1}|L\tilde{\psi}| + 2K(a\tilde{\psi}_x, \tilde{\psi}_x). \tag{19}$$

We next come to inequality (10). It follows from (11) and (12) that

$$\sum_k (\sigma^k, \psi_x)^2 = 2(a\psi_x, \psi_x) \leqq 2 \operatorname{tr} a|\psi_x|^2 \leqq 2K^3|\tilde{c}|,$$

$$\sum_k |(\sigma^k, \tilde{\psi}_x)_{(\xi)}|^2 \leq (1 + \frac{\delta}{4}) \sum_k |(\sigma^k, \psi_x)_{(\xi)}|^2 + N \sum_k |(\sigma^k, \psi\psi_x)_{(\xi)}|^2$$

$$\leq \left(1 + \frac{\delta}{4} + N\psi^2\right) \sum_k |(\sigma^k, \psi_x)_{(\xi)}|^2 + N|\tilde{c}| \cdot |\psi_{(\xi)}|^2,$$

where (and hereafter) N denotes the constants depending only on K, K_1, and δ. Also, note that, by (11) and (12), we have $|(B\xi, a\psi_x)| \leq |a\psi_x| \cdot |B\xi| \leq K^{5/2} \cdot |\tilde{c}| (B\xi, \xi)^{1/2}$.

Therefore,

$$(B\xi, (a\tilde{\psi}_x)_{(\xi)}) = (1 - 2K_1\psi)(B\xi, (a\psi_x)_{(\xi)}) - 2K_1\psi_{(\xi)}(B\xi, a\psi_x)$$

$$\leq (1 - 2K_1\psi)(B\xi, (a\psi_x)_{(\xi)}) + \frac{\delta}{4}|\tilde{c}|(B\xi, \xi) + N|\tilde{c}| \cdot |\psi_{(\xi)}|^2,$$

$$\frac{1}{2}\left(1 + \frac{\delta}{4}\right) \sum_k |(\sigma^k, \tilde{\psi}_x)_{(\xi)}|^2 + 2(B\xi, (a\tilde{\psi}_x)_{(\xi)})$$

$$\leq (1 - 2K_1\psi)\left[\frac{1}{2}\left(1 + \frac{\delta}{4}\right)\left(1 + \frac{\delta}{4} + N\psi^2\right)(1 - 2K_1\psi)^{-1} \times \right.$$

$$\left. \times \sum_k |\sigma^k, \psi_x)_{(\xi)}|^2 + 2(B\xi, (a\psi_x)_{(\xi)})\right] +$$

$$+ \frac{\delta}{2}|\tilde{c}|(B\xi, \xi) + N|\tilde{c}| \cdot |\psi_{(\xi)}|^2$$

$$\leq \frac{1}{2}(1 - 2K_1\psi)\left[\left(1 + \frac{\delta}{4}\right)\left(1 + \frac{\delta}{4} + N_1\psi^2\right)(1 - 2K_1\psi)^{-1} - \right.$$

$$\left. - (1 + \delta)\right] \sum_k |(\sigma^k, \psi_x)_{(\xi)}|^2 +$$

$$+ \left(1 - \frac{1}{2}\delta\right)|\tilde{c}|[(B\xi, \xi) + N|\psi_{(\xi)}|^2]$$

in $\Delta_{\kappa_1} D$.

Without loss of generality, we can assume that κ_1 satisfies the inequality

$$(1 + \delta/4)(1 + \delta/4 + N_1\kappa_1^2)(1 - 2K_1\kappa_1)^{-1} - (1 + \delta) < 0,$$

but then it follows from (18) that

$$\frac{1}{2}(1 + \delta/4) \sum_k |(\sigma^k, \tilde{\psi}_x)_{(\xi)}|^2 + 2(B\xi, (a\tilde{\psi}_x)_{(\xi)})$$

$$\leq \varepsilon^{-1}(1 - \delta/2)|L\tilde{\psi}|[(B\xi, \xi) + N|\tilde{\psi}_x|^2] \qquad (20)$$

in $\Delta_{\kappa_1}D$. Finally, (19) and (20) show how we should choose ε and κ_1 to obtain

$$\mathrm{tr}\, aB + \frac{1}{2} L^0\tilde{\psi} \leqq (1 - \delta/4)|L\tilde{\psi}| + N(a\tilde{\psi}_x, \tilde{\psi}_x),$$

$$\frac{1}{2}(1 + \delta/4)\sum_k |(\sigma^k, \tilde{\psi}_x)_{(\xi)}|^2 + 2(B\xi, (a\tilde{\psi}_x)_{(\xi)})$$

$$\leqq (1 - \delta/4)|L\tilde{\psi}|[(B\xi, \xi) + N|\tilde{\psi}_{(\xi)}|^2]$$

in $\Delta_{\kappa_1}D$.

To complete the proof of the theorem, we apply Remark 3. Here we encounter only two difficulties, namely D may have a form different from $\{\tilde{\psi} > 0\}$, and $\Delta_{\kappa_1}D$ may have a form different from $\{0 < \tilde{\psi} < \kappa_2\}$. In this connection we define an infinitely differentiable function $\chi(t)$ on $(-\infty, \infty)$ such that $\chi(t) = t - K_1 t^2$ for $t \leqq (8K_1)^{-1}$, $\chi(t) = (8K_1)^{-1}$ for $t \geqq (4K_1)^{-1}$, and $\chi' \geqq 0$. It is then obvious that $D = \{\chi(\psi) > 0\}$. Moreover, since $\kappa_1 \leqq (8K_1)^{-1}$, then $\Delta_{\kappa_1}D = \{0 < \psi < \kappa_1\} = \{0 < \chi(\psi) < \chi(\kappa_1)\}$ and $\chi(\psi) = \tilde{\psi}$ on $\Delta_{\kappa_1}D$. Thus, all conditions of Remark 3 hold on the set $\{0 < \chi(\psi) < \kappa_2\}$ with $\kappa_2 = \chi(\kappa_1)$ for $\chi(\psi)$, $\frac{1}{4}\delta$, $N(\delta, K, K_1)$ instead of ψ, δ, K. The theorem is proved.

The following corollary plays a dominant role in applications of the established results.

5. COROLLARY. Let the conditions of Theorem 4 hold, let $u \in C^2(\overline{\Delta_\kappa D})$ $|u_x| \leqq K$, and let $|\psi_x| \geqq \delta$ on $\Delta_\kappa D$. Then there exists a constant N such that $|u_{(\xi)(\eta)}| \leqq N$ for any $x \in \partial D$ and unit vectors $\xi \perp \psi_x(x)$, $\eta || \psi_x(x)$. Here N depends only on γ, κ, δ, K, K_1, the norm of ψ in $C^2(\Delta_\kappa D)$, and the constant bounding above the eigenvalues of the inverse matrix to $B + \frac{1}{4}\psi^{-1}\psi_x\psi_x^*$ in $\Delta_\kappa D$.

Let us prove this fact. We make use of the assertion of Theorem 4 with $\rho = \rho_0$ denoting by $\tilde{C}$ the inverse matrix to $(1 - N_3\tilde{\psi})B + \frac{1}{4}\tilde{\psi}^{-1}\tilde{\psi}_x\psi_x^*$. By Theorem 4,

$$(\tilde{C}(u/\tilde{\psi})_x, (u/\tilde{\psi})_x) + N_1(u/\tilde{\psi})^2$$

$$\leqq \max_{x:\psi(x)=\rho_0} [(\tilde{C}(u/\tilde{\psi})_x, (u/\tilde{\psi})_x) + N_1(u/\tilde{\psi})^2] \vee N_2 \qquad (21)$$

in $\Delta_{\rho_0}D$. Since $1 - N_3\rho_0 \geqq 1/2$ and $1 - 2K_1\rho_0 \geqq 1/2$, the right-hand side of (21) is bounded above by a constant depending only on the parameters specified in the corollary. In addition, as in Theorem 3.1, it can easily be proved that the function $u\tilde{\psi}^{-1}$ can be extended to a function of the class $C^1(\overline{\Delta_{\rho_0}D})$ which, by l'Hôpital's rule, is equal to $(u_x, \psi_x)|\psi_x|^{-2}$ on ∂D. Furthermore, we take $x_0 \in \partial D$, $x_n \in D$, $x_n \to x_0$, $\xi \perp \psi_x(x_0)$, $|\xi| = 1$, $\eta = \psi_x(x_0)|\psi_x(x_0)|^{-1}$, set $\zeta_n = (u\tilde{\psi}^{-1})_x(x_n)$, and select ξ_n such that $\xi_n \to \xi$, $\xi_n \perp \psi_x(x_n)$. By the Cauchy–Buniakovski inequality and (11), for every $t \in (-\infty, \infty)$, we have

$$(\tilde{C}(x_n)\zeta_n, \zeta_n) \geqq 2t(\zeta_n, \xi_n) - t^2((\tilde{C}(x_n))^{-1}\xi_n, \xi_n)$$

$$= 2t(\zeta_n, \xi_n) - t^2(1 - N_3\tilde{\psi}(x_n))(B(x_n)\xi_n, \xi_n)$$

$$\geqq 2t(\zeta_n, \xi_n) - t^2K|\xi_n|^2.$$

Let us calculate the supremum over t in this expression. We can then see that the lower limit on the left of (21) at the points x_n is no smaller than the value of

$$K^{-1}[((u_x, \psi_x)|\psi_x|^{-2})_{(\xi)}]^2$$

$$= K^{-1}[u_{(\xi)(\eta)}|\psi_x|^{-1} + (u_x, (\psi_x|\psi_x|^{-2})_{(\xi)})]^2$$

at the point x_0. This clearly proves the corollary.

We now come to the proof of Theorem 1. The hypothesis of Theorem 1 is assumed to hold until the end of this section. We make use of the method in Section 3, which enables us to reduce an equation in a domain to an equation on a manifold, and the results established in Section 5 concerning the estimates of the first derivatives of solutions to equations on a manifold. Accordingly, as in Section 3, we introduce the auxiliary space E_{d+4}, adopt the convention on the Greek superscripts, introduce notation (1.4), (3.6), (3.7), and set $V(\rho) = \{x \in E_{d+4} : 0 < \psi(x) < \rho\}$, $U(\rho) = \{x \in E_{d+4} : 0 < [x]^2 = \psi(x) < \rho\}$, $\bar{\psi}(x) = \psi(x) - [x]^2$, $\bar{u} = u\psi^{-1}$, $\bar{L} = \bar{L}(\omega) = a^{ij}(\omega)L_{ij} + b^i(\omega)L_i + c(\omega)$, $\bar{L}^0 = \bar{L} - \bar{L}1$.

We need to represent the operator $\bar{L}$ in the form (7.1.2) and formulas (3.6) and (3.7) straightforwardly yield the following objects (cf. (1.7)):

$$\bar{\sigma}^{i\nu k} = x^\nu\sigma^{ik}, \quad \bar{\sigma}^{\mu\nu k} = \delta^{\mu\nu}\,\frac{1}{2}(\sigma^k, \psi_x),$$

$$\bar{b}^i = 2a^{ij}\psi_{x^j} + [x]^2 b^i, \quad \bar{b}^\mu = \frac{1}{2} x^\mu L^0 \psi,$$

where $i = 1, \ldots, d$, $k = 1, \ldots, d_1$, $\nu, \mu = d + 1, \ldots, d + 4$. The function $\bar{c}$ is taken from (2). Note that $\bar{L}1 = L\psi = \bar{c}$, $\bar{L}^0 = \bar{L} - \bar{c}$. The following lemma can be proved in a straightforward manner.

6. LEMMA. (a) $|\bar{\psi}_x| > 0$ on $V(\kappa)$;

(b) $L_\alpha\bar{\psi} = L_\alpha(\bar{\psi}^2) = 0$ on $U(\kappa)$ with $\alpha = i, ij$, specifically, $\bar{L}\bar{\psi} = \bar{L}(\bar{\psi}^2) = 0$ on $U(\kappa)$;

(c) if $v \in C^2_{loc}(V(\kappa))$, then

$$\bar{L}v = \frac{1}{2}\delta^{\nu\mu}\delta^{kr} v_{(\bar{\sigma}^{\nu k})(\bar{\sigma}^{\mu r})} + v_{(\bar{b})} + \bar{c}v$$

on $V(\kappa)$;

(d) if $v(x^1, \ldots, x^d) \in C^2_{loc}(\Delta_\kappa D)$, then

$$\frac{1}{2}(L_{ij} + L_{ji})v = (\psi v)_{x^i x^j}, \quad L_i v = (\psi v)_{x^i}, \quad Lv = \bar{L}(\psi v),$$

$$\bar{L}^0 v = \psi L^0 v + 2(a\psi_x, v_x)$$

on $U(\kappa)$;

(e) $F\left(\frac{1}{2}(L_{ij} + L_{ji})\bar{u},\ L_i\bar{u},\ \psi\bar{u}\right) = 0$ on $U(\kappa)$.

By using the matrix $B = (B^{ij};\ i, j = 1, \ldots, d)$, we also introduce a matrix $\bar{B} = (\bar{B}^{ij};\ i, j = 1, \ldots, d + 4)$ such that $\bar{B}^{ij} = B^{ij}$ for $i, j \leq d$, $\bar{B}^{\mu\nu} = \delta^{\mu\nu}$ for $\mu, \nu = d + 1, \ldots, d + 4$, and $\bar{B}^{ij} = 0$ for the remaining values of $i, j \in \{1, \ldots, d + 4\}$. A number of results of straightforward calculations are integrated into the following lemma, where and in what follows we denote by $B\xi$ an element of E_d with the coordinates

$$(B\xi)^i = \sum_{j=1}^{d} B^{ij}\xi^j \text{ and } (B\xi, \xi) = \sum_{i,j=1}^{d} B^{ij}\xi^i\xi^j$$

for $\xi \in E_{d+4}$ or $\xi \in E_d$.

7. LEMMA. For $\xi \in E_{d+4}$ we have on $U(\kappa)$:

$$\bar{L}^0(\bar{B}\xi, \xi) = \psi L^0(B\xi, \xi) + 2(a\psi_x, (B\xi, \xi)_x),$$

$$\left(\bar{B}_{(\bar{\sigma}^{\nu k})}\xi,\ \bar{\sigma}^{\nu k}_{(\xi)}\right) = x^\nu\xi^\nu\left(B_{(\sigma^k)}\xi,\ \sigma^k\right) + \psi\left(B_{(\sigma^k)}\xi,\ \sigma^k_{(\xi)}\right).$$

$$\left(\bar{B}\bar{\sigma}^{\nu k}_{(\xi)}, \bar{\sigma}^{\nu k}_{(\xi)}\right) = [\xi]^2(B\sigma^k, \sigma^k) + 2\xi^\nu x^\nu(B\sigma^k, \sigma^k_{(\xi)}) + $$
$$+ \psi(B\sigma^k_{(\xi)}, \sigma^k_{(\xi)}) + \sum|(\sigma^k, \psi_x)_{(\xi)}|^2,$$

$$(\bar{B}\xi, \bar{b}_{(\xi)}) = (B\xi, (2a\psi_x + [x]^2 b)_{(\xi)}) + \frac{1}{2}[\xi]^2 L^0\psi +$$
$$+ \frac{1}{2}\xi^\nu x^\nu (L^0\psi)_{(\xi)},$$

$$\sum_{k=1}^{d_1}\sum_{\nu=d+1}^{d+4}(\bar{B}\xi, \bar{\sigma}^{\nu k})^2 = \psi\sum(B\xi, \sigma^k)^2 + \xi^\nu x^\nu(B\xi, \sigma^k)(\sigma^k, \psi_x) +$$
$$+ \frac{1}{4}[\xi]^2\sum_k(\sigma^k, \psi_x)^2,$$

$$\sum_{k=1}^{d_1}\sum_{\nu=d+1}^{d+4}\left(\bar{B}_{(\bar{\sigma}^{\nu k})}\xi + 2\bar{B}\bar{\sigma}^{\nu k}_{(\xi)}, \xi\right)^2 =$$
$$= \psi\sum_k\left[\left(B_{(\sigma^k)}\xi, \xi\right) + 2(B\xi, \sigma^k_{(\xi)})\right]^2 +$$
$$+ 2x^\nu\xi^\nu\left[\left(B_{(\sigma^k)}\xi, \xi\right) + 2(B\xi, \sigma^k_{(\xi)})\right]\cdot[2(B\xi, \sigma^k) +$$
$$+ (\sigma^k, \psi_x)_{(\xi)}] + [\xi]^2\sum_k[2(B\xi, \sigma^k) + (\sigma^k, \psi_x)_{(\xi)}]^2.$$

Prior to formulating the following lemma we define one more object. For $\omega \in \Omega$, $x \in \Delta_\kappa D$, $\xi \in E_d$, and $t \geq 0$, let

$$J = J(\omega, x, \xi, t) = \psi L^0(B\xi, \xi) + 2(a\psi_x, (B\xi, \xi)_x) +$$
$$+ \psi_{(\xi)}\left(B_{(\sigma^k)}\xi, \sigma^k\right) + 2\psi\left(B_{(\sigma^k)}\xi, \sigma^k_{(\xi)}\right) + \psi_{(\xi)}(B\sigma^k, \sigma^k_{(\xi)}) +$$
$$+ \psi(B\sigma^k_{(\xi)}, \sigma^k_{(\xi)}) + \sum_k|(\sigma^k, \psi_x)_{(\xi)}|^2 + 2(B\xi, (2a\psi_x + \psi b)_{(\xi)}) +$$
$$+ \frac{1}{2}\psi_{(\xi)}(L^0\psi)_{(\xi)} + t[(B\sigma^k, \sigma^k) + L^0\psi]. \quad (22)$$

8. LEMMA. Let $\rho \in (0, \kappa)$. For $\bar{L}^0$, $\bar{B}$, $\bar{\sigma}$, $\bar{b}$, $\bar{c}$, all $\omega \in \Omega$, and $(\xi^1, \ldots, \xi^{d+4}) \neq 0$ inequality (5.17) holds on $U'(\rho)$ if and only if for all $\omega \in \Omega$, $x \in \Delta_\rho D$, $\xi \in E_d$, $t \geq (1/4)|\psi_{(\xi)}|^2\psi^{-1}$, $|\xi|^2 + t > 0$ we have

$$J < 2(1 - \delta)|\bar{c}|((B\xi, \xi) + t) + K\Big[\psi\sum(B\xi, \sigma^k)^2 +$$

$$+ \frac{1}{2}\psi_{(\xi)}(B\xi, \sigma^k)(\sigma^k, \psi_x) + \frac{1}{4}t\sum_k (\sigma^k, \psi_x)^2\Big]. \tag{23}$$

Proof. Lemma 7 demonstrates that for $\bar{L}^0$, $\bar{B}$, $\bar{\sigma}$, $\bar{b}$, $\bar{c}$ inequality (5.17) is as follows: it results from (23) if $\psi_{(\xi)}$, t, and ψb in (22) and (23) are replaced by $2\xi^\nu x^\nu$, $[\xi]^2$, and $[x]^2 b$, respectively. The fact that we are interested in inequality (5.17) only on $U'(\rho)$ specifically means that x and $\xi \in E_{d+4}$ are related by the equation $\bar{\psi}_{(\xi)}(x) = 0$, i.e., $\psi_{(\xi)}(x) = 2\xi^\nu \cdot x^\nu$. This indicates that the replacement of $2\xi^\nu \cdot x^\nu$ by $\psi_{(\xi)}$ and $[x]^2 b$ by ψb is correct. Moreover, by the Cauchy–Buniakowski inequality applied at $(x, \xi) \in U'(\rho)$, we obtain $|\psi_{(\xi)}(x)| = 2|\xi^\nu x^\nu| \leq 2[\xi]\sqrt{\psi(x)}$. Therefore, regarding t as $[\xi]^2$, we can see that if (23) holds for all $t \geq (1/4)|\psi_{(\xi)}|^2\psi^{-1}$, then (5.17) holds too. On the other hand, for every $x \in \Delta_\rho D$, $\xi \in E_d$, $t \geq (1/4)|\psi_{(\xi)}|^2\psi^{-1}$ we can choose $x^{d+1}, \ldots, x^{d+4}$, $\xi^{d+1}, \ldots, \xi^{d+4}$ in a number of ways so that $\bar{\psi}(x^1, \ldots, x^{d+4}) = 0$, $\psi(x^1, \ldots, x^d) = x^\nu x^\nu$, $\xi^i \bar{\psi}_{x^i} + \xi^\nu \bar{\psi}_{x^\nu} = 0$, $\psi_{(\xi)}(x^1, \ldots, x^d) = 2\xi^\nu x^\nu$, $\xi^\nu \xi^\nu = t$. Therefore, condition (5.17) implies (23). The lemma is proved.

9. LEMMA. There exists a constant $\rho_0 \in (0, \kappa)$ depending only on γ, κ, δ, K such that inequality (23), where δ and K are replaced by $(1/2)\delta$ and 2K, respectively, holds true for all $\omega \in \Omega$, $x \in \Delta_{\rho_0} D$, $\xi \in E_d$, $t \geq (1/4)|\psi_{(\xi)}|^2\psi^{-1}$, $|\xi|^2 + t > 0$.

Proof. Both sides of (23) are linear in t, therefore, the inequality between them holds true for all $t \geq (1/4)|\psi_{(\xi)}|^2\psi^{-1}$ if it is true for $t = (1/4)|\psi_{(\xi)}|^2\psi^{-1}$ and as $t \to \infty$. As $t \to \infty$, it holds with 2K, $\delta/2$ in place of K, δ by (7). Furthermore, it follows from (7) that for $x \in \Delta_\rho D$, $t = (1/4)|\psi_{(\xi)}|^2\psi^{-1}$ we obtain

$$J(\omega, x, \xi, t) - J(\omega, x, \xi, 0) - [2(1 - \delta/2)|\bar{c}| + $$

$$+ K(a\psi_x, \psi_x)]t$$

$$\leq -\delta|\bar{c}|t \leq -\frac{1}{4}\delta\rho^{-1}|\bar{c}|\cdot|\psi_{(\xi)}|^2.$$

Therefore, to prove the lemma, it suffices to find the appropriate ρ such that

$$J(\omega, x, \xi, 0) \leq 2(1 - \delta/2)|\bar{c}|(B\xi, \xi) +$$

$$+ K\psi_{(\xi)}(B\xi, \sigma^k)(\sigma^k, \psi_x) + \frac{1}{4}\delta\rho^{-1}|\bar{c}|\cdot|\psi_{(\xi)}|^2 \tag{24}$$

for all $\omega \in \Omega$, $x \in \Delta_\rho D$, $\xi \in E_d$. Using (2) — (6), we estimate different terms involved in $J(\omega, x, \xi, 0)$. We denote by N various constants depending only on δ, K and set

$$I := (B\xi, \xi) + |\psi_{(\xi)}|^2.$$

For $x \in \Delta_\rho D$, $\rho \leqq \kappa$ we have

$$\psi L^0(B\xi, \xi) \leqq \rho^{1-\gamma}K|\bar{c}|I,$$

$$\psi_{(\xi)}\left(B_{(\sigma^k)}\xi, \sigma^k\right) \leqq N|\bar{c}|^{1/2}|\psi_{(\xi)}|\left(\sum_k |B_{(\sigma^k)}\xi|^2\right)^{1/2}$$

$$\leqq N|\bar{c}|\cdot|\psi_{(\xi)}|^2 + \frac{\delta}{8}|\bar{c}|I,$$

$$\psi\left(B_{(\sigma^k)}\xi, \sigma^k_{(\xi)}\right) \leqq \rho\left(\sum_k |\sigma^k_{(\xi)}|^2\right)^{1/2}\left(\sum_k |B_{(\sigma^k)}\xi|^2\right)^{1/2} \leqq N\rho|\bar{c}|I,$$

$$\psi_{(\xi)}(B\sigma^k, \sigma^k_{(\xi)}) \leqq |\psi_{(\xi)}|N|\bar{c}|I^{1/2} \leqq N|\bar{c}|\cdot|\psi_{(\xi)}|^2 + \frac{\delta}{8}|\bar{c}|I,$$

$$\psi(B\sigma^k_{(\xi)}, \sigma^k_{(\xi)}) \leqq N\rho|\bar{c}|I,$$

$$(B\xi, (\psi b)_{(\xi)}) \leqq N(B\xi, \xi)^{1/2}[|\bar{c}|\cdot|\psi_{(\xi)}| + \rho|\bar{c}|I^{1/2}]$$

$$\leqq N|\bar{c}|\cdot|\psi_{(\xi)}|^2 + (\frac{\delta}{8} + N\rho)|\bar{c}|I,$$

$$\psi_{(\xi)}(L^0\psi)_{(\xi)} = \psi_{(\xi)}\bar{c}_{(\xi)} - \psi_{(\xi)}(c\psi)_{(\xi)}$$

$$\leqq N|\psi_{(\xi)}|\cdot|\bar{c}|I^{1/2} + N|\bar{c}|\cdot|\psi_{(\xi)}|^2 \leqq N|\bar{c}|\cdot|\psi_{(\xi)}|^2 + \frac{\delta}{8}|\bar{c}|I.$$

(If the matrices A and B are symmetric, $B \geqq 0$ and $|(A\xi, \xi)| \leqq K(B\xi, \xi)$ for all $\xi \in E_d$, then $|A\xi|^2 \leqq K^2|B|(B\xi, \xi)$ for all $\xi \in E_d$.)

Moreover,

$$-K\psi_{(\xi)}(B\xi, \sigma^k)(\sigma^k, \psi_x) \leqq N|\psi_{(\xi)}|\cdot|B\xi|\cdot|\bar{c}| \leqq$$

$$\leqq N|\psi_{(\xi)}|\cdot|\bar{c}|I^{1/2} \leqq N|\bar{c}|\cdot|\psi_{(\xi)}|^2 + \frac{\delta}{8}|\bar{c}|I.$$

This and (8) imply that

$$J(\omega, x, \xi, 0) - K\psi_{(\xi)}(B\xi, \sigma^k)(\sigma^k, \psi_x)$$
$$\leqq \left(2 - \delta - \frac{1}{4}\delta + N\rho^{1-\gamma}\right)|\bar{c}|(B\xi, \xi) + N|\bar{c}|\cdot|\psi_{(\xi)}|^2.$$

This inequality shows which ρ should be taken to furnish inequality (24). The lemma is proved.

10. Proof of Theorem 1. By (3) and (6) for f, $\bar{c}$, $\bar{B}$, $\bar{b}$, $\bar{\sigma}$ and the appropriate $\rho_0 = \rho_0(\gamma, \kappa, \delta, K)$, not only condition (5.17) holds on $U'(\rho_0)$, but also condition (5.18) (with a constant $N(K)$ in place of K). Moreover, it follows from Theorem 1(c) that for a constant $\mu > 0$ and $\xi \in E_d$ we have $(B\xi, \xi) + (1/4)|\psi_{(\xi)}|^2\psi^{-1} \geqq \mu|\xi|^2$ in $\Delta_\kappa D$ and $\Delta_{\rho_0} D$. Since for $x \in U(\rho_0)$, $\xi \in E_{d+4}$, $\bar{\psi}_{(\xi)}(x) = 0$ the inequality $|\psi_{(\xi)}(x)| \leqq 2[\xi]\sqrt{\psi(x)}$ holds, we obtain

$$(\bar{B}\xi, \xi) \geqq [\xi]^2,$$

$$(\bar{B}\xi, \xi) \geqq \mu \sum_{i=1}^{d} |\xi^i|^2, \quad (\bar{B}\xi, \xi) \geqq \frac{1 \wedge \mu}{2} \sum_{i=1}^{d+4} |\xi^i|^2.$$

Therefore, for $\bar{\psi}$, $\bar{u}$, $\bar{B}$, $\bar{\sigma}$, $\bar{b}$, $\bar{c}$, f, $U'(\rho_0)$ the assumptions in the first sentence of the statement of Theorem 5.8 hold. We take $\psi^{-1/2}$ as the function ψ_1. Then (by Lemma 6) we obtain $\bar{L}\psi_1 = L\sqrt{\psi} \leqq (1/2)\psi^{-1/2}\cdot L^0\psi + c\psi^{1/2}$ $= 1/2(\bar{c} + c\psi)\psi^{-1/2}$ on $U(\rho_0)$, and the latter expression is negative on $U(\rho_0)$ by (3) and by reduction in ρ_0 if necessary. According to (5.26), we denote

$$U_m(\rho_0) = \{x \in U(\rho_0) : \psi_1(x) < m\}$$

for the chosen function ψ_1. Obviously, $\bar{u}$ and $\bar{u}_{x^i}$ are continuous on $\bar{U}_m(\rho_0)$, $\bar{U}_m(\rho_0) \setminus U(\rho_0) = \{x \in U(\kappa) : \psi(x) = \rho_0\}$. Furthermore, for $x \in U(\kappa)$ we denote

$$P(x) = \sup_{\xi}\{[\lambda(\bar{B}(x)\xi, \xi) + |\xi^{d+5}|^2]^{-1/2}\partial(\tilde{\xi})\bar{u}(x) : (x, \tilde{\xi}) \in U'(\kappa)\},$$

where λ is taken from Lemma 5.1 applied to $\bar{\psi}$, $\bar{B}$, $\bar{\sigma}$, $\bar{b}$, $\bar{c}$, f, $U'(\rho_0)$. The value $P(x)$ can easily be calculated if we note that

$$\partial(\tilde{\xi})\bar{u} = \sum_{i=1}^{d} (u\psi^{-1})_{x^i}\xi^i + \xi^{d+5}u\psi^{-1},$$

$$(\bar{B}\xi, \xi) = (B\xi, \xi) + [\xi]^2$$

$$\geq \sum_{i,j=1}^{d} B^{ij}\xi^i\xi^j + \frac{1}{4}\psi^{-1} \sum_{i,j=1}^{d} \psi_{x^i}\psi_{x^j}\xi^i\xi^j,$$

for every $\xi^1, \ldots, \xi^d$ we can choose $\xi^{d+1}, \ldots, \xi^{d+4}$ such that the latter inequality becomes an equality (cf. the proof of Lemma 8). It turns out that

$$P^2 = \frac{1}{\lambda} (C\bar{u}_x, \bar{u}_x) + (\bar{u})^2.$$

Finally, by Theorem 5.8 for $x \in U'(\rho_0)$

$$P^2(x) \leq N_1 \vee \max\{P^2(y) : y \in U(K), \psi(y) = \rho_0\}.$$

Since $P(x) = P(x^1, \ldots, x^d)$, we have for $x \in \Delta_{\rho_0} D$

$$P^2(x) \leq N_1 \vee \max\{P^2(y) : y = (y^1, \ldots, y^d), \psi(y) = \rho_0\}.$$

This asserts (9) for $\rho = \rho_0$ and since the above argument obviously remains valid for any $\rho \leq \rho_0$, the theorem is proved.

8.7. The Solvability of Equations Weakly Nondegenerate Along the Normal

Let us consider the situation described at the beginning of the previous section, but abandon the requirement that the lower bound in (7.2.3) has to be attained. Suppose that σ, b, c, f are defined for all $x \in \bar{D}$, continuous in $\bar{D}$, twice continuously differentiable with respect to x in D, and bounded in (ω, x). Also, let $\psi \in C^3(D)$. To clarify some ideas, we provisionally suppose that $|u| \leq K\psi$ in D for any sufficiently smooth solution to equation (6.1) in D with zero condition on ∂D. Sometimes this assumption holds in view of $|f| \leq -K\bar{c}$, where $\bar{c} := L\psi$. Needless to say that the inequality $|u| \leq K\psi$ implies $|u_x| \leq K|\psi_x|$ on ∂D and we obtain an estimate of the normal derivative of u on ∂D. Moreover, by Meusnier's theorem on the surface cross-section curvature applied to $x \in D$, $\xi \perp \psi_x(x)$, we have

$$u_{(\xi)(\xi)} = u_{(n)}\psi^{-1}_{(n)}\psi_{(\xi)(\xi)} = (\psi_x, u_x)|\psi_x|^{-2}\psi_{(\xi)(\xi)}, \tag{1}$$

where $n = \psi_x|\psi_x|^{-1}$. Therefore, an estimate of $u_{(\xi)(\xi)}$ is also available. In this section we will derive an estimate of $u_{(\xi)(n)}$ from Corollary 6.5

so that it will be possible to estimate $u_{(n)(n)}$ (on ∂D) from the equation (6.1) itself, roughly speaking, by solving it relative to $u_{(n)(n)}$. To make this possible, $u_{(n)(n)}$ should 'enter' equation (6.1). We have already repeatedly encountered the need to estimate the measure of the involvement of derivatives in an equation (e.g., see Section 2.2); when using upper estimates of pure second derivatives of a solution we had to estimate them *from below*. We now have to estimate $u_{(n)(n)}$ *from above* using the equation, formula (1), and the estimates of $u_{(\xi)(n)}$. Therefore, the 'measure of the involvement' of $u_{(n)(n)}$ in an equation is here measured by a method stemming from the following definition rather than $\mu(1)$ (see (2.2.5)).

1. Definition. Let there exist constants λ, $\nu > 0$ and $\varepsilon_0 \in (0, 1)$ such that for any $\varepsilon \in (0, \varepsilon_0)$ and any solution $u^\varepsilon \in C^2(\bar{D})$ to the problem

$$\varepsilon^2 \Delta u - \varepsilon u + F\left(u_{x^i x^j}, u_{x^i}, u, x\right) = 0 \text{ in } D, \tag{2}$$

$$u = 0 \text{ on } \partial D \tag{3}$$

the inequality

$$F\left(|\psi_x|^{-2}(u_x^\varepsilon, \psi_x)\psi_{x^i x^j} + \lambda n^i n^j, |\psi_x|^{-2}(u_x^\varepsilon, \psi_x)\psi_{x^i}, 0, x\right) \geqq \nu \tag{4}$$

holds on ∂D.

Then the operator $F\left(u_{x^i x^j}, u_{x^i}, u, x\right)$ is said to be *weakly non-degenerate* along the normal to ∂D.

Let us discuss this definition. When $c \leqq 0$, the maximum principle implies that problem (2) and (3) may have only one solution of the class $C^2(\bar{D})$. Under the conditions on σ, b, c, f which will be imposed in the existence theorems, this problem for every $\varepsilon \in (0, 1)$ will indeed have a solution of the class $C^2(\bar{D})$, although in Definition 1 it is not generally assumed that problem (2) and (3) has a solution. It is only required that inequality (4) holds on ∂D whenever problem (2) and (3) is solvable in $C^2(\bar{D})$. Sometimes it may be useful to keep in mind that, by the definitions of F and $\bar{c}$, inequality (4) can be rewritten as

$$\inf_\omega \left[|\psi_x|^{-2}(u_x^\varepsilon, \psi_x)\bar{c}(\omega) + \lambda(a(\omega)n, n) + f(\omega)\right] \geqq \nu. \tag{5}$$

Below we state some simple conditions which are sufficient for the operator F to be weakly nondegenerate along the normal.

2. Remark. If $c \leq 0$, $\varepsilon_0 \in (0, 1)$, $|f| \leq -K(1 + \varepsilon^2\Delta)\psi$ for all $\varepsilon \in (0, \varepsilon_0)$, $x \in D$, $\omega \in \Omega$, and the operator F is *uniformly nondegenerate* along the normal to ∂D, i.e., if

$$a^{ij}(\omega, x)\psi_{x^i}(x)\psi_{x^j}(x) \geq \nu_1, \tag{6}$$

where ν_1 is a positive constant, for all $x \in \partial D$, $\omega \in \Omega$, then F is weakly nondegenerate along the normal too.

Indeed, it follows from the maximum principle that $|u^\varepsilon| \leq K\psi$ in D, $|u^\varepsilon_x| \leq K|\psi_x|$ on ∂D, therefore, the left-hand side of (5) with $\lambda = 0$ is bounded on ∂D by a constant independent of ε, and for any $\lambda \geq 0$, by the property of the infimum of a sum, it is greater than the sum of its value for $\lambda = 0$ and the value

$$\lambda \inf_\omega a^{ij}n^i n^j,$$

which is greater than $\lambda\nu_1|\psi_x|^2$ by condition (6). Therefore, for a sufficiently large λ we can furnish inequality (5) (with any ν).

3. Remark. If Ω consists of a single point, i.e., we deal with linear equations, if the second derivatives of u^ε are bounded on ∂D uniformly in ε, and if the operator F is weakly nondegenerate along the normal, then it is uniformly nondegenerate along the normal too.

Indeed, if the left-hand side of (6) is zero at a point $x_0 \in \partial D$, then the left-hand side of (4) at x_0 is independent of λ and, by (1), it is equal to

$$F\left(u^\varepsilon_{x^i x^j}(x_0),\ u^\varepsilon_{x^i}(x_0),\ u^\varepsilon(x_0),\ x_0\right).$$

By (2), this coincides with $(-\varepsilon^2\Delta u^\varepsilon(x_0))$ and cannot be greater than $\nu > 0$ for all $\varepsilon \in (0, \varepsilon_0)$. We can similarly prove that in the general case the supremum of the left-hand side of (6) over $\omega \in \Omega$ is strictly positive at every point $x \in \partial D$ if the operator F is weakly nondegenerate along the normal to ∂D and the second derivatives of u^ε are bounded on ∂D uniformly in ε.

Remarks 2 and 3 show that in practically significant cases of linear equations the notions of weak and uniform nondegeneracy along the normal coincide. The case of nonlinear equations is quite different.

4. Remark. Let $f \geq 0$ and $\bar{c} \leq -\nu_1$ for $x \in \partial D$, and let $\omega \in \Omega$, where $\nu_1 > 0$ is a constant. Also, let the *a priori* estimate of u^ε: $(u^\varepsilon_x, \psi_x) \leq$

$\leqq -\nu_1$ on ∂D be available. Then the operator F is weakly nondegenerate along the normal and we can set $\lambda = 0$ in (4).

This remark is valid since the left-hand side of (4) on ∂D with $\lambda = 0$ and $f \geqq 0$ is greater than or equal to

$$-|\psi_x|^{-2}(u_x^\varepsilon, \psi_x)\inf_\omega(-\bar{c}(\omega)) \geqq |\psi_x|^{-2}\nu_1^2.$$

5. Remark. Let D be a connected domain and let there exist a smooth function $g \leqq 0$, $g \not\equiv 0$ in $\bar{D}$ and a constant $\nu_1 > 0$ such that for all $x \in D$, $u_{ij} = u_{ji}$, u_i, u the inequality

$$F(u_{ij}, u_i, u, x) \leqq \nu_1\delta^{ij}u_{ij} + g(x) \tag{7}$$

holds. Then there exists a constant $\nu_2 > 0$ such that $(u_x^\varepsilon, \psi_x) \leqq -\nu_2$ on ∂D for $\varepsilon \in (0, 1)$.

To prove this fact, we define smooth functions v^ε, $\varepsilon \in (0, 1)$, such that $(\varepsilon^2 + \nu_1)\Delta v^\varepsilon - \varepsilon v^\varepsilon + g = 0$ in D, $v^\varepsilon = 0$ on ∂D. Using elementary barriers, the smoothness of ∂D, and the connectivity of D, one can easily prove that $v^\varepsilon_{(n)} \leqq -\nu_2$ on ∂D, where ν_2 is independent of x and ε, and $\nu_2 > 0$. It follows from (7) that $0 \leqq \varepsilon^2\Delta u^\varepsilon - \varepsilon u^\varepsilon + \nu_1\Delta u^\varepsilon + g$, hence by using the maximum principle, we conclude that $u^\varepsilon \leqq v^\varepsilon$ in $\bar{D}$, $(u_x^\varepsilon, \psi_x) \leqq (v_x^\varepsilon, \psi_x) \leqq -\nu_2|\psi_x|$ on ∂D.

It follows from the above discussion that condition (7) can be substantially relaxed. For example, it suffices that there exists a point $\omega_0 \in \Omega$ such that $c(\omega_0) \leqq 0$ and the equation $(\varepsilon^2\Delta + L(\omega_0))v - \varepsilon v + f(\omega_0) = 0$ in D with zero boundary condition on ∂D has a solution of $C^2(\bar{D})$ such that $v_{(n)} \leqq -\nu_2$ on ∂D, where $\nu_2 > 0$ and ν_2 is independent of x and ε. Note that if condition (7) holds, then it is possible to expand the set Ω by adding a point $\bar{\omega} \notin \Omega$ and defining $L(\bar{\omega})u + f(\bar{\omega}) = \nu_1\Delta u + g$. By virtue of (7), the function F remains unchanged under such expansion. Remarks 4 and 5 enable us to construct a large number of examples of nonlinear operators which are weakly nondegenerate, but not uniformly nondegenerate along the normal. This can easily be done by assuming that $f = 0$ with $x \in \partial D$, $\omega \in \Omega$. Turning back to Remark 3 on linear equations, note that if Ω consists of a single point, then condition (7) readily yields the operator $L(\omega)$: $L(\omega) = \nu_1\Delta$. Nonlinear equations do not enjoy this property, which leads to the difference between the notions of weak and uniform nondegeneracy along the normal.

6. Remark. The condition $f \geqq 0$ on $\Omega \times \partial D$ in Remark 4 may seem too re-

strictive. It can be abandoned in the following case. Let Ω be a metric compactum, let σ, b, c, f be continuous in (ω, x) on $\Omega \times \partial D$, and let $\varepsilon_0 \in (0, 1)$. Denote

$$P = \{(\omega, x) : \omega \in \Omega,\ x \in \partial D,\ a^{ij}(\omega, x)n^i(x)n^j(x) = 0\},$$

$$\Gamma = \{x \in \partial D : (\omega, x) \in P \text{ for an } \omega \in \Omega\}.$$

If the closed set P were empty, then the operator F would be uniformly nondegenerate along the normal to ∂D. Therefore, we suppose that $P \neq \emptyset$. Also, let $f \geq 0$ on P and, for a constant $\nu_1 > 0$, let $\bar{c} \leq -\nu_1$ on P. Finally, suppose that for a domain G containing a closed set Γ, for any $\varepsilon \in (0, \varepsilon_0)$, and any solution $u^\varepsilon \in C^2(\bar{D})$ to problem (2) and (3) the inequality $(u^\varepsilon_x, \psi_x) \leq -\nu_1$ holds on $G \in \partial D$. It turns out that in this case the operator F is weakly nondegenerate along the normal to ∂D.

To prove this, we assume the contrary. Then inequality (5) does not hold on ∂D for any λ, ν and consequently for every sequence $\lambda_r \to \infty$ there exist $\varepsilon_r \in (0, \varepsilon_0)$, $\nu_r > 0$, $x_r \in \partial D$ such that $\nu_r \to 0$ as $r \to \infty$ and inequality (5) is violated if we substitute $\varepsilon = \varepsilon_r$, $\lambda = \lambda_r$, $\nu = \nu_r$, and $x = x_r$ in it. Of course, we may assume that the sequence x_r converges to a point $x_0 \in \partial D$. Moreover, since Ω is compact and σ, b, c, f are continuous in ω, the infimum in the definition of F is attainable. We denote by ω_r the point at which the infimum in (7.2.3) is attained when the arguments of F are taken from (4) with $\varepsilon = \varepsilon_r$, $\lambda = \lambda_r$, and $x = x_r$. Without loss of generality we assume that $\omega_r \to \omega_0$ as $r \to \infty$.

For every r we obtain

$$\nu_r \geq |\psi_x(x_r)|^{-2}\left(u^{\varepsilon_r}_x, \psi_x\right)(x_r)\bar{c}(\omega_r, x_r) + $$

$$+ \lambda_r(a(\omega_r, x_r)n(x_r), n(x_r)) + f(\omega_r, x_r). \tag{8}$$

Dividing both sides of this inequality by λ_r, letting $r \to \infty$, and using the continuity of a in (ω, x), we obtain $(\omega_0, x_0) \in P$. We next omit the second summand in (8), let $r \to \infty$, and use our hypotheses. Then we can see that

$$0 \geq |\psi_x(x_0)|^{-2}\nu_1(-\bar{c}(\omega_0, x_0)) + f(\omega_0, x_0) \geq |\psi_x(x_0)|^{-2}\nu_1^2.$$

Since this is not true, the validity of the remark is proved.

Let us consider a simple, and as the reader will apprehend far from unique, example of applying Remarks 5 and 6.

7. Example. Let $d \geqq 2$, let ψ be strictly concave on $\bar{D}$, $g(x)$ be smooth on $\bar{D}$, $g \leqq 0$, $g \not\equiv 0$, and let Ω be the set of all nonnegative symmetric $d \times d$ matrices with unit trace. Let the operator

$$\inf_{\omega \in \Omega}\left[\omega^{ij}u_{ij} + \sqrt[d]{\det \omega}\; g(x)d\right]$$

be taken as F and let us define the functions σ, b, c, f in an obvious way. Since ψ is strictly concave, we have $c \leqq -\nu_1$ for all ω and x, where $\nu_1 > 0$. Moreover, it is obvious here that $\det \omega = 0$ for $(\omega, x) \in P$, hence $f := \sqrt[d]{\det \omega}\; gd = 0$ on P. Finally, since $d^{-1}(\delta^{ij}) \in \Omega$, condition (7) holds if $\nu_1 = d^{-1}$ in it. By Remarks 5 and 6, the operator in question is weakly nondegenerate along the normal to ∂D.

The basic result of this section is presented in Theorem 9. To prove this theorem, we shall need the following lemma whose statement is preceded by recalling that the conditions formulated at the beginning of this section are assumed to hold.

8. LEMMA. Let σ, b, c, f be twice continuously differentiable with respect to x in D, bounded in ω, x, and let their first and second derivatives with respect to x be bounded in ω, x. Let $B = (B^{ij}) = B^* \geqq 0$, $B \in C^2_{loc}(D)$, $\tilde{u} \in C^2(D) \cap C(\bar{D})$, $\tilde{u} \geqq 0$, $\chi(\omega, x) \geqq 0$ be some functions and $\delta_1, \delta_2 \in (0, 1)$, $K_1 > 0$. We define the function $M(\omega, x, \xi)$ as in Remark 5.7 and suppose that for all $\omega \in \Omega$, $x \in D$ and unit vectors $\xi \in E_d$ inequalties (5.23), (5.38), (5.39), and

$$\delta_1 \leqq \tilde{u} \leqq K, \quad c \leqq 0, \quad \bar{c} := L\psi \leqq -\delta_2, \quad |f| \leqq -K\bar{c}. \tag{9}$$

hold. Also, assume that for all $x \in \Delta_\kappa D$, $\xi \in E_d$, $\omega \in \Omega$ inequalities (6.11) – (6.17) hold, $|\psi_x| \geqq \delta_1$, $(B\xi, \xi) \geqq \delta_1|\xi|^2$, and $\psi^\gamma||(B\xi, \xi)_{xx}|| \leqq K|\xi|^2$. Then there exists an $\varepsilon_0 \in (0, 1)$ such that for every $\varepsilon \in (0, \varepsilon_0)$ problem (2) and (3) has a unique solution $u \in C^2(\bar{D})$ such that

$$|u(x)| \leqq N\psi(x), \quad |u_x(x)| \leqq N, \quad |u_{(\eta)(\zeta)}(y)| \leqq N,$$
$$u_{(\xi)(\xi)}(x) \leqq N_1(t) \tag{10}$$

for $t := \sup_{\partial D}(u_{(n)(n)})_+$, $x \in D$, $y \in \partial D$, all unit vectors $\xi, \eta, \zeta \in E_d$, $\eta || \psi_x(y)$, $\zeta \perp \psi_x(y)$, where N, $N_1(t)$ depend only on the norm of ψ in $C^2(D)$ and K, K_1, δ, δ_1, γ, κ (N_1 also depends on the singled-out argument t; N and N_1 are independent of ε and δ_2).

Proof. The existence of a solution $u \in C^2(\bar{D})$ was shown in Theorem 6.4.2, the uniqueness readily follows from the maximum principle. The first inequality in (10) also follows from the maximum principle if $\varepsilon^2\Delta\psi - \varepsilon\psi + L\psi \leqq (1/2)\bar{c}$, which, by (9) and the fact that ψ has a finite norm in $C^2(D)$, is true at least for sufficiently small ε. Of course, the inequality $|u| \leqq N\psi$ implies $|u_x| \leqq N|\psi_x|$ on ∂D.

Furthermore, we will apply Theorems 5.10 and 5.11 to u. To this purpose we need, first of all, to convert equation (2) to the form (5.25), where the function F is constructed from some σ_ε, b_ε, c_ε, f_ε according to (7.2.1) — (7.2.3). This is a simple operation, it suffices to set $\sigma_\varepsilon^{ik} = \sigma^{ik}$ for $k = 1, \ldots, d_1$, $i = 1, \ldots, d$, $\sigma^{i,d_1+k} = \sqrt{2\varepsilon}\delta^{ik}$ for $i, k = 1, \ldots, d$, $b_\varepsilon = b$, $f_\varepsilon = f$, and $c_\varepsilon = c - \varepsilon$.

Furthermore, we set $\delta_3 = \frac{1}{4}(2 - \delta)\delta_1(1 - \delta_1)$, $\chi_\varepsilon = \chi + \delta_3\varepsilon$ and construct M_ε from σ_ε, b_ε, and c_ε. It is easily seen that $\delta_3\varepsilon \geqq \delta_3\varepsilon$ and since the norm of $\tilde{u}$ in $C^2(D)$ is finite, we also have $M_\varepsilon \geqq M + \delta_3 \geqq \delta_3\varepsilon$ for all sufficiently small $\varepsilon > 0$. If ε is small, inequalities (5.23), (5.38), and (5.39) hold for σ_ε, b_ε, c_ε, f_ε, χ_ε, and M_ε. If we now knew that $u \in C^3_{loc}(D)$ and if the infimum in (7.2.3) were attainable for the original σ, b, c, and f (and hence for σ_ε, b_ε, c_ε, and f_ε), then, by Theorem 5.10 (see also Remark 5.12), we would obtain the second inequality in (10) too. On the other hand, if we knew that $u \in C^4_{loc}(D)$, then by Theorem 5.11 we could write

$$u_{(\xi)(\xi)} \leqq N_1\left(\sup_{\partial D, |\eta|=1} (u_{(\eta)(\eta)})_+\right) \tag{11}$$

for $\xi \in E_d$, $|\xi| = 1$. We could then note that for sufficiently small ε conditions (6.11) — (6.17) hold with σ_ε, b_ε, c_ε, f_ε if δ, K were replaced by $(1/2)\delta$, 2K, therefore the latter formula, by Corollary 6.5 and (1), would give the third and fourth inequalities in (10).

The above discussion shows how we should choose ε_0. When this is done, the only difficulty is connected with the lack of sufficient smoothness of u and the fact that the infimum in (7.2.3) is unattainable.

To surmount this difficulty, we make use of the arguments of Section 7.3. First, using the transformations of the proof of Lemma 7.3.2 we prove the lemma in case when the set Ω is finite, σ, b, c, f are infinitely differentiable with respect to x in D. Of course, we take advantage of passage to the limit as $p \to \infty$, the function ψ_p and formulas (7.3.16) — (7.3.20), where u_t is omitted since we consider elliptic equations. Moreover, we pay attention to the fact that in such a case the constants N and the function $N_1(t)$ in (10) by no means depend on the

the number of elements in Ω.

In the next step the lemma is proved under the condition that Ω is finite. Here we use passage to the limit when we pass from the infinitely differentiable σ_n, b_n, c_n, f_n to the initial functions. Note that since $M_\varepsilon \geqq \delta_3\varepsilon$, $\chi_\varepsilon \geqq \delta_3\varepsilon$, $\psi^\gamma||(B\xi, \xi)_{xx}|| \leqq K|\xi|^2$, and B is uniformly nondegenerate in $\Delta_\kappa D$, thus it is absolutely immaterial how we construct the infinitely differentiable approximations provided that they and their derivatives of first and second order uniformly converge to the initial functions and their derivatives on D. Also, note that passage to the limit in the corresponding equations presents no difficulties since, by Theorem 6.4.2, we have the estimates of the norms in $C^{2+\alpha}(\bar{D})$ of solutions to the equations involved in the limiting process and these estimates are independent of n, and $\alpha \in (0, 1)$ (specific for every ε) is also independent of n. Passage to the limit over the number of elements in Ω asserts the lemma for countable sets Ω, whereas the fact that equation (2) is equivalent to the same equation constructed for a countable subset of Ω (cf. the proof of Theorem 6.1.5) enables us to prove the lemma in the general case. The lemma is proved.

The following theorem presents the basic results of this section.

9. THEOREM. Let the conditions of Lemma 8 hold and let the operator F be weakly nondegenerate along the normal. Then there exists a function u on $\bar{D}$ such that:

(a) $u \in C(\bar{D})$, $u = 0$ on ∂D, there exist Sobolev derivatives u_x and measures u_{xx} in D;

(b) $|u| \leqq N\psi$, $|u_x| \leqq N$ (a.e.D), $u_{(\xi)(\xi)}(dx) \leqq N\,dx$ in D for all unit vectors $\xi \in E_d$, where N depends only on γ, κ, δ, K, K_1, δ_1, the norm of ψ in $C^2(D)$, and the constants λ, ν in Definition 1;

(c) $a^{ij}(\omega)u_{x^ix^j}(dx) \ll dx$ for all $\omega \in \Omega$

$$F\left(u^{(0)}_{x^ix^j}, u_{x^i}, u, x\right) = 0 \text{ (a.e.D).}$$

The function u enjoying the above properties is unique. Moreover, if a domain D_0 belongs to D, $l \in E_d$, $|l| = 1$, and the function $\mu(l)$ defined in (7.3.30) is bounded away from zero on D_0, then the generalized derivative $u_{(l)(l)}$ is bounded on D_0. But if the function μ in (7.3.31) is bounded away from zero on D_0, then all generalized derivatives u_{xx} are bounded on D_0.

Proof. We first discuss the uniqueness. By Lemma 3.6.2, the function u satisfies the equation

$$\underset{\omega}{\wedge}\left[a^{ij}(\omega)\circ u_{x^ix^j} + \left(b^i(\omega)u_{x^i} + c(\omega)u + f(\omega)\right)\circ\Lambda\right] = 0,$$

where Λ is the Lebesgue measure in D. Since u is independent of t, the same equation can be written in $Q := (0, \infty) \times D$ for $u(t, x) := u(x)$ and we can add the derivative of u with respect to t to it. The uniqueness now follows from Corollary 3.7.10 if we set $\tilde{\psi}(t, x) = \psi(x)e^{\alpha t}$ and make use of the inequality $\tilde{\psi}_t + L\tilde{\psi} = e^{\alpha t}(\bar{c} + \alpha\psi) \leqq 0$ for small $\alpha > 0$.

We now come to the existence of a solution. We reduce ε_0 in Definition 1, if necessary, so that it is suited for Lemma 8, and for $\varepsilon \in (0, \varepsilon_0)$ denote by u^ε the function u involved in this lemma and associated with this ε. We fix a point $x_0 \in \partial D$ and choose unit vectors $\xi_i \perp \psi_x(x_0)$, $i = 1, \ldots, d - 1$ such that $(\xi_i, \xi_j) = \delta^{ij}$, $i, j \leqq d - 1$. A straightforward calculation at the point x_0 for small ε leads from (1) and (10) to

$$\begin{aligned}\varepsilon^2\Delta u^\varepsilon + Lu^\varepsilon &= |\psi_x|^{-2}(u^\varepsilon_x, \psi_x)(\varepsilon^2\Delta + L)\psi + \\ &+ \sum_{i\geqq 2} 2(a\xi_i, n)\left[u^\varepsilon_{(\xi_i)(n)} - |\psi_x|^{-2}(u^\varepsilon_x, \psi_x)\psi_{(\xi_i)(n)}\right] \\ &+ (\varepsilon^2 + (an, n))\left[u^\varepsilon_{(n)(n)} - |\psi_x|^{-2}(u^\varepsilon_x, \psi_x)\psi_{(n)(n)}\right] \\ &\geqq |\psi_x|^{-2}(u^\varepsilon_x, \psi_x)\bar{c} - \frac{\nu}{2} + \\ &+ (\varepsilon^2 + an, n))\left[u^\varepsilon_{(n)(n)} - N_0\right]\end{aligned}$$

(making use of the fact that

$$2|(a\xi_i, n)| \leqq \beta(a\xi_i, \xi_i) + \beta^{-1}(an, n)$$

for any $\beta > 0$), where N_0 depends on initial data as desired. By Definition 1 (also, see (5)) it follows that $u^\varepsilon_{(n)(n)} \leqq \lambda + N_0$ on ∂D, and this together with Lemma 8 yields the estimates $|u^\varepsilon| \leqq N\psi$, $|u^\varepsilon_x| \leqq N$, $u^\varepsilon_{(\xi)(\xi)} \leqq N$ which are uniform in $\varepsilon > 0$, $x \in D$, and unit $\xi \in E_d$. The existence of the function u enjoying properties (a) — (c) can now be proved as in Lemma 7.3.4. The last assertion of the theorem can be proved following the proof of Theorem 2.2.6. The theorem is proved.

The test of the hypotheses of Lemma 8 can be a rather complicated problem consisting of finding appropriate B and u. Therefore, in the theorem to come we present two simple cases based on specific values of B and u. For the sake of convenience the statement of the theorem involves the necessary conditions, which are implicitly meant in this section.

10. THEOREM. Let $\gamma, \delta \in (0, 1)$, let $\psi \in C^2_{loc}(E_d)$, Let $D = \{x : \psi(x) > 0\}$ be a bounded domain, let $\psi \in C^3(D)$, and let $|\psi_x| > 0$ on ∂D. Also, let σ^{ik}, b^i, c, f be continuous in $\bar{D}$ and twice continuously differentiable with respect to x in D, and let them and all their derivatives with respect to x^r and $x^r x^s$ be bounded on $\Omega \times D$. Suppose that one of the two following groups of conditions (a) or (b) holds true:

(a) for all $x \in D$, $\omega \in \Omega$ and unit vectors $\xi \in E_d$ the inequalities (5.40), $c \leqq -\delta, L\psi \leqq -\delta$, and $(a\psi_x, \psi_x) \geqq \delta$ hold on $\Omega \times \partial D$;

(b) σ, b are independent of x, the function ψ is strictly concave,

$$\psi \in C^3(D) \cap C^4_{loc}(D), \quad (b, \psi_x) + 2(1 - \delta)L\psi \leqq 0$$

on $\Omega \times \partial D$, for all $(\omega, x) \in \Omega \times D$ and unit vectors $\xi, \eta \in E_d$

$$\psi^{\gamma}|\psi_{(\xi)(\xi)(\eta)(\eta)}| \leqq K, \quad c \leqq 0, \quad L\psi \leqq -\delta, \tag{12}$$

and moreover, the operator F is weakly nondegenerate along the normal to ∂D.

Then all assertions of Theorem 9 are true without specifying the initial parameters on which N depends in assertion (b) on this theorem.

Proof. We first consider the case when condition (a) holds true. We set $\tilde{u} \equiv 1$, $\chi = 1$. Then (5.23) turns out to follow from (5.40), inequalities (5.38), (5.39), and (9) hold true with some $\delta_1, \delta_2 > 0$. Moreover, it is almost obvious that for a $\kappa > 0$ we can find, first, a large constant K_1, and next a large constant N such that in $\Delta_\nu D$ the inequalities $(a\psi_x, \psi_x) \geqq (1/2)\delta$, and (6.11) — (6.17) hold with B^{ij} and K in the latter inequalities being replaced by δ^{ij} and N respectively. Finally, in case (a) the operator F is uniformly nondegenerate along the normal (cf. Remark 2) and thus all assumptions of the theorem hold true.

If conditions (b) hold, then we take $\tilde{u} = \psi + 1$, $\chi = 1$ and set $B = (1/2)\psi_{xx}$. Then we can take $\delta_2 = \delta$ in (9), while the left-hand side of (5.23) is equal to zero and conditions (5.38) and (5.39) are obviously satisfied. Furthermore, for $K_1 = 0$ in (6.11) we have $\tilde{c} = \bar{c} \leqq -\delta$, and since $(B\xi, \xi) = -(1/2)\psi_{(\xi)(\xi)} \geqq \nu|\xi|^2$ for a constant $\nu > 0$ in view of the strict concavity of ψ, therefore inequalities (6.12) — (6.15) hold for an appropriate constant K. The first inequality in (6.11) can easily be furnished in $\Delta_\kappa D$ by choosing a sufficiently small $\kappa(|\psi_x| > 0$ on $\partial D)$. Moreover, the left-hand side of (6.16) is equal to $(1/2)(b, \psi_x)$ which is smaller than $(1 - \delta)|\bar{c}|$ on ∂D, by hypothesis, and smaller than $(1 - (1/2)\delta)|\bar{c}|$ in $\Delta_\kappa D$ if κ is small. Finally, the left-hand side of (6.17) is $\delta(a\psi_{x(\xi)}, \psi_{x(\xi)})$, which in $\Delta_\kappa D$ can be made smaller than $(1 - \delta)\bar{c}\psi_{(\xi)(\xi)}$ by choosing small δ in (6.17). Thus in case (b) the conditions of Theorem 9 hold also true. The theorem is proved.

In both cases (a) and (b) the above theorem enables us to examine a large number of specific examples. In particular, it is possible to add lower terms (if appropriate conditions hold) to the examples of elliptic equations in Section 2. Here we consider only the example of the simplest Monge–Ampère equation which is most significant from the historical point of view.

11. Example. Let ψ be strictly concave, let $\psi \in C^4(D)$, let f be twice continuously differentiable in $\bar{D}$, let $f \geq 0$, and let $d \geq 2$. Then the simplest Monge–Ampère equation (2.2) with zero boundary data on ∂D has a unique solution in $\bar{D}$, which is concave and has bounded generalized derivatives u_x and u_{xx}. This can be shown by reducing equation (2.2) to the form (2.3), using Example 7 and the fact that $\bar{c} \leq -\nu_1$ for a constant $\nu_1 > 0$ as mentioned in the discussion of Example 7. If desired, the smoothness of the solution can be improved as shown in Example 2.2.

12. Remark. In the previous example the conditions on ψ can be replaced by those on D to obtain a somewhat stronger result. Indeed, if D is a strictly convex domain whose boundary belongs to the class $C^{4-\gamma}$, then we can construct a strictly concave function ψ such that $\psi \in C^{4-\gamma}(\bar{D}) \cap C^4_{loc}(D)$, $D = \{\psi > 0\}$, and the first condition in (12) holds (with the same γ, although we could restrict ourselves to the case when γ in (12) is changed; in this connection see Lemma 1.1.1).

13. Remark. Conditions (a) and (b) of Theorem 10 characterize two extreme situations. One can easily imagine that there are intermediate situations when conditions (6.11) — (6.17) are furnished on a part of the boundary of D by uniform nondegeneracy along the normal, whereas on some other part they hold because the boundary is 'outward' convex. The version of Theorem 10 with conditions (a) can be found in Remark 8.8.

14. Remark. Theorem 10 is far from exhausting all possible applications of Theorem 9. For instance, in the proof of Theorem 10 we always set $\chi = 1$. Using different functions χ, we can examine the set of problems with 'free boundaries'. Here we point out only one such problem, but first let us note that Lemma 8 is also true when its conditions are replaced by the following (*a priori*) assumptions:

(a) any solution $u^{\varepsilon} \in C^2(\bar{D})$ to problem (2) and (3) for any sufficiently small $\varepsilon > 0$ also satisfies the equation

$$\varepsilon^2 \Delta u - \varepsilon u + \hat{F}\left(u_{x^i x^j}, u_{x^i}, u, x\right) = 0 \text{ in } \Delta_K D, \tag{13}$$

where $\hat{F}$ is constructed from some $\hat{\sigma}$, $\hat{b}$, $\hat{c}$, $\hat{f}$ defined on $\Omega \times \Delta_K D$;

(b) $|u^{\varepsilon}| \leq K\psi$ in D, the original σ, b, c, f, B, $\tilde{u}$, χ, δ_1, δ_2 satisfy the conditions of Lemma 8 listed at the beginning of its statement prior to conditions (9); moreover, $\delta_1 \leq \tilde{u} \leq K$, $c \leq 0$;

(c) $\hat{\sigma}$, $\hat{b}$, $\hat{c}$, $\hat{f}$ are twice continuously differentiable with respect to x in $\Delta_K D$, bounded in ω, x in $\Omega \times \Delta_K D$, and their first and second derivatives with respect to x are also bounded in $\Omega \times \Delta_K D$.

(d) for all $x \in \Delta_K D$, $\xi \in E_d$, $\omega \in \Omega$ for $\hat{\sigma}$, $\hat{b}$, $\hat{c}$, $\hat{f}$ inequalities (6.11) – (6.17) hold as well as the inequalities $|\psi_x| \geq \delta_1$, $(B\xi, \xi) \geq \delta_1|\xi|^2$, $\psi^{\gamma}||(B\xi, \xi)_{xx}|| \leq K|\xi|^2$, $\hat{c} \leq 0$, and the expression for $\tilde{c}$ in (6.11) constructed from K_1, $\hat{\sigma}$, $\hat{b}$, $\hat{c}$ is smaller than $(-\delta_2)$.

Theorem 9 is also true if the conditions of Lemma 8 are replaced by conditions (a) – (d), and the condition of the weak degeneracy of F should be preserved (it can readily be replaced by an equivalent condition on $\hat{F}$).

Let us prove this fact. The manner of deriving Theorem 9 from Lemma 8 remains unchanged, therefore it suffices to discuss Lemma 8. The existence and uniqueness of $u^{\varepsilon} \in C^2(\bar{D})$ can be obtained as in Lemma 8. It follows from (b) that $|u^{\varepsilon}_x| \leq K|\psi_x|$ on ∂D. Hence for small $\varepsilon > 0$ the second inequality in (10) and inequality (11) for $u = u^{\varepsilon}$ can be derived using Theorems 5.10 and 5.11 again as in Lemma 8. For sufficiently small $\varepsilon > 0$ conditions (6.11) – (6.17) hold for equation (13) in view of (d), if δ and K in these conditions are replaced by $(1/2)\delta$ and 2K respectively. Consequently, if we knew that $u^{\varepsilon} \in C^3_{loc}(\Delta_K D)$ and the infimum in the definition of F is attainable, then by Corollary 6.5 we could derive from (11) the remaining estimates in (10).

We can surmount this difficulty following the proof of Lemma 8. Namely, let us replace Ω in the definition of $\hat{F}$ by a finite subset $\{\omega_1, \ldots, \omega_r\}$, then transform the infimum over $\{\omega_1, \ldots, \omega_r\}$ using the function Ψ_p and formulas (7.3.16) – (7.3.20). We thus construct equations with parameters r and p, which 'converge' to equation (13) in $\Delta_K D$, as $r \to \infty$, $p \to \infty$. The equations involved in the limiting process are considered in $\Delta_\kappa D$ with the boundary data $u = u^\varepsilon$ on $\partial\Delta_\kappa D$. By Theorem 6.4.2, the solutions $u^{\varepsilon rp}$ to the equations involved in the limiting process, as well as u^ε, lie in $C^{2+\alpha}(\overline{\Delta_\kappa D})$ for an $\alpha \in (0, 1)$ (which, of course depends on ε), their norms in $C^{2+\alpha}(\Delta_\kappa D)$ are bounded by a common constant.A $p \to \infty$, $r \to \infty$, the functions $u^{\varepsilon rp}$ converge to u^ε, along with their first and second derivatives with respect to x, uniformly on $\Delta_\kappa D$ in view of the uniqueness ($\hat{c} \leq 0$) of the solution of (13). Furthermore, as in Lemma 1.3.2, it can be shown that $u^{\varepsilon rp} \in C^3_{loc}(\Delta_\kappa D)$ (here we need not smooth $\hat{\sigma}$, $\hat{b}$, $\hat{c}$, $\hat{f}$). Corollary 6.5 now yields an upper bound for the values $|u^{\varepsilon rp}_{(\xi)(\eta)}(x)|$ for $x \in \partial D$, $\xi, \eta \in E_d$, $|\xi| = |\eta| = 1$, $\xi \perp \psi_x(x)$, $\eta || \psi_x(x)$, which for large p depends only on γ, κ, δ, K, K_1, δ_1, the norms of ψ in $C^2(D)$, and the suprema of the functions $|u^{\varepsilon rp}\psi^{-1}|$ and $|u^{\varepsilon rp}_x|$ over $\Delta_\kappa D$. These suprema converge to the already estimated suprema of $|u^\varepsilon\psi^{-1}|$ and $|u^\varepsilon_x|$ as $p \to \infty$, $r \to \infty$ (the functions $u^{\varepsilon rp}\psi^{-1}$ can be studied as, for example, in the proof of Theorem 3.1). Thus we have derived the desired estimates of $|u^\varepsilon_{(\xi)(\xi)}|$ on ∂D, which completes the proof of estimates (10).

15. Example. Let ψ be strictly concave, $\psi \in C^4(D)$, and let $g \in C^2(D) \cap C(\bar{D})$ be a function such that $g > 0$ near ∂D and $g < 0$ at some points of D. We consider the problem of stretching a convex envelope over the graph of g and contour ∂D which most tightly enclose the graph of g, i.e., such that the solid composed of this envelope and the 'lid' of D has the minimum volume. This problem reduces to that of finding a function u which is equal to zero on ∂D and satisfies the conditions

$$(u_{xx}) \geq 0, \quad u \leq g, \quad (g - u)\det(u_{xx}) = 0 \text{ (a.e.D)}. \tag{14}$$

Putting aside the question of the equivalence between these two problems, we prove the existence and uniqueness of a solution to system (14) in the class of functions which are equal to zero on ∂D and have bounded generalized derivatives u_x and u_{xx}.

To this purpose we fix an integer $p \geq 2$ and set

$$F(u_{ij}, u_i, x) = \inf_{\omega\in\Omega,\ \theta\in(0,1-p^{-1})} \left[\omega^{ij}u_{ij} + \frac{\theta}{1-\theta}(g - u)\right],$$

where Ω is the set in Example 7. The function F falls within our scheme if $\sigma^k(\omega, \theta)$ is the kth column of the matrix $\sqrt{2\omega}$, $b = 0$, $c(\omega, \theta) = -\theta(1 - \theta)^{-1}$, $f(\omega, \theta) = \theta(1 - \theta)^{-1}g$. We also define $\kappa \in (0, 1)$ such that the inequalities $g > 0$ and $|\psi_x| > 0$ hold in $\overline{\Delta_\kappa D}$, and let $\hat{\sigma}^k(\omega, \theta) =$

$\hat{\sigma}^k(\omega)$ be the kth column of the matrix $\sqrt{2\omega}$, $\hat{b}^i = \hat{c} = \hat{f} = 0$. Let us check that the introduced objects satisfy conditions (a) — (d) of Remark 14.

Condition (c) obviously holds, the check of condition (d) (with $\gamma = 1/2$, $B = -(1/2)\psi_{xx}$, and some constants K, K_1, δ, δ_1, δ_2) is carried out as in the proof of Theorem 10. For our further consideration we take a $\zeta \in C_0(D)$ such that $0 \le \zeta \le 1/2$, $\zeta g \le 0$, $\zeta g \not\equiv 0$, and define v^ε as a solution to the problem $\varepsilon^2 \Delta v - \varepsilon v + d^{-1}\Delta v + \zeta(1-\zeta)^{-1}(g - v) = 0$ in D, $v = 0$ on ∂D.

It is clear that $v^0 \in C^2(\bar{D})$, $v^0 \le 0$ in D, $(v^0_x, \psi_x) \le -2\nu_1$ on ∂D, where ν_1 is a positive constant. It is also clear that for all sufficiently small $\varepsilon > 0$ we have $v^\varepsilon \le 0$ in D, $(v^\varepsilon_x, \psi_x) \le -\nu_1$ on ∂D. Since $F \le d^{-1}\delta^{ij}u_{ij} + \zeta(1-\zeta)^{-1}(g - u)$, we obtain (cf. Remark 5) $u^\varepsilon \le v^\varepsilon$ in D and $(u^\varepsilon_x, \psi_x) \le -\nu_1$ on ∂D.

It follows from Remark 4 that the operator F is weakly nondegenerate along the normal. Moreover, for small ε, $u^\varepsilon \le v^\varepsilon \le 0$ in D, $g - u^\varepsilon > 0$ in $\Delta_\kappa D$, the infimum of $\theta(1-\theta)^{-1}(g - u^\varepsilon)$ over $\theta \in (0, 1 - p^{-1})$ is equal to zero on $\Delta_\kappa D$ and u^ε satisfies equation (13) in $\Delta_\kappa D$, i.e., condition (a) holds. Also, note that the left-hand side of (2) calculated for $u = -N_1\psi$ is clearly positive if $g + N_1\psi > 0$ in D. By the maximum principle, $u^\varepsilon \ge -N_1\psi$, $|u^\varepsilon| \le N_1\psi$ for all sufficiently small ε. To check the remaining conditions in (b), it suffices to set $\tilde{u} = 1$, $\chi = \theta(1-\theta)^{-1}$, and to note that in such a case the function M is greater than $(1/2)(1-\delta_1)(2-\delta)\chi$, the left-hand side of (5.23) is equal to zero, and inequalities (5.38) and (5.39) can easily be furnished by choosing a sufficiently large K.

Note that the desired constants γ, κ, δ, K, K_1, δ_1, λ, ν could be chosen independent of p. By Theorem 9 for every integer $p \ge 2$ there exists a function u_p satisfying properties (a) and (b) of Theorem 9, possessing bounded generalized derivatives u_{px} and u_{pxx}, and such that

$$\inf_{\omega\in\Omega,\theta\in(0,1-p^{-1})} \left[(1-\theta)\omega^{ij}u_{px^ix^j} + \theta(g - u_p) \right] = 0 \quad (\text{a.e.}D). \tag{15}$$

It follows from the latter relationship and the upper bounds on $u_{p(\xi)(\xi)}$, which are uniform in p, that (e.g., see Theorem 2.2.6) $||u_{pxx}||$ are uniformly bounded on D. Furthermore, following, for example, the proof of Theorem 7.4.4 and using Theorem 3.5.9 (or Theorem 3.6.3), we can easily pass to the limit in (15) as $p \to \infty$ and to find a function $u \in C(\bar{D})$ with bounded generalized derivatives u_x and u_{xx}, which is equal to zero on ∂D and such that

$$\inf_{\omega\in\Omega,\theta\in(0,1)}\left[(1-\theta)\omega^{ij}u_{x^ix^j}+\theta(g-u)\right]=0 \text{ (a.e.D)}. \tag{16}$$

To prove the existence of a solution to system (14), it remains for us to note that it is equivalent to the single equation (16). The uniqueness of the solution to (16) immediately follows from Theorem 3.4.7 (or as in Theorem 9, it can be derived from Corollary 3.7.10).

8.8. The Solvability of Degenerate Equations in a Domain

The results established in Section 7 turn out to be effective only if the normal derivatives of second order are directly estimated on ∂D from an equation via other derivatives. In this section we develop the methods of estimating the second derivatives in the case when the equation in question does not have the above property. The basic result of this section is Theorem 1 which essentially reduces the solvability of an equation to the solvability of some inequalities. These inequalities are of rather complicated form, which gives rise to Theorems 4 and 7 concerning the solvability of equations with constant coefficients and equations with a 'large' coefficient of u.

Theorems 4 and 7 are obtained from Theorem 1 as a result of solving the mentioned inequalities in specific cases.

Let $\delta \in (0, 1)$, $K \geq 0$, let $d, d_1 \geq 1$ be integers, let $\psi \in C^3_{loc}(E_d)$, let $D := \{x \in E_d : \psi(x) > 0\}$ be a nonempty bounded domain, and let $\psi \in C^4(D)$, $|\psi_x| > 0$ on ∂D. Also, suppose that Ω is a set and, for $\omega \in \Omega$, $i = 1, \ldots, d$, $k = 1, \ldots, d_1$, $x \in D$, real-valued functions $\sigma^{ik}(\omega, x)$, $b^i(\omega, x)$, $c(\omega, x)$, and $f(\omega, x)$ are defined which are twice continuously differentiable with respect to x in D and bounded on $\Omega \times D$ along with their first and second derivatives with respect to x. As always, we shall use notation (7.2.1) — (7.2.3) (where z is replaced by x). We need a matrix function $B(x) = (B^{ij}(x);\ i, j = 1, \ldots, d)$ such that $B = B^* \geq 0$, $B \in C^2(D)$.

Prior to stating Theorem 1, recall that the function $J = J(\omega, x, \xi, t)$ was introduced in (6.22).

1. **THEOREM.** For all x, ω, ξ, t satisfying the conditions

$$x \in D,\quad \omega \in \Omega,\quad 4\psi(x)t \geq |\psi_{(\xi)}(x)|^2,\quad (B\xi, \xi) + t = 1,$$

let the inequality

$$J(\omega, x, \xi, t) + \frac{1}{2}\psi\sum_k\left[\left(B_{(\sigma^k)}\xi, \xi\right) + 2(B\xi, \sigma^k_{(\xi)})\right]^2 +$$

$$+ \frac{1}{2}\psi_{(\xi)}\left[\left(B_{(\sigma^k)}\xi, \xi\right) + 2(B\xi, \sigma^k_{(\xi)})\right]\cdot\left[2(B\xi, \sigma^k) + \right.$$

$$+ (\sigma^k, \psi_x)_{(\xi)}\Big] + \frac{1}{2} t \sum_k [2(B\xi, \sigma^k) + (\sigma^k, \psi_x)_{(\xi)}]^2$$

$$\leqq -(1 - \delta)L\psi + K\Big[\psi \sum_k (B\xi, \sigma^k)^2 +$$

$$+ \frac{1}{2} \psi_{(\xi)} (B\xi, \sigma^k)(\sigma^k, \psi_x) + \frac{1}{4} t \sum_k (\sigma^k, \psi_x)^2\Big] \tag{1}$$

hold.

In addition, let $|\psi_x|^2 + \psi \geqq \delta^2$, $(B\xi, \xi) \geqq \delta|\xi|^2$, $L\psi \leqq -\delta$ for all $\omega \in \Omega$, $\xi \in E_d$, $x \in D$. Then all assertions of Theorem 7.9, where the parameters affecting N in assertion (b) are not specified, hold true.

To prove this theorem, we need the following.

2. LEMMA. Let the hypotheses of Theorem 1 (and the assumptions in the beginning of this section) hold true and suppose that the infimum in (7.2.3) is attainable. Then if $u \in C^3(D) \cap C(\bar{D}) \cap C^4_{loc}(D)$, u is a solution to equation (6.1) in D, u = 0 on D, then $|u| \leqq N\psi$, $|u_x| \leqq N$, $u_{(\xi)(\xi)} \leqq N$ on D for all unit vectors $\xi \in E_d$, where N depends only on d, K, δ, and the constants bounding above the norm of ψ in $C^4(D)$, the norms of B^{ij}, b^i, c, f, in C(D), and the sums of squared σ^{ik} and their derivatives with respect to x^r, $x^r x^s$ on D.

Proof. We denote by N various constants depending only on the initial objects specified in the statement of the lemma. Note that the estimate $|u| \leqq N\psi$ immediately follows from the maximum principle. We now repeat the construction carried out after the proof of Corollary 6.5. Here, however, there is no hope of pasting local estimates into a 'global' one, therefore we take $\kappa = 1 + \max\{\psi, D\}$ and omit κ in the denotation of $U(\kappa)$ and $U'(\kappa)$. Note that $|\bar{\psi}_x|^2 = |\psi_x|^2 + 4[x]^2$ and this expression is greater than δ^2 on U since $|\psi_x|^2 + 4\psi \geqq \delta^2$ on D by hypothesis. To use the results of Section 5, it is formally required that $|\bar{\psi}_x| > 0$ on V, and since this is not the case for V from Section 6, we set

$$V := \{x \in E_{d+4} : 0 < \psi(x),\ |\psi(x) - [x]^2| < \delta_1\},$$

and the constant $\delta_1 \in (0, 1)$ is chosen so that $|\bar{\psi}_x| > 0$ on V.

Of course, all calculations which lead us to Lemmas 6.6 and 6.7 remain in force. Furthermore, we would like to apply Theorem 5.9 to $\bar{\psi}$, $\bar{\sigma}$, $\bar{b}$, $\bar{c}$, f, $\bar{B}$, and the function $\bar{u} = u\psi^{-1}$ satisfying the equation

$$\inf_{\omega}[\bar{L}(\omega)\bar{u} + f(\omega)] = 0$$

on U in the case when Remark 5.6 is used to check the conditions of Lemma 5.3. The first hypothesis of Remark 5.6 holds true for $\bar{\psi}$, $\bar{\sigma}$, $\bar{b}$, $\bar{c}$, f, and $\bar{B}$ in view of our assumptions if the constant K in this hypothesis if replaced by a constant N. Condition (5.22) obviously holds U' for $\delta_1 = \delta$. Inequality (5.21) with $(\bar{B}\xi, \xi) = 1$ can, by Lemma 6.7, be rewritten in terms of the initial σ, b, c, B in the form (1) on U' if the functions $\psi_{(\xi)}$, t, and ψb in the definition (6.22) of J and in (1) are replaced by $2x^\nu\xi^\nu$, $[\xi]^2$, and $[x]^2 b$. Since we consider inequality (5.21) only for $(x, \xi) \in U'$, i.e., when $\bar{\psi}_{(\xi)} = 0$, $\psi_{(\xi)} - 2x^\nu\xi^\nu = 0$, $|\psi_{(\xi)}|^2 \leqq 4[x]^2[\xi]^2 = 4\psi[\xi]^2$, it follows from (1) that condition (5.21) for $\bar{\psi}$, $\bar{\sigma}$, $\bar{b}$, $\bar{c}$, and $\bar{B}$ holds on $U' \cap \{(\bar{B}\xi, \xi) = 1\}$. In view of homogeneity in ξ this inequality holds for any $(\bar{B}\xi, \xi)$.

We now check the other conditions of Theorem 5.9. We have already noted that $|\bar{u}| \leqq N$. By Theorem 3.1 and the inequalities $|\bar{\psi}_x| \geqq \delta$ and $(\bar{B}\xi, \xi) \geqq \delta|\xi|^2$, which hold true on U', the inequalities $|\bar{u}_{(\xi)}|^2$, $\tilde{\partial}^2(\xi)\bar{u} \leqq M(\bar{B}\xi, \xi)$ also hold on this set for a (finite) constant M. Let us take $\psi^{-\alpha}$ as ψ_1 in Theorem 5.9 and choose $\alpha \in (0, 1)$ such that

$$\bar{L}\psi^{-\alpha} = L\psi^{1-\alpha} = (1 - \alpha)\psi^{-\alpha}\left(\frac{\alpha}{1 - \alpha} c\psi + L\psi\right) -$$

$$- (1 - \alpha)\alpha\psi^{-\alpha-1}(a\psi_x, \psi_x) \leqq 0$$

on U (see Lemma 6.6d)). For such a function ψ_1 the sets $\bar{U}_m \setminus U$ are empty and, by Theorem 5.9, we conclude

$$\left|\sum_{i=1}^{d+1} \xi^i\tilde{\partial}_i\bar{u}\right|^2, \quad \sum_{i,j=1}^{d+4} \xi^i\xi^j\tilde{\partial}_i\tilde{\partial}_j\bar{u} \leqq N \sum_{i=1}^{d+4} |\xi^i|^2 \tag{2}$$

on U'.

The vector $(\tilde{\partial}_1\bar{u}(x), \ldots, \tilde{\partial}_{d+4}\bar{u}(x))$ with $x \in U$ results from the orthogonal projection of the gradient of u on the tangent plane to U at x. Specifically, this vector lies in the tangent plane, and since the first estimate in (2) is valid for any $\xi \perp \bar{\psi}_x(x)$, it follows that the squared length of this vector does not exceed N. The squared length can easily be found by the Pythagoras theorem if it is taken into account that $\bar{u}$ is independent of $x^{d+1}, \ldots, x^{d+4}$ and the first d coordinates of the unit normal to the surface U at x are determined as $\psi_{x^i}(x) \times$

$(|\psi_x(x)|^2 + 4\psi(x))^{-1/2}$. Hence we obtain

$$N \geq \sum_{i=1}^{d+4} |\partial_i \bar{u}|^2 = \sum_{i=1}^{d} |\bar{u}_{x^i}|^2 - (\bar{u}_x, \psi_x)^2(|\psi_x|^2 + 4\psi)^{-1}$$

$$= \psi^{-2}(|\psi_x|^2 + 4\psi)^{-1}\Big[4\psi|u_x - \frac{u}{\psi}\psi_x|^2 + |\psi_x|^2 \cdot |u_x|^2 -$$

$$- (u_x, \psi_x)^2\Big], \tag{3}$$

and it follows that $|u_x - u\psi^{-1}\psi_x|^2 = |u_x - \bar{u}\psi_x|^2 \leq N\psi$, and since $|\bar{u}| \leq N$, we have $|u_x| \leq N$ on U or equivalently on D.

To derive consequences from the second inequality in (2), we first apply Lemma 6.6(d) and (b) and then Definition (3.7) and Lemmas 4.2 and 4.5. Thus, for any $\xi \in E_d$ we obtain

$$\sum_{i,j=1}^{d} \xi^i\xi^j u_{x^ix^j} = \sum_{i,j=1}^{d} \xi^i\xi^j L_{ij}\bar{u} = \sum_{\nu=d+1}^{d+4} \sum_{i,j=1}^{d+4} \xi^i_\nu\xi^j_\nu \tilde{\partial}_i\tilde{\partial}_j\bar{u} +$$

$$+ 2\sum_{i,j=1}^{d} \xi^i\xi^j\psi_{x^i}\tilde{\partial}_j\bar{u} +$$

$$+ \frac{1}{2}\psi_{(\xi)(\xi)} \sum_{\nu=d+1}^{d+4} x^\nu\tilde{\partial}\nu\bar{u} + \bar{u}\psi_{(\xi)(\xi)}$$

on U, where the vector $\xi_\nu = (\xi^1_\nu, \ldots, \xi^{d+4}_\nu)$ with $\nu = d + 1, \ldots, d + 4$ is defined as $\xi^i_\nu = x^\nu\xi^i$ for $i \leq d$, $\xi^\mu_\nu = \delta^{\mu\nu} \cdot \frac{1}{2}\psi_{(\xi)}(x)$ for $\mu = d + 1, \ldots,$ $d + 4$. Obviously, $\xi_\nu \perp \bar{\psi}_x$ and it follows from (2) and (3) that $u_{(\xi)(\xi)} \leq N|\xi|^2$ on U and on D. The lemma is proved.

3. <u>Proof of Theorem 1</u>. Theorem 7.9 asserts the uniqueness and the boundedness of some derivatives under certain conditions. In the case at hand these assertions are proved by the same arguments which were developed in the proof of Theorem 7.9.

It remains to prove the existence of the function u satisfying the assertions (a) — (c) of Theorem 7.9. As we always do in such situations, let us take an $\varepsilon > 0$ and consider problem (7.2) and (7.3). To prove its solvability, we will apply Theorem 6.4.2 keeping in mind the situation described in Example 6.1.8. Formally, we cannot do this since we have no assumption concerning the negativity of c and, thus, condition (6.1.8) can be violated (for small ε). Of course, the assumption $c \leq 0$ could not strongly affect the generality of the result, however, this assumption would be helpful only in this place, while we have already encountered the difficulty in question in Examples 6.4.7 and 2.2. Following Example 2.2, we change the unknown function in (7.2) as $v = u(\gamma + \psi)^{-1}$, where $\gamma > 0$. It can easily be shown that, by the inequality $L\psi \leqq -\delta$ for all sufficiently small ε and γ, the equation for v is the case studied in Example 6.1.8. Therefore, the equation for v and, consequently, equation (7.2) with the boundary condition (7.3) for sufficiently small ε have solutions of $C^{2+\alpha}(\bar{D})$ where $\alpha > 0$.

Let u^{ε} be a solution to problem (7.2) and (7.3) of the class $C^{2+\alpha}(\bar{D})$. The next step consists of applying Lemma 2 to u^{ε}. Here we again encounter difficulties caused by the fact that equation (7.2) is not of the form (6.1) and u^{ε} do not generally belong to $C^3(D) \cap C^4_{loc}(D)$. However, for sufficiently small ε these difficulties can be surmounted as in the proof of Lemma 7.8. We do not repeat the discussion again.

Finally, as in the proof of Theorem 7.9, we obtain the desired function u by passing to the limit as $\varepsilon \downarrow 0$ and using the arguments of the proof of Lemma 7.3.4. The theorem is proved.

Let us consider a simple example of applying Theorem 1.

4. THEOREM. Let $D = \{x \in E_d : |x| < 1\}$, let $\psi = 1 - |x|^2$, let a^{ij} and b^i be independent of x, and let $c \leqq 0$, $2|b| \leqq (1 - \delta)\text{tr } a$, $\delta \leqq \text{tr } a \leqq K$ for all $\omega \in \Omega$. Then all assertions of Theorem 7.9 hold except for the indication of the parameters on which the constant N depends in assertion (b) of this theorem.

Proof. We take the kth column of the matrix $\sqrt{2a}$ as σ^k, $k = 1, \ldots, d$. Note that $L\psi = -2\text{tr } a - 2(x, b) + c\psi \leqq -(1 + \delta)\text{tr } a$. Also, let $B = (\delta^{ij})$. Then the left-hand side of (1) is

$$-2[(x, \xi)(\xi, b) + t(x, b)] \leqq 2(|\xi|^2 + t)|b|$$
$$\leqq (|\xi|^2 + t)(1 - \delta)\text{tr } a \leqq -(|\xi|^2 + t)(1 - \delta)L\psi.$$

Therefore, condition (1) for $(B\xi, \xi)$ holds true. The other conditions of Theorem 1 obviously hold and the theorem is proved.

5. Remark. Theorem 4 extends Theorem 1.1 and makes it possible to study equations with lower terms in a ball. We shall present an extension of Theorem 1.1 to arbitrary smooth strictly convex domains in Remark 9.

6. Remark. The results of this chapter and the previous one can be applied to linear equations, and it is possible to study some cases which are not covered by the theory presented in Oleinik and Radkevich [1]. As an example one can consider the problem

$$u_{x^1} + u_{x^2x^2} + \ldots + u_{x^dx^d} = f \text{ in } D, \quad u = 0 \text{ on } \partial D, \tag{4}$$

where $f \in C^2(D)$, $D = \{|x| < 1\}$. When $d \geq 4$, Theorem 4 guarantees the existence of a solution of problem (4) whose all second derivatives are bounded above and, by linearity of the problem, below. Also, note that in case of $d = 2$, $f(-1, 0) \neq 0$, u_{x^1} of the solution of problem (4) is unbounded near the point $(-1, 0)$.

For a slightly modified problem (4), i.e., for the problem $-|u_{x^1}| + u_{x^2x^2} + \ldots + u_{x^dx^d} = f$ in D, $u = 0$ on ∂D with the same f, D and $d \geq 4$, Theorem 4, by the formula $(-|u_{x^1}|) = \min(u_{x^1}, -u_{x^1})$, guarantees the existence of a solution whose second derivatives are bounded above and the derivatives u_{x^i}, $i = 1, \ldots, d$, $u_{x^ix^j}$, $i, j \geq 2$, are bounded.

In conclusion we perform the analysis of the conditions of Theorem 1 in the case when $c \leq 0$ and the value $|c|$ can be chosen arbitrarily large. It turns out that inequality (1) has to be checked only on the boundary.

7. THOEREM. Let Ω be a metric compactum, let σ, b, B, and their first derivatives with respect to x be continuous on $\Omega \times \bar{D}$. Suppose that $(B\xi, \xi) \geq \delta|\xi|^2$ for all $\xi \in E_d$, $x \in D$. Also, for all $\omega \in \Omega$, $x \in \partial D$, $t \geq 0$, $\xi \in E_d$ such that $\xi \perp \psi_x(x)$, $(B(x)\xi, \xi) + t = 1$, $(a(\omega, x)\psi_x(x), \psi_x(x)) = 0$, let the inequalities

$$L^0\psi \leq -\delta, \tag{5}$$

$$\sum_k |(\sigma^k, \psi_x)_{(\xi)}|^2 + 4(B\xi, (a\psi_x)_{(\xi)}) + $$
$$+ t\Big[(B\sigma^k, \sigma^k) + L^0\psi + \frac{1}{2}\sum_k (2(B\xi, \sigma^k) + (\sigma^k, \psi_x)_{(\xi)})^2\Big]$$
$$\leq -(1 - \delta)L^0\psi \tag{6}$$

hold. (If there are no $\omega \in \Omega$, $x \in \partial D$ satisfying $(a(\omega, x)\psi_x(x), \psi_x(x)) = 0$, then inequalities (5) and (6) are not required to hold for any ω, x, t, ξ).

Then there exists a constant m_0 depending only on D, Ω, ψ, σ, b, δ, d, c such that for every $m \geq m_0$ all assertions of Theorem 7.9 hold if we omit the parameters on which N depends and $c(\omega, x)$ in the definition of F is replaced by $c(\omega, x)-m$.

Proof. We define an infinitely differentiable function $r(s)$ on $[0, \infty)$ such that $r(s) = s - s^2$ for small s and $\frac{1}{2} \leq r'(s) \leq 2$, $r''(s) < 0$ for all

$s \geq 0$. For $n = 1, 2, \ldots$ we set $\psi_n = n^{-1}r(n\psi)$ and prove that $L\psi_n - n\psi_n \leq -(1/4)\delta$ on $\Omega \times \bar{D}$ for sufficiently large n. If we assume the contrary, then there exist infinitely many values of n such that some points (ω_n, x_n) furnish the opposite inequality $r'(n\psi)L^0\psi + c\psi_n + nr''(n\psi) \times (a\psi_x, \psi_x) \geq n\psi_n - (1/4)\delta$.

The left-hand side of this inequality is bounded above; therefore, $n\psi_n(x_n)$ is bounded. Since $\psi \leq 2\psi_n \leq 4\psi$, it follows that $n\psi(x_n)$ is bounded, $nr''(n\psi(x_n)) \to -\infty$; and for every limit point (ω_0, x_0) of the sequence (ω_n, x_n) we obtain $\psi(x_0) = 0$, $x_0 \in \partial D$;

$$(a(\omega_0, x_0)\psi_x(x_0), \quad \psi_x(x_0)) = 0,$$

$$L^0(\omega_0, x_0)\psi(x_0) \geq \varliminf_{n\to\infty}\left(-\frac{1}{4}\delta\right)(r'(n\psi(x_n)))^{-1} \geq -\frac{1}{2}\delta.$$

The latter is impossible in view of (5). Moreover, it is easily seen that for $x \in \partial D$, $\xi \perp \psi_x(x)$ we have $(\sigma^k, \psi_x)_{(\xi)} = (\sigma^k, \psi_{nx})_{(\xi)}$, $-L^0\psi_n \geq -L^0\psi$.

Therefore, reducing δ and passing, if necessary, from c, ψ to c - n, ψ_n for an appropriate n, we can and will assume that the second hypothesis of Theorem 1 holds in addition to the hypotheses of the current theorem. Of course we can assume that $c \leq 0$ in which case the consideration of the same function ψ_n enables us to preserve the results obtained and shows that, by (6) for large n, the inequality

$$2(a\psi_{nx}, (B\xi, \xi)_x) + \sum_k |(\sigma^k, \psi_{nx})_{(\xi)}|^2 + 4(B\xi, (a\psi_{nx})_{(\xi)})$$
$$\leq -(1 - \delta/2)L^0\psi_n,$$

holds whenever $\xi \perp \psi_x(x)$, $x \in \partial D$, $(B\xi, \xi) = 1$. In short, we can assume that the second condition of Theorem 1 and the hypotheses of the current theorem hold and

$$2(a\psi_x, (B\xi, \xi)_x) + \sum_k |(\sigma^k, \psi_x)_{(\xi)}|^2 + 4(B\xi, (a\psi_x)_{(\xi)})$$
$$\leq -(1 - \delta)L^0\psi \tag{7}$$

with $x \in \partial D$, $\xi \perp \psi_x(x)$, $(B(x)\xi, \xi) = 1$.

It now remains for us to show that if c, δ, K in (1) are replaced by c - m, $(1/2)\delta$, and m, then for sufficiently large m the first hypothesis of Theorem 1 holds. Again, assume the contrary. Then for every $m = 1, 2 \ldots$ there exist $x_m \in D$, $\omega_m \in \Omega$, $\xi_m \in F_d$, and $t_m \geq 0$ such that for $(\omega, x, \xi, t) = (\omega_m, x_m, \xi_m, t_m)$ we obtain $4\psi t \geqq |\psi_{(\xi)}|^2$, $(B\xi, \xi) + t = 1$ and the left-hand side of (1) is greater than or equal to

$$-\left(1 - \frac{\delta}{2}\right)L\psi + \left(1 - \frac{\delta}{2}\right)m\psi +$$
$$+ m\left[\psi \sum_k (B\xi, \sigma^k)^2 + \frac{1}{2}\psi_{(\xi)}(B\xi, c^k)(\sigma^k, \psi_x) + \frac{1}{2}t(a\psi_x, \psi_x)\right].$$

Since this sequence is, in particular, bounded, it follows that $m\psi(x_m)$ is bounded, $\psi(x_m) \to 0$, $4\psi(x_m)t_m \geqq |\psi_{(\xi_m)}(x_m)|^2 \to 0$ as $m \to \infty$, and for any limit point $(\omega_0, x_0, \xi_0, t_0)$ of the sequence $(\omega_m, x_m, \xi_m, t_m)$ we have

$$t_0(a(\omega_0, x_0)\psi_x(x_0), \psi_x(x_0)) = 0, \quad \psi(x_0) = 0, \quad \psi_{(\xi_0)}(x_0) = 0,$$

$$(B(x_0)\xi_0, \xi_0) + t_0 = 1,$$

$$2(a\psi_x, (B\xi_0, \xi_0)_x) + \sum_k |(\sigma^k, \psi_x)_{(\xi_0)}|^2 + 4(B\xi_0, (a\psi_x)_{(\xi_0)}) +$$
$$+ t_0\left[(B\sigma^k, c^k) + L^0\psi + \frac{1}{2}\sum_k (2(B\xi_0, \sigma^k) + (\sigma^k, \psi_x)_{(\xi_0)})^2\right]$$
$$\geqq -\left(1 - \frac{\delta}{2}\right)L^0\psi. \tag{8}$$

If $t_0 = 0$, then $(B\xi_0, \xi_0) = 1$ and (8) contradicts (7). But if $t_0 > 0$, then $(a\psi_x, \psi_x) = 0$ and (8) contradicts (6). The theorem is proved.

8. Remark. Suppose that Ω is a metric compactum, σ, b, and their first derivatives are continuous on $\Omega \times \bar{D}$, $L^0\psi \leqq -\delta$ on $\Omega \times \partial D$, and for every $\omega \in \Omega$ the set $\Gamma(\omega) := \{x \in \partial D : (a(\omega, x)\ \psi_x(x), \psi_x(x)) = 0$ is the closure of the set of its relatively (in the topology on ∂D) interior points. Then the assertion of Theorem 7 still holds true.

Indeed, by definition, $(\sigma^k, \psi_x) = 0$, $a\psi_x = 0$ on $\Gamma(\omega)$. Consequently, at any relatively interior point $x \in \Gamma(\omega)$ for $\xi \perp \psi_x(x)$ we obtain $(\sigma^k, \psi_x)_{(\xi)} = 0$, $(a\psi_x)_{(\xi)} = 0$. By continuity, these equalities can be extended to any points $x \in \Gamma(\omega)$ and $\xi \perp \psi_x(x)$. Specifically, the left-hand side of (6) with $B = (\delta^{ij})$, $x \in \partial D$, $\xi \perp \psi_x(x)$, $(a\omega, x)\psi_x(x), \psi_x(x)) = 0$ is equal to $t2\text{tr } a + tL^0\psi + 4t(a\xi, \xi)$ and it can be made smaller than $(-\frac{1}{2})L^0\psi$ merely by substituting $N\psi$ for ψ, where N is a sufficiently large constant.

9. Remark. Suppose that Ω is a metric compactum, σ, b are independent of x and continuous on Ω, ψ is strictly concave on $\bar{D}$, $(b, \psi_x) \leq -(1 - \delta)L^0\psi$, $L^0\psi \leq -\delta$ for $x \in \partial D$, and $\omega \in \Omega$. Then the assertion of Theorem 7 remains in force.

To prove this fact it suffices to note that for $B := -\frac{1}{2}\psi_{xx}$ the left-hand side of (6) is equal to $t(b, \psi_x)$. This remark is readily applicable to example (4) considered in a domain other than the ball. The fact that in this case $c \equiv 0$ makes no difference since the equation is parabolic and the change $u \to u \exp(\lambda x^1)$ allows any c. Also, note that if $b \equiv 0$, then all conditions in the statement of the remark can obviously be replaced by the following three: σ is independent of x and continuous on the compactum Ω, $\operatorname{tr} a > 0$, and D is strictly convex and belongs to the class C^4.

8.9. Comments

Sections 1 — 3. The ideas, results, and some examples of these sections are taken from Krylov [17]. This paper, however, was written in probabilistic language and the estimates of solutions was established in C^2. The strongest results on the solvability of degenerate parabolic Bellman equations in a smooth cylinder of general form were obtained by Safonov [1 - 3], where it is required that the operators under consideration are uniformly nondegenerate along the normal to the boundary (see Comments in Chapter 5, Section 1). The arguments in Remark 1.3 are typical of the theory of controlled diffusion processes (see Krylov [10]). Example 2.6 is taken from Krylov [23]. Equation (2.6) with $d = 2$ was earlier considered by Pogorelov [1] who assumed that A and f depend on u_x. The Monge—Ampère parabolic equation was introduced by Krylov [8, 9, 17]. Example 2.2 will be discussed later.

Sections 4 — 6. The material of these sections is taken from Krylov [22, 23]. Many of the results presented do not require that the equation under consideration has the structure of Bellman's equation, specifically, the function F in (5.25) may be nonconvex (nonconcave) in u_{ij} (see Krylov [22]). As compared with Krylov [23], Lemma 5.3 is substantially strengthened and now its validity for a function B depends only on the manner in which B acts on the tangent vectors.

Section 7. The notation of the weak nondegeneracy along the normal is first introduced in this book, although it was implicitly used in Krylov [22, 23]. Theorem 9 is an extension of the corresponding result in [23]. Theorem 10 under condition (a) was established by Safonov [1] (see also Lions [2] where the uniform nondegeneracy of operators on ∂D seems to be required). If the operators $L(\omega)$ are uniformly nondegenerate on ∂D, then the solution near ∂D belongs to $C^{2+\alpha}$ (see Chapter 6) and the interior estimates of derivatives can be obtained from Theorems 5.10 and 5.11. The $C^{2+\alpha}(\bar{D})$ solvability of the simplest Monge—Ampère equation in a smooth strictly convex domain was proved by Krylov [20, 22]. The same result in a ball in the class of functions whose second derivatives are bounded was earlier established by probabilistic techniques (see Krylov [17]). A large number of papers are devoted to investigating into the Monge–Ampère equation (e.g., see Aleksandrov [2], Pogorelov [1 - 3], Bakel'man [1], Sabitov [1], Cheng and Yau [1], Ivochkina [1], and the references there). Except for the two-dimensional case, the solvability of these equations was proved, prior to the results in Krylov [20 - 22], either in the generalized sense (Aleksandrov [2]), or in the class $C^n_{loc}(D)$, which was first done by Pogorelov [2, 3]. One can now obtain a great deal of information on the solvability of the Monge–Ampère equation in the classes $C^{n+\alpha}(\bar{D})$ with $f = f(u_x, u, x) > 0$ in Ivochkina [1], Caffarelli, Nirenberg, and Spruk [1 - 3], Caffarelli, Kohn, Nirenberg, Spruck [1], Urbas [1].

<u>Section</u> 8. The results of this section are taken from Krylov [23], where some other applications of Theorem 1 are described. In Krylov [28] it is proved if Ω is a metric compactum, $b = 0$, $c = 0$, σ is independent of x and continuous in ω, $\operatorname{tr} a > 0$, D is strictly convex and belongs to $C^{3,1}$, then the assertion of Theorem 7 holds true for $m_0 = 0$. Further results concerning degenerate equations in a domain can be found in Krylov [27].

Appendix 1

PROOF OF LEMMA 4.1.6

In Section 4.1 we made use of a fact concerning measure theory in Euclidean space. This fact was formulated in Lemma 4.1.6. Let us recall that in this lemma $\Gamma \subset C_{1,1}$ is a measurable set, $q, \eta, \zeta \in (0. 1)$, $\mathbf{B}$ is the family of all sets Q of the form $(t_0, x_0) + C_{\rho^2,\rho}$ which lie in $C_{1,1}$ and such that $\Lambda(Q \cap \Gamma) \geqq q\Lambda(Q)$. If $Q = (t_0, x_0) + C_{\rho^2,\rho} \in \mathbf{B}$, then we assume that $Q^0 = Q$.

$$Q^1 = (t_0, x_0) - C_{\eta^{-1}\rho^2,\rho}, \quad Q^2 = (t_0 - \eta^{-1}\rho^2, x_0) + C_{\eta^{-1}\rho^2\zeta^2,\rho\zeta}.$$

Moreover, let

$$\Gamma^i = \bigcup_{Q \in \mathbf{B}} Q^i, \quad i = 0, 1, 2.$$

The proof of Lemma 4.1.6 consists of combining the folloiwng Lemmas 1 — 4 which are proved as in Krylov and Safonov [1, 2] and Safonov [4].

1. LEMMA. $\Lambda(\Gamma \setminus \Gamma^0) = 0$ and if $\Lambda(\Gamma) \leqq q\Lambda(C_{1,1})$, then $\Lambda(\Gamma^0) \geqq (1 - \frac{1-q}{3^{d+1}})^{-1} \times \Lambda(\Gamma)$.

Proof. The first assertion immediately follows from the theorem asserting that almost every point of a measurable set is its Lebesgue point and from the fact that $q < 1$ (see Guzmàn [1]). Let us prove the second assertion.

Denote

$$\mathbf{B}^0 = \{Q \in \mathbf{B} : \Lambda(Q \cap \Gamma) = q\Lambda(Q)\}.$$

We take a cylinder $Q \in \mathbf{B}$ and for $\theta \in [0, 1]$ somehow define expanding cylinders $Q(\theta)$ of standard form such that $Q(0) = Q$, $Q(1) = C_{1,1}$ and $Q(\theta)$ is continuous in θ. Then the function $\varphi(\theta) := \Lambda(Q(\theta) \cap \Gamma)\Lambda^{-1}(Q(\theta))$ is continuous on $[0, 1]$, $\varphi(0) \geqq q$, $\varphi(1) \leqq q$. Therefore, $\varphi(\theta) = q$ for a θ and $Q \subset Q(\theta) \in \mathbf{B}^0$. Since the cylinder $Q \in \mathbf{B}$ is arbitrarily chosen, it follows that

$$\Gamma^0 = \bigcup_{Q \in \mathbf{B}^0} Q. \tag{1}$$

Furthermore, we construct a (countable or finite) sequence of cylinders Q_i in the following way. For $Q = (t_0, x_0) + C_{\rho^2,\rho} \in \mathbf{B}^0$ we denote $\rho(Q) = \rho$ and let $\rho_1 = \sup\{\rho(Q) : Q \in \mathbf{B}^0\}$. It readily follows from the dominated convergence theorem that there exists a cylinder $Q_1 \in \mathbf{B}^0$ such that $\rho(Q_1) = \rho_1$. Suppose that Q_i with $i = 1, 2, \ldots, k$ have already been constructed for a $k \geqq 1$. Denote

$$\mathbf{B}^k = \{Q \in \mathbf{B}^0 : Q \cap Q_i = \emptyset, \quad i = 1, \ldots, k\}.$$

If $\mathbf{B}^k$ is nonempty, we denote $\rho_{k+1} = \sup\{\rho(Q) : Q \in \mathbf{B}^k\}$. As above, we find that the supremum is attained for a $Q_{k+1} \in \mathbf{B}^k$. If the sets $\mathbf{B}^i$ are nonempty for all $i = 1, 2, \ldots$, we obtain a countable sequence of cylinders Q_i, $i = 1, \ldots$. But if $\mathbf{B}^i \neq \emptyset$ for $i = 1, 2, \ldots, k$, while $\mathbf{B}^{k+1} = \emptyset$, a finite sequence of cylinders Q_i, $i = 1, 2, \ldots, k+1$, results. In this latter case, by definition, we set $\rho_{k+2} = \ldots = 0$. Note that, by construction, the cylinders Q_i do not intersect and $\rho_i \downarrow 0$ as $i \to \infty$.

We now denote by $\tilde{Q}_i$ the cylinder which is obtained by the three-fold stretching of Q_i in E_{d+1} with the centre of stretching at the geometrical centre of Q_i. We assert that

$$\Gamma^0 \subset \bigcup_{i \geqq 1} \tilde{Q}_i. \tag{2}$$

Assume the contrary. Then by (1) for a $Q \in \mathbf{B}^0$ we obtain $Q \not\subset \tilde{Q}_i$, $i = 1, 2, \ldots$. For $\rho_i \geqq \rho(Q)$ this obviously implies $Q \cap Q_i = \emptyset$. Moreover, for a k it is clear that $\rho_1 \geqq \ldots \geqq \rho_k \geqq \rho(Q) > \rho_{k+1}$. Therefore, $Q \cap Q_i = \emptyset$ for $i = 1, \ldots, k$, $Q \in \mathbf{B}^k$ and then, by the definition of ρ_{k+1}, we obtain the inequality $\rho(Q) \leqq \rho_{k+1}$. This contradicts the inequality $\rho(Q) > \rho_{k+1}$ and proves (2).

It follows from (2) that

$$\Lambda(\Gamma^0) \leqq 3^{d+1} \sum_i \Lambda(Q_i).$$

Moreover, since the cylindres Q_i are pairwise nonintersecting and

$\Lambda((\Gamma^0 \setminus \Gamma) \cap Q_i) = \Lambda(Q_i) - \Lambda(Q_i \cap \Gamma) = (1 - q)\Lambda(Q_i)$, it follows that

$$\Lambda(\Gamma^0 \setminus \Gamma) \geqq \sum_i \Lambda((\Gamma^0 \setminus \Gamma) \cap Q_i) = (1 - q) \sum_i \Lambda(Q_i).$$

Hence

$$\Lambda(\Gamma^0) = \Lambda(\Gamma) + \Lambda(\Gamma^0 \setminus \Gamma) \geqq \Lambda(\Gamma) + (1 - q)3^{-d-1}\Lambda(\Gamma^0).$$

The lemma is proved.

The estimates of $\Lambda(\Gamma^0)$ in terms of $\Lambda(\Gamma^1)$ and $\Lambda(\Gamma^2)$ are obtained by use of the following lemma.

2. LEMMA. Let $\kappa \geqq 1$, let $-\infty \leqq t_1 < t_2 \leqq \infty$, let Σ be the set of all open subintervals of the interval (t_1, t_2), let $\sigma \subset \Sigma$, and let a function $g : \Sigma \to \Sigma$ be defined such that $|g(I)| \leqq \kappa|I|$ for $I \in \Sigma$, where $|g(I)|$, and $|I|$ are (one-dimensional) Lebesgue measures of $g(I)$ and I. Also, let $g(I_1) \subset g(I_2)$ if $I_1 \subset I_2$. Then

$$\left|\bigcup_{I\in\sigma} g(I)\right| \leqq \kappa\left|\bigcup_{I\in\sigma} I\right| .$$

Proof. The set $\cup\{I : I \in \sigma\}$ is open and can be represented as a sum of nonintersecting intervals I_n. Therefore,

$$\left|\bigcup_{I\in\sigma} g(I)\right| \leqq \left|\bigcup_n \bigcup_{I\in I_n} g(I)\right| \leqq \left|\bigcup_n g(I_n)\right| \leqq \sum_n \kappa|I_n| = \kappa\left|\bigcup_{I\in\sigma} I\right|$$

as we wished to prove.

3. LEMMA. $\Lambda(\Gamma^0) \leqq (1 + \eta)\Lambda(\Gamma^1)$.

Proof. It is obviously sufficient to prove that $\Lambda(\Gamma^0 \cup \Gamma^1) \leqq (1 + \eta) \times \Lambda(\Gamma^1)$ to which purpose, by Fubini's theorem, it is sufficient to prove that for any x the one-dimensional Lebesgue measure of the set $\{t : (t, x) \in \Gamma^1 \cup \Gamma^0\}$ does not exceed the product of $(1 + \eta)$ and the measure of $\{t : (t, x) \in \Gamma^1\}$. To prove this, we set $\kappa = 1 + \eta$, $t_1 = -t_2 = -\infty$, $\sigma = \{\{t : (t, x) \in Q^1\} : Q \in \mathbf{B}\}$ in Lemma 2 and define $g(I)$ as the result of the κ-fold stretching of I centred at the left extremity of I. The desired assertion now follows from Lemma 2 and the fact that

$$g(\{t : (t, x) \in Q^1\}) = \{t : (t, x) \in Q^1 \cup Q\},$$

$$\bigcup_{I\in\sigma} I = \{t : (t, x) \in \Gamma^1\}.$$

The lemma is proved.

4. LEMMA. $\Lambda(\Gamma^1) \leqq \zeta^{-d-2}\Lambda(\Gamma^2)$.

Proof. The set Γ^2 can be obtained by consecutively contracting Γ^1 d+1 times, first by a factor of ζ^{-2} along the t-axis, and by a factor of ζ^{-1} along each of the remaining axes separately. The first contraction reduces the volume of Γ^1 no more than ζ^{-2} times. This can be shown as in the previous lemma. Every contraction along a coordinate axis reduces the volume by no more than a factor of ζ^{-1}. This can be proved using Lemma 2. By Fubini's theorem, it suffices to prove that if σ is a set of intervals lying in (-1, 1) and g(I) is an interval centred at the same point as I, but ζ^{-1} times greater, then $\left|\bigcup_{I\in\sigma} g(I)\right| \leqq \zeta^{-1}\left|\bigcup_{I\in\sigma} I\right|$. This obviously follows from Lemma 2. The lemma is proved.

Appendix 2

ALEKSANDROV—BUSEMANN—FELLER THOEREM

In this book we have repeatedly used Theorem 1.2.1 and its extensions to functions depending on t. The Aleksandrov—Busemann—Feller theorem and its extensions are available only in journal papers, so for the reader's convenience we present the corresponding proof following Krylov [15 - 16].

Let a bounded Borel function $u(t, x)$ convex in x be defined in the domain $Q = (-1, 1) \times S_1$. We set $v(t, x) = \mathrm{Var}\{u(s, x) : s \in (-1, t]\}$ for $t < 1$, $v(1, x) = v(1-, x)$ and assume that $v(1, x)$ is fintie and summable with respect to S_1 (in terms of Section 3.6, $u \in W^{bv,+}(Q)$). Note that the latter condition makes sense since $v(t, x)$ increases and is continuous in t on the right, $v(t + \frac{1}{n} -, x) \to v(t, x)$, $v(t-, x)$ is lower semicontinuous in (t, x) and, therefore, $v(t-, x)$, $v(t, x)$ are Borel functions of (t, x). For functions u, which are independent of t, Theorem 1.2.1 follows from the theorem to come where the parameters involved in its statement do exist in view of Lemma 3.6.1.

1. THEOREM. Let

$$P(t, x, s, y) = u(t, x) + u_t^{(0)}(t, x)s + u_{x^i}(t, x)y^i + \frac{1}{2} u^{(0)}_{x^i x^j}(t, x)y^i y^j.$$

Then for almost all $(t, x) \in Q$ we obtain $u(t + s, x + y) = P(t, x, s, y) + o(|s| + |y|^2)$ as $(s, y) \to 0$.

To prove this theorem we need some auxiliary propositions. The first one is a version of the well-known Lebesgue theorem.

2. LEMMA. Let $Q(r) = (-r^2, r^2) \times S_r$ and let μ be a measure in Q. Then for almost all (in the sense of Lebesgue) points $(t, x) \in Q$ we have $|\mu - \mu^{(0)}(t, x)\Lambda|((t, x) + Q(r)) = o(r^{d+2})$ as $r \to 0$.

This proposition is presumably well known in the theory of measure differentiation, although the author cannot give an exact reference. Nevertheless, we merely sketch the proof here. It should be explained that $\mu^{(0)}(t, x)\Lambda$ is the Lebesgue measure in Q multiplied by the constant $\mu^{(0)}(t, x)$. If we took $Q'(r) = (-r, r) \times S_r$ instead of $Q(r)$ and $o(r^{d+1})$ instead of $o(r^{d+2})$, then we would obtain the exact assertion of the Lebesgue theorem. So the lemma sounds quite natural. However, it is known that if we take $(-r', r') \times S_r$ in place of $Q(r)$, where r' is *any* number in the segment $[r^2, r^{1/2}]$, replacing $o(r^{d+2})$ by $o(r'r^d)$, then the resulting assertion is false (see Appendix IV in Guzmàn [1]). Therefore the lemma deserves attention. The result can be derived from Theorems 7.2, 6.12, and 8.12 of Morse [1], whose results are formulated in Theorems 6.2.2, 4.2.5, and 2.3.2 of Hayes and Pauc [1] so that our lemma is straightforward. In the author's opinion, the simplest way to prove Lemma 2 is to repeat almost literally the proofs of Theorems 3.12.6, 3.12.8 and Lemma 3.12.5 of Dunford and Schwartz [1], the latter being proved by using, rather than Theorem 3.12.3 of this book, simple arguments of the proof of Theorem 3.1 in Chapter 1 of Guzmàn [1] (see Remark 1 in Section 1.3 there).

3. LEMMA. Let $n \geqq 1$ be an integer, let $h(x)$ be a locally summable function in S_1 which has all Sobolev derivatives up to the $(n - 1)$th ones, and such that all its derivatives $h_{x^{i_1}\ldots x^{i_n}}$ in the sense of the theory of generalized functions are measures in S_1. We denote

$$P(x, y) = h(x) + \ldots + \frac{1}{(n-1)!} h_{x^{i_1}\ldots x^{i_{n-1}}}(x)y^{i_1}\ldots y^{i_{n-1}} +$$

$$+ \frac{1}{n!} h^{(0)}_{x^{i_1}\ldots x^{i_n}}(x)y^{i_1}\ldots y^{i_n}.$$

Then for almost all $x_0 \in S_1$ we have $r^{-n}[h(x_0 + ry) - P(x_0, ry)] \to 0$ in $L_1(S_2)$ as $r \downarrow 0$.

Proof. We consider only one case, $n = 2$. The other cases when $n \geqq 1$ can be proved similarly.

Let us recall a notion related to Lebesgue measure. A point x_0 is said to be Lebesgue of a measure μ in S_1 if

$$|\mu - \mu^{(0)}(x_0)\Lambda|(x_0 + S_r) = o(r^d).$$

If the measure μ is defined as $z(x)dx$, then the Lebesgue points of the function z are described. By Lebesgue's theorem, for every measure μ almost every point is Lebesgue. Therefore, it suffices to prove the lemma for every point which is a common Lebesgue point for h, h_{x^i}, and $h_{x^ix^j}$. Let us fix such a point x_0. Without loss of generality we can assume that $x_0 = 0$. Moreover, since we can subtract an appropriate polynomial from h, it follows that we can assume that $h(x_0) = h_{x^i}(x_0) = h^{(0)}_{x^i x^j}(x_0) = 0$. Furthermore, we take a function $\zeta \geqq 0$, $\zeta \in C_0^\infty(E_d)$ such that $\int \zeta\, dx = 1$, $\zeta(x) = \zeta(|x|)$. For $\tau > 0$ let $h^{(\tau)}(x) = h(x) * \tau^{-d}\zeta(x\tau^{-1})$ and let $h^{(\tau)}_{x^i}$ and $h^{(\tau)}_{x^i x^j}$ be defined in a similar manner. Note that $(h_{x^i x^j})^{(\tau)} = (h^{(\tau)})_{x^i x^j}$. Finally, let $w^\tau(y) := h^{(\tau)}(y) - h^{(\tau)}(0) - h^{(\tau)}_{x^i}(0)y^i$.

For a fixed y the expression of $w^\tau(ry)$ is the remainder in the Taylor theorem for $h^{(\tau)}(ry)$. Writing it in Lagrange's integral form, next integrating with respect to y, and applying Fubini's theorem, we obtain

$$\int_{|y|\leqq 2} |h(ry)|\,dy = \lim_{\tau\downarrow 0} \int_{|y|\leqq 2} |w^\tau(ry)|\,dy$$

$$\leqq \overline{\lim_{\tau\downarrow 0}} \int_{|y|\leqq 2p\leqq 2r} (r-p)p^{-d-2}\left|h^{(\tau)}_{y^i y^j}(y)y^i y^j\right| dp\,dy$$

$$\leqq \overline{\lim_{\tau\downarrow 0}}\, 4r \sum_{i,j} \int_{|y|\leqq 2p\leqq 2r} p^{-d}\left|h_{y^i y^j}\right|^{(\tau)} dp\,dy$$

$$= \overline{\lim_{\tau\downarrow 0}}\, 4r \sum_{i,j} \int_{|y|\leqq 2r} \left(\left|\frac{1}{2}y\right|^{1-d} - r^{1-d}\right)\left|h_{y^i y^j}\right|^{(\tau)} dy$$

$$\leqq \overline{\lim_{\tau\downarrow 0}}\, 2^{d+1} r \int \eta_r^{(\tau)}(y)\left|h_{y^i y^j}\right|(dy), \tag{1}$$

where $\eta_r(y) = |y|^{1-d}$ for $|y| \leqq 2r$, $\eta_r(y) = 0$ for $|y| > 2r$. Note that $\eta^{(\tau)}(y) \leqq |y|^{1-d} * \tau^{-d}\zeta(y\tau^{-1})$, the latter being equal to $\tau^{1-d}|y_1|^{1-d} * \zeta(y_1)$ at the point $y_1 = y\tau^{-1}$. Since $|y_1|^{1-d} * \zeta(y_1) \leqq N(d)|y_1|^{1-d}$, it follows that $\eta^{(\tau)}(y) \leqq N|y|^{1-d}$. Thus, we obtain from (1)

$$\int_{|y|\leqq 2} |h(ry)|\,dy \leqq Nr \sum_{i,j} \int_{|y|\leqq 4r} |y|^{1-d}\left|h_{y^i y^j}\right|(dy)$$

$$= Nr \sum_{i,j} \int_{|y|\leqq (4r)\wedge p} p^{-d}\left|h_{y^i y^j}\right|(dy)\,dp$$

$$= Nr \sum_{i,j} \int_0^\infty p^{-d}\left|h_{y^i y^j}\right|(S_{(4r)\wedge p})\,dp$$

$$\leq Nr^2 \sum_{i,j} \sup_{p\in(0,4r]} p^{-d} \left| h_{y^i y^j} \right| (S_p).$$

The latter is $o(r^2)$ by choice of x_0. The lemma is proved.

In what follows we make use of one more standard fact of measure theory.

4. LEMMA. For every $t \in (-1, 1)$, let a measure $\mu(t, dx)$ be defiend in S_1 such that $\int f(x)\mu(t, dx)$ is a Borel function of t, summable on $(-1, 1)$ for any $f \in C_0^\infty(S_1)$. Then there exists the measure $\nu(dt\ dx) = \mu(t, dx)dt$ in Q. For this measure we have $\nu^a(dt\ dx) = \mu^a(t, dx)dt$ and $|\nu(dt\ dx)| = |\mu(t, dx)|dt$.

We now turn to studying the functions u and v introduced before Theorem 1. For every $x \in S_1$ the measure $\mu(dt, x)$ is introduced in $(-1, 1)$ in a natural way using the increasing function $v(t, x)$ which is continuous on the right. By Lemma 4, where t and x should be interchanged, the measure

$$\nu(dt\ dx) = \mu(dt, x)dx$$

is defined in Q.

5. LEMMA. $\nu^{(0)}(t, x) = |u_t^{(0)}(t, x)|$ almost everywhere in Q in the Lebesgue measure.

Proof. For every $x \in S_1$ the generalized derivative of u with respect to t in the sense of the theory of generalized functions is a measure on $(-1, 1)$. We denote by $\bar{u}_t(dt, x)$ the differential of this measure. Let $\bar{u}_t^a(dt, x)$ and $\mu^a(dt, x)$ be absolutely continuous parts of the measures $\bar{u}_t(dt, x)$ and $\mu(dt, x)$ in one-dimensional Lebesgue measure. Their Radon—Nikodym derivatives are known to coincide, almost everywhere on $(-1, 1)$, with the ordinary derivative of $u(t, x)$ with respect to t and its absolute value respectively. Therefore, $\mu^a(dt, x) = |\bar{u}_t^a(dt, x)|$. Moreover, it can easily be obtained from definitions and Fubini's theorem that $u_t(dt, dx) = \bar{u}_t(dt, x)dx$. By Lemma 4, it follows that $\nu^a(dt\ dx) = |\bar{u}_t^a(dt, x)|dx = |u_t^a(dt\ dx)| = |u_t^{(0)}|dt\ dx$, as desired.

By Lemma 2, it follows from Lemma 5 that

$$\lim_{r\downarrow 0} \Lambda^{-1}(Q)r^{-(d+2)} \int_{|y-x|<r} \operatorname*{Var}_{(t-r^2,\ t+r^2)} u(s, y)dy = |u_t^{(0)}(t, x)| \tag{2}$$

for almost all $(t, x) \in Q$. Of course this assertion remains in force if $u(s, y)$ in it is replaced by $u(s, y) - \xi s$ and $u_t^{(0)}(t, x)$ by $u_t^{(0)}(t, x) - \xi$, where ξ is any scalar. The set of (t, x), where (2) holds, now depends on ξ. However, the expressions involved in the limiting process and the right-hand side of the version of (2) under consideration obviously satisfy a Lipschitz condition in ξ with a unit constant, therefore there exists one and the same set of full Lebesgue measure which is suited for all ξ. Hence for $\xi = u_t^{(0)}(t, x)$ we obtain the following result.

6. LEMMA. For almost all $(t, x) \in Q$

$$\int_{|y-x|<r} \operatorname*{Var}_{(t-r^2,t+r^2)} [u(s, y) - su_t^{(0)}(t, x)]dy = o(r^{d+2}) \qquad (3)$$

as $r \downarrow 0$.

Before formulating the next lemma, we denote

$$w(t, x, s, y) = u(t + s, x + y) - P(t, x, s, y),$$

$$G = (-4, 4) \times S_2.$$

7. LEMMA. For almost all $(t, x) \in Q$ we have $r^{-2} w(t, x, r^2 s, ry) \to 0$ as $r \downarrow 0$ in $L_1(G)$.

Proof. We set $w_1(t, x, s, y) = u(t + s, x + y) - u(t, x + y) - u_t^{(0)}(t, x)s$, $w_2(t, x, y) = u(t, x + y) - P(t, x, 0, y)$ so that $w = w_1 + w_2$.

Note that the function w_2 is independent of s. Moreover, for a fixed t the function $u(t, x)$ is convex in x. Therefore, there exist the measures $u_{x^i x^j}(t, dx)$ in S_1. By Fubini's theorem, $u_{x^i x^j}(dt\, dx) = u_{x^i x^j}(t, dx)dt$ and, by Lemma 4, $u_{x^i x^j}^{(0)}(t, x)$ coincides with $u^a_{x^i x^j}(t, dx)/dx$ for almost all (t, x). Therefore, for almost every t we have this coincidence for almost all x and this can be taken into account for constructing $P(t, x, 0, y)$. Then by Lemma 3, for almost every t we have $r^{-2} w_2(t, x, ry) \to 0$ in $L_1(S_2)$ and $L_1(G)$ as $r \downarrow 0$ for almost all x. In other words, $r^{-2} w_2(t, x, ry) \to 0$ in $L_1(G)$ as $r \downarrow 0$ for almost all $(t, x) \in Q$.

Furthermore, we take $\zeta \geq 0$, $\zeta \in C_0^\infty(E_{d+1})$ such that $\int \zeta\, dt\, dx = 1$ and set $u^{(\tau)}(t, x) = u(t, x) * \tau^{-d-1} \zeta(t\tau^{-1}, x\tau^{-1})$. Since the function u is bounded on Q, we have $u^{(\tau)} \to u$ in $L_1(\Gamma)$ as $\tau \downarrow 0$ for every compactum

$\Gamma \subset Q$. It follows that there exists a sequence $\tau_n \downarrow 0$ such that $u^{(\tau_n)}(t, \cdot) \to u(t, \cdot)$ in $L_1(\Gamma)$ for any compactum $\Gamma \subset S_1$ for almost all t. To prove the lemma, it remains for us to show that $r^{-2}w_1(t_0, x_0, r^2s, ry) \to 0$ in $L_1(G)$ if the point (t_0, x_0) is such that

$$|u_t - u_t^{(0)}(t_0, x_0)\Lambda|((t_0, x_0) + Q(r)) = o(r^{d+2}),$$

$$\int_{|y|\leq 1/2} \left|u^{(\tau_n)}(t_0, y) - u(t_0, y)\right| dy \to 0.$$

For the sake of convenience let $(t_0, x_0) = 0$. Moreover, we may assume that $u(t_0, x_0) = u_t^{(0)}(t_0, x_0) = 0$ since we can subtract the corresponding linear function from u. It is clear that

$$\int_G |w_1(0, 0, r^2s, ry)|\,ds\,dy$$

$$= \lim_{n\to\infty} \int_G \left|u^{(\tau_n)}(r^2s, ry) - u^{(\tau_n)}(0, ry)\right| ds\,dy$$

$$= \overline{\lim_{n\to\infty}}\, r^{-d-2} \int_{Q(2r)} \left|\int_0^s u_t^{(\tau_n)}(p, y)\,dp\right| ds\,dy$$

$$\leq \overline{\lim_{n\to\infty}}\, r^{-d-2} \int_{Q(2r)} (4r^2 - |s|)\left|u_t^{(\tau_n)}(s, y)\right| ds\,dy$$

$$\leq 4r^{-d}|u_t|(Q(2r))$$

for small r. By choice of (t_0, x_0) the latter expression of $o(r^2)$, which proves the lemma.

8. Proof of Theorem 1. We need to prove that $w(t, x, s, y) = o(|s| + |y|^2)$ as $(s, y) \to 0$ for almost all $(t, x) \in Q$. It is easily seen that equivalently $r^{-2}w(t, x, r^2s, ry) \to 0$ uniformly in $(s, y) \in Q$ for almost all $(t, x) \in Q$. By Lemmas 6 and 7, it suffices to prove the latter uniform convergence only for those (t, x) satisfying the hypotheses of these lemmas. Let us fix an appropriate point (t, x) and denote

$$P_0(y) = \frac{1}{2} u^{(0)}_{x^ix^j}(t, x)y^iy^j.$$

Let $r_n \downarrow 0$ so that

$$\overline{\lim_{r\downarrow 0}} \sup_Q r^{-2} w(t, x, r^2 s, ry) =$$

$$= \lim_{n\to\infty} \sup_Q r_n^{-2} |w(t, x, r_n s, r_n y)|. \tag{4}$$

By choice of the point (t, x) for the function

$$u_n(s, y) := r_n^{-2}\left[u(t + r_n^2 s, x + r_n y) - u(t, x) - \right.$$

$$\left. - u_t^{(0)}(t, x) r_n^2 s - u_{x^i}(t, x) r_n y^i\right]$$

$$= r_n^{-2} w(t, x, r_n^2 s, r_n y) + P_0(y)$$

we have

$$\rho_n(s) := \int_{|y|\leq 2} |u_n(s, y) - P_0(y)| dy \to 0 \text{ in } L_1(-4, 4). \tag{5}$$

By virtue of (3), the variation of $\rho_n(s)$ in $s \in (-4, 4)$ tends to zero. This and (5) show that $\rho_n(s) \to 0$ uniformly on $(-4, 4)$. Since $u_n(s, y)$ are convex in y, it follows that the boundedness of $\rho_n(t_n)$ for all $t_n \in (-4, 4)$ implies the uniform boundedness of $u_n(t_n, y)$ on $S_{3/2}$. In the smaller domain S_1 this yields the uniform boundedness of Lipschitz constants in y for the functions $u_n(t_n, y)$ and $u_n(t_n, y) - P_0(y)$ since uniformly bounded convex functions defined on a bounded domain satisfy a Lipschitz condition with the same constant in every interior subdomain. This and the convergence of $u_n(t_n, \cdot) - P_0$ to zero in $L_1(S_1)$ readily yield the uniform convergence $u_n(t_n, y) - P_0(y) \to 0$ for $y \in S_1$. Since t_n can be chosen arbitrarily, we conclude that $u_n - P_0 \to 0$ uniformly on Q. Therefore the right-hand side of (4) is zero, which asserts the theorem.

9. Remark. If we wish to prove only the Aleksandrov—Busemann—Feller theorem 1.2.1, then Lemmas 2 and 4 — 7 are superfluous. Having proved Lemma 3, it suffices to repeat the previous proof — simplifying it in an obvious manner. Note that the same scheme can be adjusted for proving Rademacher's theorem on the differentiability a.e. of Lipschitz continuous functions (one should apply Lemma 3 for $n = 1$ and note that $r^{-1}u\ \ (x + ry)$, as a function of y, satisfies the same Lipschitz condition as u).

10. Remark. Assertions resembling Lemma 3 with similar proofs can be found in Calderon and Zygmund [1].

11. Remark. Let $u(x)$ be a convex function in S_1. Then at every point $x \in S_1$ it has the right derivative with respect to any x^i. Denote this derivative by $u_i(x)$. Aleksandrov [1] proved that for almost all $x \in S_1$ the equality $u_i(x + y) = u_i(x) + u^{(0)}_{x^i x^j}(x)y^j + o(|y|^2)$ holds as $y \to 0$. Specifically, $u^{(0)}_{x^i x^j}(x)$ is an ordinary partial derivative of u_i with respect to x^j and this yields a 'direct' method of calculating $u^{(0)}_{x^i x^j}$ from u.

The Aleksandrov result mentioned can be obtained in the following way. It is well known that if functions which are monotone on a segment converge to a continuous function, then the convergence is uniform. In the one-dimensional case the right derivative of a convex function defines the right tangent line to the graph of this function, and if there is a sequence of convex functions $f_n(t)$ converging on a segment $[a, b]$ to a continuously differentiable function $f(t)$, then, of course the tangent lines to the graphs of $f_n(t)$ at the point $t \in (a, b)$ converge to the tangent line for $f(t)$, while the right derivatives of $f_n(t)$, as monotone functions, thus uniformly converge on every segment $[a', b']$ properly contained in $[a, b]$ to the derivative of $f(t)$. Considering an expression of the form $u^n_1(t, x^2_n, \ldots, x^d_n)$ and assuming the contrary, we obtain that if a sequence of convex functions u^n uniformly converges to a continuously differentiable function u on the cube D_2, then $u^n_1 \to u_1$ uniformly in D_1.

Also, note that u_i is not only the right derivative of u with respect to x^i, but also the Sobolev derivative of the form u_{x^i}, as easily follows from Fubini's theorem.

We now take any point $x_0 \in S_1$ such that

$$u^{(0)}_{x^i x^j}(x_0) = u^{(0)}_{x^j x^i}(x_0),$$

$$u(x_0 + y) = u(x_0) + u_i(x_0)y^i + \frac{1}{2} u^{(0)}_{x^i x^j}(x_0)y^i y^j + 0(|y|^2)$$

as $y \to 0$ (Note that

$$u_{x^i x^j}(dx) = u_{x^i x^j}(dx), \; u^{(0)}_{x^i x^j} = u^{(0)}_{x^j x^i} \text{ a.e.)}$$

Then the functions $r^{-2}[u(x_0 + ry) - u(x_0) - u_i(x_0)ry^i]$, which are convex in y, uniformly converge to the continuously differentiable function $\frac{1}{2} u^{(0)}_{x^i x^j}(x_0)y^i y^j$ on the cube D_2 as $r \downarrow 0$. Their right derivatives with respect to y_1, i.e., $r^{-1}[u_1(x_0 + ry) - u_1(x_0)]$, uniformly converge to $u^{(0)}_{x^1 x^j}(x_0)y^j$ on D. Therefore, $u_1(x_0 + y) = u_1(x_0) + u^{(0)}_{x^1 x^j}(x_0)y^j + o(|y|)$ as $y \to 0$. The cases $i \geqq 2$ can be studied in a similar manner.

An extension of Aleksandrov's result to arbitrary monotone mappings can be found in Krylov [18].

BIBLIOGRAPHY

Agmon, S., A. Douglis, and L. Nirenberg

1. Estimates near the boundary for solutions of elliptic partial differential equations satisfying general boundary conditions. I. Comm. Pure Appl. Math. 12, 623-727 (1959).

Aleksandrov, A.D.

1. Almost everywhere existence of the second differential of a convex function and some related properties of convex surfaces. Uchenye Zapiski Leningrad Univ. 37, No. 6, 3-35 (1939) (Russian).

2. Dirichlet's problem for the equation Det $||z_{ij}|| = \varphi(z_1, \ldots z_n, z, x_1, \ldots, x_n)$. Vestnik Leningrad Univ. 13 No. 1, 5-24 (1958) (Russian).

3. Research into maximum principle. YI. Izvestiya Vysshikh Uchebnykh Zavedenii. Ser. Math. No. 1, 3-20 (1961) (Russian).

4. Uniqueness conditions and estimates for solution of the Dirichlet problem. Vestnik Leningrad Univ. 18, No. 3, 5-29 (1963) (Russian). (English translation in Amer. Math. Soc. Transl. (2) 68, 89-119 (1968).)

5. Majorization of solutions of second-order linear equations. Vestnik Leningrad Univ. 21 No. 1, 5-25 (1966) (Russian). (English translation in Amer. Math. Soc. Transl. (2) 68, 120-143 (1968).)

Bakel'man, I.YA.

1. Geometrical methods for solving elliptic equations. Moscow: Nauka (1965) (Russian).

Belopol'skaya, YA.I., and YU.L. Daletskii

1. Markov processes related to nonlinear parbolic systems. Dokl. Akad. Nauk SSSR 250, No. 3, 521-524 (1980) (Russian).

Bismut, J.M.

1. Martingales, the Malliavin calculus and hypoelepticity under general Hörmander's conditions. Z. Wahrsch. 56, 469-505 (1981).

Blagoveshchenskii, YU.N., and M.I. Freidlin

1. Some properties of diffusion processes depending on a parameter. Dokl. Akad. Nauk SSSR 138, 508-511 (1961) (Russian).

Bony, J.M.

1. Principe du maximum dans les espaces de Sobolev. C.R. Acad. Sci. Paris 265, No. 12, 333-336 (1967).

Brezis, H., and L.C. Evans

1. A variational inequality approach to the Bellman—Dirichlet equation for two elliptic operators. Arch. Rational Mech. Anal. 71, 1-13 (1979).

Caffarelli, L., L. Nirenberg, and J. Spruck

1. The Dirichlet problem for nonlinear second order elliptic equations, I. Monge-Ampère equations. Comm. Pure Appl. Math. 37, No. 3,

360-402 (1984).

2. III. Functions of the eigenvalues of the Hessian. Acta Math., v.155, N3-4, 261-301 (1985).

3. IV. Starshaped compact Weingarten hypersurfaces. Preprint

Caffarelli, L., J.J. Kohn, L. Nirenberg, and J. Spruck

1. The Dirichlet problem for nonlinear second order equations. II. Complex Monge—Ampère, and uniformly elliptic equations. Comm. Pure Appl. Math. 38, 209-252 (1985).

Calderón, A.P., and A. Zygmund

1. Local properties of solutions of elliptic partial differential equations. Studia Math. 20, 171-225 (1961).

Cheng, S.Y., and S.T. Yau

1. On the regularity of the Monge—Ampère equation det $(\partial^2 u/\partial x_i \partial x_j) = F(x, y)$. Comm. Pure Appl. Math. 30, 41-68 (1977).

Conway, E., and E. Hopf

1. Hamilton's theory and generalized solutions of the Hamilton—Jacobi equation. J. Math. Mech. 13, 939-986 (1964).

Crandall, M.G., and P.-L. Lions

1. Viscosity solutions of Hamilton—Jacobi equations. Trans. Amer. Math. Soc. 277, No. 1, 1-42 (1983).

De Giorgi, E.

1. Sulla differenziabilità e l'analiticità della estremali degli integrali multipli regolari. Mem. Accad. Sci. Torino Cl. Sci. Fis. Mat. Natur. (3) 3, 25-43 (1957).

Dunford, N., and J.T. Schwartz

1. Linear operators, Part 1. New York: Interscience (1958).

Evans, L.C.

1. A convergence theorem for solutions of nonlinear second order elliptic equations. Indiana University Math. J. 27, 875-887 (1978).

2. Classical solutions of fully nonlinear convex, second order elliptic equations. Comm. Pure Appl. Math. 25, 333-363 (1982).

3. Classical solutions of the Hamilton—Jacobi—Bellman equation for uniformly elliptic operators. Trans. Amer. Math. Soc. 275, 245-255 (1983).

Evans, L.C., and A. Friedman

1. Optimal stochastic switching and the Dirichlet problem for the Bellman equation. Trans. Amer. Math. Soc. 253, 365-389 (1979).

Evans, L.C., and P.L. Lions

1. Résolution des équations de Hamilton—Jacobi—Bellman pour des operateurs uniformement elliptiques. C.R. Acad. Sci. Paris 290, 1049-1052 (1980).

2. Fully nonlinear second order elliptic equations with large zeroth order coefficient. Ann. Inst. Fourier (Grenoble) 31, fasc. 2, 175-191 (1981).

Feller, W.

1. An introduction to probability theory and its applications, Vol. 2. New York: Wiley (1966).

Fleming, W.H., and R.W. Rishel

1. Deterministic and stochastic optimal control. New York: Springer-Verlag (1975).

Freidlin, M.I.
1. On the smoothness of solutions of degenerate elliptic equations. Izvestiya Acad. Nauk SSSR. Ser. Math. 32, No. 6, 1391-1413 (1968) (Russian).
2. On factorization of positive semi-definite matrices. Teoriya Veroyatnostei I Ee Primeneniya 13, No. 2, 374-378 (1968) (Russian).

Friedman, A.
1. Partial differential equations of parabolic type. Englewood Cliffs, New Jersey: Prentice-Hall (1964).

Genis, I.L., and N.V. Krylov
1. An example of a one-dimensional controlled process. Teoriya Veroyatnostei I Ee Primeneniya 21, No. 1, 147-151 (1976) (Russian).

Gikhman, I.I.
1. On the theory of differential equations of random processes, Part 1. Ukrainskii Matematicheskii Zhurnal 2, No. 4, 37-63 (1950); Part 2, 3, No. 3, 317-339 (1951) (Russian).

Gilbarg, D., and N.S. Trudinger
1. Elliptic partial differential equations of second order. Berlin: Springer-Verlag (1983).

Guzmàn, M.
1. Differentiation of integrals in R^n. Berlin: Springer-Verlag (1975).

Hausman, U.
1. On the integral representation of functionals of Ito processes. Stochastics 3, 17-27 (1979).

Hayes, C.A., and C.Y. Pauc
1. Derivation and martingales. Berlin: Springer-Verlag (1970).

Ivanov, A.V.
1. A priori estimates of the second derivatives of solutions to non-linear second order equations on the boundary of the domain. Zap. Naucn. Sem. Leningrad. Otdel. Mat. Inst. Steklov (LOMI) 69, 65-76 (1977) (Russian). (English translation in J. Soviet Math. 10, 44-53 (1978).)
2. Quasilinear degenerate and nonuniformly elliptic and parabolic equations of second order. Trudy Mat. Inst. Steklov, Leningrad: Nauka (1982) (Russian).

Ivochkina, N.M.
1. The classical solvability of the Dirichlet problems for Monge-Ampère equations. Zap. Nauchn. Sem. Leningrad. Otdel. Mat. Inst. Steklov (LOMI) 131, 72-79 (1983) (Russian).

Il'in, A.M.
1. Degenerate elliptic and parabolic equations. Nauchn. Dokl. Vyssh. Shkoly 1, No. 2, 48-53 (1958) (Russian).
2. Degenerate elliptic and parabolic equations. Mat. Sbornik 50, No. 4, 443-498 (1960) (Russian).

Kamynin, L.I.
1. The maximum principle and the boundary α-estimates of the solution to the first boundary-value problem for a parabolic equation in a noncyclindric domain. Zhurn. Vychisl. Mat. I Mat. Fiz. 7, No. 3, 551-567 (1967) (Russian).

Kamynin, L.I., and B.N. Khimchenko
1. On local behaviour of the solution of a parabolic equation of second order near the lower lid of the parabolic boundary. Sibirsk. Mat. Zhurn. 20, No. 1, 69-94 (1979) (Russian).

Kohn, J.J., and L. Nirinberg

1. Degenerate elliptic-parabolic equations of second order. Comm. Pure Appl. Math. 20, 551-585 (1967).

Krasnosel'skii, M.A., P.P. Zabreiko, Pustyl'nik E.I., and P.E. Sobolevskii

1. Integral operators in spaces of summable functions. Moscow: Nauka (1966) (Russian).

Kruzhkov, S.N.

1. Nonlinear partial differential equations, Part 2. Moscow: Moscow University Press (1970) (Russian).

2. Generalized solutions of the Hamilton-Jacobi equation of the eikonal type. Mat. Sbornik 98, 450-493 (1975) (Russian).

Kruzhkov, S.N., and O.A. Oleinik

1. Quasilinear parabolic equations of second order with many independent variables. Uspekhi Mat. Nauk 16, No. 5, 115-155 (1961) (Russian).

Krylov, N.V.

1. On equations of minimax type in the theory of elliptic and parabolic equations on a plane. Mat. Sbornik 81, No. 1, 3-22 (1970) (Russian).

2. Boundedly inhomogeneous elliptic and parabolic equations on a plane. Mat. Sbornik 82, No. 2, 99-110 (1970) (Russian).

3. On the theory of nonlinear degenerate elliptic equations. Dokl. Akad. Nauk SSSR 201, No. 6, 1279-1281 (1971) (Russian).

4. On controlling the solution of a stochastic integral equation. Teoriya Veroyatnostei I Ee Primeneniya 17, No. 1, 111-127 (1972) (Russian).

5. On controlling the solution of a stochastic integral equation under degeneracy. Izv. Akad. Nauk SSSR, Ser. Mat. 36, No. 1, 248-261 (1972) (Russian).

6. On the control of diffusion type processes. Second Japan—USSR Symp. on Probability Theory, Kyoto, 1, 126-137 (August, 1972).

7. Some estimates in the theory of the stochastic integral. Teoriya Veroyatnostei I Ee Primeneniya 18, No. 1, 56-65 (1973) (Russian).

8. Some estimates of the distribution density of the stochastic integral. Izv. Akad. Nauk SSSR, Ser. Mat. 38, No. 1, 228-248 (1974) (Russian).

9. Sequences of convex functions and estimates of the maximum of the solution to a parabolic equation. Sibirsk. Mat. Zhurn. 17, No. 2, 290-303 (1976) (Russian).

10. Controlled diffisuion processes. Moscow: Nauka (1977) (Russian). (English Translation: Berlin: Springer-Verlag (1980).)

11. On the maximum principle for nonlinear parabolic and elliptic equatoins. Izv. Akad. Nauk SSSR, Ser. Mat. 42, No. 4, 1050-1062 (1978). (Russian).

12. On passage to the limit in Bellman parabolic equations. Izv. Akad. Nauk SSSR, Ser. Mat. 42, No. 6, 1417-1425 (1978) (Russian).

13. On passage to the limit in degenerate Bellman equations, 2. Mat. Sbornik, 107, No. 1, 56-68 (1978) (Russian).

14. Some new results in the theory of controlled diffusion processes. Mat. Sbornik 109, No. 1, 146-164 (1979) (Russian).

15. On traditional derivation of the Bellman equation for controlled diffusion processes. Lit. Mat. Sbornik 21, No. 1, 59-68 (1981) (Russian).

16. On controlled diffusion processes with unbounded coefficients. Izv. Akad. Nauk SSSR, Ser. Mat. 45, No. 4, 734-759 (1981) (Russian).

17. On control of a diffusion process up to the time of first exit from a region. Izv. Akad. Nauk. SSSR, Ser. Mat. 45, No. 5, 1029-1048(1981) (Russian). (English Translation in Math. USSR Izv. 19, 297-313 (1982).)

18. Certain properties of monotone mappings. Lit. Mat. Sbornik 22, No. 2, 80-87 (1982) (Russian).

19. Boundedly inhomogeneous elliptic and parabolic equations. Izv. Akad. Nauk SSSR, Ser. Mat. 46, No. 3, 487-523 (1982) (Russian).

20. On nonlinear equations in a domain. Uspekhi Mat. Nauk 37, No. 4, 91 (1982) (Russian).

21. Boundedly inhomogeneous elliptic and parabolic equations in a domain. Izv. Akad. Nauk SSSR, Ser. Mat. 47, No. 1, 75-108 (1983) (Russian).

22. On degenerate nonlinear elliptic equations . Mat. Sbornik 120, No. 3, 311-330 (1983) (Russian).

23. On degenerate nonlinear elliptic equations, 2. Mat. Sbornik 121, No. 2, 211-232 (1983) (Russian).

24. On estimates of solutions to nonlinear parabolic equations. Dokl. Akad. Nauk SSSR 274, No. 1, 23-26 (1984) (Russian).

25. On the G-convergence of elliptic operators in nondivergence form. Mat. Zametki 37, No. 4, 522-527 (1985) (Russian).

26. On the estimates of the maximum of the solution of parabolic equations and the estimates of the distribution of semimartingales, Matem. Sbornik, 130, No. 2, 207-221 (1986). (In Russian).

27. Some new results in the theory of nonlinear elliptic and parabolic equations, International Congres of Math., Berkely (1986).

Krylov, N.V., and M.V. Safonov

1. An estimate of the probability that a diffusion process hits a set of positive measure. Dokl. Akad. Nauk SSSR 245, No. 1, 18-20 ((1979) (Russian). (English Translation in Soviet Math. Dokl. 20, 253-255 (1979)).

2. Certain properties of solutions of parabolic equations with measurable coefficients. Izv. Akad. Nauk SSSR 44, No. 1, 161-175 (1980) (Russian).

Ladyzhenskaya, O.A., and N.N. Ural'tseva

1. Linear and quasilinear elliptic equations. Moscow: Nauka, 1964 (Russian). (English Translation: New York: Academic Press (1968). 2nd Russian ed. (1973).)

2. Hölder estimates for solutions of second order quasilinear elliptic equations in general form. Zap. Nauchn. Sem. Leningrad. Otdel. Mat. Inst. Steklov. (LOMI) 96, 161-168 (1980) (Russian).

3. On an estimate of the Hölder constant for solutions of quasi-linear parabolic equations in nondivergent form. Uspekhi Mat. Nauk 36, No. 4, p.230 (1981) (Russian).

4. Estimates of Hölder constants for bounded solutions of second order quasilinear parabolic equations of nondivergent form. LOMI Pre-prints, E-11-81-Leningrad (1981).

Ladyzhenskaya, O.A., V.A. Solonnikov, and N.N. Ural'tseva

1. Linear and quasilinear parabolic equations. Moscow: Nauka (1967) (Russian).

Landis, E.M.

1. Elliptic and parabolic equations of second order. Moscow: Nauka (1971) (Russian).

Lenhart, S.

1. Bellman equation for optimal stopping time problems. Indiana

University Math. J. 32, No. 3, 363-375 (1983).

Lieberman, G.M. and N.S. Trudinger

1. Nonlinear oblique boundary value problems for nonlinear elliptic equations, Trans. Amer. Math. Soc., 295, No. 2, 509-546 (1985).

Lions, P.-L.

1. Contrôle de diffusions dans R^N.C.R. Acad. Sci. Paris, Ser. A-B 288, 339-342 (1979).

2. Équations de Hamilton—Jacobi—Bellman dégénérées. C.R.Acad. Sci. Paris, Ser. A, 289, No. 5, 329-332 (1979).

3. Control of diffusion processes in R^N. Comm. Pure Appl. Math. 34, 121-147 (1981).

4. Resolution analytique des problemes de Bellman—Dirichlet. Acta Math. 146, Nos. 3 and 4, 151-166 (1981).

Lions, P.-L., and J.-L. Menaldi

1. Optimal control of stochastic integrals and Hamilton—Jacobi—Bellman equations. 1, 2. SIAM J. Control and Optimization 20, No. 1, 58-95 (1982).

Mamedov, I.T.

1. On an a priori estimate of the Hölder norm of solutions of quasilinear parabolic equations with discontinuous coefficients. Dokl. Akad. Nauk SSSR 252, No. 5, 1052-1054 (1980) (Russian). (English Translation in Soviet Math. Dokl. 21, 872-875 (1980).)

Morse, A.P.

1. A theory of covering and differentiation. Trans. Amer. Math. Soc. 55, No. 2, 205-235 (1944).

Moser, J.

1. A new proof of de Giorgi theorem concerning the regularity problem for elliptic differential equations. Comm. Pure Appl. Math. 13, 457-468 (1960).

2. A Harnack's inequality for parabolic differential equations. Comm. Pure Appl. Math. 17, 101-134 (1964).

Nash, J.

1. Continuity of solutions of parabolic and elliptic equations. Amer. J. Math. 80, 931-954 (1958).

Nazarov, A.I., and N.N. Ural'tseva

1. Convex-monotone hulls and estimates of the maximum of the solution of parabolic equations. Zap. Nauchn. Sem. LOMI, 147, 71-86 (1985). (In Russian.)

Novruzov, A.A.

1. Estimates of the Hölder norm of solutions to quasilinear elliptic equations with discontinuous coefficients. Dokl. Akad. Nauk SSSR, 253, No. 1, 31-33 (1980). (English Translation in Soviet Math. Dokl. 22, 25-28 (1980).)

2. On an approach to studying qualitative properties of solutions to elliptic equations of second order in nondivergence form. Mat. Sbornik 122, No. 3, 360-387 (1983) (Russian).

Nisio, M.

1. Remarks on stochastic optimal controls. Japan J. Math. 1, 159-183 (1975).

2. On nonlinear semigroup associated with optimal stopping for Markov processes. Appl. Math. Optimization, No. 4, 143-169 (1978).

Oleinik, O.A., and E.V. Radkevich

1. Second order equations with nonnegative characteristic form. (Mathematical Analysis 1969). Moscow: Itogi Nauki 1971 (Russian). (English Translation: Providence, R.I., American Math. Soc. (1973).)

Phillips, R.S., and L. Sarason

1. Elliptic-parabolic equations of second order. J. Math. Mech. 17

Pogorelov, A.V.

1. Monge–Ampère equations of elliptic type. Groningen: Noordhoff (1964).

2. The Dirichlet problem for the n-dimensional analogue of the Monge–Ampère equation. Dokl. Akad. Nauk SSSR 201, No. 4, 790-793 (1971) (Russian). (English Translation in Soviet Math. Dokl. 12, 1727-1731 (1971).)

3. The Minkowski multidimensional problem. New York: Wiley (1978).

Pragarauskas, G.

1. On the Bellman equation in the measure lattice for general controlled random processes, 1, 2. Lit. Mat. Sbornik 21, No. 4, 169-184 (1981); 22, No. 1, 138-145 (1982) (Russian).

2. On the uniqueness of a solution to the Bellman equation in case of general controlled processes. Lit. Mat. Sbornik 22, No. 2, 137-156 (1982) (Russian).

Sabitov, I.KH.

1. The regularity of convex domains with a metric which is regular in Hölder classes. Sibirsk. Mat. Zhurn. 17, No. 4, 907-915 (1976) (Russian).

Safonov, M.V.

1. On the control of diffusion processes in a multidimensional cylindrical domain. Abstr. Comm. Intern. Symp. Stoch. Diff. Eq. Vilnius: Institute Math. Cyb. Lit. SSR Akad. Nauk, 168-172 (1978) (Russian).

2. On the Dirichlet problem for the Bellman equation in a multidimensional domain. Dokl. Akad. Nauk SSSR 253, No. 3, 535-540 (1980) (Russian).

3. The control of diffusion processes in a bounded domain. Cand. of Sci. Thesis, Moscow University (1980) (Russian).

4. Harnack inequalities for elliptic equations and Hölder continuity of their solutions. Zap. Nauchn. Sem. Leningrad. Otdel. Mat. Inst. Steklov (LOMI) 96, 272-287 (1980) (Russian).

5. An example of the diffusion processes with singular distribution at a fixed time instant. Abstr. Comm. Third Intern. Vilnius Conf. Probability Theory and Math. Statist. 2, 133-134 (1981) (Russian).

6. Boundary $C^{2+\alpha}$-estimates for solutions of nonlinear elliptic equations. Uspekhi Mat. Nauk 38, No. 5, 146-147 (1983) (Russian).

7. On the classical solution of the Bellman equation. Dokl. Akad. Nauk SSSR, 278, No. 4, 810-813 (1984) (Russian).

8. On the classical solutions of nonlinear parabolic equations of the Bellman type, Uspekhi Mat. Nauk., 41, No. 4, 174-175 (1986). (In Russian.)

Skrypnik, I.V.

1. On the topological characteristic of general nonlinear elliptic operators. Dokl. Akad. Nauk SSSR 239, No. 3, 538-541 (1978) (Russian).

Trudinger, N.S.

1. Local estimates for subsolutions and supersolutions of general second order elliptic quasilinear equations. Invent. Math. 61, 67-79 (1980).

2. Fully nonlinear, uniformly elliptic equations under natural structure conditions. Trans. Amer. Math. Soc. 278, No. 2, 751-769 (1983).

Urbas, J.I.E.

1. Elliptic equations of Monge–Ampère type. Ph.D. Thesis, The Australian National University (1984).

Zhikov, V.V., and M.M. Sirazhudinov

1. On G-compactness of a class of second order elliptic operators in nondivergence form. Izv. Akad. Nauk SSSR Ser. Mat. 45, No. 4, 718-733 (1981) (Russian).

Zhikov, V.V., S.M. Kozlov, and O.A. Oleinik

1. On G-convergence of parabolic operators. Uspekhi Mat. Nauk 36, No. 1, 11-58 (1981) (Russian).

SUBJECT INDEX